Low-Power CMOS Circuits

Technology, Logic Design and CAD Tools

Low-Power CMOS Circuits

Technology, Logic Design and CAD Tools

Christian Piguet
CSEM
Neuchatel, Switzerland

CRC Taylor & Francis
Taylor & Francis Group
Boca Raton London New York

A CRC title, part of the Taylor & Francis imprint, a member of the Taylor & Francis Group, the academic division of T&F Informa plc.

This material was previously published in *Low-Power Electronics Design.* © CRC Press LLC 2004

Published in 2006 by
CRC Press
Taylor & Francis Group
6000 Broken Sound Parkway NW, Suite 300
Boca Raton, FL 33487-2742

Printed in the United States of America on acid-free paper
10 9 8 7 6 5 4 3 2 1

International Standard Book Number-10: 0-8493-9537-2 (Hardcover)
International Standard Book Number-13: 978-0-8493-9537-6 (Hardcover)
Library of Congress Card Number 2005050631

Library of Congress Cataloging-in-Publication Data

Piguet, Christian.
Low-power CMOS circuits : technology, logic design, and CAD tools / Christian Piguet.
p. cm.
Includes bibliographical references and index.
ISBN 0-8493-9537-2 (alk. paper)
1. Metal oxide semiconductors, Complementary--Computer-aided design. 2. Low voltage integrated circuits. I. Title.

TK7871.99.M44P56 2005
621.39'5--dc22 2005050631

informa
Taylor & Francis Group
is the Academic Division of Informa plc.

Visit the Taylor & Francis Web site at
http://www.taylorandfrancis.com

and the CRC Press Web site at
http://www.crcpress.com

Preface

Purpose and Background

This book is a part of *Low-Power Electronics Design*, edited by C. Piguet, published in November 2004. It contains only the chapters that describe the design of low-power circuitry, from technology aspects to transistors and logic gates, including some CAD tools to design these circuits. All the other chapters, describing microcontrollers, microprocessors, DSP cores and systems on chips (SoCs), are also included in another smaller book entitled *Low Power Processors and Systems on Chips.*

The goal of this book is to cover all the low-level aspects of the design of low-power integrated circuits (ICs) in deep submicron technologies. Today, the power consumption of ICs is considered one of the most important problems for high-performance chips, as well as for portable devices. For the latter, the problem is due to the limited cell battery lifetime, while it is the chip cooling for the first case. As a result, for any chip design, power consumption has to be taken into account very seriously. Before 1993–1994, only speed and silicon area were important in the design of integrated circuits, and power consumption was not an issue. Later, it was recognized that power consumption must be taken into account as a main design parameter. Many papers and books were written to describe all the first design methodologies to save power limited to circuit design. However, today, we have to cope with many new problems implied by very deep submicron technologies, such as leakage power, interconnect delays and robustness.

Today, we are close to designing one billion transistor chips down to 0.10 μm and below, supplied at less than half a Volt and working at some GHz. This is due to an unexpected evolution of the microelectronics technologies. This evolution is not yet at its end, so the next decade will also see some spectacular improvements in the design of integrated circuits. However, it is sure that the microelectronics evolution will slow down in the future, but as pointed out by Gordon Moore, "No exponential is forever, but we can delay 'forever'."

Organization

The first part of the book covers some of the history of low-power electronics. It then describes some existing and future very deep submicron technologies, as well as some completely different technologies that could be used for the design of integrated circuits, i.e., nanoelectronics and optical chips. This look at the past, as well as into the future, of microelectronics is a very good introduction to low-power design.

The second part of the book comprises a set of chapters describing many interesting techniques to reduce power consumption at low levels, i.e., at transistor and gate levels. They have been written by different authors who are recognized as leading specialists in their domains. Successively, starting with delay and power models of logic gates and low-power standard libraries, the next chapters present the design of dynamic logic, arithmetic circuits, pipelining and parallelization for low power and VHDL for low power. The next chapters present more specific contributions to the reduction of power consumption, i.e., clocking, leakage reduction, interconnecting and communication on chips and finally, adiabatic circuits. The last two chapters conclude this second part of the book by presenting weak inversion logic and robustness of integrated circuits, a main issue today.

The third section of the book presents some CAD tools used to design low-power integrated circuits, starting at high level with the two first chapters and then presenting the tools and low-power issues of three major companies providing logic synthesizers.

The key benefits for readers will be this complete picture of what is done today for reducing power at the logic level, while also looking into the future of integrated circuits to be fully aware of what is going on for the design of chips 10 or 15 years from now.

Locating Your Topic

Several avenues are available through which to access desired information. A complete table of contents is presented at the front of the book, and each of the chapters is also preceded with an individual table of contents. Each contributed chapter contains comprehensive references, including books, journal and magazine papers and sometimes Web pointers.

Acknowledgments

The value of this book is completely based on the many excellent contributions of experts. I am very grateful to them, as they spent considerable time to write excellent texts without receiving compensation. Their sole motivation was to provide readers with their excellent insights. I would like to thank all these authors very much, as I am sure that this book will be a useful text for many readers and students interested in low-power design. I am indebted to Prof. Vojin G. Oklobjzija for asking me to edit this book and trusting me with this project. I would like to thank very much Nora Konopka and Allison Taub of CRC Press for their excellent work in combining all this material into its present form. The work of all has made this book.

Editor

Christian Piguet, a native of Nyon, Switzerland, received his M.S. and Ph.D. degrees in Electrical Engineering from the Ecole Polytechnique Fédérale de Lausanne, Switzerland, in 1974 and 1981 respectively.

Dr. Piguet joined the Centre Electronique Horloger S.A., Neuchâtel, Switzerland, in 1974. He worked on CMOS digital integrated circuits for the watch industry, and on low-power embedded microprocessors, as well as on CAD tools based on a gate matrix approach. He is now head of the Ultra-Low-Power Sector at the CSEM Centre Suisse d'Electronique et de Microtechnique S.A., Neuchâtel. He is presently involved in the design and management of low-power and high-speed integrated circuits in CMOS technology. His main interests include the design of very low-power microprocessors and DSPs, low-power standard cell libraries, gated clock and low-power techniques, as well as asynchronous design.

Dr. Piguet, who is a professor at the Ecole Polytechnique Fédérale Lausanne (EPFL), also lectures in VLSI and microprocessor design at the University of Neuchâtel and in the ALaRI master's program at the University of Lugano, Switzerland. He is also a lecturer for many postgraduate courses in low-power design.

Dr. Piguet holds some 30 patents in digital design, microprocessors and watch systems. He is author and co-author of more than 170 publications in technical journals, as well as books and book chapters on low-power digital design. He has served as a reviewer for many technical journals, and also served as guest editor for the July 96 *JSSC* issue. He is a member of the steering and program committees of numerous conferences and served as program chairman of PATMOS '95 in Oldenburg, Germany, co-chairman at FTFC '99 in Paris, chairman of the ACiD '2001 workshop in Neuchâtel, co-chair of VLSI-SOC 2001 in Montpellier and co-chair of ISLPED 2002 in Monterey. He was chairman of the PATMOS executive committee during 2002 and low-power topic chair at DATE 2004 and 2005.

Christian Piguet
CSEM SA
Jaquet-Droz 1
2000 Neuchâtel, Switzerland
Christian.piguet@csem.ch

Contributors

Amit Agarwal
Purdue University
West Lafayette, Indiana

Amara Amara
ISEP
Paris, France

Daniel Auvergne
LIRMM, University of Montpellier
Montpellier, France

Nadine Azémard
LIRMM, University of Montpellier
Montpellier, France

Marc Belleville
CEA-LETI
Grenoble, France

Olivier Faynot
CEA-LETI
Grenoble, France

Antoni Ferré
UPC
Barcelona, Spain

Joan Figueras
UPC
Barcelona, Spain

Jerry Frenkil
Sequence Design
Santa Clara, California

Frédéric Gaffiot
Ecole Centrale de Lyon
Lyon, France

Domenik Helms
OFFIS
Oldenburg, Germany

Ed Huijbregts
Magma Design Automation
Eindhoven, the Netherlands

Chris H. Kim
Purdue University
West Lafayette, Indiana

Lars Kruse
Magma Design Automation
Eindhoven, the Netherlands

Mark Lundstrom
Purdue University
West Lafayette, Indiana

Enrico Macii
Politecnico di Torino
Torino, Italy

Philippe Maurine
LIRMM, University of Montpellier
Montpellier, France

Renu Mehra
Synopsys Inc.
Mountain View, California

Wolfgang Nebel
Oldenburg University
Oldenburg, Germany

Ian O'Connor
Ecole Centrale de Lyon
Lyon, France

Vojin G. Oklobdzija
University of California-Davis
Davis, California

Barry Pangrle
Synopsys Inc.
Santa Clara, California

Christian Piguet
CSEM & LAP-EPFL
Neuchâtel, Switzerland

Massimo Poncino
Universita di Verona
Verona, Italy

Kaushik Roy
Purdue University
West Lafayette, Indiana

Philippe Royannez
Texas Instruments
Villeneuve Loubet, France

Eric Seelen
Magma Design Automation
Eindhoven, The Netherlands

Dimitrios Soudris
Democritus University of Thrace
Xanthi, Greece

Thad E. Starner
Georgia Institute of Technology
Atlanta, Georgia

Christer Svensson
Linköping University
Linköping, Sweden

Lars Svensson
Chalmers University
Göteborg, Sweden

Arnaud Tisserand
INRIA LIP Arénaire
Lyon, France

Harry Veendrick
Philips Research Laboratories
Eindhoven, The Netherlands

Eric A. Vittoz
CSEM
Neuchâtel, Switzerland

Jing Wang
Purdue University
West Lafayette, Indiana

Jiren Yuan
Lund University
Lund, Sweden

Contents

Part I

Technologies and Devices

1

History of Low-Power Electronics

Christian Piguet
CSEM & LAP-EPFL

1.1 Introduction

Power consumption awareness began worldwide around 1990–1992. Before that, only niche markets required low-power integrated circuits (ICs). Today, every circuit has to face the power consumption issue, for both portable devices aiming at longer battery life and high-end circuits avoiding cooling packages and reliability issues that are too complex.

It was not anticipated that microprocessors would consume 100 watts today and perhaps 300 watts in 2016, as predicted by the International Technology Roadmap for Semiconductors (ITRS). Looking at a 10-year prediction proposed in 1986 (Figure 1.1), the frequency and throughput were accurate as well as the number of transistors based on Moore's law; however, no prediction was made for power consumption, while a simple calculation would yield 40 watts. If 40 watts had been predicted in 1986, it is conceivable that the awareness about the increase of microprocessor power would have been much better.

Roughly speaking, power was not an issue during the development of microelectronics from the invention of early computers in 1940 and of the transistor in 1947 to the early 1990s; however, many ideas proposed during this period for improving electronic circuits have been rediscovered and reused in the last 10 years, focusing on power consumption reduction [1]. This chapter begins with a brief history of early computers [2,3,4] to continue by the invention of the transistor and of the IC. Besides the mainstream microelectronics evolution, some aspects of low-power consumer applications are also described before the dramatic power increase during 1990–1992.

This first chapter is far from being an exhaustive history of low-power electronics. It contains some flashes on some well-known or ignored events and provides some considerations or interpretations

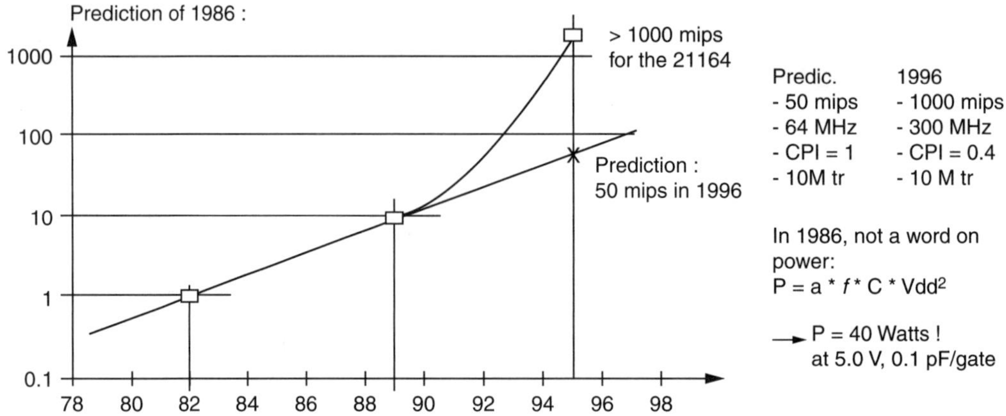

FIGURE 1.1 Predictions.

regarding low power. The history of techniques is not of general interest today, even if more and more companies try to tell their own histories within the context of the "good old days."

1.2 Early Computers

The first computer architecture, which we owe to Charles Babbage (1791–1871), was a mechanical analytical engine [5]. Although this machine was never completed, the estimated power required would certainly be very high. The first electronic computers or calculating machines were designed with vacuum tubes to significantly improve the speed over electromechanical machines.

1.2.1 Power Consumption of Early Computers

The ENIAC (1944) is generally considered to be the first electronic computer. It was programmed manually by using wires and connections between the execution units; therefore, it was very fast, achieving 100 kHz. Designed by Mauchly (1907–1980) and Eckert (1919–1995), it required 18,000 vacuum tubes and weighed 20 tons, and it was more of a huge calculator than a computer. The power consumption was 150,000 watts.

Such huge power consumption was not the highest achieved for a computer: the Whirlwind, designed by IBM in 1952 for the Semi-Automatic Ground Environment (SAGE) network (75,000 tubes, 275 tons), consumed 750,000 watts.

The introduction of transistors in the design of computers, although not really aiming at power reduction, nevertheless achieved a significant decrease of their power consumption. A transistor consumes roughly 1000 times less than a vacuum tube. Among the first transistorized computers, the TX0 designed by the Lincoln Laboratory in 1957 was an 18-bit machine containing 3500 transistors and consuming 1000 watts; however, huge mainframe computers still consumed a very large amount of power. For instance, the IBM 360 Model 91, announced in 1964, consumed a significant fraction of 1 MW [9]. The 12-bit PDP 8 minicomputer from Digital, designed in 1965, consumed 780 watts.

1.2.2 Reused Concepts for Low Power

Many concepts and ideas have been introduced for the design of early computers. Some of these ideas were rediscovered recently and used to reduce the power consumption of systems on chip (SoC). For instance, instruction formats of the early computers [6] were based on one-word instructions that could be read in one step or one clock cycle. This is much more energy efficient than the multi-byte instruction formats so common in Complex Instruction Set Computers (CISC) microprocessors. Multi-byte instruc-

FIGURE 1.2 Baby Computer of Manchester University, the world's first stored-program computer, running for the first time on June 21, 1948 (From the University of Manchester, Manchester, U.K. With permission.).

tions require several memory fetches as well as program counter updates that consume much power. It is quite interesting that the first computers were all designed with Reduced Instruction Set Computers (RISC)-like instruction sets.

Another "low-power" feature is the Harvard architecture, which comes from the name of the Harvard Mark I, designed in 1939 by Howard Aiken (1900–1973). This architecture is well-known today for providing two separate data and instruction memories, contrary to the "Von Neumann architecture" that contains only one memory (or a unified cache memory) for both instructions and data [7,8]. It results in a high sequencing of instruction execution (and a large number of clocks per instruction [CPI]), as successive instruction and operand fetches have to be performed. Today, it is well-known that this higher sequencing significantly increases power consumption.

The same applies for bit–serial architectures used for the first computers (e.g., EDVAC, Ferranti Mark I), due to the use of serial delay line memories. Consequently, many clock cycles were necessary to execute a single instruction [10]. A few years later, however, bit–parallel architectures (e.g., Von Neumann's IAS) featured a much simpler control unit and a reduced sequencing, although the execution unit was more complex. Such an observation is also valid for low power; a higher sequencing always results in larger power consumption.

Pipelined computers were introduced in the 1960s. For instance, the well-known IBM 360 Model 91, announced in 1964, was the first 360-pipelined computer with 20 stages. For scientific code, the number of clocks per instruction was about 1 (CPI = 1). Such a low CPI is very beneficial for reducing power consumption (see Section 1.5). Superscalar and parallel architectures were also introduced early in the history of computers, for instance, the 1964 CDC 6600, with 10 parallel execution units and the Illiac IV with 64 parallel processors. Parallelism is also beneficial for power consumption reduction (see Section 1.5).

1.3 Transistors and Integrated Circuits

The history of low-power electronics really starts with the invention of the bipolar transistor in late 1947. Compared to a vacuum tube, which consumes several watts, the transistor, with a range in the tens of milliwatts, is really a low-power device.

FIGURE 1.3 On the right are submini tubes used in a Zenith Royal hearing aid. The 201 date code represents week 1, 1952. On the left are examples of CK718 junction germanium transistors produced by Raytheon and used in the Zenith Royal "T" hearing aid, with 252 representing week 52, 1952. In less than 1 year, transistors had replaced the dominant vacuum tube technology in hearing aids. (From the Transistor Museum, Jack Ward, curator. With permission. [35])

1.3.1 Invention of the Transistor

When Bell Telephone Laboratories announced the invention [34] of the transistor on June 30, 1948, the general press was almost indifferent. The *New York Times* only published four short paragraphs on a back page of the paper. The inventors of the transistor were William Shockley, John Bardeen, and Walter H. Brattein. In 1956, they received the Nobel Prize for Physics for the invention of the transistor on December 23, 1947. The first working transistor was a germanium point-contact transistor; it was difficult to produce, not very reliable, and consisted of two wires pressed onto a small block of germanium. William Shockley proposed a junction or bipolar transistor as early as January 1948.

Even technical people were reluctant to recognize the benefits of the transistor. Its direct competitor was the vacuum tube, a strongly established commercial product; but transistors had very long lives, were small, and required no filament current. Despite all these advantages, however, Bell Labs decided to license it freely, and publicized it extensively in seminars and papers [14]. Imagine a free license for the transistor. This means that nobody really understood what this device was.

Since the beginning of the century, more studies have been devoted to solid-state physics, metals, and semiconductors. In 1929, a patent on a metal-oxide semiconductor (MOS)-like transistor was issued to Julius Lillienfeld (i.e., insulated material such as glass coated with a metal film having unidirectional conductivity); however, this device was impossible to fabricate with the available materials, and the world was in the midst of the Great Depression. In 1935, a patent was issued to Oscar Heil for a field-effect triode, although he was not able to explain how it worked. It is ironic that the concept of field-effect transistors, so marvelously simple, provides practical implementations after the invention of the far more complex bipolar transistor [14]. The latter was just much simpler to fabricate.

The first market pull came from telephone switching equipment and military computers, but also from hearing aids and portable radios, for which miniaturization and low power were a must. Compared to vacuum tubes, the power could be reduced by a factor of 1000; therefore, the introduction of the transistor was really a significant first step to lower power devices. Sonotone announced the first transistorized hearing aid in February 1953; it contained five transistors, but still required a pair of miniaturized tubes for the input and driver stages. Figure 1.3 and Figure 1.5 show such transistors and mini tubes. It is likely that power consumption was already an issue for a product like a hearing aid.

The second transistorized product to be introduced was a portable four-transistor radio. This is why people simply call "transistors" commercial transistorized radios, for instance, those produced by RCA and Sony. It was really the first consumer market for transistors. Figure 1.4 shows such a "radio" transistor.

The first bipolar silicon transistors were introduced in 1954 by Texas Instruments. One million transistors were produced in 1953, 3.5 million in 1955, and 29 million in 1957. Companies producing these transistors were Raytheon, Western Electric, RCA, Philco, General Electric, Texas Instruments (TI), and Fairchild. The average price was about $4 per transistor.

FIGURE 1.4 RCA introduced the 2N109 in 1955 (Germanium PNP Alloy Junction). It was an affordable and reliable germanium audio transistor used in many transistorized radios. In 1956, the 2N109 cost a little over $2 and had dropped to approximately $1 by the early 1960s (From the Transistor Museum, Jack Ward, curator. With permission. [35]).

FIGURE 1.5 CK722 is one of the well-known transistors from the 1950s and 1960s. Raytheon introduced this device in early 1953. The CK722 was the first mass-produced germanium alloy junction transistor. Raytheon was the major manufacturer of hearing aid transistors, and those units that were not quite "good enough" for the demanding hearing aid market (i.e., not enough gain or too noisy) were sold to hobbyists as the CK722 (From the Transistor Museum, Jack Ward, curator. With permission. [35]).

Finally, the significance of the invention and the introduction of the transistor became larger year after year as more transistors and integrated circuits were embedded in equipment and devices.

1.3.2 Invention of the IC

The second major step in low-power electronics history was the invention of the IC because on-chip interconnects consume much less power than off-chip connections. The IC was invented in 1958 [15] by Jack Kilby of TI, who won the 2000 Nobel Prize, and Robert Noyce, of Fairchild, and then co-founder of Intel. In October 1958, Kilby began the design of a flip-flop on a monolithic germanium chip. The device was completed in early 1959 and was revealed "as the most significant development by Texas Instruments since … the commercial silicon transistor." The announcement was widely reported in the press, but engineers were skeptical about the devices regarding optimization and yield [14]. By February 1960, Noyce, at Fairchild, announced that his company was manufacturing resistor–transistor logic (RTL) as an integrated circuit family named Micrologic. According to a recent book about this story [15], we completely ignore these famous people today: "Do you have a PC? "Yes, I do." "Do you know who Jack Kilby is?" "No, I don't."

Even before the introduction of metal oxide semiconductor (MOS) and complementary metal oxide semiconductor (CMOS) technologies, several low-power principles were understood in the 1960s [28], such as the reduction of the supply voltage, the use of analog circuits replacing digital ones, and the design of fast and parallel circuits allowing a supply voltage reduction just satisfying to the speed constraints. A few years later, these ideas would be applied to CMOS design.

1.3.3 MOS Transistors

In 1958, the first field-effect transistor became operational. Its creator, Stanislas Teszner, working in France, called it "Tecnitron." Even before Teszner, many studies about the possibilities of such a device

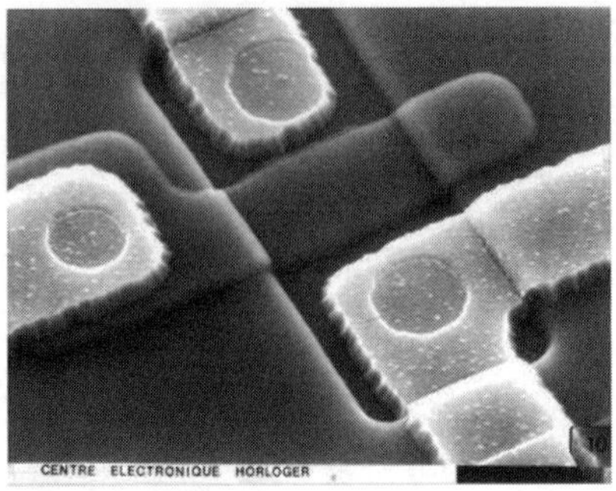

FIGURE 1.6 MOS transistor (1972).

were under way in the U.S. In 1959, RCA was already working on fluid-effect transistors (FETs) to implement logic circuits; but Dr. John T. Wallmark of the RCA Laboratories, although he had a patent, never achieved success. Two years later, Paul Weimer, another researcher from RCA, succeeded in obtaining a working FET. In 1962, RCA was fabricating multipurpose logic blocks comprising 16 MOSFETs on a single chip. By 1963, RCA had fabricated large arrays of several hundred MOS devices; however, these devices were extremely sensitive to static charge and oxide defects, and they were slower. In mid-1965, only two companies were producing MOS ICs: General Microelectronics and General Instruments. The other companies were simply waiting. Fairchild offered a 64-bit random-access memory (RAM) MOS memory in 1967, but even the first electronic watch IC was designed with bipolar transistors in 1967 [16].

Nevertheless, the move toward MOS technology was on its way. It was mainly for packing more transistors on a single chip and not for power consumption considerations. Intel developed the first microprocessor in 1971 and the first MOS EPROM memories in 1972. This first microprocessor was the famous 4004 in P-channel silicon-gate technology (Figure 1.7). The clock was 750 kHz, with an average CPI of 10 (10 clocks per instruction on average). With 0.075 MIPS (million instructions per second) performance, it consumed 0.3 watts, resulting in 0.25 MIPS /watt. (Today, 8-bit microcontrollers reach several tens of thousands of MIPS/watt.)

N-channel technology was faster than P-channel due to mobility, however. The first VLSI textbooks, such as the famous Mead/Conway [17], presented only N-MOS circuits, which were considered a dominant technology that would be used for a very long time. Figure 1.6 is an MOS transistor from 1972 in 6 μm technology. CMOS technology was, at that time, only used for special consumer markets, such as electronic watches in the early 1970s [16]; however, 10 years later, the move toward CMOS was achieved due to heat dissipation, electromigration, and reliability problems. In low-power electronics history, the introduction of the CMOS technology was the third major step after the invention of the transistor and the IC. Today, CMOS technology is clearly the dominant technology, but the CMOS power increase in the 1990s had simply followed the bipolar transistor power increase with a 10-year time shift [30].

1.3.4 Early Microprocessors

After the 4004, the technology pace could be measured by the new microprocessors introduced year after year (Figure 1.7). In 1974, the N-channel technology was chosen by Intel to produce the 8080, which was 10 times faster than the software-compatible 8008, designed by Masatoshi Shima, who later designed the Z80. The Intel 8080 was roughly equivalent to the mainframes of the 1950s. The CMOS technology was used for the RCA 1802. In 1974, Texas introduced its 4-bit TMS 1000. Motorola kept a low profile and introduced its first 8-bit microprocessor in 1974, the MC 6800. Rockwell proposed the 8-bit PPS-8 and 4-bit PPS-4 microprocessors. National Semiconductor presented a 16-bit machine called PACE, while Signetics proposed its 2650 8-bit microprocessor. By Fall 1975, almost 40 different microprocessors crowded the market. The addition time was 10.8 μsec for the 4004, 20 μsec for the 8008, 2 μsec for the 8080, 1.3 μsec for the 8085, 0.375 μsec for the 8086, 0.25 μsec for the 80296, and 0.125 μsec for the

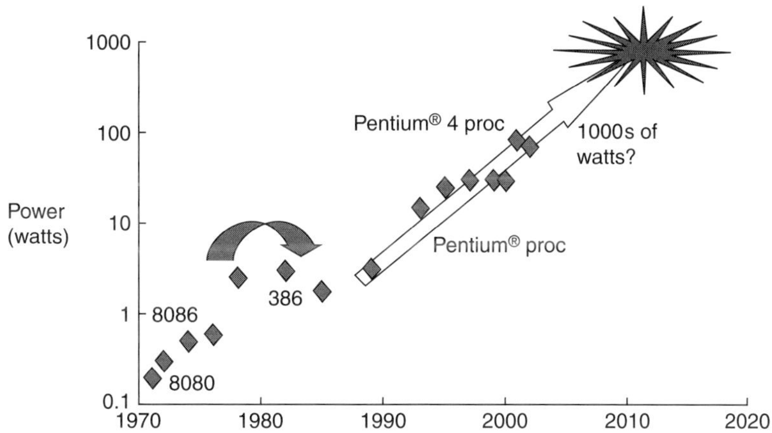

FIGURE 1.7 Power dissipation of Intel microprocessors (Courtesy of Intel).

TABLE 1.1 First Intel Microprocessors [18]

Year	μP	Technology	Nb of MOS	Address
1971	4,004	P-MOS 8 μm	2,300	4K
1971	8,008	P-MOS 8 μm	3,500	16K
1974	8,080	N-MOS 6 μm	5,000	64K
1976	8,085	N-MOS 4 μm	6,000	64K
1978	8,086	N-MOS 3 μm	29,000	1M
1982	80,286	N-MOS 2.3 μm	130,000	16M
1985	80,386	CMOS 2 μm	275,000	4G

TABLE 1.2 First Motorola Microprocessors [19]

Year	μP	Nb of MOS	Technology	Frequency
1974	6,800	5,000	N-MOS 6 μm	2 MHz
1979	68,000	68,000	N-MOS 4 μm	8 MHz
1984	68,020	200,000	CMOS 2 μm	16 MHz
1987	68,030	275,000	CMOS 1.3 μm	20 MHz
1989	68,040	2,000,000	CMOS 0.8 μm	25 MHz

80386. Table 1.1 presents some data about the first Intel microprocessors. It is interesting to note that CMOS was introduced in Intel microprocessors in 1985. Table 1.2 is a similar presentation for Motorola microprocessors, with a shift to CMOS in 1984.

1.3.5 RISC Machines

In 1981, two opposite approaches to designing future computers were possible [20]:

1. Continue the mainstream trend to design increasingly complex CISC machines.
2. Take the opposite direction and build simpler processors (i.e., RISC machines).

The alternative to complexity was obviously simplicity: less instructions, instructions executed in one clock cycle, load/store architectures, and hardware control units [20]. The 1975 IBM 801 is the first RISC machine, but the RISC concept was made popular by the RISC I and RISC II architectures from the University of California-Berkeley in 1980. The concept of RISC machines, however, rediscovered from Harvard Mark I and EDVAC early computers, is very beneficial for reducing power consumption.

The dramatic increase in the number of transistors per chip, as well as architectural advances, including the use of RISC ideas, pipelining, and caches, have produced an improvement in the performance of microprocessors at a rate of 1.5 to 2 times per year between 1980 and 1990 [21]. At that time, however, power consumption was still not an issue, the supply voltage was 5.0 volts, and no one predicted any change, as confirmed by Moore's law [22].

> "It's a 5 volt world, and to change to 1.5 volt would mean that the whole world would have to change!"
>
> **Gordon Moore**

1.4 Low-Power Consumer Electronics

Until 1990, power consumption was not an issue for a huge majority of ICs. Only a few niche applications had to take care of power, such as electronic watches, hearing aids, pacemakers, pocket calculators, pagers, and some battery-less applications. It was already for "portable" products, but only wristwatches, hearing aids, and pacemakers were, at that time, considered "portable." The history of each of these niche products would be quite interesting, but, for the most part, no documents are available. Recently, a book was published about the first electronic watch designed in Switzerland [23].

1.4.1 First Electronic Wristwatch

The Horological Electronics Center or Centre Electronique Horloger (CEH) developed the first electronic watch, a Swiss quartz watch named Beta [16, 23]. Such research was performed at that time without any public support, but pushed by some visionary people. Fortunately, this research was quite successful, producing the first quartz electronic watch. Commercialization was quite difficult, however, due to the structure of the Swiss watch industry. The impact of this research went far beyond the watch industry because it opened the way to very low-power ICs and microprocessors developed by CEH and, from 1984, by CSEM.

In the late 1950s, the Swiss watch industry was very successful in the production of mechanical watches. The president of the Swiss Horological Federation, Gérard Bauer, a visionary and powerful president, however, was under the impression that electronics could be a source of trouble for the watch industry. He was not an engineer, and he had been Swiss ambassador in Paris, so he was not able to support his contention with technical evidence, but his vision was that electronics or microelectronics would be a very dynamic discipline, producing many inventions and new products. He succeeded in convincing the Swiss watch industry to create a new laboratory known as CEH. CEH was officially created in Neuchâtel, Switzerland, on January 30, 1962, with the mission "to develop an electronic wristwatch with at least one advantage over existing watches." It was clear that few watchmakers believed that electronic watches were a major threat. It was decided that the CEH scientific results should be kept secret to prevent one company from using them before the other companies.

A silicon process was clearly required. Kurt Hübner, coming from Shockley Semiconductors (inventor of the transistor), was hired to set up a microelectronic technology. In 1963, the bipolar technology was more reliable than the MOS technology, and the first CEH circuits were bipolar circuits (Figure 1.8). It was the first silicon process installed in Switzerland. The goal was to have 1.0 to 1.3 volts and a 10-μA power consumption for the complete circuit. CMOS technology was chosen by CEH only a few years later — 10 μm in 1964 and 6 μm in 1966 — largely before the main semiconductor companies, which only switched to CMOS around 1984 (Table 1.1 and Table 1.2).

Under the direction of Max Forrer, several design projects were also defined, such as a 8.2-kHz quartz resonator (considered as a very risky project) quartz oscillators, frequency dividers in bipolar technology (consuming a few microwatts at 1.3 volts), and a vibrating motor at 256 Hz. So, from 8192 Hz, only five frequency dividers by 2 were necessary.

The first electronic watch was presented to the CEH board members in August 1967. Because all the research projects had been carried out secretly, it was a major shock for the Swiss watchmakers. A seminar

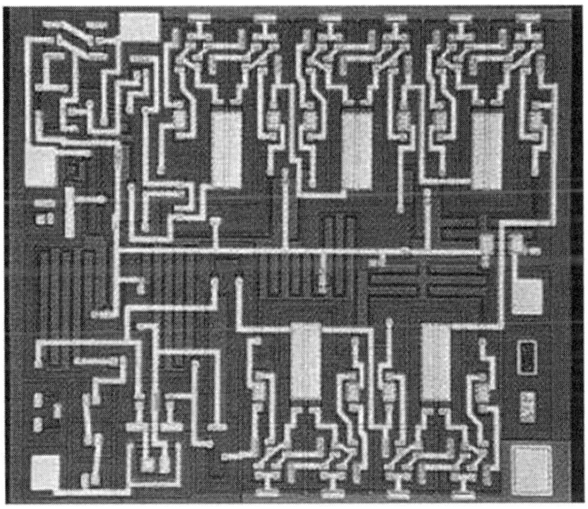

FIGURE 1.8 The bipolar circuit of the Beta 21.

was organized for CEH shareholders in December 1967, and this was the beginning of CEH's very strong reputation in low power.

The Beta wristwatch was able to reach a 1-year autonomy with the chosen battery. There were several bipolar chips: a 8192-Hz quartz oscillator and other circuits consisting of five frequency dividers to provide a 256-Hz signal used for the mechanical motor and the hands. The technology was a 10 μm bipolar process. Each divider by 2 consumed approximately 1 μA. The frequency divider was a flip-flop designed by Eric Vittoz with four NPN transistors and some integrated resistors and capacitors. Two frequency dividers (12 elements) were integrated on the same die of 2.1 mm^2. The motor control circuit was also a digital circuit, producing the right motor pulses. The total chip power consumption was 15 to 30 μA at 1.3 volts, and the chosen battery cell was supposed to deliver 18 μA during 1 year; therefore, the lifetime of 1 year was satisfied. Figure 1.9 is the Beta 2, with the electromechanical part on the left of the photograph and the printed circuit with the IC and the quartz crystal on the right.

FIGURE 1.9 The first Beta electronic watch.

Ten Beta electronic watches were presented to the "Observatoire de Neuchâtel" in December 1967. The result was 10 CEH watches at the first 10 places with only a few tenths of a second offset per day, followed by four other quartz watches from Seiko, Japan. This competition was suspended the next year. The improvement was a factor of 10 in precision compared with the mechanical wristwatches presented during the prior year.

1.4.2 Electronic Watches in Japan [24]

In the development of a quartz wristwatch, Seiko was close to CEH. Seiko began to look at quartz timekeeping in 1958 with the development of a quartz crystal clock. In 1959, Seiko was developing a quartz watch. Obviously, they had to reduce the size of a quartz-based clock to that of a quartz wristwatch. The result of this project was the world's first analog quartz watch to reach the market, the Seiko 35SQ Astron, introduced at Christmas 1969. The next Seiko, model 36SQC, was introduced in 1970; it was the first quartz watch to use a CMOS chip. Seiko and other Japanese watch companies, such as Citizen and Casio, quickly and successfully switched to electronics. As a result, Japan took the lead in worldwide watch production in 1978.

1.4.3 Electronic Watches in the U.S. [14]

The first electronic watch from the U.S. was announced in 1970 and introduced to the market by Hamilton in Fall 1971: a digital model called Pulsar. A button was necessary to display the time using light-emitting diodes (LEDs). The company bought the chip from RCA, which was the first U.S. company to produce CMOS chips.

Nevertheless, other U.S. companies were thinking of developing CMOS watch chips. For instance, Motorola offered the first integrated electronic watch kit to manufacturers in early 1972. The CMOS circuit, the quartz crystal, and a miniature microwatt motor were offered for only $15, thus beginning a move toward very cheap electronic watches.

In 1974, National Semiconductor introduced six watches. American Microsystems came out with a digital watch module that was smaller and consumed less power. By February 1975, about 40 companies were offering electronic watches with digital displays. Industry experts believed that solid-state digital watches would soon occupy a large part of the market, if not all of it.

In 1976, TI introduced its plastic-cased, five-function LED watch that sold for $19.95. As a result, National Semiconductor reduced its watch prices. Six months later, watchmakers were predicting a $9.95 digital watch would be on the market by Christmas. As liquid-crystal display (LCD) prices dropped, TI introduced a watch with liquid-crystal hands in August 1976.

By 1977, the price of digital watches had fallen from more than $100 to less than $10, in just two years. Profits evaporated. As with calculators, in 1977 only three real survivors remained: again, two Japanese competitors, Casio and Seiko, and TI [25]. Twenty years later, Intel chairman Gordon Moore was still wearing his ancient Microna watch ("My $30 million watch," he called it) to remind him of that lesson.

The U.S. watch market rise and fall was largely due to a brilliant but dangerous strategy by TI. By so dramatically reducing the TI electronic watch prices, they succeeded in eliminating all their U.S. competitors. Ultimately, however, TI found itself with watch prices that had been driven so low that the company did not make any profit. Worse than that, Japan's Seiko and Casio, with relatively low labor costs at that time, were fierce competitors. These firms soon did to TI what TI had done to its American competitors [25].

1.5 The Dramatic Increase in Power

Until 1990, CMOS integrated circuits, showing only moderate power consumption, were not designed with a serious interest in saving power [27, 28]. In 1992, however, the first very fast microprocessor, Alpha, which ran at 200 MHz, consumed about 30 watts [26]. It was a big shock for the semiconductor industry to see that power consumption was higher than expected. The second issue was the market growth of portable devices beyond the classical wristwatches, pacemakers, and pocket calculators (i.e., personal digital assistants (PDA), cellular phones, global positioning system (GPS) receivers, and palmtop computers and notebooks). New types of applications, such as ad hoc networks, did require extremely low power consumption for each network node. The reduction of power consumption was also in line with the global awareness of environmental issues.

Due to the exponential microprocessor frequency increase in the early 1990s, the increase of power dissipation, if it did not result in panic, was suddenly a major issue. A third low-power constraint was suddenly added to the well-known speed and silicon area constraints in the design of integrated circuits. It was also a major issue for chip cooling techniques [30]. In the early 1990s, therefore, engineers and managers were eager to learn how to reduce power consumption. Consequently, any course or conference labeled "low-power" was sure to be successful. Low-power postgraduate courses, such as Mead Education courses [29], attracted many participants in the early 1990s, perhaps more in the U.S. than in Europe.

1.5.1 Low-Power Workshops

The first "Low-Power" conferences and workshops were organized around 1993. Before the famous ISLPED (International Symposium on Low Power Electronics and Systems), some U.S. workshops were organized, such as the 1993 Low-Power Electronics Conference in Arizona and the 1994 Workshop on Low-Power Design in Napa, CA. These workshops were merged to create the first ISLPED conference that was held in 1995 at Dana Point, and is now regularly organized each year. The conference was held for the first time in Europe in 2000 (Rapallo, Italy), and was held for the first time in Asia (Seoul, Korea) in 2003. In 2001 and 2002, ISLPED was held at Huntington Beach, CA, and Monterey, CA, respectively.

Interestingly enough, it was in Europe that the first low-power workshops appeared, such as the PATMOS (Power and Timing Modeling Optimization and Simulation) project, followed by PATMOS workshops organized in 1993 in Montpellier, France, and in 1994 in Barcelona, Spain, with some attendees from the U.S. The PATMOS conference was originally a European project about timing and power modeling (1990–1993); however, it was decided to continue the organization of annual meetings on timing with more information on low-power issues. The PATMOS conference was then organized in 1995 at Oldenburg, Germany, Bologna, Italy, Louvain la Neuve, Belgium, Lyngby, Denmark, Kos Island, Greece, Göttingen, Germany, Yverdon, Switzerland, and Sevilla, Spain. It was organized in Torino, Italy, in 2003, in Santorini Island, Greece, in 2004, and will be in Leuven, Belgium, in 2005.

In 1999, the IEEE Alessandro Volta Memorial Workshop on Low-Power Design (VOLTA '99) was organized in Como, Italy. It was dedicated to low power as well as to recall that Volta invented the electric battery 200 years earlier in that town. A French-speaking conference, called Faible Tension Faible Consommation (FTFC), is also organized in Paris every 2 years since 1997, and the fourth edition was organized in May 2003.

One-day low-power workshops were also organized by Dimes Delft University, Netherlands, within the framework of the ESD-LPD (Electronic System Design — Low-Power Design) from 1997 to 2001. These 1-day workshops with invited speakers were usually held on the day before or after PATMOS, VOLTA, and ISLPED conferences. Some aspects of the projects of this European Low-Power Initiative are described in three books [31,32,33].

1.5.2 Low-Power Design Techniques

Many of the techniques described in the first low-power conferences were not really new ideas or concepts, but often the reuse of old techniques for achieving low power, such as asynchronous, pipelined or parallel machines (see Section 1.2), state assignment of finite state machines, reduced swing, and transistor sizing.

Pipelining and parallelism were proposed [11,13] to reduce power consumption by increasing the throughput of logic blocks and processors to reduce frequency and supply voltage. Pipelining is used today to reduce power consumption, as illustrated in Figure 1.10. A pipelined execution unit presents a shorter

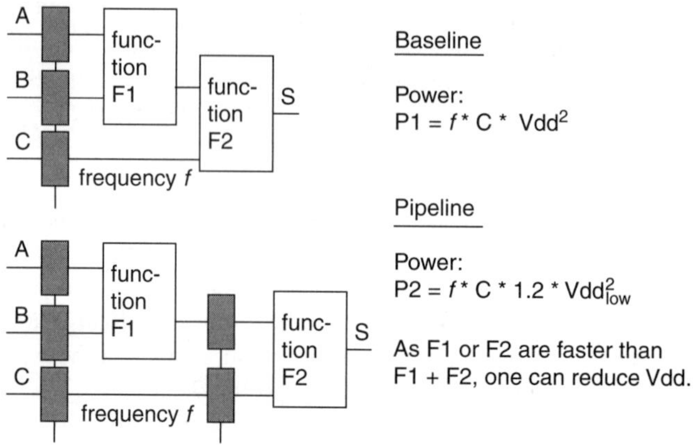

FIGURE 1.10 Pipeline for low power.

stage delay than a nonpipelined execution unit. It is therefore possible to work at the same operating frequency while reducing the supply voltage [11,13]. A lower Vdd helps to save a lot of dynamic power.

Parallelism has also been proposed [11] to lower frequency and supply voltage while maintaining the same throughput. Examples are parallel datapaths, memories (Figure 1.11), and shift registers (Figure 1.12). Figure 1.11 presents, for instance, interleaved memories accessed in an overlapped fashion at the

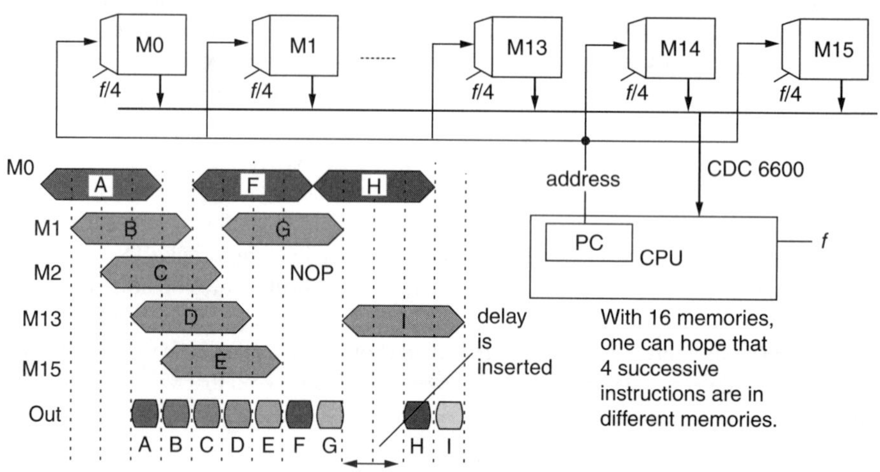

FIGURE 1.11 Interleaved memories.

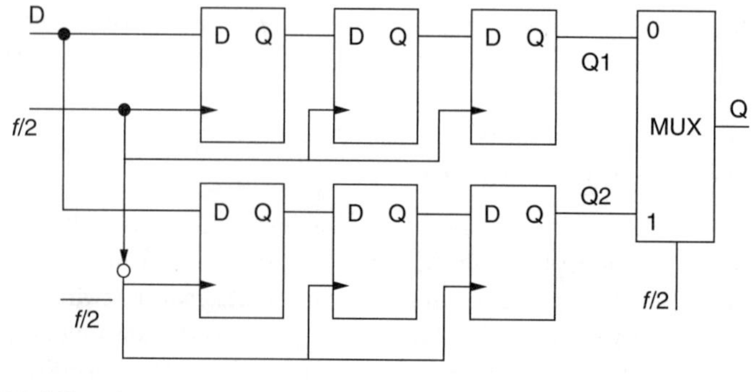

FIGURE 1.12 Parallel shift register.

frequency f/N. Each of the N memories (size 1/N) provides one instruction executed at the frequency f. The supply voltage of the memory modules can therefore be reduced because the access time is N times larger; however, this architecture comes from the early CDC 6600 computer [6].

The same idea has been proposed for parallel shift registers, as presented in Figure 1.12 [13]. The input is successively provided to the upper or to the lower half shift register at a reduced frequency, while the output multiplexer restores the output at the frequency f. The total number of D-flip-flops is the same as in the nonparallelized shift register. Parallelized shift registers with the same throughput present a reduced power consumption according to $P = f/2 \cdot C \cdot Vdd^2$. Furthermore, as flip-flops are working at a reduced frequency ($f/2$), the supply voltage can be reduced to save dynamic power. This idea was first proposed for bubble memories with major and minor loops [12].

Asynchronous or self-timed architectures have been proposed to reduce power consumption by removing the clock tree known to be a large consumer. Today, some asynchronous microprocessors are available, such as the Amulet, Titac, MiniMIPS, ASPRO, or Philips 8051. John Von Neumann proposed this idea for the first time for the Institute of Advanced Studies (IAS) computer designed at Princeton University. The IAS execution units worked at their own speed, and each unit had to send a completion signal to indicate when it had finished.

Some new but obvious ideas were also proposed, such as gated clock and activity reduction, which are applied today to a majority of ICs. Adiabatic logic was also proposed, which was an old idea rediscovered and adapted to modern technologies. Although supply voltage reduction had already been proposed in the 1960s [28], in the early 1990s, it was considered a major shift in the way circuits had to be designed. The trend to very low supply voltages down 0.2 volt is still a major issue. New schemes proposed include dynamic voltage scaling (i.e., supply voltage variation depending on the application load).

In 2003, the major themes in low-power conferences deal with the increasing complexity of SoCs in very deep submicron technologies. This complexity has to be considered at all design levels because most of the power can be saved at the highest levels. At the system level, which is strongly application dependent, designers have to consider many design parameters, such as partition, activity, number of executed steps, simplicity, data representation, locality, cache memories, and distributed or centralized memories. Furthermore, many new dramatic issues result from the use of very deep submicron technologies, such as leakage, variable V_T, very low supply voltages, interconnect delays, networks on chip, cross talk, and soft errors, and require very innovative design techniques. Today, leakage in standby and active modes is certainly a main issue in very deep submicron technologies, pioneered by some authors in Japan [36,37] and now studied all over the world.

1.6 Conclusion

The design of nearly 1 billion transistor chips, down to 0.10 μm and below, and supplied at less than 1 volt but working at several GHz, is a very challenging task. It was certainly considered an impossible task a few years ago.

The microelectronics revolution is fascinating: the transistor was invented only 55 years ago, and today we are using 130 and 90 nanometers technologies. For 2016, the 2001 SIA Roadmap predicts a 0.022-μm CMOS process (probably silicon-on-insulator (SOI)) with 16 billion transistors for high-performance chips, with 0.4 volts, 288 watts, and 28 GHz as the local frequency. As a result, more transistors exist in the world today (10^{17}) than ants (10^{16}).

What is the future of microelectronics? Are we close to the end of this marvelous story? Does the future belong to nanotechnologies that could completely replace microelectronics (although 0.015 μm transistors are 15 nanometers long)? Nanodevices have been constructed, and they are capable of switching a current or single electrons with a ratio between the on/off current of 1000 to 1 million. Such elements could be promising because their size of a few nanometers and their extremely low power consumption are very attractive. Carbon nanotubes, quantum dots, single-electron devices, or molecular switches are the most promising nanodevices. For instance, depending on its diameter, a carbon nanotube is a semiconductor device (otherwise, it is a conductor, and is not usable as a switch). If one of 10 nanotubes is a semiconductor, however, how do we select and interconnect the semiconducting ones to provide a useful logic function?

Quantum dots are based on the Coulomb blockade effect, and electrons are moved one by one from dot to dot. They have been constructed atom by atom by atomic force microscopes. Due to noise, it is better to construct cellular automata with several dots, and to define a given state of the automata as the logic "0" and another state as "1." Majority gates have been demonstrated as well as AND/OR gates. The main problem is still how to interconnect these gates to provide useful functions. Furthermore, it is hard to construct a complete chip atom by atom with several billion elements.

Design methods could be completely different from today because nanodevices could be constructed randomly, without any predefined schematic or layout; however, a useful function could emerge from this huge number of nanodevices, or some auto-organization could occur. It is somewhat similar to natural selection, for which only the useful functions will survive, but it will be hard to design a predefined and very complex function like a Pentium microprocessor.

Most likely, microelectronics will be used until about 2020. Nanoelectronics will probably not replace microelectronics because the two technologies will coexist with possibly different applications.

References

[1] C. Piguet, Are early computer architectures a source of ideas for low-power? Invited paper, Volta '99, Como, Italy, March 4–5, 1999.

[2] C. Piguet, Histoire des ordinateurs. Invited paper at FTFC '99, Paris, May 26–28, 1999, pp. 7–16.

[3] M. R. Williams, *History of Computing Technology, 2nd Ed.*, IEEE Computer Society Press, Los Alamitos, CA, 1997.

[4] H. D. Huskey, and V. R. Huskey, Chronology of computing devices, *IEEE Trans. on Computers,* Vol. C-35, No. 12, December 1976, pp. 1190–1199.

[5] C. Piguet, Babbage, l'inventeur de l'ordinateur. Invited paper at FTFC '01, Paris, May 30–31, June 1, 2001.

[6] J. P. Hayes, *Computer Architecture and Organization,* McGraw-Hill Book Company, New York, 1978.

[7] W. Aspray, *John von Neumann and the Origins of Modern Computing,* MIT Press, Cambridge, MA, 1990.

[8] H. H. Goldstine, *The Computer from Pascal to von Neumann*, Princeton University Press, Princeton, NJ, 1972.

[9] M. J. Flynn, *Computer engineering 30 years after the IBM Model 91, IEEE Computer,* Vol. 31, No. 4, April 1998, pp. 27–31.

[10] R. F. Krick and A. Dollas, The evolution of instruction sequencing, *IEEE Computer,* Vol. 24, No. 4, April 1991, pp. 5–15.

[11] A. P. Chandrakasan, S. Sheng, and R. W. Brodersen, Low-power CMOS digital design *IEEE J. of Solid-State Circuits,* Vol. 27, No. 4, April 1992, pp. 473–484.

[12] D. Rutland, *Why Computers Are Computers,* Wren Publishing, 1995.

[13] W. Nebel and J. Mermet, eds., Low-Power Design in Deep Submicron Electronics, NATO ASI Series, E 337, 1997, Kluwer Academic Publishers, Dordrecht, Chapters 4.2 and 9.1.

[14] Editors of Electronics, *An Age of Innovation, The World of Electronics 1930–2000,* McGraw-Hill, New York, 1981.

[15] R. Reid, *The CHIP: How Two Americans Invented the Microchip and Launched a Revolution, 2nd ed.,* Random House Trade Paperbacks, New York, 2001.

[16] C. Piguet, The First Quartz Electronic Watch. Invited talk at PATMOS, Sevilla, Spain, September 11–13, 2002, pp. 1–15.

[17] C. Mead and L. Conway, *Introduction to VLSI Systems,* Addison-Wesley, Reading MA, 1980.

[18] G. J. Myers et al., Microprocessors technology trends, *Proc. IEEE,* Vol. 74, No. 12, December 1986, pp. 1605–1622.

[19] *IEEE MICRO Issue,* December 1996.

[20] R. Bernhard, ed., More hardware means less software, *IEEE Spectrum,* December 1981, pp. 30–37.
[21] J. L. Henessy and N. P. Jouppi, Computer technology and architecture: an evolving interaction, *Computer,* September 1991, pp. 18–29.
[22] C. Freeman, *The Economics of Industrial Innovation,* Penguin Books, 1974.
[23] M. Forrer et al., *L'Aventure de la Montre à Quartz, Mutation Technologique Initiée par le Centre Electronique Horloger, Neuchâtel,* O. Attinger, ed., Neuchâtel, Switzerland, 2002.
[24] Smithsonian Museum, http://www.si.edu/lemelson/Quartz/index.html.
[25] M. S. Malone, *The Microprocessor. A Biography,* Springer-Verlag, New York, December, 1995.
[26] D. W. Dobberpuhl et al., A 200 MHz 64b dual issue CMOS microprocessor, *IEEE JSSC,* Vol. 27, No. 11, November 1992, pp. 1555–1567.
[27] J. D. Meindl, A history of low power electronics: how it began and where it's headed," *Proc. 1997 Int. Symp. On Low Power Electronics and Design, ISLPED '97,* August 18–20, pp. 149–151.
[28] J. D. Meindl, Low-*Power Microelectronics: Retrospect and Prospect, Proc. IEEE,* Vol. 83, No. 4, April 1995, pp. 619–635.
[29] Mead Education, http://www.mead.ch.
[30] A. Kaveh, The history of power dissipation, *Electronics Cooling,* January 2000, Vol. 6, No. 1.
[31] F. Catthoor, ed., *Unified Low-Power Design Flow for Data-Dominated Multi-Media and Telecom Applications,* Kluwer Academic Publishers, Dordrecht, 2000.
[32] J. Sparso, S. Furber, eds., *Principles of Asynchronous Circuit Design, A Systems Perspective,* Kluwer Academic Publishers, Dordrecht, 2001.
[33] D. Soudris, C. Piguet, and C. Goutis, eds., *Designing CMOS Circuits for Low Power,* Kluwer Academic Publishers, Dordrecht, 2002.
[34] W. F. Brinkman, D. E. Haggan, and W. W. Troutman, A history of the invention of the transistor and where it will lead us, *IEEE JSSC,* Vol. 32, No. 12, December 1997, pp. 1858–1865.
[35] Transistor Museum, http://semiconductormuseum.com/Museum_Index.htm.
[36] T. Sakurai, Perspectives on power-aware electronics, plenary talk 1.2, *Proc. ISSCC 2003,* San Francisco, CA, Feb. 9–13, pp. 26–29.
[37] K. Itoh, *VLSI Memory Chip Design, Springer Series in Advanced Microelectronics,* Springer-Verlag, New York, 2001.

2

Evolution of Deep Submicron Bulk and SOI Technologies

Marc Belleville
Olivier Faynot
CEA-LETI

2.1 Introduction

Metal-oxide semiconductor (MOS) transistor behavior has already been demonstrated at the research level, down to a 6-nm gate length on fully depleted silicon on insulator (SOI) [1]. As complementary metal oxide semiconductor (CMOS) technologies continue to shrink, however, new physical phenomena are becoming increasingly important in the device behavior, setting up new challenges especially for low-power design. Some authors are even suggesting that power consumption will set the limits of scaling on an application dependent way [2]. Compared to traditional CMOS bulk technologies, SOI technologies are foreseen as alternative technologies that could lead to a better trade-off between active and leakage power. After a brief overview of the various scenarios proposed by the International Technology Roadmap for Semiconductors (ITRS) [3], the four main causes of limitations are discussed:

1. Voltage limits and subthreshold leakage
2. Tunneling currents

3. Statistical dispersions
4. Poly depletion and quantum effects

In addition, for each of those limitations, design challenges and proposed solutions are briefly presented, for bulk and SOI technologies. Finally, new innovative transistor architectures and technologies are described, and their relevance regarding the previous problems discussed.

2.2 Overview of ITRS Roadmap

2.2.1 Major Evolutions

Moving a design from an old technology to a newer one, with smaller design rules, has always been, up to now, an interesting way to lower the power consumption. Indeed, the overall parasitic capacitances (i.e., gates and interconnects) are decreased, the available active current per device is higher, and, consequently, the same performance can be achieved with a lower supply voltage. Moving to a new technology generation, however, induces a scale down of the power supply voltage (Vdd), the threshold voltage (VT), and the gate oxide thickness (T_{ox}). Beginning with the 0.18-μ technologies, it appeared that building a transistor with a good active current (I_{on}) and a low leakage current (I_{off}) was becoming more difficult. Therefore, two families of transistors were introduced: high-speed transistors and low-leakage transistors. The threshold voltages of the two families are tuned differently, thanks to a different channel doping. When moving to more advanced technologies, those two families are not sufficient anymore, regarding technological constraints. The ITRS introduces three main groups of transistors:

1. High performance (HP)
2. Low operating power(LOP)
3. Low standby power (LSTP)

At this stage, the channel doping is not only different, but also the gate oxide thickness.

2.2.2 Bulk CMOS Technologies

Table 2.1 summarizes the main parameters required for the next generations of bulk metal-oxide semiconductor field-effect transistor (MOSFET) devices, in case of HP, LOP, and LSTP technology options. The HP technology uses the shortest gate lengths in order to achieve the higher drive current. A higher leakage current is also allowed in the technology. For the LOP technology, the main target is to reduce the operating power of the circuit. Compared to the HP technology, the LOP one uses a longer physical gate length, a thicker gate oxide in order to achieve a leakage current hundred of times lower, for a given node.

The main purpose of LSTP technology is to achieve transistors with a very low leakage current (roughly five orders of magnitude smaller than the HP technology). To satisfy this criteria, gate length and gate oxide scalings are relaxed, compared to both HP and LOP technologies. In addition, threshold voltage values must be significantly increased to lower the leakage current. As discussed in the following sections, many key issues have no available solution today.

How to shrink the gate length and achieve good performances
How to shrink the gate oxide thickness and match the leakage current targets
How to reduce the supply voltage, while keeping operational circuits and low leakage current

2.2.3 SOI Technologies

For several years, SOI technologies have been developed to improve the performance of bulk technologies. The main difference between bulk and SOI substrates is the buried oxide layer located below the active silicon layer (i.e., layer where the MOSFET devices are processed). Therefore, each transistor can be electrically isolated from the others. Depending on the silicon thickness used for the SOI wafer, the transistor can operate in partially depleted or fully depleted modes [4]. When the SOI film is thick enough

TABLE 2.1 Main Device Characteristics for HP, LOP, and LSTP Technologies, Based on ITRS 2002 Update [3]

Year	2001	2002	2003	2004	2005	2006	2007	2010	2013	2016
LOP										
Physical L_G (nm)	90	75	65	53	45	37	32	22	16	11
Physical EOT (nm)	2.0–2.4	1.8–2.2	1.6–2.0	1.4–1.8	1.2–1.6	1.1–1.5	1.0–1.4	0.8–1.2	0.7–1.1	0.6–1
V_{DD} (V)	1.2	1.2	1.1	1.1	1	1	0.9	0.8	0.7	0.6
I_{ON} (μA/μm)	600	600	600	600	600	600	700	700	800	900
I_{OFF} (pA/μm)	100	100	100	300	300	300	700	1000	3000	10000
LSTP										
Physical L_G (nm)	100	90	75	65	53	45	37	28	20	16
Physical EOT (nm)	2.4–2.8	2.2–2.6	2.0–2.4	1.8–2.2	1.6–2	1.4–1.8	1.2–1.6	0.9–1.3	0.8–1.2	0.7–1.1
V_{DD} (V)	1.2	1.2	1.2	1.2	1.2	1.2	1.1	1	0.9	0.9
I_{ON} (μA/μm)	300	300	400	400	400	400	500	500	600	700
I_{OFF} (pA/μm)	1	1	1	1	1	1	1	3	7	10
HP										
Physical L_G (nm)	65	53	45	37	32	28	25	18	13	9
Physical EOT (nm)	1.3–1.6	1.2–1.5	1.1–1.6	0.9–1.4	0.8–1.3	0.7–1.2	0.6–1.1	0.5–0.8	0.4–0.6	0.4–0.5
V_{DD} (V)	1.2	1.1	1	1	0.9	0.9	0.7	0.6	0.5	0.4
I_{ON} (μA/μm)	900	900	900	900	900	900	900	1200	1500	1500
I_{OFF} (pA/μm)	10,000	30,000	70,000	100,000	300,000	700,000	1E + 06	3E + 06	7E + 06	1E + 07

Note: L_G = gate length; EOT = equivalent oxide thickness.

(thicker than the depletion region), a neutral floating body region exists below the channel, inducing the partially depleted electrical behavior. The floating body effects increase the speed of the circuits. When the SOI film thickness is thinner than the depletion region, the entire film is depleted, inducing the fully depleted electrical behavior. In this case, the floating body effects are suppressed, and an ideal subthreshold swing of 60 mV/decade is theoretically achievable thanks to the constant depletion charge. This is a strong advantage compared to bulk devices. To achieve such ideal performances, ultra-thin silicon layers have to be used (with a ratio of 3 to 5 between the gate length and the SOI film thickness), which induce many technological issues, such as implantation-induced amorphization layer, thin SOI layer uniformity control, and silicon epitaxy growth.

SOI technologies are providing a complete set of transistors (HP, LOP, LSTP) just like bulk technologies. In the next sections, the advantages of SOI technologies compared with bulk technologies will be discussed.

2.3 Transistors Saturation and Subthreshold Currents

2.3.1 Subthreshold Leakage and Voltage Limits

The subthreshold current of a transistor is typically described by the following equation:

$$I_{OFF} \approx a\frac{1}{L_{EFF}}\exp\left(\frac{q(V_G - V_T)}{kT}\right) \tag{2.1}$$

where a is a constant, L_{EFF} is the effective gate length, V_G is the gate voltage, V_T is the threshold voltage, and kT/q is the thermal voltage.

In a typical scaling scenario, the electric fields are kept constant in the device by shrinking all the voltages and dimensions by the same factor. All doping levels are increased by the same scaling factor. As I_{off} increases exponentially when V_T decreases, however, static power consumption sets a lower limit to the scaling down of threshold voltages of the transistors. As the dynamic performance is directly related

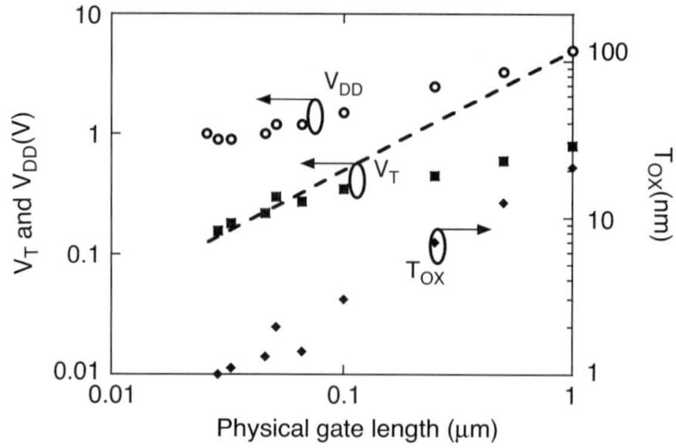

FIGURE 2.1 Supply voltage, threshold voltage, and gate oxide evolutions vs. gate length.

to the V_{dd}/V_T ratio, the power supply voltage also does not scale down easily. Consequently, in the ITRS roadmap scenario [3] (Figure 2.1) supply voltages do not shrink as rapidly as device dimensions. This results in a higher electric field in the device that has to be handled at the device level. Another consequence is the lower benefit granted to the dynamic power consumption which is proportional to Vdd^2. This is mainly because it is getting impossible to have simultaneously good active and leakage currents that several sets of transistors are required in advanced technologies.

To minimize the active power consumption, trading extra transistors against lower clock frequencies has been proposed by many authors and is well used today. The strong increase of the static power in respect to the active power in the upcoming technologies could lead to an opposite scenario: the number of transistors will have to be as small as possible regarding the targeted performance.

2.3.2 SOI Benefits

A significant active power reduction can be achieved by using SOI devices. Indeed, SOI devices are well-known for achieving the same performances than bulk devices, but with a lower power supply. This is achieved thanks to (a) lower parasitic capacitances (thick buried oxide for electrical isolation instead of junctions) and (b) lower threshold voltage in dynamic mode in case of partially depleted SOI devices. Active power reduction up to 50% can then be achieved with SOI [5, 6].

Another interesting advantage of single and multiple gates fully depleted SOI is their capability to achieve a nearly ideal (i.e., meaning 60 mV/decade at 300 K) subthreshold swing, compared to other devices. For a given OFF current, the fully depleted devices should achieve a higher drive current (due to a smaller threshold voltage) than its bulk counterpart.

To achieve subthreshold swing lower than 60 mV/decade (at 300 K), a new device structure is proposed [7]. According to the authors, 10 mV/decade can be achieved, which makes this type of device an excellent candidate for low leakage technology. The drawback is that it is a new kind of transistor, for which all the technology and design expertise have to be rebuilt.

2.3.3 Bulk CMOS Design Solutions for Subthreshold Leakage

For logical operators that can be stopped for a while, various techniques are used or foreseen to minimize standby power consumption. Those techniques can be used with the different kinds of CMOS transistors (HP, LOP, LSTP).

Triple well, or equivalent insulating technologies, are allowing an individual biasing of each independent pwell and nwell. Therefore, it is possible to tune the N-channel MOSFET (Nmos) and P-channel MOSFET (Pmos) substrate potentials to the required activity: a positive substrate potential (for an Nmos) will lower the threshold voltage of the transistors, therefore increasing its dynamic characteristics; on the contrary, a negative substrate voltage will increase the threshold voltage, consequently minimizing the subthreshold leakage. This technique, sometimes called variable threshold CMOS (VTCMOS) [8], requires efficient DC-to-DC converters. For a given technology, there is an optimum in reverse body bias, as the improvement in subthreshold leakage is compensated by an increase in source/drain to body junction leakage. Unfortunately, this technique is getting less effective with technologies scaling down [9].

With high-VT and low-VT transistors simultaneously available, other techniques are proposed: using low-VT transistors only in critical paths, or in multi-threshold CMOS (MTCMOS), introducing high-VT power switches to limit leakage current in standby mode [10]. A further level of optimization introduces multiple V_{dd} in a design [11].

More advanced concepts are now proposed to help minimize this subthreshold leakage. For instance, Abdollahi [12] is setting up, during sleep mode, the logical internal states so that the total leakage current of the circuit is minimized.

2.3.4 SOI CMOS Design Solutions for Subthreshold Leakage

Each SOI transistor has its own individual substrate usually called the "body." In fully depleted SOI technologies, the body potential roughly follows the source potential. No special design techniques are foreseen regarding subthreshold leakage in fully depleted SOI technologies; the advantage of this technology will be directly related to the better subthreshold swing of the devices. Regarding partially depleted SOI technologies, the body cannot be directly accessed in floating body SOI transistors. Only when using body-contacted SOI transistors, a body electrical node is available; the drawback is a much larger layout and because of that, their use is reserved to seldom situations when a floating body implementation is not possible.

Circuits based only on body-contacted transistors can use all the subthreshold leakage reduction mechanisms that are proposed for CMOS Bulk technologies. In MTCMOS SOI implementations, mixing floating body transistors for the logic and body contacted transistors for the standby current control can lead to a very efficient solution: driving independently the body of this transistor allows a low VT (and high current) in active mode, and a high VT in standby mode [13].

Regarding floating body transistors, substrate biasing, such as in triple-well technologies, is not possible. Other mechanisms have to be used to control the body potential and consequently the threshold voltage of the transistors. One very interesting feature of the partially depleted SOI transistors is the dynamic threshold modulation: the gate-to-body and drain-to-body capacitive couplings dynamically control body potential. Regarding subthreshold leakage, one key point is the ratio between those two capacitances. Figure 2.2 presents the current flowing through an inverter for three different values of this ratio. When the drain to body capacitance is lower than the gate to body capacitance, the leakage current of the inverter is, for a period after the input transition, lower than the DC leakage value. Then the body potential of the Nmos transistor comes back to a higher value, through the influence of DC phenomenon (impact ionization, gate tunneling current), and the leakage current rises back to the DC point. This low leakage state depends on the technology, but is frequently in the 100 microseconds–1 millisecond range.

Consequently, leakage control in SOI has to be preferably considered in a dynamic way and not in a static way. A first approach, for a block in standby mode, is to refresh the leakage current of the logic gates to lower values simply by generating one transition in the largest possible number of gates. Scan path, already inserted for test purpose can be used for this generation process.

Another dynamic approach to SOI standby was proposed by Morishita et al. [14] using a different way to control the body potential of SOI transistors. In this technique, the body potential is lowered by pulling out accumulated carriers from the floating body region with the forward biased current of the body source diode; consequently, the threshold voltage is increased. The extraction of the carriers occurs when the source line voltage is pulled down to a negative voltage; the source line is then pulled back up to 0 V. This technique requires additional switching transistors to pull up and down the power and ground lines; it also requires on chip DC-to-DC converters to generate the two extra power supply voltages. Notice however that as this occurs during standby mode, those power supplies have only to handle the leakage currents and the power line transitions. Refresh control is again required. Considering

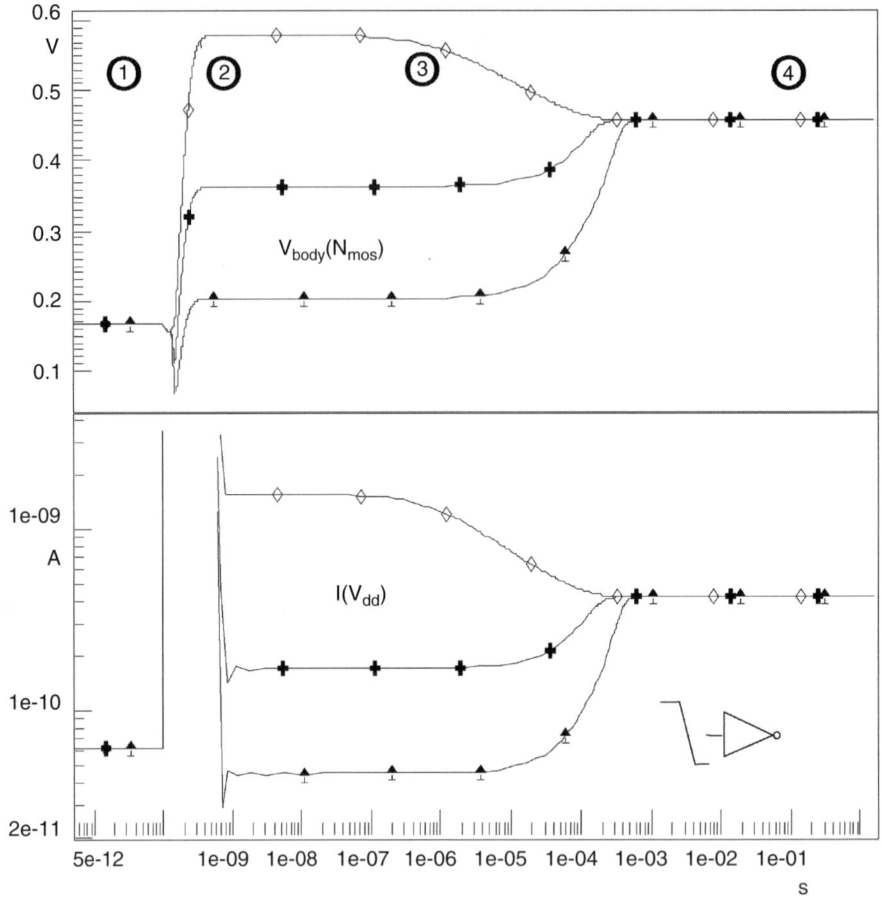

FIGURE 2.2 Power supply current of an inverter and body potential of the Nmos, before and after an input transition, for three different gate-to-body and drain-to-body capacitive ratios: 1 corresponds to the Pmos DC leakage; 2 to the input transition; 3 to the posttransition leakage evolution; and 4 to the Nmos DC leakage.

circuit architecture and layout, using this technique in memories like static random access memories (SRAMs) and standard-cell blocks is promising. The only constraint is that functions with independent shutdown modes cannot share the same rows.

2.4 Gate and Other Tunnel Currents

2.4.1 Tunneling Effects

Quantum mechanical tunneling of carriers through the energy barriers becomes more important as the dimensions of the transistors are scaled down to the nanometer range. Three main forms of leakage appear: the tunneling through the gate oxide, the band-to-band tunneling between body and drain and the direct drain to source tunneling current. The band-to-band tunneling current between the body and the drain of the transistor is strongly dependent on the electric field in the junction. The increased doping level of the body region, required by the scaling laws, emphasizes this tunneling current. Because direct band-to-band tunneling current depends on conduction-band states lining up with valence-band states, a forward bias of the body (with $V_{BS}>V_{DS}$) should limit this tunneling current. Nevertheless, this forward bias will induce a drastic increase of the direct junction leakage, which will not be acceptable. The easiest way to limit this tunneling current is to use intrinsic body devices with metal gate.

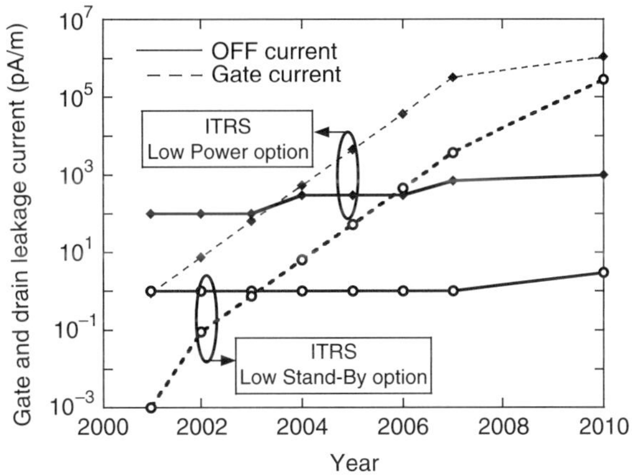

FIGURE 2.3 Evolution of drain and gate leakage current for LOP and LSTP technologies.

The second tunneling current is the direct source to drain tunneling, through the channel barrier. Recent analyses have shown that it becomes problematic for gate length smaller than 10 nm [15].

2.4.2 Gate Current

The third source of leakage current is the tunneling current through the gate oxide. The interface between silicon and silicon dioxide is still considered perfect in terms of abruptness and in terms of electrical properties. Silicon dioxide has carried us this far without limitations by extrinsic factors such as defect density, surface roughness or large-scale thickness, and uniformity control. Interface trap and fixed charge densities are so low that it corresponds to less than one surface defect in 10^5 surface silicon atoms [16]. This has made silicon dioxide become the only insulator used for MOS transistors. Based on the scaling laws, the gate oxide thickness must be few percent of the device gate length. This law implies that oxides thinner than 1.0–1.5nm must be employed for 35-nm gate lengths. These thicknesses comprise only a few layers of atoms and approach the fundamental limits. With such thicknesses, direct tunneling currents become very large, as shown in Figure 2.3 (data based on [17]). The basic equation describing this phenomenon is:

$$J_n \propto e^{\frac{-8\pi\sqrt{2m_{ox}}\Phi_B^{3/2}\left[1-\left(1-\frac{V_{ox}}{\Phi_B}\right)^{3/2}\right]}{3hqV_{ox}/T_{ox}}} \tag{2.2}$$

where Φ_B is the tunneling barrier height, m_{ox} is the oxide effective mass, T_{ox} is the gate oxide thickness, and V_{ox} is the voltage applied to the oxide [18].

This current is an exponential function of the gate oxide thickness and the applied voltage. As gate oxide decreases rapidly with the new upcoming technologies, the gate leakage current will become larger than the required leakage current of the transistor after 2004. Direct-tunneling current depends on a combination of tunneling probability and on the number of tunneling carriers. Due to the higher oxide tunneling barrier for holes (4.5 eV) than for electrons (3.1 eV) and due to the heavier effective mass of holes, the hole tunneling current is roughly one order of magnitude less than the electron tunneling current.

2.4.3 Design Issues and Possible Solutions

Several authors suggest that the upper limit of acceptable gate leakage current is in the range of 1–10 A/cm2 or even 100 A/cm2 [2] in very high performance chips. These currents become too large to accommodate the standby power requirements of integrated circuit (IC) applications. For low-power applications, the minimum oxide thickness is thicker (by 0.3–0.4nm) than for high performance. This thicker oxide reduces by two orders of magnitude the gate leakage current (for the same node).

When compared with subthreshold leakage for which several circuit design techniques are proposed to minimize it, circuit designers are facing with gate leakage a new problem with only limited solutions foreseen. A first possibility is to lower the power supply voltage in standby mode, taking advantage of the exponential relationship described in Equation 2.2. Another proposal [19] takes advantage of the difference between Pmos and Nmos gate leakage (one order of magnitude less for Pmos). For instance, by changing in MTCMOS logic Nmos gating to Pmos gating, leakage reduction between 41 and 60% are achieved [19]. This can be an interesting solution at some point; nevertheless, as gate leakage will increase exponentially from one generation to the next, solutions have to be found at the device level.

2.4.4 High-K Materials and Other Device Options

Many options are available to overcome the gate leakage problem. The first approach is to replace the silicon dioxide by a higher permittivity gate insulator. In such a case, due to the higher K, we can achieve a small electrical thickness with a thicker material. Up to now, the only successful insulator used is a silicon oxide/nitride composite. Higher-K insulators are currently studied (HfO_2, HfSiON), but even the most advanced works are still research works [20]. Nevertheless, as the dielectric constant of those binary insulators increases, their bandgap tends to decrease, as illustrated in Campbell et al. [21]. Because of the linear dependence on insulator thickness, and of the square root dependence on the barrier height shown in Equation 2.2, a large bandgap (hence a larger barrier height) is desirable when the aim is to reduce the gate tunneling current; however, other constraints exist on high-K materials. The two main constraints are: increased short channel effects and mobility degradation.

Due to its higher K value, the physical thickness of the insulator is larger than the one of silicon dioxide, and the ratio between their thicknesses is equal to the ratio between their permitivities. This thicker physical thickness increases the drain penetration under the gate, increasing drastically the short channel effects. In reference [22], it is shown that a permitivity ratio larger than 20 degrades significantly the scaling of the transistor. Therefore, the upper limit for the high-K permittivity and thickness is driven by the control of short channel effects, while the lower limit is driven by the tunneling current through the insulator [2].

The second drawback of using high-K material is related to the mobility degradation observed on MOSFET devices. Based on literature [23], from 10 to 40% of mobility degradation has been evaluated with high-K dielectric. Lots of analyses and modeling are under progress to explain this degradation. The high-K quality, the interface states density, as well as the fixed charges in the thin oxide localized between the silicon and the high K can be considered to be responsible for this degradation: the interface quality is not as good as the one obtained with silicon dioxide. Many experiments are in progress to explain and reduce this mobility degradation, which is one of the main problems to be solved for a use of high K in the semiconductor industry.

The second approach to limit the gate tunneling current is to stop the scaling of the gate oxide. Unfortunately, the use of a thicker gate oxide reduces the control of the gate on the channel conduction, leading to higher short channel effects and DIBL (drain induced barrier lowering) effects. Furthermore, the subthreshold swing expression is proportional to $1.0 + C_{si}/C_{ox}$, where C_{si} is the depletion layer capacitance, and C_{ox} is the gate capacitance. As C_{si} is proportional to $N^{1/2}$ (where N is the doping level), and C_{ox} proportional to $1/T_{ox}$, the gate oxide thickness is the most effective way to control the subthreshold leakage current. Thicker gate oxide will induce higher subthreshold swing value, and other parameters need to be scaled, in order to compensate this thicker gate oxide. Increased body doping and/or highly doped pockets can be used to optimize the device without reducing the gate oxide thickness, but the increase of the body doping reduces simultaneously the carrier mobility and the depletion depth of the channel region (i.e, reduces short channel effects). Forward biasing the body-to-source diode can also achieve this depletion depth reduction. The main drawback is an increase of the diode leakage currents and a degradation of the subthreshold swing of the device (and hence of I_{off}).

The third approach is the use of novel architectures, such as Double gate, FinFET, Triple gate, or Gate All Around, which are well known to improve the scaling of short transistors, compared to planar CMOS. These new types of architecture will be commented in the last section. With multiple gate devices, the vertical control is ensured by 2, 3, or 4 gates, enabling the gate control to be better than planar CMOS for a given gate oxide thickness. We could thus imagine using a thicker gate oxide in order to achieve the same short channel effects.

2.5 Statistical Dispersion of Transistor Electrical Parameters

2.5.1 Dopant Fluctuation

As the device dimensions shrink, the doping in the channel must be increased, in order to achieve acceptable short channel effects and VT, compatible with standard CMOS technologies. One of the main sources of variation in MOSFETs at the limit of scaling is randomness in the exact location of dopant atoms. Ion implantation leads to randomness at the atomic scale in the form of spatial fluctuations in the local doping concentration. This leads to dramatic variations of the transistor parameters. The number of dopant atoms is decreasing with scaling, as illustrated in Figure 2.4. For gate lengths smaller than 50 nm, the total number of dopant atoms is less than 100 in the device depletion region. As the standard deviation of fluctuations is equal to the square root of the number of dopant atoms, the $\pm 3\sigma$ boundaries shown in Figure 2.4 becomes extremely large when the gate length is scaled down to 30 nm.

Many publications have investigated the impact of the dopant fluctuations on the parameters of the transistors. Most of the publications use stochastically placed dopants in three-dimensional (3D) simulations [24–26]. Figure 2.5 illustrates an example of the discrete distribution of dopant atoms in the channel region. The fluctuation of the number of atom dopants induces a large variation of the parameters of the device (leakage current, subthreshold swing, etc.).

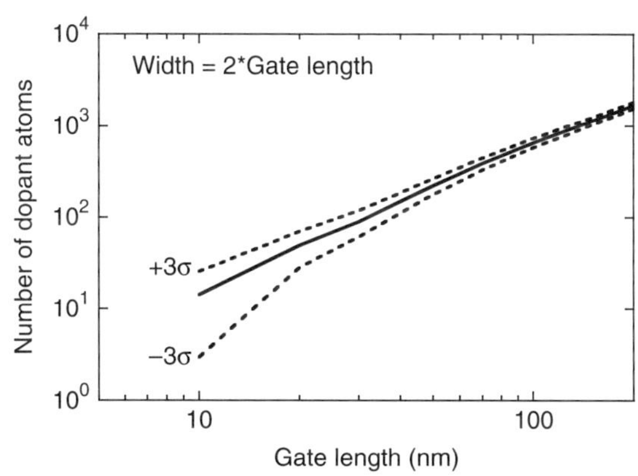

FIGURE 2.4 Variation of the number of dopant atoms and the associated fluctuations (in dashed lines) vs. gate length for the polysilicon gate.

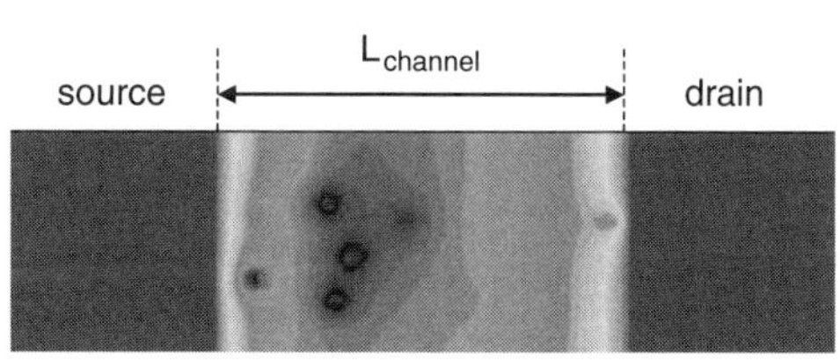

FIGURE 2.5 2D simulation of the discrete distribution of the dopant atoms in the channel region of a MOSFET transistor.

TABLE 2.2 Variation of the Number of Dopant Atoms and Its Associated Fluctuation Standard Deviation as a Function of Doping Level for a Transistor with L/W = 20 nm/40 nm

Doping level (at/cm^3)	10^{19}	10^{18}	10^{17}	10^{16}	10^{15}
N = Number of dopant atoms	93	28	8	2	0.7
$\sigma = \sqrt{N}$	9	5	2.8	1.5	0.8

It is obvious that such a device will be useless from a circuit point of view. Two solutions can be investigated to overcome this problem. First, the dopants can be moved in the body, back away from the surface. This will lead to retrograde channel doping profiles, for which the VT uncertainty is significantly reduced [26]. The second solution is the elimination of the dopants, combined with the use of metal gates. In such case, VT will be set by the gate work function, and not by the dopants. This trend is confirmed by the calculations presented in Table 2.2.

It is interesting to note that when metal gates are used with intrinsic devices, the number of dopant atoms can be smaller than one.

2.5.2 Design Issues and Possible Solutions

When considering digital circuit behavior, the major parameter to consider is the induced threshold voltage variation. An approximate expression for the delay variation of a logic gate is given as [27]:

$$\sigma Tpd \approx \frac{\sigma VT}{(Vdd - VT)^{2.5}} \tag{2.3}$$

When moving to advanced technologies, VT does not scale down as fast as Vdd; this is especially true for low-power devices. Consequently, the impact of the statistical VT variation on the gate delays will be very large. Eisele et al. [28] studied the impact of local delay variations due to VT variations on path delays in low-voltage circuits. For a given nominal VT, the relative delay variation increases with reduced logic depth, reduced supply voltages, and smaller device dimensions. In synchronous digital design, keeping a good yield consequently implies to increase the ratio "nominal path delay over clock period." One consequence is that the extra speed (or the lower active power) brought by a new technology will be partly lost in compensating for delay variations. Furthermore, all sensitive signals, like clock trees, will require large enough devices to minimize the statistical variations. Kishor et al. [27] compared the robustness of various CMOS logic design families regarding threshold voltage statistical variations; overall, static CMOS logic performs the best. It has also been demonstrated that the stability of memory cells is affected [29]. Self-Timed Logic, thanks to its inherent robustness and lower sensitivity to such variations could be a solution to overcome those limitations.

Regarding subthreshold leakage, dopant fluctuation induces variations on the threshold voltage, and on the subthreshold swing. As those variations are strongly non linear, the average standby current of a set of transistors will be higher than the sum of all the nominal standby currents. Figure 2.6 illustrates the impact of VT fluctuation on the static power consumption of a large number of inverters. The VT estimations are based on a doping level stable over the generations; as, when scaling down, doping levels are usually increased, the effective degradation will be worse. Another additive degrading factor will be the subthreshold swing fluctuation.

2.6 Physical and Electrical Gate Oxide Thickness

2.6.1 Poly Depletion

Polysilicon has been the widely used material for the gate electrode for more than 25 years. Its main advantage is that it can be doped n-type or p-type, leading to suitable work function for PMOS and

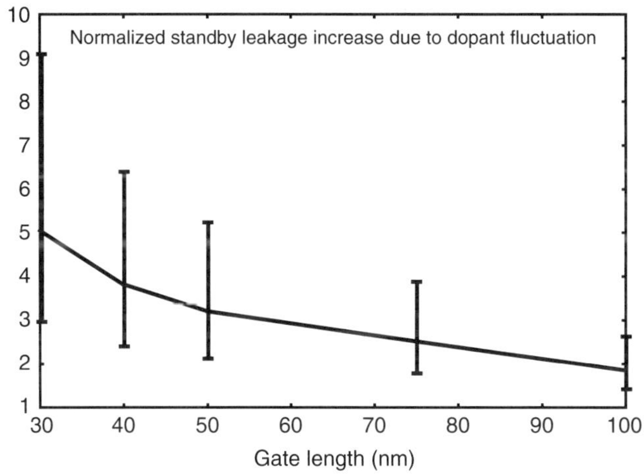

FIGURE 2.6 Normalized standby leakage increase for a set of 100 inverters (LOP), due to the impact of dopant fluctuation on VT (data from [30]).

NMOS devices. The polysilicon depletion effect occurs when the device is biased toward inversion. The applied voltage begins to deplete the highly doped polysilicon region near its interface with the gate oxide. The result of this effect is an apparent increase of the oxide thickness by a few angstroms.

As gate oxide thickness is reduced, the poly depletion effect becomes more severe. To prevent those effects, the increase of the active doping level near the gate oxide interface is required, but this increase is limited by two factors. The first factor is the maximum activation level that is not as good as the one of single crystal (maybe mainly due to the grain boundaries of polysilicon). The second one is the dopant outdiffusion from the polysilicon to channel through the gate insulator. This second effect is enhanced by the use of thinner and thinner gate dielectric, that makes the optimization of the polysilicon doping profile very difficult. This is particularly true with P+ gates because boron diffuses rapidly through SiO_2. The obvious solution to overcome this problem is the use of a metal gate instead of polysilicon. In such a case, the depletion is completely suppressed. Another important advantage of metal is related to the resistivity of the gate, which is significantly reduced compared to silicided polysilicon. This can be a strong benefit, especially for radio frequency (RF) applications. As the work function of metal is close to midgap, the use of very low doping levels in the channel region is required, and many efforts must be made on work function engineering to achieve multiple VT for NMOS and PMOS.

2.6.2 Quantum Effects

Quantum effects occur in the silicon because of carrier quantum confinement in a potential field. The carrier concentration is low at the interface, and the peak of carrier concentration moves deeper into the silicon, by a few angstroms [31] (Figure 2.7). If d is the effective distance of carriers below the interface, and ε_{SI} the silicon dielectric constant, the main contribution of quantum effects to the gate capacitance is the addition of an extra capacitance (ε_{SI}/d) in series with the gate oxide capacitance. This results in a thicker effective gate oxide thickness. For 1-nm oxide thickness, the increase of the electrical thickness can reach a few tenths of a percent.

2.6.3 Circuit Dynamic Performances

Combining poly depletion and quantum effects leads to an effective gate oxide electrical thickness larger; consequently, the saturation current will be smaller. For a given I_{off}, the circuit dynamic characteristics will be degraded. For a chain of unloaded inverters in 25-nm technology, the delay is expected to be slightly degraded (~5%), mainly because the input capacitance of the next stage also decreases; for heavily loaded inverters, delay degradation approaches those of the saturation currents (10 to 20%) [32].

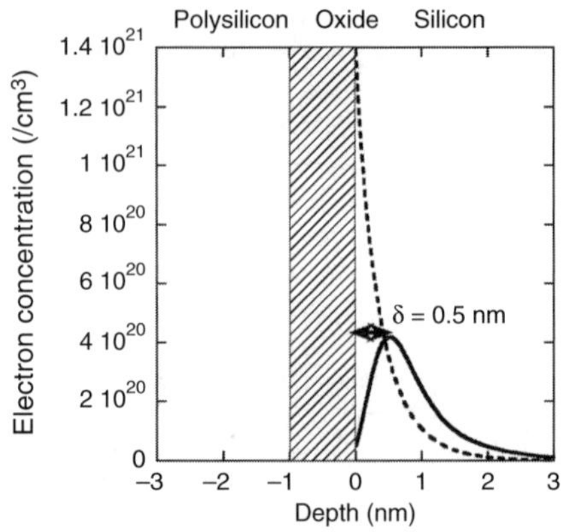

FIGURE 2.7 Quantum simulation of the inversion charge distribution with classical model (dashed line) and with quantum effects (solid line) for 1-nm gate oxide.

2.7 Innovative Transistor Architectures

This section presents innovative solutions for transistor architectures, in an attempt to solve most of the problems listed in the previous section.

2.7.1 Strained Silicon

To overcome the reduction of carrier mobility encountered in advanced devices (mainly due to the increasing doping level), two main types of strained material are under development. Strained SiGe substrates are the first option for the improvement of the mobility. By increasing the Ge concentration in the $SiGe_x$ alloy, the effective mass of holes is significantly reduced, leading to a strong improvement of the hole mobility in the case of long channel devices [33]. The drawback of such substrate is the decrease of the electron mobility. On short-channel PMOS, no significant improvement has been demonstrated, as illustrated on Figure 2.8 [34].

The second type of investigated material is strained silicon on relaxed SiGe. In such a case, both electron and hole mobilities can be slightly improved. As for the strained SiGe case, the gain in mobility decreases

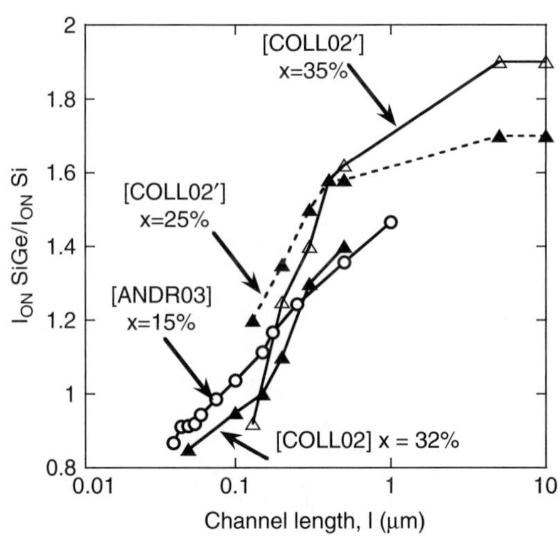

FIGURE 2.8 PMOS current gain as a function of channel length in case of strained SiGe substrates.

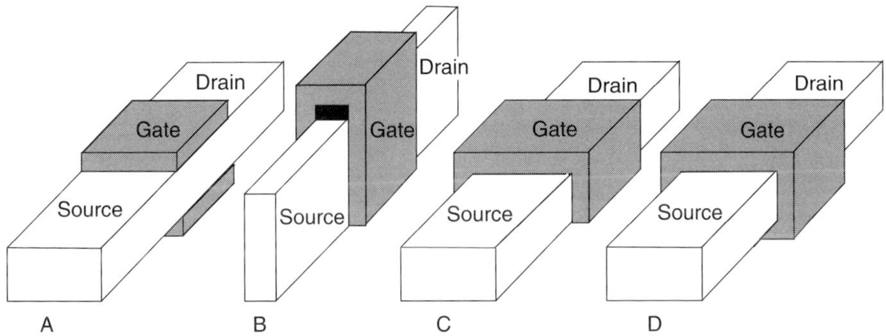

FIGURE 2.9 Scheme of multiple gate SOI devices. Planar double gate (a) FinFET, (b) triple gate, (c) quadruple gate, or (d) gate all around.

as the gate length is reduced. A 20–35% current gain has been obtained on NMOS for 70-nm gate length [35], but no significant gain has been published on short PMOS transistors. Only Intel has recently incorporated such material for its new 90-nm node.

Based on those two families of strained material, many other combinations of SiGe alloys on more or less strained layers are under analysis. Up to now, however, no real product has been fabricated with those materials.

2.7.2 Multiple Gate Devices

The use of SOI material makes the fabrication of multiple gate devices easier. Different kinds of transistors are developed: double gate [36], triple gate [37], FinFET [38] or gate all around [39]. Those architectures are summarized on Figure 2.9. Two, three, or four gates are used to control the channel regions, leading to an increase of the electric field induced by the gate. With multiple gates, the transverse electric field (i.e., gate-to-channel electric field) is reinforced compared to lateral electric field (i.e., source-to-drain electric field). The main purpose of these architectures is to:

- Improve the scaling of the transistors (i.e., by limiting DIBL and short channel effects for a given drive current) for gate length below 50 nm
- Achieve levels of performance that are as good as classical planar CMOS devices.

Planar Double gate devices [36] are the architecture that is the closest to classical planar devices. The channel conduction is achieved in the same <100> crystal orientation, compared to planar CMOS devices. The two channels (the top and the bottom ones) are controlled by two gates. In the ideal case, the current level and the gate capacitance are doubled, compared to classical devices. The main drawback of this device is the process [36] that has to be used to fabricate such a transistor: the bonding of SOI and bulk wafers, and the alignment of both top and bottom gates makes its manufacturability difficult to achieve with the equipment used today. Non-planar double gate (i.e., FinFET) and triple gate have significant easier processes compared to the previous transistor. The technology used is close to planar devices, and this makes these types of transistors very good candidates for the next generations. The main drawback is that most of conduction is ensured in vertical gates, for which the mobility and the interface quality are not well-known. Many publications are now focused on those aspects [41] to achieve drive currents that are as good as classical planar devices. The gate all around structure is the ideal case for short channel effect control: four gates control the body region potential. The gate all around process is comparable to the planar double gate process, in terms of complexity: the oxide below the channel region must be etched and the deposited gate must have the same length all around the transistor to ensure the ideal performances of the transistor (i.e., current level and gate capacitance multiplied by 4). If this is not the case (i.e., bottom and sidewall gates larger than the top one), the multiplication factor of the capacitance will be more than 4, while that of the current level will be less than 4.

With the improved performances achieved by those multiple gate SOI devices, we can imagine:

1. Having a lower I_{OFF} for a given VT criterion of a bulk transistor, possibly thanks to its steeper subthreshold swing
2. Having a lower gate tunneling current
3. Having smaller performance variations, with the combination of intrinsic devices and metal gate material

2.8 Conclusion

This chapter presented an overview of the problems generated by the scaling of MOSFETs transistors, outlining their impacts on circuit design. For each limitation, the investigated solutions have been presented. High-K materials are proposed to reduce the gate leakage current, metal gate is used to suppress the polysilicon gate depletion, and SOI technologies with single or multiple gate transistors offer opportunities for further scaling down of the transistor dimensions. Most of the proposed solutions for the transistor performance enhancement are new processes (high K, metal gate, SOI technologies). It is also clear that technology alone will not be able to solve all the problems foreseen in the coming technologies. Innovative circuit design techniques will be required to team with advanced devices in order to make circuits with acceptable dynamic performances and power consumption. Statistical dispersions and DC leakage currents, combined with low-voltage design, will be two challenges for circuit designers in the coming years.

References

[1] B. Doris et al. Extreme scaling with ultra-thin Si channel MOSFETs, *IEDM '02, Tech. Dig.*, pp. 267–270, 2002.

[2] D. J. Frank et al. Device scaling limits of SiMOSFETs and their application dependencies, *Proc. IEEE*, Vol. 89, No. 3, March 2001.

[3] The International Technology Roadmap for semiconductors, 2001 and 2002 update, http://public.itrs.net.

[4] C. Raynaud, SOI Process Integration, Short Course, *IEEE SOI Conf.*, Durango, Colorado, October 2001.

[5] L. E. Thon et al. 250–600 MHz 12b digital filters in 0.8–0.25 μm bulk and SOI CMOS technologies, *Int. Symp. on Low-Power Electronics*, pp. 89–92, 12–14 August 1996.

[6] M. Itoh et al. Fully depleted SIMOX SOI process technology for low-power digital and RF device, Silicon on Insulator Technology and Devices, *Proc. 10th Int. Symp. Electrochemical Society*, Washington, D.C., March 25–29, 2001, pp. 331—336.

[7] K. Gopalakrishnan et al. I-MOS: a novel semiconductor device with a subthreshold slope lower than kT/q, *IEDM Tech. Dig.*, pp. 289–292, 2002.

[8] T. Kuroda et al. "Variable threshold-voltage CMOS technology, *IEICE Trans. Electron.*, Vol. E83-C, No. 11, November 2000.

[9] A. Keshavarzi et al. Technology scaling behavior of optimum reverse body bias for standby leakage power reduction in CMOS IC's, *ISLPED '99*, San Diego, California, August 16–17, 1999, pp. 252–254.

[10] S. Mutoh et al. 1-V high-speed digital circuit technology with 0.5 μm multi-threshold CMOS, *ASIC Conf. and Exhibit*, September 27–October 1, 1993, pp. 186–189.

[11] T. Kuroda et al. Optimization and control of VDD and V_{th} for low-power, high-speed CMOS design, *ICCAD '02*, November 2002, pp. 28—34.

[12] A. Abdollahi et al. Runtime mechanisms for leakage current reduction in CMOS VLSI circuits, *ISPLED '02*, August 12–14, Monterey, CA, pp. 213—218.

[13] T. Douseki et al. A 0.5-V SIMOX–MTCMOS circuit with 200 ps logic gate, *Solid-State Circuits Conf., 1996. Digest of Technical Papers*, pp. 84–85, February 1996.

[14] F. Morishita et al. Dynamic floating body control SOI CMOS for power-managed multimedia ULSIs, *IEICE Trans. Electron.*, Vol. E84-C, No. 2, pp. 253–259, February 2001.
[15] Y. Naveh et al. Modeling of 10-nm scale ballistic MOSFETs, *IEEE Electron. Device Lett.*, 21, 242–244, 2000.
[16] J. D. Plummer et al. Material and process limits in silicon VLSI technology, *Proc. IEEE,* Vol. 89, No. 3, March 2001.
[17] S. H. Lo et al. Quantum mechanical modeling of electron tummeling current from the inversion layer of ultra-thin-oxide nMOSFETs, *IEEE Electron. Device Lett.,* 18, 209–211, 1997.
[18] W. C. Lee et al. Modeling gate and substrate currents due to conduction- and valence-band electron and hole tunneling, *VLSI Symp., 2000,* pp. 198–199.
[19] F. Hamzaoglu et al. Circuit-level techniques to control gate leakage for sub-100nm CMOS, *ISPLED '02,* August 12–14, Monterey, CA, pp. 60–63.
[20] A. L. P. Rotondaro et al. Advanced CMOS transistors with a novel HfSiON Gate dielectric, *VLSI Symp. 2002, Tech. Dig.*, p. 148.
[21] S. A. Campbell et al. MOSFET transistors fabricated with high permitivity Ti02 dielectrics, *IEEE Trans. on Electron. Devices,* Vol. 44, p. 104, 1997.
[22] D. J. Frank et al. Generalized scale length for two-dimensional effects in SOI MOSFETs, *IEEE Electron. Device Lett.*, Vol. 19, pp. 385–387, October 1998.
[23] B. Guillaumot et al. 75-nm damascene metal gate and high K integration for advanced CMOS devices, *IEDM '02, Tech. Dig.*, pp. 355–358, 2002.
[24] H. S. Wong et al., Three-dimensional atomistic simulation of discrete microscopic random dopant distributions effects in nanometer-scale MOSFETs, *Microelectron. Reliability,* 38, 1447–1456, 1998.
[25] H. S. Wong et al., Discrete random dopant distribution effects in nanometer-scale MOSFETs, *Microelectron. Reliability,* 38, 1447–1456, 1998.
[26] D. J. Frank et al. Monte Carlo modeling of threshold variation due to dopant fluctuations, *Symp. on VLSI Technol., Dig. of Tech. Papers,* pp. 1169–170, 1999.
[27] M. Kishor et al. Threshold voltage and power supply tolerance of CMOS logic design families, *Symp. on Defect and Fault Tolerance in VLSI Systems,* October 25–27, 2000, Yamanashi, Japan, pp. 349–357.
[28] M. Eisele et al. The impact of intra-die device parameter variations on path delays and on the design for yield of low voltage digital circuits, *IEEE Trans. on VLSI Systems,* Vol. 5, No. 4, pp. 360–368, December 1997.
[29] D. Burnett et al. Implications of fundamental threshold voltage variations for high-density SRAM and logic circuits, *Symp. on VLSI Technol., 1994,* pp. 15–16.
[30] A. Asenov, Efficient 3D "atomistic" simulation technique for studying of random dopant induced threshold voltage lowering and fluctuations in decanano MOSFETs, *6th Int. Workshop on Computational Electronics, 1998,* pp. 263–266.
[31] J. D. Plummer et al. Material and process limits in silicon VLSI technology, *Proc. IEEE,* Vol. 89, No. 3, pp. 240–258, 2001.
[32] Y. Taur et al. 25-nm CMOS design considerations, *Int. Electron. Devices Meeting, IEDM '98,* December 6–9, 1998, pp. 789–792.
[33] M. V. Fischetti et al. Band structure, deformation potentials, and carrier mobility in strained Si, Ge, and SiGe alloys, *J. Appl. Phys.*, Vol. 80, No. 4, pp. 2234–2252, 1996.
[34] F. Andrieu et al. "SiGe channel p-MOSFETs scaling-down, to be published at *ESSDERC '03,* Estoril, Portugal, September 16–18, 2003, pp. 267–270.
[35] K. Rim et al. Strained Si MOSFETs for high-performance CMOS technology, *Symp. on VLSI Technol.* pp. 59–60, 2001.
[36]] H. S. P. Wong et al. Self-Aligned (Top and Bottom) Double Gate MOSFET with a 25nm Thick Silicon Channel, *IEDM Tech. Dig.*, 1997, pp. 427–430.
[37] J. T. Park et al. Pi-gate SOI MOSFET, *IEEE Electron. Dev. Lett.*, Vol. 22, No. 8, pp. 405–408, 2000.
[38] D. Hisamoto et al. FinFET- A self-Aligned Double-Gate MOSFET Scalable to 20nm, *IEEE Trans. Electron Dev.*, Vol. 47, No. 12, pp. 2320–2325, 2000.

[39] J. P. Colinge et al. Silicon-on-Insulator Gate-All-Around device, *IEDM Tech. Dig.*, pp. 595-598, 1990.

[40] J. H. Lee et al. Super self-aligned double-gate (SSDG) MOSFETs utilizing oxidation rate difference and selective epitaxy, *IEDM Tech. Dig.*, 1999.

[41] Y. K. Choi et al. FinFET process refinements for improved mobility and gate work function engineering, *IEDM Tech. Dig.*, pp. 259–262, 2002.

3

Leakage in CMOS Nanometric Technologies

Antoni Ferré
Joan Figueras
UPC

3.1 Introduction

This chapter is devoted to the characterization of the different sources of leakage current that appear in complementary metal-oxide semiconductor (CMOS) devices; physical origins of I_{LEAK} are outlined and data on how voltage and temperature affect the various components of I_{LEAK} are provided. In addition, the impact of technology scaling is presented. This characterization is used in order to predict the circuit level impact of I_{LEAK} of a CMOS circuit. The characterization of leakage components in nanometric technologies is an important issue: Whereas old technology, long channel transistors had basically one leakage mechanism, the well-known reverse-biased pn junction leakage, deep submicron, and nanometric transistors may have different leakage sources depending on the fabrication process.

3.2 I_{LEAK} Components of MOSFET Devices

Different physical phenomena contribute to the leakage currents causing the static consumption when one or more transistors in the V_{DD} to *GND* paths are in OFF-state. These currents are listed next, and are separated into five classes according to the physical origin of the current:

1. Tunneling currents of electrons across the thin gate oxide between the gate and the substrate I_G [1, 6], due to the high electric field in the gate oxide. The responsible mechanism in nanometric devices is direct tunneling through the oxide bands.
2. Subthreshold conduction producing leakage currents I_{SUBTH}, which flow from the drain to source [1, 2]. When the MOSFET has a gate voltage below the threshold voltage, the device surface is in weak inversion or depletion. When gate-to-source voltage V_{GS} is applied, even below the device

threshold voltage, sufficient charge carriers are on the surface region that can still create a significant current flow [3].

3. Gate-induced drain leakage I_{GIDL} currents flowing from the drain to the substrate. These currents are due to the tunneling of electrons from the valence to conduction band in the transition zone of the drain-substrate junction below the gate-to-drain overlap region where a high electric field exists [4, 5].
4. Reverse-biased pn junctions in the circuit. The leakage currents I_D of reverse-biased pn junctions are due to various mechanisms such as diffusion and thermal generation in the depletion region of the junctions [7, 8]. In nanometric technologies, junction-tunneling current due to bulk band-to-band tunneling (BTBT) current I_{BTBT} may appear [5, 9].
5. Bulk punchthrough current I_P from the source to the drain due to lateral bipolar transistor formed by the source (emitter), the bulk (base), and the drain (collector) [3]. If the drain voltage is large enough to deplete the neutral base region, a direct current I_P flows between the source and drain.

In general, CMOS technologies have one dominant OFF-state leakage mechanism and may have some secondary OFF-state leakage mechanisms. These mechanisms have evolved due to technological changes in MOSFET fabrication. The evolution is illustrated in Figure 3.1 for an OFF-state NMOS transistor.

Old technologies using long channel transistors, approximately defined as those above 0.7–1 μm channel length, had normally reverse-biased pn junction leakage current of the drain–substrate (well)

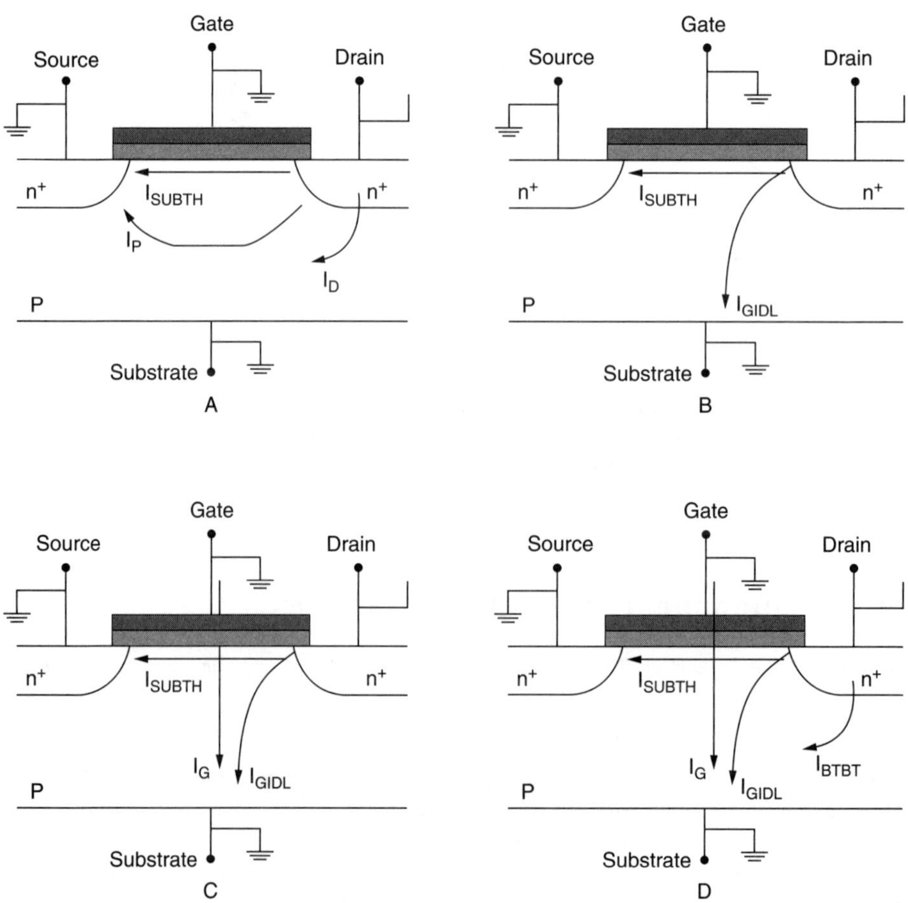

FIGURE 3.1 An NMOS transistor with its leakage currents depending on technology: (a) $L \geq 500$ nm; (b) 500 nm $\geq L \geq 100$ nm; (c) 100 nm $\geq L \geq 50$ nm; (d) 50 nm $\geq L$.

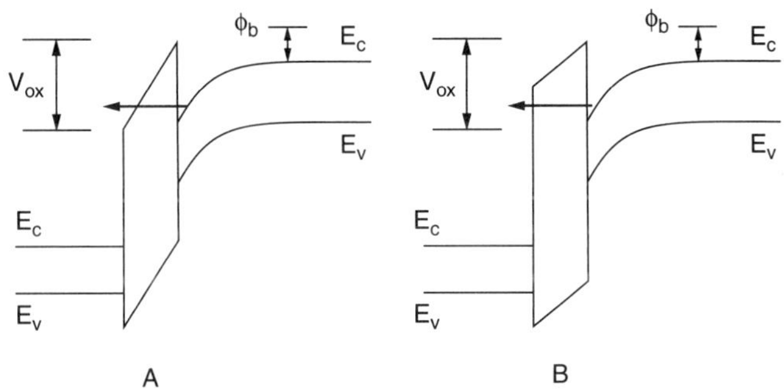

FIGURE 3.2 Gate Tunneling Mechanisms in MOSFETs: (a) Fowler–Nordheim Tunneling; (b) Direct Tunneling.

and substrate–well pn junctions as the dominant mechanism. The contribution from subthreshold leakage currents, the secondary mechanism, was usually negligible [10]. As the technology reached the 0.5 μm, the dominant mechanism changed to subthreshold leakage current [11, 12]. The punchthrough current was present also as secondary mechanism in some technologies. It is usually negligible in present technologies. This component is well controlled by raising the impurity concentration in the bulk channel region [3].

For submicron technologies below 0.5 μm, the dominant mechanism is the subthreshold leakage current. As a secondary leakage mechanism, reverse-biased pn junction leakage and gate-induced drain leakage current have been reported [10, 13].

For nanometric technologies, below 100 nm, the decrease in the gate oxide thickness needed to achieve a high current drive capability and to reduce the short-channel effects causes the magnification of nonideal effects such as gate tunneling currents. For ultrathin gate oxides, the direct tunneling increases and becomes one of the dominant leakage current mechanisms.

For sub-50-nm MOSFETs, the body-to-drain junction tunneling current is expected to become one of the dominant mechanisms due to high doping concentration [1].

Let us review the main characteristics of these currents and quantify the evolution when scaling down the technology.

3.2.1 Gate Tunneling Currents

For nanometric technologies, tunneling currents become a major issue. These currents are also greatly enhanced when scaling down the technology. Gate direct-tunneling current is produced by the quantum mechanical wave function of a charged carrier through the gate oxide potential barrier into the gate, which depends not only on the device structure, but also on its bias conditions [23–26].

The high electric field in the gate oxide may cause tunneling currents through the gate by means of two mechanisms: direct tunneling or Fowler–Nordheim tunneling through the oxide bands as illustrated in Figure 3.2. For the voltages and structures of modern MOSFETs, direct tunneling is the dominant component. Fowler–Nordheim tunneling typically appears when the oxide layer is thicker than 6 nm, and the applied field is higher than the electric field found at present day technologies.

The gate-tunneling current from the Si inversion layer to the poly-Si gate has been traditionally computed using an independent electron approximation and an elastic tunneling process. Because the exact form of the electronic tunneling barrier is not generally known, the potential barrier was commonly assumed as triangular for potentials higher than the Si/SiO_2 barrier voltage $\phi_b = 3.2V$. Whenever oxide voltage is lower than 3.2 V the electron-tunneling barrier changes from being triangular to trapezoidal. The silicon surface is strongly inverted or strongly accumulated, and the surface electric field on the silicon side and the potential barrier on the SiO2 side confine electrons at the Si/SiO2 interface. In these cases, the contribution to the leakage due to direct tunneling is given by [27]:

$$J_G = J_0 \cdot E_{ox}^2 \cdot e^{-k \cdot tox} \quad (3.1)$$

where J_0 is a technology dependent parameter adjusted to match experimental data, E_{ox} is the oxide electric field, and t_{ox} is the gate oxide thickness. The imaginary part of the wave vector k when a V_G voltage is applied is given by [27]:

$$k = \frac{2k_0}{3} \frac{\phi_b}{V_G} \left[1 - \left[1 - \min\left(1, \frac{V_G}{\phi_b} \right) \right] \right] \quad (3.2)$$

where V_G is a technological parameter. These expressions show the high sensitivity of direct tunneling on oxide thickness and power supply voltage. In Figure 3.3, the tunneling current density is plotted as a function of gate bias for several oxide thicknesses.

Temperature variations have a low impact on gate tunneling. The gate leakage current depends on temperature through the energy ground level, which leads to a reduction of the effective barrier height ϕ_b. The temperature also affects the mean-free path of electrons in the oxide conduction band. Consequently, when the ambient temperature is increased, the gate tunneling increases very slightly [28].

Note that the tunneling leakage in current SiO_2 dielectrics will be dominating in NMOS devices because PMOS has a higher barrier for hole tunneling and, therefore, a lower leakage current. With high-k dielectrics, such as Si_3N_4 dielectric, however, the gate current is higher in p^+ poly-Si PMOS than in n^+ poly-Si NMOS, and the scaling limit due to excessive tunneling leakage current will be first reached for PMOS.

In MOSFETs having ultrathin gate oxide thicknesses (1.4–2.4 nm), a direct current of electrons from n^+ poly-Si to underlying n-type drain extension in off-state n-channel contributes also to the gate leakage current [29]. This effect was reported experimentally by Henson *et al.* [30] and Yang et al. [31]. They found that the gate current weakly depends on channel length. The off-state bias configuration $V_{GS} = 0$ V and $V_{DS} = V_{DD}$ exhibited a non-negligible gate current but not as significant as the case when $V_{GS} = V_{DD}$ and $V_{DS} = 0$. This means that the tunneling current is localized in the edge region for the off-state condition. Figure 3.4 illustrates these various gate-tunneling components in a scaled NMOS; the gate-to-channel current I_{go}, and the direct tunneling current appearing between the source drain extension (SDE) and the gate overlap, usually called the edge direct tunneling (EDT) currents (I_{gso} and I_{gdo}) and become dominant in front of the gate-to-channel current I_{go}. In long-channel devices, EDT currents are less important than because the gate overlap length is small compared to the channel length. In very short channel devices, the portion of the gate overlap compared to the total gate length increases.

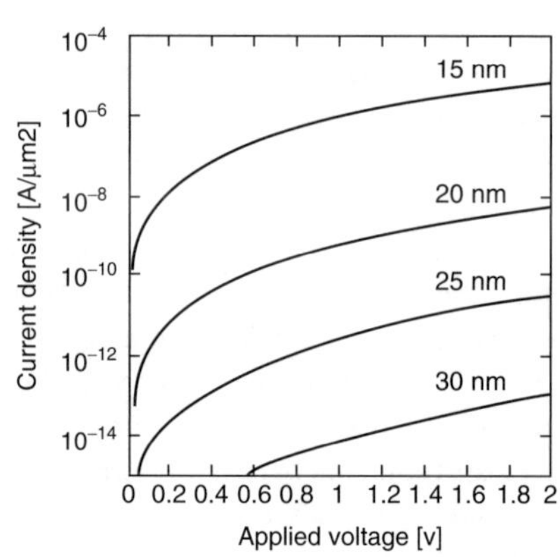

FIGURE 3.3 Gate-tunneling current as a function of applied voltage and oxide thickness.

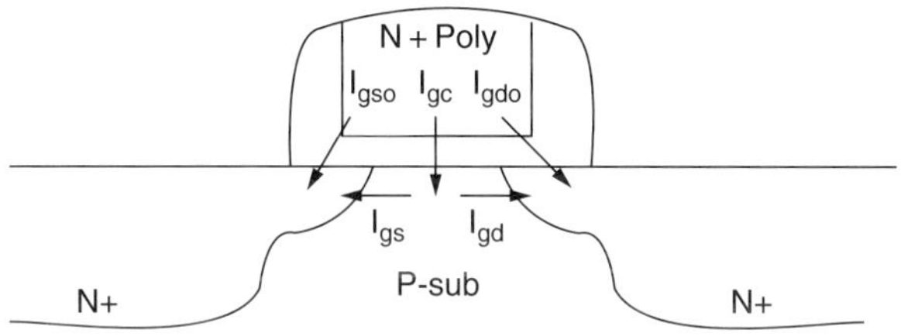

FIGURE 3.4 Gate-tunneling current components.

Different models have been proposed for EDT currents. Yu *et al.* [32] presented a model where gate direct tunneling currents are described using voltage-dependent current sources as a function of the terminal voltages. The partitioning of channel gate current is modeled by using variable resistances in each part of the channel. The channel currents of each region are obtained by adjusting the BSIM3-model parameters to fit the current-voltage curves obtained from simulation at device level.

Lee and Hu [33] proposed an accurate dielectric leakage model for metal-oxide semiconductor (MOS) capacitors based on modeling the electron conduction band (ECB), electron valence band (EVB), and hole valence band (HVB) currents. A physical source-drain current partition model is introduced using these contributions. This model has been implemented into the BSIM4 transistor model [34].

We propose a simpler but sufficiently accurate model. The use of two antiparallel current sources with an exponential (diode-like) dependence connected between the poly gate and each edge of the channel allows the modeling of the current flowing in both directions (see Figure 3.5). Currents flowing through D_{GS} and D_{SG} account for $I_{gso} + I_{gs}$, while currents flowing through D_{GD} and D_{DG} account for $I_{gdo} + I_{gd}$. The voltage drop by poly-Si gate is also taken into account by introducing a series resistance (R_{poly} in Figure 3.5). A similar model is used for PMOS with the only difference that the current driven by the antiparallel current sources is smaller than for NMOSs according to experimental data for transistors using SiO_2 as dielectric. Similar models can be derived for high-k dielectrics. The parameters of each current source are adjusted to match as close as possible the direct tunneling currents.

Comparison performed between the values obtained from expression and the simplified model shows an error lower than 2.5% through all the input gate voltage range $[0, V_{DD}]$ for $V_{DD} = 1.5$V. Figure 3.6 and Figure 3.7 show the simulation results of this model for a device with $L = 70$ nm at different V_{DS} values. The observed “dips” in the measured gate tunneling currents in these Figures are due to the combined

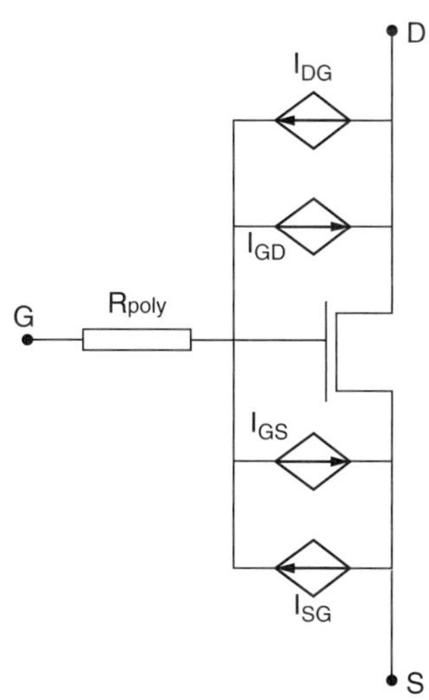

FIGURE 3.5 Gate-tunneling current model using antiparallel current sources.

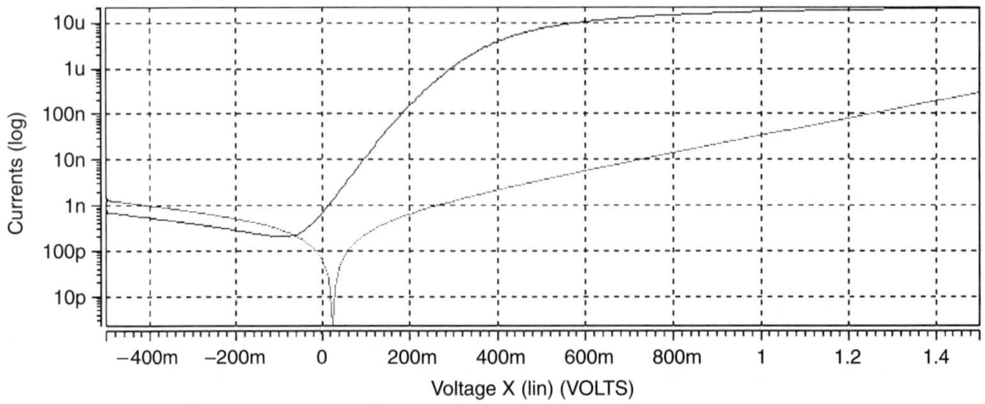

FIGURE 3.6 Leakage Current Simulation using MOSFET model including antiparallel current sources. W = 140 nm, L = 70 nm, V_{DS} = 0.05 V.

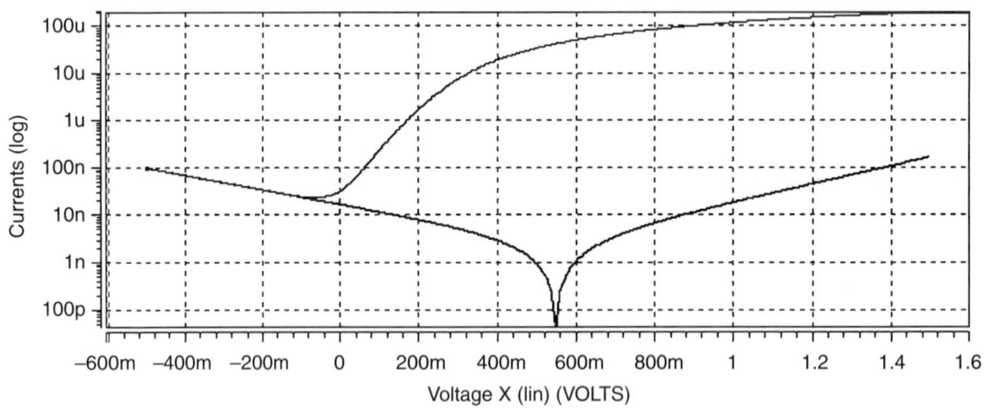

FIGURE 3.7 Leakage Current Simulation using MOSFET model including antiparallel current sources. W = 140 nm, L = 70 nm, V_{DS} = 1.5 V.

effect of source-to-channel tunneling and drain-to-gate tunneling currents in opposite directions. This effect shows the importance of the EDT currents, especially for small transistors.

This model is used later (see Section 3.4) to estimate the leakage of CMOS circuits.

3.2.2 Subthreshold Leakage Currents

When gate voltage is lower than the threshold voltage and there is a voltage applied between drain and source of a MOS transistor, a diffusion current appears due to the different carrier concentrations at the inversion layer in source and drain terminals. This current depends exponentially on gate-to-source voltage V_{GS} and drain-to-source voltage V_{DS} through the carrier concentrations. For an NMOS transistor, the subthreshold current is given by [14]:

$$I_{SUBTH} = \mu_N C_{ox} \frac{W_N}{L_N} V_t^2 \exp\left[\frac{V_{GS} - V_{TH}}{nV_t}\right]\left[1 - \exp\left[-\frac{V_{DS}}{V_t}\right]\right] \tag{3.3}$$

where μ_N is the electron carrier mobility, C_{ox} is the gate capacitance per unit area, W_N is the channel width, L_N is the channel length, V_t is the thermal voltage, and V_{TH} is the threshold voltage. The inverse slope of the subthreshold current n is given by [14]:

$$n = 1 + \frac{C_D}{C_{ox}} \tag{3.4}$$

where C_D is the depletion channel region capacitance per unit area. The subthreshold parameter n is related to the subthreshold swing S — the gate voltage change needed to raise the subthreshold current by one decade:

$$S = \ln 10 \cdot n \cdot V_t \tag{3.5}$$

At room temperature, the minimum theoretical value for S is about 60 mV/dec ($n = 1$). For present CMOS technologies, S takes values in the range of 80-90 mV/dec.

The threshold voltage for an NMOS long channel transistor with substrate bias V_{BS} is expressed by [14]:

$$V_{TH} = V_{TH0} + k_1\sqrt{\phi_S - V_{BS}} - k_2 V_{BS} \tag{3.6}$$

where V_{TH0} is the long-channel threshold voltage without substrate-bias, ϕ_S is the surface potential. The body effect and the nonuniform doping effect are modeled using the parameters k_1 and k_2.

In short-channel transistors, V_{TH} is further modified as a function of the channel length and the drain-to-source bias: these are the so-called short-channel and drain-induced barrier lowering (DIBL) effects [15].

Figure 3.8 illustrates the origin of short-channel effect. In a long-channel device, the depth of the depletion region in source and drain regions is relatively unimportant. As the channel length is reduced, however, these depletion regions occupy more space of the channel region. The depletion regions near the source and drain edges are shared with the channel. This effect produces a reduction of the threshold voltage when decreasing channel length and, therefore, increases subthreshold current.

The short-channel effect may be modeled following the Phillips model by reducing the effective threshold voltage as a function of the effective channel length L_{eff} [16]:

$$\Delta V_{TH}^{SCE}(L_{eff}) = \frac{u_{L1}}{L_{eff}} - \frac{u_{L2}}{L_{eff}^2} \tag{3.7}$$

where u_{L1} and u_{L2} are technology dependent parameters.

The drain-induced barrier lowering (DIBL) effect consists of lowering the energy barrier between the source and the channel. This causes excess injection of charge barriers into the channel and gives rise to an increased subthreshold current. Figure 3.9 presents qualitatively the band diagram at the interface of the channel, for short- and long-channel transistors. At the interface, the channel consists of three regions: the source-channel junction, the middle region, and the drain-channel junction. For the long-channel

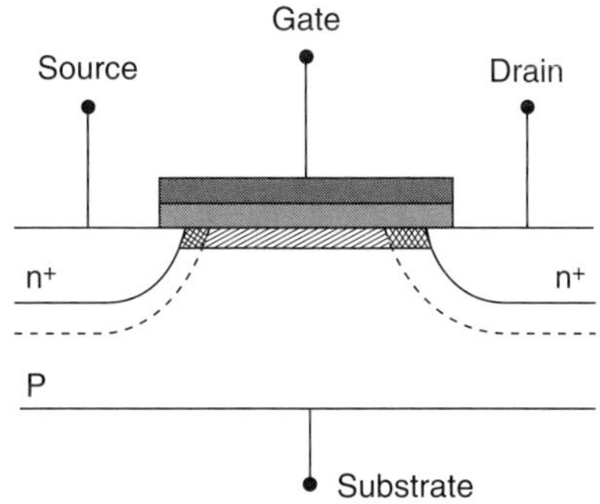

FIGURE 3.8 Physical origin of short-channel effect.

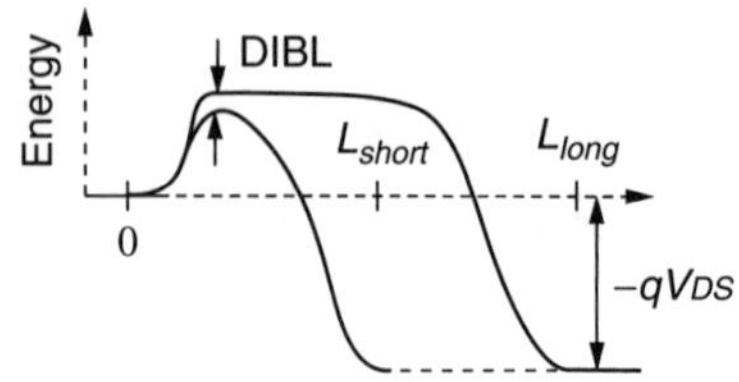

FIGURE 3.9 Physical origin of the DIBL effect.

transistor, the energy bands in the central part of the channel can be taken to be approximately constant because the voltage almost drops at the drain-channel junction. As channel length is reduced, however, this situation is no longer true. Consequently, a reduction in the interface energy barrier occurs at the source-channel junction where the maximum of the barrier is reached. This is the so-called DIBL effect [15]. The DIBL effect may be modeled, following the Phillips model, by additionally reducing the effective threshold voltage as a function of the drain to source voltage V_{DS} in the following amount [16]:

$$\Delta V_{TH}^{DIBL}(L_{eff}, V_{DS}) = \frac{S_L}{L_{eff}^2} \cdot (\phi_S + V_{SB})^{1/2} \cdot V_{DS} \tag{3.8}$$

where S_L is a constant for a given technology, ϕ_S is the surface potential, and V_{SB} is the source-to-bulk voltage.

The leakage current I_{LEAK} also depends on the power supply voltage V_{DD} through the dependence on the drain-to-source voltage V_{DS}. Two main factors are responsible for this:

1. Carrier concentration at the drain. This factor refers to the term

   $$\left[1 - \exp(-V_{DS}/V_t)\right]$$

   in Equation 3.3. For

   $$V_{DS} \geq 4 \cdot V_t$$

   this effect becomes negligible.
2. DIBL effect. For reduced channel lengths, the leakage current I_{LEAK} depends exponentially on V_{DS}. If the channel length increases, the DIBL effect reduces. For large enough channel lengths, I_{LEAK} will be almost independent on V_{DS}.

Another important issue is the variation on the subthreshold current due to the temperature. MOS transistor characteristics are strongly dependent on temperature. One of the main parameters responsible for this is the effective mobility, which is known to decrease with temperature [17]:

$$\mu(T) = \mu(T_r)\left(\frac{T}{T_r}\right)^{-\kappa_1} \tag{3.9}$$

where T is the absolute temperature [in Kelvin], T_r is room absolute temperature [in degrees Kelvin], and κ_1 is a constant technology-dependent. This value varies usually from 1.2–2.0 [17]. Other temperature-dependent parameters are the surface potential ϕ_S (through the variation of the intrinsic concentration) and the flat-band voltage V_{FB} (trough the variation in the work-function ϕ_{MS}). These effects are manifested in the value of the threshold voltage, V_{TH}, as an almost straight-line decrease with temperature [17]:

$$V_{TH}(T) = V_{TH}(T_r) - \kappa_2(T - T_r) \tag{3.10}$$

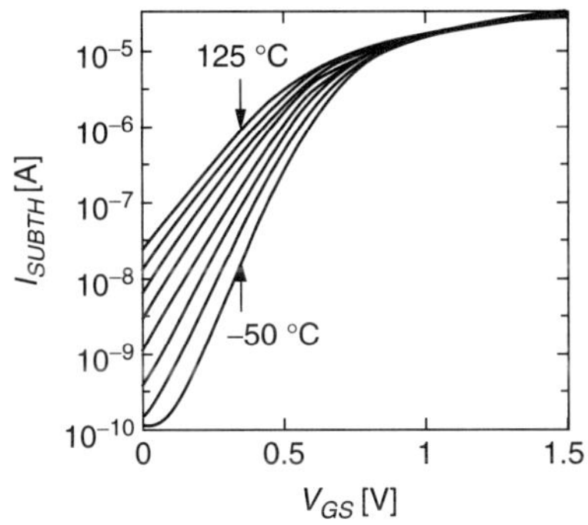

FIGURE 3.10 Subthreshold leakage current in an NMOS (L = 70 nm) depending on temperature.

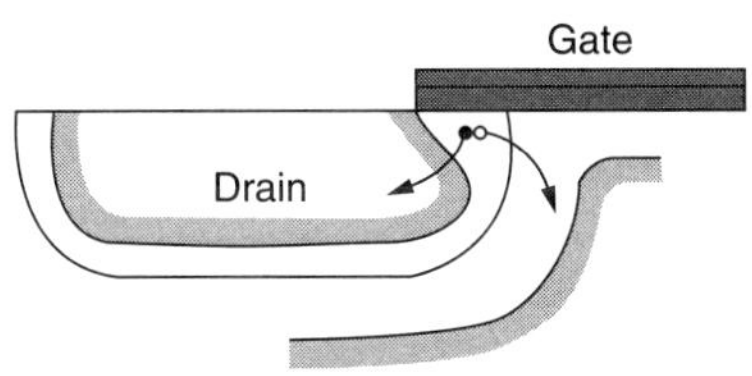

FIGURE 3.11 Physical origin of GIDL current.

where κ_2 is usually between 0.5 and 3 mV/K [17], with larger values in this range corresponding to heavier doped substrates, thicker oxides, and higher values of V_{BS}.

Thus, a temperature increase tends to increase the drain current (exponentially in the subthreshold region) through the threshold voltage variation and to decrease it through the mobility variation. At the subthreshold region, the decrease of the threshold voltage dominates. Therefore, increasing the temperatures produces an exponential increase in subthreshold current.

The temperature also affects the slope of the leakage current curves through the thermal voltage. Figure 3.10 plots the variation of leakage current of a 70-nm NMOS transistor as found by HSPICE simulation.

3.2.3 Gate-Induced Drain Leakage Currents

In some nanometric technologies, gate-induced drain leakage (GIDL) current I_{GIDL} may appear, usually for relatively high power supply voltages [2, 11]. I_{GIDL} current of a NMOS transistor flows from the drain to the substrate. This is caused by the effects of the high electric field region under the gate in the region of the drain overlap as illustrated in Figure 3.11. In this region, pair creation can occur.

Several possible mechanisms contribute to this current, which are presented in Figure 3.12. These include thermal emission, trap-assisted tunneling, and band-to-band tunneling [2, 18, 19]. It is the BTBT that has the metal-oxide semiconductor fluid-effect transistor (MOSFET) relevance at the voltages and structures of modern devices. This current is due to the direct tunneling of electrons from the valence to conduction band in the gate-to-drain overlap region where a high normal electric field E_n exists. This current may be further enhanced because the generated carriers are accelerated by the longitudinal electric field E_l in the drain-substrate junction, and this causes impact ionization [20].

The expression to estimate this leakage component as a function of the longitudinal E_l and normal E_n components of the electric field in the gate drain overlap area is [20]:

$$I_{GIDL} = A_{b1} \cdot W \cdot E_n \cdot e^{\frac{-B_{b2}}{E_n}} \cdot E_l \cdot e^{\frac{-B_{b2}}{E_l}} \tag{3.11}$$

where A_{b1}, B_{b1}, and B_{b2} are technology dependent parameters. W is the width of the device. In Figure 3.13, the contributions of subthreshold and GIDL currents in a 70-nm NMOS with V_{DD} = 1.5 V are

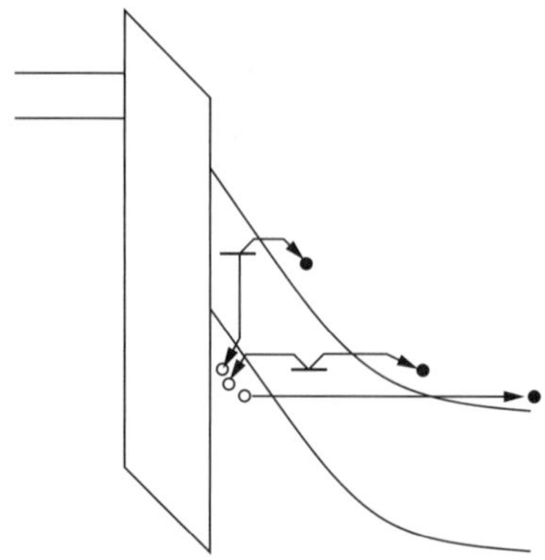

FIGURE 3.12 Tunneling mechanisms in GIDL: BTBT and trap-assisted tunneling.

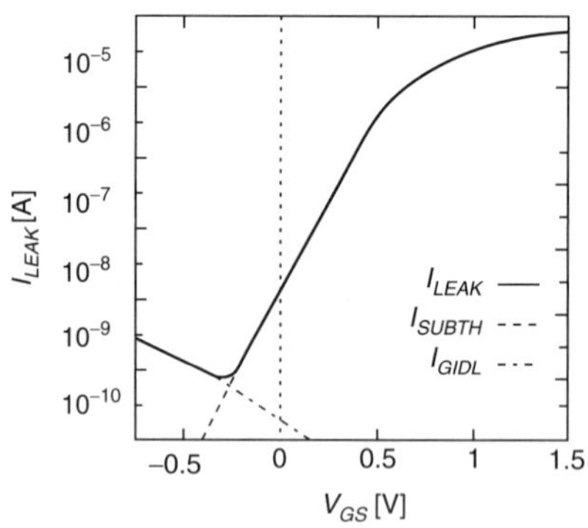

FIGURE 3.13 Subthreshold and GIDL leakage currents in an NMOS (L = 70 nm).

plotted shown as well as the total leakage current I_{LEAK}. Notice the significant effect of the I_{GIDL} leakage as the gate voltage in an NMOS transistor is brought to high negative levels.

The dependence on power supply voltage also depends on power supply. The increase in V_{DD} implies an increase of the normal electric field and, therefore, an exponential increase of I_{GIDL}.

GIDL currents may be especially important in buried channel devices. Comparison of GIDL in surface channel PMOS devices (p^+ poly gate) and buried channel PMOS devices (n^+ poly gate) have shown that buried channel PMOS has higher GIDL than the equivalent surface device for a given supply voltage [21]. This is due to the large flat band voltage between the n^+ poly gate and the p^+ junction in the overlap region of the drain. In this case, the GIDL current may become the dominant component of leakage current as illustrated in [22]. GIDL may be also a limiting factor when applying leakage reduction techniques such as body bias control (BBC) [66].

3.2.4 Junction Leakage Currents

In micrometric technologies, the leakage currents of reverse-biased pn junctions I_D are due to two mechanisms: diffusion of carriers and thermal generation currents in the depletion region of the silicided junctions [7]. Generation and diffusion currents depend strongly on temperature, through the intrinsic carrier concentration n_i. For low to moderate temperatures, I_D is dominated by generation mechanisms and increases with temperature at a rate proportional to n_i. At high temperatures, I_D is determined primarily by diffusion mechanisms and increases more rapidly at a rate proportional to n_i^2 [35]. For present day technologies, reverse-biased pn junction leakage current is lower than subthreshold leakage current and can be neglected.

For sub-50-nm technologies with highly doped pn junctions, the narrow depletion region produce tunneling currents I_{BTBT} [1, 67]. In addition, if V_{DD} increases and approaches the junction breakdown voltage, avalanche current appears from impact ionization in the depletion region [1, 27]. An important contribution to the total leakage in these devices is observed in [67].

3.2.5 Punchthrough Currents

In CMOS circuits, parasitic lateral bipolar transistors formed by the source (emitter), the bulk or the well (base) and the drain (collector) of MOS transistors are formed. If the drain voltage is large enough to deplete the neutral base region, the potential barrier height between the source and the channel region is lowered not only by the gate bias but also by the drain bias. Therefore, a punchthrough current I_P flows between the source and drain. The punchthrough current causes a large leakage current not controllable by the gate bias voltage. This component is well controlled by raising the impurity concentration in the bulk channel region [3].

3.3 Scaling

For the last four decades, silicon technology has been progressively reducing the channel length of MOSFETs from 25 μm at 5–10 V supply voltage to nanometric lengths and power supplies below 1 V in current production technologies. To maintain the transistor performance at lower voltages, the oxide thickness has also been reduced from 100 to a few nanometers in accordance with Dennard constant-field scaling law. [36–44].

As discussed in the previous section, as technology evolves and channel length becomes nanometric, total leakage current increases. This fact is mainly due to: (a) the lowering of threshold [1], which increases the subthreshold current, (b) the increased short-channel effects when reducing the channel length [45], which also increases the subthreshold current, and (c) the reduction of oxide thickness, which increases the gate tunneling current.

3.3.1 Scaling of V_{TH} and its Impact on Subthreshold Current

To avoid reliability degradation effects such as hot-carrier injection or oxide breakdown due to the high electrical fields, reduction of the power supply voltage is required. Reduction of the supply voltage has a negative effect on circuit performance: propagation delays may increase as the supply voltage decreases. To reduce these undesirable effects, the threshold voltage should be reduced. The impact of threshold voltage reduction on subthreshold current and, consequently, on leakage current consumption is illustrated in Figure 3.14. I_{TH} is defined to be the drain current when the gate voltage is equal to the threshold voltage. If the threshold voltage is reduced from V_{TH} to V'_{TH}, the OFF-state leakage current I_{OFF}, defined as the drain current when the gate voltage is zero, increases exponentially. Because S is about 80–100 mV/dec at room temperature for present technologies, a reduction of 80–100 mV in V_{TH} implies an increase of I_{OFF} by an order of magnitude.

3.3.2 Short-Channel Effects

Short-channel effects are another main factor responsible of the increase for leakage of current when downsizing the technology, as we have seen before. These undesirable effects may be reduced by scaling the gate length, the source-drain junction depth, the bulk doping concentrations and the gate oxide thickness. All these quantities must be scaled together. The use of shallow drain/source junctions reduces the charge shared by the junction and the channel; however, these ultrashallow junctions increase the resistance of the device and, therefore, degrades its performance. To improve the performance, the use of elevated source and drain structures has been introduced [46].

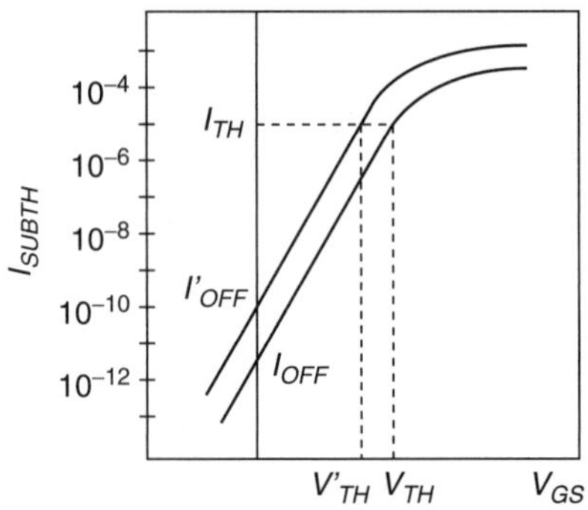

FIGURE 3.14 Impact of scaling on subthreshold leakage current.

In CMOS technologies, new doping profiles have been introduced to reduce short-channel and DIBL effect. By counter-doping the channel surface, threshold voltages can be fixed while still obtaining acceptable V_{TH} roll-off characteristics [47]. In addition, the locally high doping concentration in the channel near source-drain junctions (also known as *pocket* or *halo* implants) improves the short-channel effects [48]; however, many trade-offs exist between the improvement of short-channel immunity and the other device electrical performance that needs further research. These doping techniques may enhance I_{GIDL} and BTBT in general.

Scaled technologies have thin oxides and high channel doping concentrations. If the power supply voltage is kept at a high value, high normal electric fields may appear and, therefore, GIDL currents, even using LDD [22]. In addition, Guo *et al.* [19] have shown that the difference of GIDL between single-diffused drain (SD) devices and diffused drain (DD) devices is reduced when forward substrate bias is applied. Applying a band-trap-band tunneling model, they found the lateral electric field E_l due to the V_{BD} voltage and the ratio of lateral field to total field (E_l/E_{ox}) as the two key factors responsible for the tunneling barrier lowering and enhancement of I_{GIDL} [49].

3.3.3 Gate-Tunneling Currents

The exponential increase of direct-tunneling current should be considered due to the reduction of the oxide thickness in order to better control the inversion layer by the gate should be considered. Previous results show the high sensitivity of direct tunneling on oxide thickness and power supply voltage. Furthermore, as mentioned in previous section, scaled gate oxide thickness approaches the direct-tunneling regime, the EDT of electron from n^+ poly-Si to underlying n-type drain dominates the gate leakage. This phenomenon is more pronounced for thinner oxide thicknesses. At 1-V operation, the direct-tunneling current remains high for a gate oxide thickness of about 2 or 2.5 nm [2]. Every 0.2-nm reduction between 2 and 1 nm implies an increase of an order of magnitude of gate-tunneling current. For devices with 2-, 1.5-, and 1-nm oxide depths, the direct-tunneling current would be around 5 pA/μm, 2 nA/μm, and 50 μA/μm, respectively. A direct-tunneling current of 5 pA/μm is acceptable for high performance circuits, and 2 nA/μm is approaching the upper limit of OFF-state subthreshold leakage current (I_{OFF}); however, 50 μA/μm is certainly not acceptable.

Considering operation at high temperatures (105°C), the tunneling current is found to be ten times lower than the subthreshold leakage, since the latter increases rapidly with temperature while the former does not. Therefore, it is projected that gate oxide can be scaled to 1.5–2.0 nm before running into such a limit. Below these limits, the gate tunneling current quickly becomes problematic. When the gate becomes thinner, extremely larger direct tunneling hinders the formation of the inversion layer, and the drain current will not increase. Therefore, the limit of gate oxide thickness would be between 1.2–1.5 nm. This sets a limit for bulk CMOS scaling [45]. A solution is the use of high dielectric insulators [6, 50–53, 67]. If a good high-k dielectric insulator is developed and the direct tunneling suppressed, the gate oxide thickness may be reduced to 0.7–0.5 nm [45]. In conclusion, Figure 3.15 plots data on the

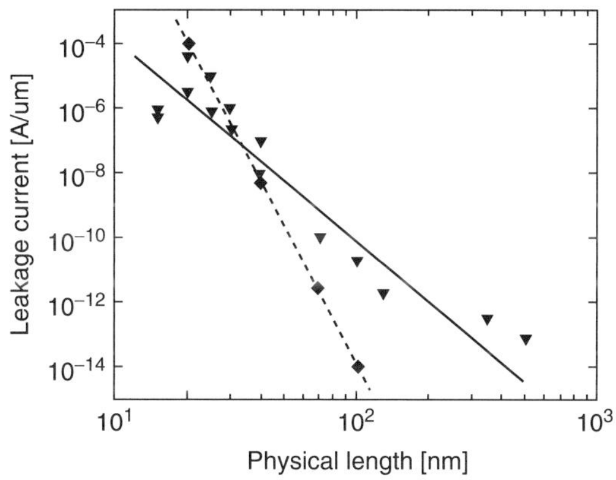

FIGURE 3.15 Trends in leakage current components as technology scales down based on published data: gate current (♦), subthreshold current (▼).

increasing leakage (subthreshold leakage current and gate tunneling current) of deep submicron transistors from different sources [10, 57, 65] and the trends computed using the expressions presented in previous sections. Notice, however, that prediction on future leakage current values, which is always based o the Semiconductor Industry Association (SIA) Roadmap [56], may be inaccurate because equal geometry transistors may have different current characteristics.

For future sub-40-nm technologies, with highly doped pn junctions, the narrow depletion region may produce worse problems due to the already mentioned BTBT currents [1, 27]. The very thin depletion depths needed in future CMOS require very high doping concentration, perhaps into the 5×10^{18} cm^{-3} range for sub-40-nm MOSFETs. At these doping levels, junction-tunneling current may appear [1, 39, 67]. Further, the combination of these effects with other effects associated to controllability and reliability of the MOSFET devices at these very small dimensions are becoming an issue. For instance, lithography variation and doping fluctuation in channel regions affecting V_{TH} control [54, 55]. To avoid these problems, novel three-dimensional (3D) double-gate transistor structures are emerging [39, 50].

3.4 Circuit Level

This section addresses the estimation and computation of leakage current in basic CMOS gates. CMOS circuits are built by series-parallel combination networks of MOS transistors. This implies that I_{LEAK} is also dependent on the input vector and circuit state: as the input vector applied to the circuit changes, the configuration of the transistors also changes. For circuits driving only subthreshold leakage currents, this dependence has been studied extensively [59–61], including the well-known stack effect. When gate currents are taken into account, the estimation of the leakage current is complicated by the state dependence of both the gate and subthreshold currents. I_{SUBTH} through OFF transistors and I_G through both ON and OFF transistors combine at internal nodes. In general, these currents are interdependent and must be analyzed simultaneously. This topic is currently an active area of research [62–64].

The models presented in Section 3.2 have been used to analyze the behavior and compute the leakage current of basic CMOS gates. At cell level, the current consumption depending on the input vector for each kind of cell in the circuit is obtained using SPICE simulation of the circuit with BSIM3 or BSIM4 (without I_G) transistor models and the added antiparallel current sources with an exponential dependence. The model is also very useful in order to highlight the current sharing between gates. Other models such as Lee et al. [63] or Mukhopadhyay et al. [64] produce similar results.

For instance, let us consider several inverters in series as illustrated in Figure 3.16. Two possible situations are shown: input V_G low (second inverter in the chain) and input V_G high (second inverter in the chain). All tunneling paths are also illustrated. In this case, for inverters with high V_G, the subthreshold current of NMOS transistor combines with the gate current of the previous transistor, while for inverters

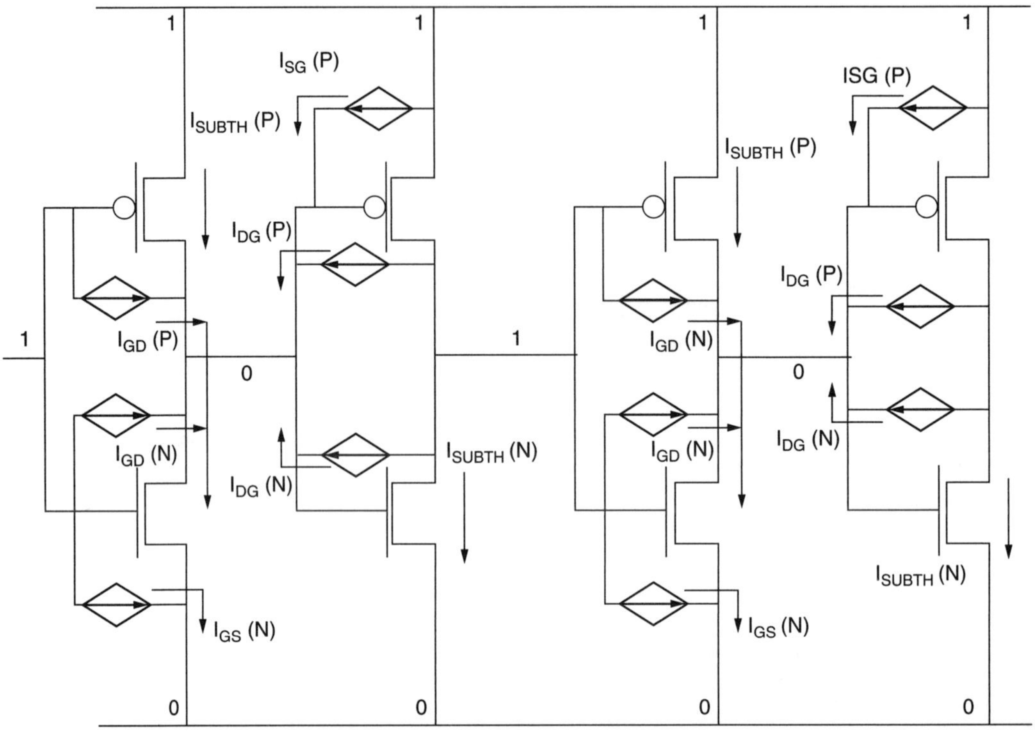

FIGURE 3.16 Inverter chain showing all gate tunneling and subthreshold leakage current using the MOSFET model including antiparallel contributing current sources.

with high V_G the subthreshold current of PMOS transistor combines with the gate current of the next transistor. In this case, gate and subthreshold currents can be computed independently.

Let us analyze the behavior of the NAND cell (Figure 3.17). The analysis of the NOR cell is similar. The input 11 produces an output equal to 0, and the leakage current is simply the sum of the subthreshold leakage current of the PMOS transistors and the SDE-to-gate tunneling current of PMOS and NMOS transistors. The leakage current flows from V_{DD} to GND internally through the NAND gate.

The inputs 00-01-10 produce an output of 1. In these cases, the internal path to GND is blocked by one or two transistors. Let us examine these cases in more detail:

1. AB = 10. Transistor N1 (connected to ground) is ON while transistor N2 (connected to VDD through the PMOS) is OFF. In this case, tunneling currents and subthreshold currents may be computed separately again. Notice that the current is shared between the NAND gate and the driving gate.
2. AB = 01. Transistor N1 (connected to ground) is OFF, while transistor N2 (connected to V_{DD} through the PMOS) is ON. In this case, the drain of the N1 transistor is held at V_{DD}–V_{TH}. Therefore, the voltage applied to the current source I_{GS} (N2) is one order of magnitude smaller than the previous case while the I_{DG} (N1) has an applied voltage equal to V_{DD}–V_{TH} and, although smaller than the previous case, cannot be neglected.
3. AB = 00. Both transistors are OFF. In this case, the subthreshold current exhibits the stack effect and the internal nodes has a voltage in the range of $\eta \cdot V_{DD}$ = 100–200mV (η models the DIBL effect). The tunneling currents I_{DG} (N1) and I_{SG} (N2) are one order of magnitude smaller than case a) and can be neglected.

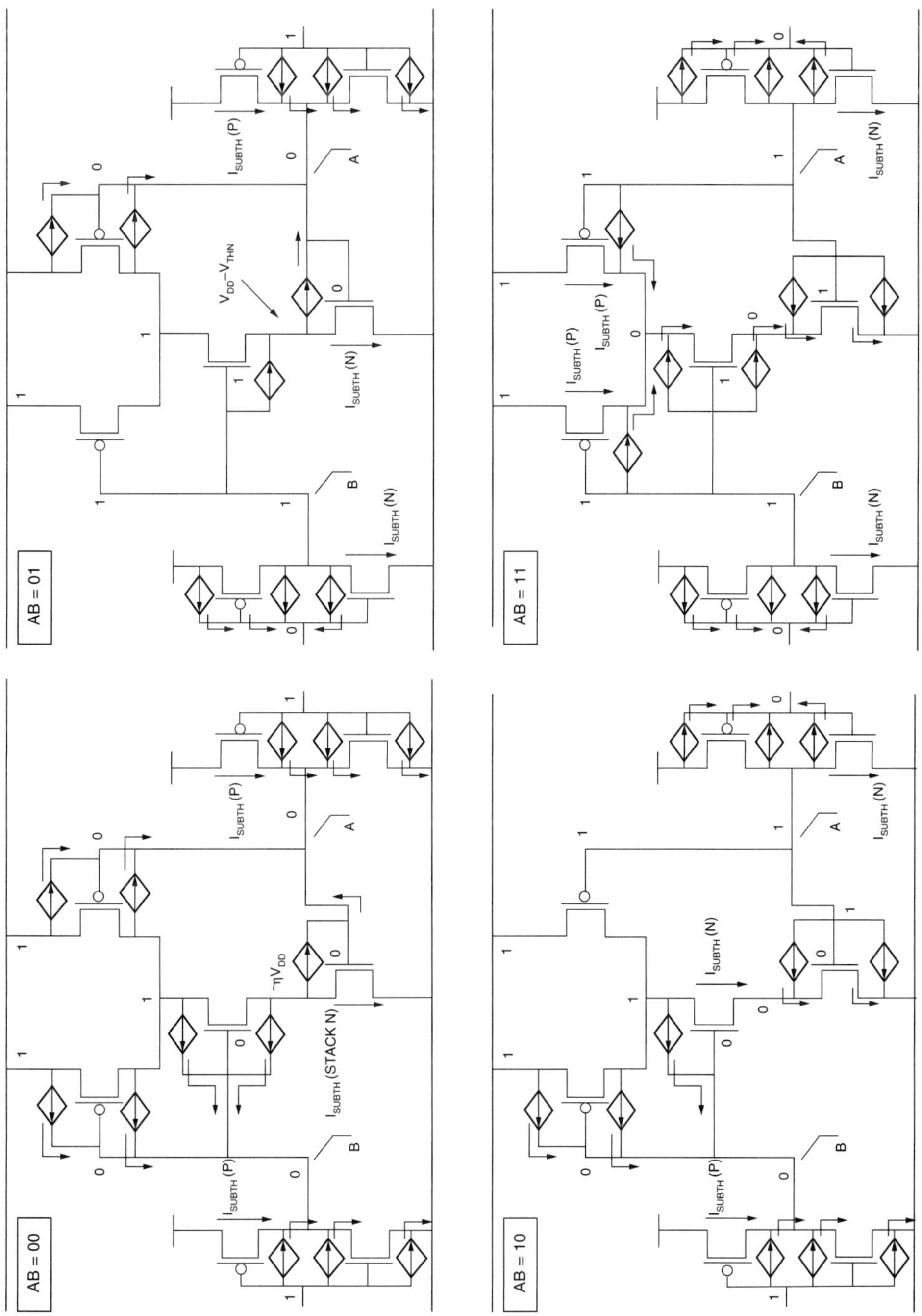

FIGURE 3.17 Two-input NAND showing all gate tunneling and subthreshold leakage current using the MOSFET model including antiparallel current sources.

In three input NAND gates, the different cases are very similar to these except the case 010. In this case, well discussed in Lee et al. [63], the current flowing from the SDE of the MOSFET in the middle of the stack increases the internal voltages and produces a reduction of the subthreshold leakage current. Therefore, the gate tunneling current therefore effectively displaces the subthreshold current, leaving the total leakage current relatively unchanged.

From these examples, it is clear that the EDT currents can degrade the performance of CMOS gates by increasing the OFF-state driving current. Therefore, an analysis of the leakage current and its bounds including gate-tunneling current are mandatory to avoid undesirable and unexpected functioning of the circuit.

For circuits with subthreshold leakage current, different methods to estimate bounds on total leakage consumption depending on the state have been presented [61]. In general, for very large circuits, it was found that the structural dependence between cells, and therefore, between the leakage current of these cells, decreased with the distance (in terms of levels) between the gates. Then, for large enough circuits, leakage current distributions were found to be Gaussian or nearly Gaussian. In other words, for large enough circuits with subthreshold leakage current as dominant leakage, some parts may be set independently in a certain logic state and control the leakage consumption [61].

When gate current is taken into account, some new issues appear. The first difference between the state dependence of I_{SUBTH} and I_G is that the value of I_{SUBTH} depends of the number of stacked transistors, while I_G depends strongly on the position of the OFF transistors in the stack [63]. Second, the current flowing in actual circuits with complex gates and the structural dependence between cells have not been deeply investigated.

These issues should be addressed to find maximum or minimum I_{LEAK} and the vectors producing these extremes. Because only in small circuits and in some special cases, it is possible to find exactly the maximum and minimum I_{LEAK}, for large circuits, the use of heuristics in order to find input vectors producing near maximum and minimum leakage currents is requested. The feasibility of using previously developed methodologies should be investigated and new methods should be introduced.

3.5 Conclusions

As CMOS technologies are scaled to nanometric ranges, the power consumption caused by leakage currents is becoming a significant part of the global power consumption. This fact has motivated a growing interest from technologists to very large scale integration (VLSI) designers to the leakage mechanisms influencing the different leakage currents and its impact on the future of CMOS.

This chapter has attempted to provide the necessary physical concepts to understand the causes of leakage and the main technology factors used to quantify the degree of leakage. The impact of scaling on the different components is expected to help in predicting the future evolution of leakage power at device and circuit levels.

The different leakage components of MOS transistors — gate tunneling, subthreshold conduction, GIDL, junction leakage, and punchthrough leakage — have been analyzed. For each component, the dependence on technology parameters, power supply, and temperature has been quantified to assess its importance as technologies evolve.

The impact of gate tunneling currents and its future trends has been studied. It is interesting to note that current available data show that for technologies with thin gate dielectrics below 40 nm for SiO_2 the contribution of the tunneling currents may become dominant over the subthreshold leakage.

The estimation of leakage at circuit level is of prime importance for VLSI designers to explore adequate solutions in the design space. The estimation of the leakage power in nanometric CMOS is a challenging problem due to the fact that gate tunneling provides new consumption paths involving driver and load gates. A model to help in estimating these currents has been proposed and used to estimate the leakage in simple circuits. Current research efforts in this area were reported.

References

[1] H-S. P. Wong *et al.* Nanoscale CMOS. *Proc. IEEE,* Vol. 87, April 1999.

[2] S. M. Sze, Ed. *Modern Semiconductor Device Physics.* John Wiley & Sons, New York, 1998.

[3] S. M. Sze, Ed. *High-Speed Semiconductor Devices.* John Wiley & Sons, New York, 1990.

[4] K-F. You and C-Y. Wu. A new quasi-2-D model for hot-carrier band-to-band tunneling current. IEEE *Trans. Electron. Devices,* Vol. 46, June 1999.

[5] M-J. Chen *et al.* Back-Gate Bias Enhanced Band-to-Band Tunneling Leakage in Scaled MOSFETs. *IEEE Electron. Device Lett.,* Vol. 19, April 1998.

[6] C.T. Liu. Circuit requirement and integration challenges of thin gate dielectrics for ultra small MOSFETs. In *IEDM Tech. Dig.,* pp. 747–750, 1998.

[7] H-D. Lee and J-M. Hwang. Accurate extraction of reverse leakage current components of shallow silicided p^+-n junction for quarter- and sub-quarter-micron MOSFETs. *IEEE Trans. Electron. Devices,* Vol. 45, August 1998.

[8] Y. Murakami and T. Shingyouji. Separation and analysis of diffusion and generation components of pn junction leakage current in various silicon wafers. *J. Applied Physics,* Vol. 75, April 1994.

[9] Y. Taur *et al.* CMOS scaling into the nanometer regime. *Proc. IEEE,* Vol. 85, April 1997.

[10] A. Keshavarzi, K. Roy, and C. F. Hawkins. Intrinsic IDDQ: origins, reduction, and applications in deep sub-um low power CMOS ICs. *Proc. Int. Test Conf. (ITC),* pp. 167–176, 1997.

[11] A. Keshavarzi, K. Roy, and C. F. Hawkins. Intrinsic leakage in deep submicron CMOS ICs. Measurement-based test solutions. *IEEE Trans. VLSI Syst.,* Vol. 8, December 2000.

[12] D. Josephson, M. Storey, and D. Dixon. Microprocessor IDDQ testing: a case study. *IEEE Design & Test of Computers,* Vol. 12, Summer 1995.

[13] P. C. Maxwell and J. R. Rearick. A simulation-based method for estimating defect-free IDDQ. *IEEE Int. Workshop on IDDQ Testing, Digest of Papers,* pp. 80–84, 1997.

[14] G. Massobrio and P. Antognetti. *Semiconductor Device Modeling with SPICE.* McGraw-Hill, New York, 1993.

[15] T. A. Fjeldly and M. Shur. Threshold voltage modeling and the subthreshold regime of operation of short-channel MOSFETs. *IEEE Trans. Electron. Devices,* Vol. 40, January 1993.

[16] R. Velghe, D. Klaassen, and F. Klaassen. MOS Model 9, Level 902. Technical report, also available at http://www.semiconductors.philips.com/Philips_Models/Retrieval date:2004.

[17] Y. P. Tsividis. *Operating and Modeling of the MOS Transistor.* McGraw-Hill, New York, 1999.

[18] M. Rosar, B. Leroy, and G. Schweeger. A new model for the description of gate voltage and temperature dependence of gate-induced drain leakage (GIDL) in the low electric field region. *IEEE Trans. Electron. Devices,* Vol. 47, January 2000.

[19] J-Y. Guo *et al.* A three-terminal band-trap-band tunneling models for drain engineering and substrate bias effect on GIDL in MOSFET. *IEEE Trans. Electron. Devices,* Vol. 45, July 1998.

[20] M. Tanizawa *et al.* A complete substrate current model including band-to-band tunneling current for circuit simulation. *IEEE Trans. Computer-Aided Design,* Vol. 12, November 1993.

[21] N. Lindert *et al.* Comparison of GIDL in p^+-poly PMOS and n^+-poly PMOS devices. *IEEE Electron. Device Lett.,* Vol. 17, June 1996.

[22] R. Ghodsi, S. Sharifzadeh, and J. Majjiga. Gate-induced drain leakage in buried-channel PMOS — a limiting factor in developement of low-cost, high-performance 3.3-V, 0.25-mm technology. *IEEE Electron. Device Lett.,* Vol. 19, September 1998.

[23] M. Stadele, B. R. Tuttle, and K. Hess. Tunneling through ultrathin sio2 gate oxides from microscopic models. *J. Applied Physics,* Vol. 89, January 2001.

[24] S. T. Ma and J. R. Brews. Comparison of deep-submicrometer conventional and retrograde n-MOSFETS. *IEEE Trans. Electron. Devices,* Vol. 47, August 2000.

[25] B. Majkusiak and M. H. Badri. Semiconductor thickness and back-gate voltage effects on the gate tunnel current in the MOS/SOI system with an ultrathin oxide. *IEEE Trans. Electron. Devices,* Vol. 47, December 2000.

[26] S.-H. Lo, D. A. Buchanan, and Y. Taur. Modeling and characterization of quantization, poly-Si depletion, and direct tunneling effects in MOSFETs with ultrathin oxides. *IBM J. Research Dev.*, Vol. 43, No. 3, 1999.

[27] C. A. Mead. Scaling of MOS technology to submicrometer feature sizes. *Analog Integrated Circuits-Signal Processing*, Vol. 6, No. 1, 1994.

[28] L. Larcher, A. Paccagnella, and G. Ghidini. Gate current in ultrathin MOS capacitors: a new model of tunnel current. *IEEE Trans. Electron. Devices*, Vol. 48, February 2001.

[29] K. N. Yang et al. Characterization and modeling of edge direct tunneling (EDT) leakage in ultrathin gate oxide MOSFETs. *IEEE Trans. on Electron. Devices*, Vol. 48, June 2001.

[30] W. K. Henson et al. Analysis of leakage currents and impact on OFF-state power consumption for CMOS technology in the 100-nm regime. *IEEE Trans. Electron. Devices*, Vol. 47, July 2000.

[31] N. Yang, W. K. Henson, and J. J. Wortman. A comparative study of gate direct tunneling and drain leakage currents in n-MOSFETs with sub-2-nm gate oxides. *IEEE Trans. Electron. Devices*, Vol. 47, August 2000.

[32] Z. Yu et al. Impact of gate direct tunneling current on circuit performance: a simulation study. *IEEE Trans. Electron. Devices*, Vol. 48, December 2001.

[33] W-C. Lee and C. Hu. Modeling CMOS tunneling currents through ultrathin gate oxide due to conduction- and valence-band electron and hole tunneling. *IEEE Trans. Electron. Devices*, Vol. 48, July 2001.

[34] K. M. Cao et al. BSIM4 gate leakage model including source-drain partition. *IEDM Tech. Dig.*, pp. 815–818, 2000.

[35] G. W. Neudeck. *The PN Junction Diode, 2nd ed.*, John Wiley & Sons, New York, 1988.

[36] M. T. Bohr. Nanotechnology goals and challenges for electronic applications. *IEEE Trans. Nanotechnology*, Vol. 1, March 2002.

[37] Y-S. Lin et al. Leakage scaling in deep submicron CMOS for SoC. *IEEE Trans. Electron. Devices*, Vol. 49, June 2002.

[38] Y-S. Lin, et al. On the SiO2-based gate dielectric scaling limit for low-standby power applications in the context of a 0.13-μm CMOS logic technology. *IEEE Trans. Electron. Devices*, Vol. 49, March 2002.

[39] D. J. Frank et al. Device scaling limits for Si MOSFETs and their application dependencies. *Proc. of the IEEE*, Vol. 89, March 2001.

[40] A. O. Adan and K. Higashi. OFF-state leakage current mechanisms in BulkSi and SOI MOSFETs and their impact on CMOS ULSIs standby current. *IEEE Trans. Electron. Devices*, Vol. 48, September 2001.

[41] K. A. Bowman et al. A circuit-level perspective of the optimum gate oxide thickness. *IEEE Trans. Electron. Devices*, Vol. 48, August 2001.

[42] R. D. Isaac. The future of CMOS technology. *IBM J. Research Dev.*, Vol. 44, No. 3, 2000.

[43] A. J. Bhavnagarwala et al. A minimum total power methodology for projecting limits on CMOS GSI. *IEEE Trans. VLSI Syst.*, Vol. 8, May 2000.

[44] F. Assad et al. On the performance limits for Si MOSFETs: a theoretical study. *IEEE Trans. Electron. Devices*, Vol. 47, January 2000.

[45] H. Iwai. CMOS technology — year 2010 and beyond. *IEEE J. Solid-State Circuits*, Vol. 34, March 1999.

[46] J. J. Sun et al. The effect of the elevated source/drain doping profile on performance and reliability of deep submicron MOSFETs. *IEEE Trans. Electron. Devices*, Vol. 44, June 1997.

[47] D. Hisamoto et al. A low-resistance self-aligned T-shaped gate for high-performance sub-0.1-mm CMOS. *IEEE Trans. Electron. Devices*, Vol. 44, June 1997.

[48] B. Yu et al. Short-channel effect improved by lateral channel-engineering in deep-submicronmeter MOSFETs. *IEEE Trans. Electron. Devices*, Vol. 44, April 1997.

[49] S-C. Lin *et al.* A closed-form back-gate-bias related inverse narrow-channel effect model for deep-submicron VLSI CMOS devices using shallow trench isolation. *IEEE Trans. Electron. Devices*, Vol. 47, April 2000.

[50] B. Doyle *et al.* Transistor elements for 30nm physical gate lengths and beyond. *Intel Technol. J.*, Vol. 06, June 2002.

[51] S. Mudanai *et al.* Modeling of direct tunneling current through gate dielectric stacks. *IEEE Trans. Electron. Devices*, Vol. 47, October 2000.

[52] I. C. Kizilyalli *et al.* MOS transistors with stacked SiO_2-Ta_2O_5-SiO_2 gate dielectrics for giga-scale integration of CMOS technologies. IEEE Electron Device Letters, Vol. 19, November 1998.

[53] Y-C. Yeo *et al.* MOSFET gate leakage modeling and selection guide for alternative gate dielectrics based on leakage considerations. *IEEE Trans. Electron. Devices*, Vol. 50, April 2003.

[54] M. Koh *et al.* Limit of gate oxide thickness scaling in MOSFETs due to apparent threshold voltage fluctuation induced by tunnel leakage current. *IEEE Trans. Electron. Devices*, Vol. 48, February 2001.

[55] K. Takeuchi, R. Koh, and T. Mogami. A study of the threshold voltage variation for ultra-small bulk and SOI CMOS. *IEEE Trans. Electron. Devices*, Vol. 48, September 2001.

[56] National Technology Roadmap for Semiconductors 2002. Available at http://public.itrs.net/, 2002.

[57] B. Davari, R.H. Dennard, and G. G. Shahidi, CMOS scaling for high performance and low power-the next ten years. Proc. of the IEEE, Vol. 83, April 1995.

[58] K. Roy, S. Mukhopadhyay, and H. Mahmoodi-Meimand. Leakage current mechanisms and leakage reduction techniques in deep-submicrometer CMOS circuits. *Proc. of the IEEE,* Vol. 91, February 2003.

[59] R. X. Gu and M. I. Elmasry, Power dissipation analysis and optimization of deep submicron CMOS digital circuits. *IEEE J. Solid-State Circuits,* Vol. 31, May 1996.

[60] R. M. C. Johnson, D. Somasekhar, and K. Roy, Models and algorithms for bounds on leakage in CMOS circuits. *IEEE Trans. Computer-Aided Design,* Vol. 18, June 1999.

[61] A. Ferré and J. Figueras. Leakage power bounds in CMOS digital technologies. *IEEE Trans. Computer-Aided Design,* Vol. 21, June 2002.

[62] A. Ferré and J. Figueras. Leakage power analysis considering gate tunneling currents. DEE - UPC Internal Report, No. 03/06, April 2003.

[63] D. Lee *et al.* Analysis and minimization techniques for total leakage considering gate oxide leakage. *Proc. of the DAC,* pp. 175–180, 2003.

[64] S. Mukhopadhyay, A. Raychowdhury, and K. Roy. Accurate estimation of total leakage current in scaled CMOS logic circuits based on compact current modeling. *Proc. of the DAC,* pp. 169–174, 2003.

[65] T. Mak, Leakages and its implication to test, Private Communication, June 2002.

[66] Y-F. Tsai *et al.* Implications of technology scaling on leakage reduction techniques, *Proc. of the DAC,* pp. 187–190, 2003.

[67] Brian Doyle *et al.* Transistor elements for 30-nm physical gate lengths and beyond, *Intel Technol. J.*, May 2002. Available online.

4

Microelectronics, Nanoelectronics, and the Future of Electronics

Jing Wang
Mark Lundstrom
Purdue University

4.1 Introduction

Silicon technology continues to progress rapidly, with current generation technologies having physical gate lengths well below 100 nm. At the same time, remarkable advances in nonsilicon nano- and molecular technologies are occurring. It is time to think seriously about the role that nanoelectronics and nontraditional technologies could play in future electronic systems. Moore's law describes device scaling-down in integrated circuits, which has led an unprecedented growth of the semiconductor industry. At the same time, it also carried device researchers into the nano world. Well-established concepts from mesoscopic physics [1] are now entering the working knowledge of device physicists and engineers as silicon transistors enter the nanoscale [2]. At the micrometer scale, transistors were well described by drift-diffusion equations, but now people are beginning to use a new language to describe nanoscale transistors. In addition, several interesting new devices that may have important applications are also being developed [3–6].

Nanoelectronics can play an important role in future electronic systems, if the design community is engaged to exploit the opportunities that nanoelectronics offers. Therefore, we appreciate this opportunity to give an overview of the current developments of nanoscale transistors. The chapter begins by defining nanotechnology and discussing how a metal-oxide-semiconductor field-effect transistor (MOSFET) performs in the nanometer regime (Section 4.2), then examines the ultimate scaling limit and practical limits of the silicon MOSFET (Section 4.3 and Section 4.4). After that, several new types of field-effect transistors (FETs) are introduced, which may become the substitutes for the silicon MOSFET (Section 4.5) and other nanotransistors beyond the FET (Section 4.6). Several important issues in the research of nanoelectronics are also discussed (Section 4.7). To be concise, we do not include the detailed mathematical formalism of the device theory, but the references are listed to help the reader who has particular interests find the sources.

4.2 The Silicon MOSFET as a Nanoelectronic Device

4.2.1 What Is Nanotechnology?

Nanotechnology has been defined as work at the 1–100-nm length scale to produce structures, devices, and systems that have novel properties because of their nanoscale dimensions [7]. Some insist that two dimensions lie in the 1–100 nm regime, which would rule out traditional technologies such as thin films. A key part of the definition is that new phenomena occur (caused, for example, by the dominance of interfaces and quantum mechanical effects), and that these new phenomena may be exploited to improve the performance of materials, devices, and systems. Nanotechnologies also involve the manipulation and control of matter at the nanoscale. Semiconductor technology does much of this with a "top-down" approach that lithographically imposes a pattern, and then etches away bulk material to create a nanostructure. Some argue that self-assembly is an essential component of nanotechnology. The hope is that nanostructures can be self-assembled from the "bottom up," molecule by molecule. We argue that current-day silicon technology meets the definition of nanoelectronics, that future silicon technologies will meet it even better, and that nontraditional technologies could play an important role in future electronic systems by complementing the capabilities of nanoscale silicon technology, rather than by attempting to replace it.

4.2.2 Silicon MOSFETs in the Nanometer Regime

The International Technology Roadmap for Semiconductors (ITRS) [8] calls for 9-nm physical gate lengths for integrated circuit (IC) transistors in 2016. At the same time, major IC manufacturers have reported transistors with 10-nm (or shorter) gate lengths on IEDM 2002 [9,10], which demonstrate the promise of pushing IC technology to the 10-nm regime.

To scale silicon transistors down to the 10-nm scale, new device structures are needed to suppress the short channel effects [11]. Figure 4.1 is a schematic illustration of a fully depleted, double-gate (DG) MOSFET, a device that offers good prospects for scaling silicon transistors to their limits [12]. Other approaches (e.g., the FinFET [9] and the tri-gate MOSFET [13]) are also being explored. At a 9-nm gate length, acceptable short-channel effects require a fully depleted silicon body thickness of 3 nm or less, and an equivalent gate oxide thickness of less than 1 nm. At such dimensions, the properties of the silicon material will be affected by quantum confinement (e.g., the bandgap will increase), and device properties will be influenced by quantum transport.

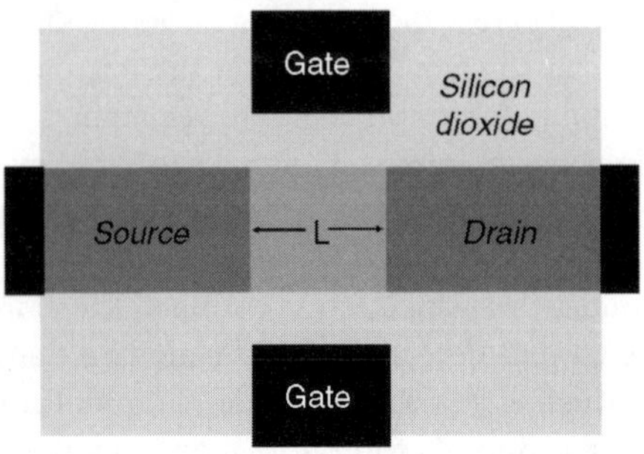

FIGURE 4.1 The double-gate MOSFET structure.

Traditional device equations are based on the drift-diffusion theory [11], which assumes that the device scale is much larger than the electron wavelength (~ 8nm at room temperature) and the electron mean-free-path (the average distance an electron travels between two collisions, ~ 10nm for electrons in the inversion layer). The first assumption allows us to treat electrons as classical particles with zero size, and the second one justifies the "local transport" property (the electron velocity at a position is solely determined by the local electric field and mobility). Unfortunately, at the nanoscale, neither of these assumptions is well satisfied. As a result, to capture the new physical effects that occur at the nanoscale, the old device theory must be modified or even completely replaced by a new quantum transport theory.

Four important phenomena need to be properly treated in the modeling of nanotransistors:

1. Quantum confinement
2. Gate tunneling
3. Quasi-ballistic transport
4. Source-to-drain (S/D) tunneling

The first two effects occur in the confinement direction (normal to the gate electrode(s)) of the MOSFET. As silicon technology entered the sub-100-nm regime (the corresponding oxide thickness < 3nm), those effects became significant and began to affect the MOSFET threshold voltage and leakage currents in the "OFF-state." Extensive work has been done to explore the physics of the first two effects, and numerous device models have been developed to capture them in device and circuit simulations [14–17]. (Here, we will not give the details of those models. Readers with particular interests should refer to the related references.) In contrast to the quantum confinement and gate tunneling, quasi-ballistic transport and source-to-drain tunneling begin to significantly affect the device performance of the silicon MOSFET when the gate length scales down to 10 nm or less [18]. Therefore, the exploration of these mesoscopic transport effects is important for the description of silicon MOSFETs at their scaling limit, as well as the understanding of device physics of other nanoscale devices (to be discussed later). In the following paragraph, a simple description of the ballistic/quasi-ballistic transport [19–23] is presented, which gives us the upper performance limit of nanoscale transistors. Section 4.3 discusses the source-to-drain tunneling in silicon MOSFETs at the scaling limit.

In a conventional MOSFET, the channel length is much longer than the electron mean-free-path, so an electron will experience numerous collisions during its travel from the source to the drain. Nevertheless, when the channel length shrinks to less than the mean-free-path, an electron may go through the channel with no or little scattering, which is called ballistic/quasi-ballistic transport. According to the quasi-ballistic transport theory [2,22,23], the current under low drain bias can be written as (assuming nondegenerate statistics),

$$I_{DS} = \frac{\lambda}{L+\lambda} W Q_i(0) \frac{\upsilon_T}{2k_BT} V_{DS} \tag{4.1}$$

where λ is the electron mean-free-path, L is the channel length of the MOSFET, $Q_i(0)$ is the sheet electron density at the beginning of the channel, $V_T \sqrt{2k_BT/\pi m^*}$ is the unidirectional thermal velocity of nondegenerate electrons, and other symbols have their common meanings. For a long channel device, $L >> \lambda$, so Equation 4.1 becomes

$$I_{DS} = W Q_i(0) \frac{\upsilon_T \lambda}{2k_BT} \frac{V_{DS}}{L} \tag{4.2}$$

Because the mobility for nondegenerate electrons can be defined as $\mu_0 = \upsilon_T \lambda / (2k_BT)$ [21], Equation 4.2 is simply the well-known classical device equation based on the drift-diffusion theory [11]. When $L << \lambda$, the current approaches its upper (ballistic) limit,

$$I_{DS} = I_{ballistic} = WQ_i(0)\frac{\upsilon_T}{2k_BT}V_{DS} \tag{4.3}$$

The point is that the conventional device equations are an approximation valid when $L >> \lambda$. As MOSFET channel lengths approach the nanoscale, the classical MOSFET equations must be modified to capture quasi-ballistic transport. It is important for both device simulation and the development of circuit models for nanotransistors.

4.3 Ultimate Limits of the Silicon MOSFET

As the gate lengths of Si MOSFETs continue to shrink, the two-dimensional (2D) electrostatics become increasingly important, which causes the well-known short-channel effects (SCEs). At the same time, for the MOSFET with a gate length < 10nm, the quantum mechanical tunneling from source to drain may also be significant. It will degrade the subthreshold slope and increase the leakage current in the OFF-state. According to our previous work [18], the ultimate scaling limit of Si MOSFETs is determined by both the semiclassical SCEs (i.e., DIBL, V_T roll-off) and the S/D tunneling.

In Wang and Lundstrom [18], S/D tunneling has been extensively examined using the nonequilibrium Green's function (NEGF) approach [24], a general and rigorous quantum model for nanoscale transistors. (The 2D quantum simulator for double-gate Si MOSFETs, nanoMOS-2.5, is available at http://nano-hub.purdue.edu.) The main conclusions are summarized next:

1. For the well-designed devices (with very thin silicon body and oxide layers that provide good electrostatics), S/D tunneling sets an ultimate scaling limit that is well below 10 nm.
2. S/D tunneling dominates OFF-current in the devices at scaling limit, and it may play an important role in the ON-state of ballistic devices.
3. Due to S/D tunneling, the sub-threshold slope saturates at low temperature (see Figure 4.2). Therefore, the leakage current in the OFF-state may still be high even at low temperature.

We also found that for a double-gate MOSFET with a 1-nm-thick silicon body and 0.6-nm(equivalent)-thick oxide layers, S/D tunneling sets a scaling limit of L = 5 nm if we require that *the subthreshold swing*

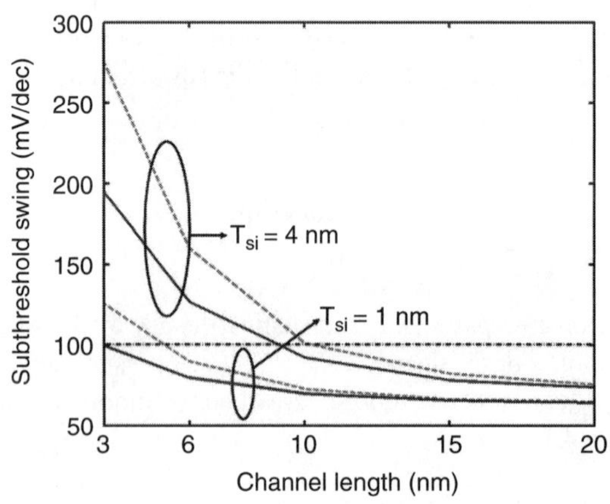

FIGURE 4.2 The subthreshold swing vs. temperature. The simulated device structure is a double-gate MOSFET with 0.6-nm(equivalent)-thick oxide layers. Two silicon body thicknesses (1 nm and 4 nm) are adopted in this simulation. The solid curves are for the semiclassical Boltzmann simulation (without S/D tunneling), while the dashed curves are for the quantum NEGF simulation (with S/D tunneling). (Obtained from J. Wang and M. Lundstrom, Does source-to-drain tunneling limit the ultimate scaling of MOSFETs? *IEEE Int. Electron. Devices Meeting (IEDM), Tech. Dig.*, pp. 707–710, San Francisco, CA, Dec. 2002. With permission.)

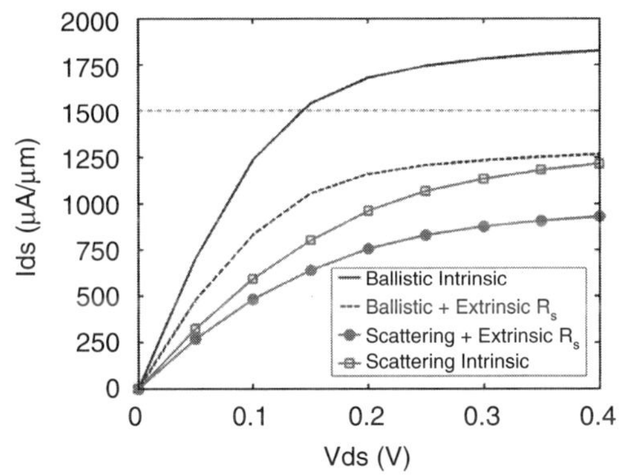

FIGURE 4.3 The top curve is the intrinsic ballistic current I^{i}_{ball}, and the dashed curve is I^{e}_{ball}, so the top two curves represent the ballistic intrinsic device. The bottom two are for intrinsic device with scattering. The curve with square markers represents I^{i}_{scatt}, and the fourth curve is I^{e}_{scatt}. (Obtained from S. Hasan, J. Wang, and M. Lundstrom, Device design and manufacturing issues for 10nm-scale MOSFETs: a computational study, *Solid State Electronics,* 48, 6, 867–875, 2004. With permission.)

is smaller than 100 mV/dec and the ON-OFF current ratio is larger than 100. Obviously, we could have different criteria to determine the ultimate scaling limit of a MOSFET. Likharev [25] proposed a criterion that the voltage gain of a CMOS inverter is larger than one, and used it to find a scaling limit of L = 2 nm. So two very important questions arise: How is the scaling limit of a MOSFET determined? What is the *worst* performance of a transistor that can be accepted by a very large scale integration (VLSI) circuit designer to build an IC chip? Clear answers to these questions require cooperation between device researchers and circuit engineers.

4.4 Practical Limits of the Silicon MOSFET

Section 4.3 discussed the ultimate scaling limit of a MOSFET. In practice, some technical issues (e.g., the source/drain series resistances, process variations, and power dissipation) may greatly affect device performance and set practical limits for MOSFETs.

In Hasan et al. [26], a computational study of the end-of-roadmap (L_G = 9nm) MOSFETs (high-performance) was presented. It was found that:

1. With a double-gate structure and a 3-nm-thick silicon body, the 10-nm-scale MOSFET can be realized but the ON-current is ~ 40% below the ITRS prediction. S/D series resistance and low gate overdrive ($V_{GS} - V_T$) were identified as limiting factors for the ON-current (see Figure 4.3 for details).
2. Process variations will seriously affect the device performance for the 10-nm-scale MOSFET. For example, a single monolayer (~ 0.3nm) variation in the silicon body thickness will cause more than 50% variation in the OFF-current (see Figure 4.4).

In summary, as the silicon MOSFET approaches its scaling limit, maintaining drive current at low supply voltages (~ 0.5V) will be very difficult, and device parasitics will be much more important than for current technology. Devices will be extremely sensitive to manufacturing variations. New design techniques will be needed to make use of devices with low drive current, high leakage, and large process variations.

4.5 Beyond the Silicon MOSFET

This section discusses several new types of FETs that are being explored by device physicists and engineers. Those devices could become either substitutes for the silicon MOSFET or complementary circuit elements that might be implemented into silicon IC circuits to improve their performance in the nanometer regime.

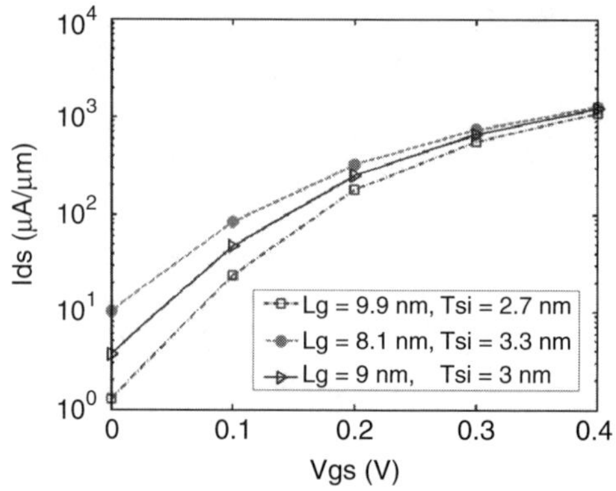

FIGURE 4.4 Intrinsic transfer characteristics of three different transistors. The top curve represents the worst transistor in terms of SCE, with L_G 10% smaller and t_{si} 10% larger, the middle one is the nominal device, and the bottom one represents the best device, with L_G 10% larger and t_{si} 10% smaller. (Obtained from S. Hasan, J. Wang, and M. Lundstrom, Device design and manufacturing issues for 10nm-scale MOSFETs: a computational study, *Solid State Electronics,* 48, 6, 867–875, 2004. With permission.)

4.5.1 Carbon Nanotube Transistors

One can think of a carbon nanotube as a 2D sheet of graphene (in which carbon atoms in a hexagonal lattice are bonded to three nearest neighbors as illustrated in Figure 4.5) that is rolled up into a tube. Depending on how the sheet is rolled up to produce a tube (in a "zigzag" pattern, "armchair," or in between (chiral), the nanotube can be either metallic or semiconducting). For semiconducting tubes, the bandgap is inversely proportional to the nanotube diameter. A diameter of 1 nm (a typical value) gives a bandgap of about 0.8 eV.

The interest in carbon nanotubes arises from the unique material properties they display. The one-dimensional (1D) energy band structure suppresses scattering, so ballistic transport can be achieved over relatively long distances. The thermal conductivity is exceptional, even higher than diamond, and nanotubes display excellent resistance to electromigration. These properties make nanotubes interesting for interconnects and heat removal in gigascale systems. Semiconducting nanotubes also display excellent transport properties, and the absence of dangling bonds may make it easier to incorporate high-K gate dielectrics into carbon nanotube field effect transistors (CNTFETs). Because the valence and conduction bands are mirror images of each other, n-type and p-type transistors should display essentially identical

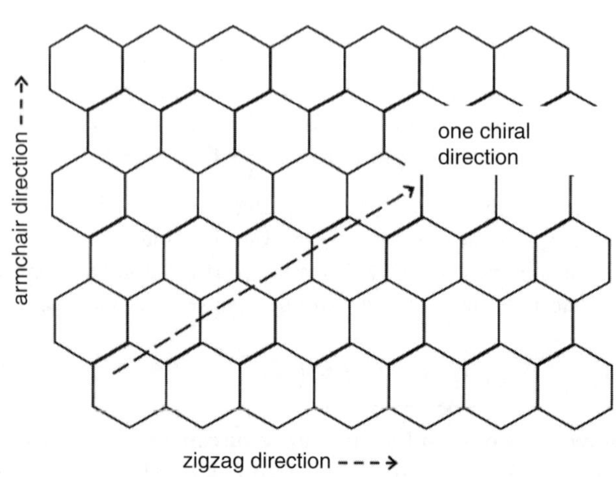

FIGURE 4.5 A 2D sheet of graphene showing the roll-up directions for different nanotubes.

characteristics, a significant advantage for complementary metal-oxide semiconductor (CMOS) circuits. Initially, CNTFETs suffered from high series resistance and low gate capacitance. Improved contacts are being developed, and new structures employ high-K gate dielectrics. The ITRS calls for an ON-current of 750 μA/μm (0.75 μA/nm) for PMOS transistors in 2016, which will be very difficult to meet by the silicon material at the low supply voltages needed (V_{DD} ~ 0.5 V). Experimental CNTFETs have already achieved over 7 μA/nm at 0.9 V [27].

It is clear that carbon nanotubes have great promise, but what are the challenges? The growth of CNTs with well-defined electronic properties is a critical issue. Growth from a catalytic seed can be used to control the CNT diameter, but it is more difficult to control the CNTs chirality (i.e., how it is rolled up). For applications in terascale systems, we will need to grow at least 10^{12} CNTs — all semiconducting with well-controlled diameters. Device structures and process flows are still primitive. One approach is to produce planar FETs with arrays of CNTs to provide sufficient current for conventional digital applications [28]. This approach aims to replace the silicon CMOS transistor with a higher-performance device. Another approach would be to explore the use of single nanotube electronics in dense locally interconnected architectures that could complement silicon CMOS. As CNT materials and device work proceeds, work at the system design level is needed to identify the most promising opportunities.

4.5.2 Organic Molecular Transistors

The organic molecular transistor is another possibility for post-CMOS devices. Figure 4.6 shows a schematic structure of the molecular FET. Compared with silicon MOSFETs and other nanotransistors, molecular FETs might have advantages on both fabrication and device performance.

1. The fabrication of molecular FETs could be with low cost, high controllability, and reproducibility. As we know, to fabricate a silicon MOSFET at the nanoscale, lithography and etching technology with extremely high resolution (< 10nm) is required, which may greatly increase the cost of IC fabrication. Moreover, the variations in lithography and etching can seriously affect the device performance. For CNTFETs (see previous paragraph), although the high-resolution lithography may not be needed for the device fabrication, the variations (i.e., the chirality and diameter of a CNT) from tube to tube could affect the controllability and reproducibility of the circuits. In contrast to silicon MOSFETs and CNTFETs, a molecular FET with numerous identical molecules might be realized at quite low cost by using the self-assembly technology [29]. The FET channel length is naturally equal to the length of the molecules so that the process variations would be effectively suppressed.
2. The molecular transistor has special physical properties that may be exploited to enhance the device performance. First, there could be no dopants in a molecular transistor. The type (n or p) of the FET can be determined by the gate work function [30]. As a result, the scattering inside the channel would be reduced. Considering the extremely short channel length of a molecular FET, transport inside the channel could be ballistic. Second, molecules are flexible and tunable,

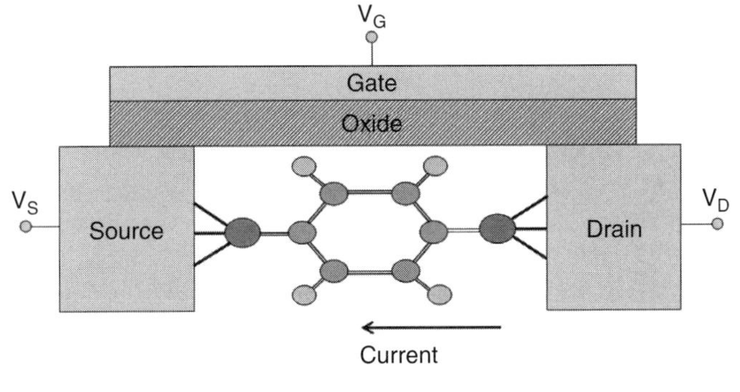

FIGURE 4.6 A schematic structure of the organic molecular transistor.

so it may be possible to control the molecules' shape (conformation) by a gate voltage [31]. This opens up the possibility of a molecular relay with a subthreshold swing better than the thermal-emission limit, $2.3k_BT/q$ (60 mV/dec at room temperature) [31]. Initial studies, however, show that thermal fluctuations of the flexible molecule are a serious issue [31].

As for any other nanotransistor, the organic molecular FET has its own challenges. Due to its extremely short channel, the molecular FET may seriously suffer the 2D electrostatics (so-called short channel effects, SCEs) (e.g., a 3-nm channel length may require a 0.2-nm equivalent oxide thickness to achieve good electrostatics, which is very difficult to realize in practice) [32]. The relatively low drive current may also limit its application as a logic circuit element. Therefore, the optimization of device performance becomes important for the future application of molecular transistors.

4.5.3 MOSFETs with New Channel Materials and Semiconductor Nanowire Transistors

A well-designed transistor should have an efficient gate control and good transport property (high channel mobility), so to improve the device performance of silicon MOSFETs, researchers are trying to exploit new channel materials and new gate geometry configurations.

Extensive experimental work [33–35] has been done on germanium and strained silicon, promising new channel materials that could provide higher mobility for both electrons and holes. On the other hand, silicon nanowire transistors are also being explored [6,36,37]. Such a 1D structure provides a possibility to make tri-gate or gate-all-around transistors that offer the best gate control. (Because the device physics of those transistors is similar to that of the silicon MOSFET, we do not discuss the details here.) With new channel materials or new gate geometry configurations, it may be possible to scale the MOSFET beyond the scaling limit of the planar silicon MOSFET.

4.6 Beyond the FET

Nanotechnology will not only provide the fabrication techniques to build nanoscale FETs, but also make it possible to realize some quantum-effects devices with special applications in the future electronics. Indeed, the most promising applications of molecular electronics may not be to replace Si MOSFETs but, instead, to complement CMOS with new capabilities. This section discusses two examples: the single electron transistor and the spin transistor.

4.6.1 Single-Electron Transistors

For future nanoscale transistors, the total number of electrons in the channel may be approximately 10, but a single-electron transistor (SET) is not just a smaller version of the same device. To produce a single electron transistor, the size of the "island" between the source and drain must be small enough so that the change in voltage due to a single electron is large compared to the thermal energy:

$$q^2/2C_G >> k_BT \tag{4.4}$$

Reliable room temperature operation requires an island size of less than about 1 nm, the size of a small molecule. In addition, we also require that the source and drain be weakly coupled to the gated island, which is usually accomplished by introducing tunnel barriers at the two contacts. When these conditions are met, some unique I-V characteristics result [38]. For example, the number of electrons on the island changes in discrete steps as the gate voltage increases, and a "Coulomb blockade" prevents current flow until V_{DS} exceeds a critical value. The critical voltage for conduction is periodic in gate voltage. Single electron transistors have been investigated for applications in digital systems, but they have several limitations [38]. The voltage gain is low and so is the drive current (because the tunnel junctions introduce a large series resistance). As might be expected, they are also extremely sensitive to stray background charges. Certain hybrid SET/MOSFET circuits, however, combine single (or few) electron devices and CMOS transistors and have interesting possibilities for memory [38].

4.6.2 Spin Transistors

The operation of a conventional transistor is based on the charge that electrons carry, but electrons also carry spin, a fundamental unit of magnetic moment. The electron's spin is the basis for magnetic memories, but it is also conceivable that spin could be modulated by a gate to realize new types of devices [39]. For example, if the source and drain were ferromagnetic, then spin-polarized electrons might be injected into a semiconductor. If they retain their spin as they propagate across the channel, they could easily exit the ferromagnetic drain, but it may be possible to rotate the electron spins by a gate voltage thereby preventing them from exiting through the drain and contributing to the drain current.

Devices of this type have not yet been demonstrated, but current research is examining how to combine ferromagnetic metals and semiconductors, how to inject spin-polarized electrons into the semiconductor, and how to maintain the spin polarization once the electrons are in the semiconductor [40]. If devices of this type could be realized, they promise faster switching and lower switching energy than conventional electrostatic MOSFETs. Eventually, it may be possible to manipulate the spins of individual electrons (single electron spin transistors), which could lead to the realization of quantum computers.

4.7 From Microelectronics to Nanoelectronics

Nanoelectronics is not simply a smaller version of microelectronics; things change at the nanoscale. At the device level, silicon transistors may give way to new materials such as organic molecules or inorganic nanowires [41]. At the interconnect level, microelectronics uses long, fat wires, but nanoelectronics seeks to use short nanowires [41]. Fundamentally new architectures will be needed to make use of simple, locally connected structures that are imperfect and are comprised of devices whose performance varies widely.

We believe that 21st-century silicon technology has evolved into a true nanotechnology. Critical dimensions are already below 100 nm. The materials used in these silicon devices have properties that differ from the bulk. Nanoscale silicon transistors have higher leakage, lower drive current, and exhibit more variability from device to device. New circuits and architectures will need to be developed to accommodate such devices. It matters little whether the material is silicon or something else; the same issues face any nanoelectronics technology. It is likely that many of the advances and breakthroughs at the circuits and systems levels that will be needed to make nanoelectronics successful will come from the silicon design community.

Developing an understanding of how devices operate at the nanoscale is a good reason to support nanoscience research. Another reason is that devices to complement silicon technology might be discovered. For example, carbon nanotube FETs could be exquisite singlemolecule detectors, and SETs could be integrated with MOSFETs for high-density memory applications. Another possibility is molecular structures that improve the performance of a CMOS platform. For example, ballistic CNTs could be high performance interconnects and efficient at heat removal. Nanowire thermoelectric cooling could lower chip temperature and increase performance [42]. Therefore, research on nanoelectronics will prove to be a good investment for several reasons.

The successful development of nanoelectronics will require a partnership between science and engineering. It was the same for semiconductor technology. The scientific community developed the understanding of semiconductor materials and physics and the engineering community used this base to learn how to design devices, circuits, and systems. Figure 4.7 summarizes this partnership. Science works in the nanoworld with individual atoms, molecules, nanoscale structures and devices, and assembly processes. Systems engineers work in the macroworld on complex systems with terascale device densities. In the middle are the device and circuit engineers. They must learn to think and work at the nanoscale to build devices and circuits that can connect to the macroworld. Their job is to hide the complexity of

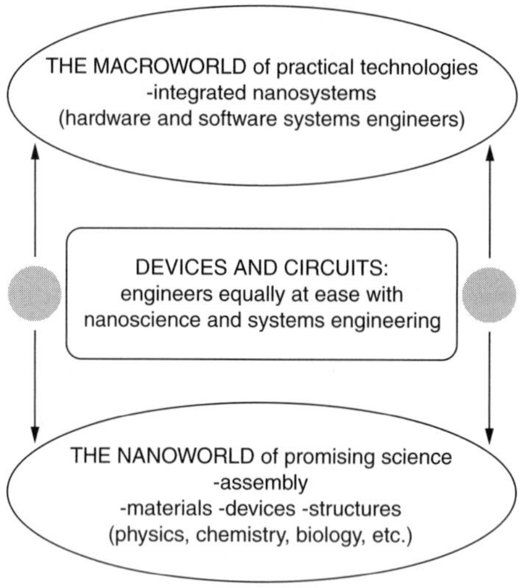

FIGURE 4.7 Science, engineering, and nanoelectronics.

the nanoscale device by packaging it in a form that systems engineers can use (e.g., a compact circuit model). To turn the promise of nanoscience into practical technologies, it is essential that the systems engineering community be engaged in the effort.

4.8 Conclusion

This chapter has introduced the emerging field of nanoelectronics — the new concepts and physical phenomena in the nanoscale MOSFET, the scaling limits of silicon MOSFETs, novel nanoscale FETs, quantum-effects devices with special applications, and the future of the electronics research. The scenario that we have outlined is an evolutionary one, but exponential evolution for another two to three decades would have a revolutionary impact on society. It is also true that it is hard to predict the future. Remember that the transistor was developed for a very specific purpose — to replace the vacuum tube; the integrated circuit was an unexpected bonus. The march of science and technology has carried us to the nanoscale; it is where the important questions are and where unforeseen breakthroughs may occur. Our march toward nanoelectronics is unstoppable. Who knows where it may lead.

4.9 Acknowledgments

It is our pleasure to acknowledge the contributions of Dr. Supriyo Datta, whose insights have deepened our understanding of conduction at the molecular scale. We also thank Sayed Hasan for providing the figures for Section 4.4. Our thanks also go to the sponsors of our work: the Semiconductor Research Corporation, the National Science Foundation, the Army Research Office (ARO) Defense University Research Initiative in Nanotechnology, and the Microelectronics Advanced Research Corporation (MARCO) Focused Research Center in Materials, Structures, and Devices (which is funded at the Massachusetts Institute of Technology, in part by MARCO under contract 2001-MT-887 and Defense Advanced Research Projects Administration (DARPA) under grant MDA972-01-1-0035).

References

[1] S. Datta, *Electronic Transport in Mesoscopic Systems*, Cambridge University Press, Cambridge, UK, 1999.

[2] M.S. Lundstrom, Elementary scattering theory of the Si MOSFET, IEEE *Electron. Dev. Lett.*, 18, 361–363, 1997.
[3] C. Collier et al. Electronically configurable molecular-based logic gates, *Science,* 285, 391–394, 1999.
[4] J. Chen, M.A. Reed, A.M. Rawlett, and J.M. Tour, Large on-off ratios and negative differential resistance in a molecular electronic device, *Science,* 286, 1550, 1999.
[5] P.L. McEuen, M.S. Fuhrer, and H. Park, Single-walled carbon nanotube electronics, IEEE *Trans. Nanotechnol.,* 1, 78–85, 2002.
[6] Y. Cui, and C.M. Lieber, Functional nanoscale electronic devices assembled using silicon nanowire building blocks, *Science,* 291, 851–853, 2001.
[7] National Nanotechnology Initiative, http://www.nano.gov, 2004.
[8] Semiconductor Industry Association (SIA), *The International Technology Roadmap for Semiconductors,* 2001.
[9] B. Yu, L. Chang, S. Ahmed, H. Wang, S. Bell, C.-Y. Yang, et al. FinFET Scaling 10-nm gate length, *IEEE Int. Electron. Devices Meeting (IEDM), Tech. Dig.*, pp. 251–254, San Francisco, CA, Dec. 2002.
[10] B. Doris, M.I.E. Ong, T. Kanarsky, Y. Zhang, R.A. Roy, O. Dokumaci, et al. Extreme scaling with ultra-thin silicon channel MOSFET's (XFET), *IEEE Int. Electron. Devices Meeting (IEDM), Tech. Dig.*, pp. 267–270, San Francisco, CA, Dec. 2002.
[11] Y. Yaur and T.H. Ning, *Fundamentals of Modern VLSI Devices,* Cambridge University Press, Cambridge, UK, 1998.
[12] D.J. Frank, R.H. Dennard, E. Nowak, P.M. Solomon, Y. Taur, and H.S.P. Wong, Device scaling limits of Si MOSFETs and their application dependencies, *Proc. IEEE,* 89, 259–288, 2001.
[13] B. Doyle, B. Boyanov, S. Datta, M. Doczy, S. Hareland, B. Jin, et al. Tri-gate fully depleted CMOS transistors: fabrication, design, and layout, *2003 Symp. on VLSI Technol.,* Kyoto, Japan, June 2003.
[14] S.H. Lo, D.A. Buchanan, Y. Taur, and W. Wang, Quantum-mechanical modeling of electron tunneling current from the inversion layer of ultra-thin-oxide nMOSFET's, *IEIEEE Electron. Dev. Lett.,* 18, 5, 209–211, 1997.
[15] W.K. Shih, E.X. Wang, S. Jallepalli, F. Leon, C.M. Maziar, and A.F. Taschjr, Modeling gate leakage current in nMOS structures due to tunneling through an ultra-thin oxide, *Solid-State Electron.,* 42, 6, 997–1006, 1998.
[16] N. Yang, W.K. Henson, J.R. Hauser, and J.J. Wortman, Modeling study of ultrathin gate oxides using direct tunneling current and capacitance-voltage measurements in MOS devices, *IEEE Trans. Electron Dev.,* 46, 7, 1464–1471, 1999.
[17] J. Wang, Y. Ma, L. Tian, and Z. Li, Modified airy function method for modeling of direct tunneling current in metal-oxide-semiconductor structures, *Applied Physics Letters,* 79, 12, 1831–1833, 2001.
[18] J. Wang and M. Lundstrom, Does source-to-drain tunneling limit the ultimate scaling of MOSFETs? *IEEE Int. Electron. Devices Meeting (IEDM), Tech. Dig.*, pp. 707–710, San Francisco, CA, Dec. 2002.
[19] K. Natori, Ballistic metal-oxide-semiconductor field effect transistor, *J. Appl. Phys.*, 76, 8, 4879–4890 1994.
[20] Z. Ren, R. Venugopal, S. Datta, M. Lundstrom, D. Jovanovic, and J.G. Fossum, The ballistic nanotransistor: a simulation study, *IEEE Int. Electron. Devices Meeting (IEDM), Tech. Dig.*, pp. 715–718, Dec. 2000.
[21] A. Rahman and M. Lundstrom, A compact scattering model for the nanoscale double-gate MOSFET, *IEEE Trans. Electron. Dev.*, 49, 3, 481–489, 2002.
[22] A. Rahman, J. Guo, S. Datta, and M. Lundstrom, Theory of ballistic nanotransistors, *IEEE Trans. Electron. Dev.,* 50, 9, 1853–1864, 2003.
[23] J. Wang and M. Lundstrom, Ballistic transport in high electron mobility transistors, *IEEE Trans. Electron. Dev.,* 50, 7, 1604–1609, 2003.
[24] S. Datta, Nanoscale device modeling: the Green's function method, *Superlattices and Microstructures,* 28, 253–278, 2000.

[25] K. Likharev, Electronics below 10nm, *Giga and Nano Challenges in Microelectron.*, J. Greer, A. Korkin, and J. Lanbanowski, Eds., North-Holland, 2003.
[26] S. Hasan, J. Wang, and M. Lundstrom, Device design and manufacturing issues for 10nm-scale MOSFETs: a computational study, *Solid State Electronics,* 48, 6, 867–875, 2004.
[27] S. Rosenblatt, Y. Yaish, J. Park, J. Gore, V. Sazonova, and P.L. McEuen, High-performance electrolyte-gated carbon nanotube transistors, *Nano Letters,* 2, 869–872, 2002.
[28] R. Saito, G. Dresselhaus, and M. Dresselhaus, *Physical Properties of Carbon Nanotubes,* Imperial College Press, London, 1998.
[29] J.M. Tour, A.M. Rawlett, M. Kozaki, Y.X. Yao, R.C. Jagessar, S.M. Dirk, et al. Synthesis and preliminary testing of molecular wires and devices, *Chemistry-A European J.*, 7, 23, 5118–5134, 2001.
[30] A. Ghosh, personal communication, May 2003.
[31] A.W. Ghosh, T. Rakshit, and S. Datta, Gating of a molecular transistor: electrostatic and conformational, *Nano Letters,* 4, 4, 565–568, 2004.
[32] P. Damle, T. Rakshit, M. Paulsson and S. Datta, Current-voltage characteristics of molecular conductors: two versus three terminal, *IEEE Trans. Nanotechnology,* 1, 3, 145–153, 2002.
[33] C-O. Chui, S. Ramanathan, B.B. Triplet, P.C. McIntyre, and K.C. Saraswat, Germanium MOS capacitors incorporating ultrathin high-κ gate dielectric, *IEEE Electron. Device Lett.*, 23, 8, 473–475, 2002.
[34] J. Hoyt, H. Nayfeh, S. Eguchi, I. Aberg, G. Xia, T. Drake, et al. Strained silicon MOSFET technology. Invited paper, *IEEE Int. Electron. Devices Meeting (IEDM), Tech. Dig.*, pp. 23–26, San Francisco, CA, Dec. 2002.
[35] K. Rim, S. Narasimha, M. Longstreet, A. Mocuta, and J. Cai, Low field mobility characteristics of sub-100nm unstrained and strained Si MOSFETs, *IEEE Int. Electron. Devices Meeting (IEDM), Tech. Dig.*, pp. 43–46, San Francisco, CA, Dec. 2002.
[36] T. Saito, T. Saraya, T. Inukai, H. Majima, T. Nagumo, and T. Hiramoto, Suppression of short-channel effect in triangular parallel wire channel MOSFETs, *IEICE Trans. Electron.*, E85-C (5), 2002.
[37] H. Majima, Y. Saito, and T. Hiramoto, Impact of Quantum mechanical effects on design of nano-scale narrow channel n- and p-type MOSFETs, *IEEE Int. Electron. Devices Meeting (IEDM), Tech. Dig.*, pp. 733–736, Washington, D.C., Dec. 2001.
[38] K. Likharev, Sub-20-nm electron devices, *Advanced Semiconductor and Organic Nano-Technologies, Part 1,* H. Morkoc, Ed., Academic Press, New York, 2003.
[39] S. Datta, and B.A. Das, Electronic analog of the electro-optic modulator, *Appl. Phys. Lett.*, 56, 665–667, 1990.
[40] D. Awschalom, M.E. Flatte, and N. Samarth, Spintronics, *Scientific American,* 67–73, June 2002.
[41] C.M. Lieber, The incredible shrinking circuit, *Scientific American,* 285, 3, 58–65, 2001.
[42] A. Seabaugh, T. Blake, B. Brar, T. Broekaert, R. Lake, F. Morris, and G. Frazier, Transistors and tunnel diodes for analog/mixed signal circuits and embedded memory, *IEEE Int. Electron. Devices Meeting (IEDM), Tech. Dig.*, pp. 429–432, San Francisco, Dec. 1998.

5

Advanced Research in On-Chip Optical Interconnects

Ian O'Connor
Frédéric Gaffiot
Ecole Centrale de Lyon

5.1 The Interconnect Problem

Due to continually shrinking feature sizes, higher clock frequencies, and the simultaneous growth in complexity, the role of interconnect as a dominant factor in determining circuit performance is growing in importance. The 2001 International Technology Roadmap for Semiconductors (ITRS) [1] shows that by 2010, high-performance integrated circuits (ICs) will count up to 2 billion transistors per chip and work with clock frequencies of the order of 10 GHz. Coping with electrical interconnects under these conditions will be a formidable task. Timing is already no longer the sole concern with physical layout: power consumption, cross talk, and voltage drop drastically increase the complexity of the trade-off problem. With decreasing device dimensions, it is increasingly difficult to keep wire propagation delays acceptable. Whereas dielectric constants below 2 (around 1.7–1.8) can be achieved using nanoporous silicon oxycarbide (SiOC)-like or organic (SiK-type) materials with an "air gap" integration approach, integration complexity is higher and mechanical properties are weaker. In addition, the use of ultra low-k materials is physically limited by the fact that no material permittivity can be less than 1 — that of air. Thus, even with the most optimistic estimates for resistance-capacitance (RC) time constants using low-resistance metals, such as copper and low-k dielectrics, global interconnect performance required for future generations of integrated circuits (ICs) cannot be achieved with metal. Furthermore, because IC power dissipation is strongly linked to switching frequency, tomorrow's architectures will require power over the 100-W mark to be able to operate in the 10-GHz range and above. At this level, thermal problems will jeopardize system performance if not strictly controlled.

5.1.1 Analysis of Electrical Interconnect Performance

The overall device scaling factor *s* makes it possible to determine the performance of interconnects. Each process shrink has a large impact on electrical parameters of metallic interconnections. Before deep submicron (DSM) nodes (the threshold is widely accepted as being around the 0.35-μm technology node), the gate delay was higher than the interconnect delay, such that each shrink led to an improvement of the maximum working speed of a system by a factor of *1/s*. In the present DSM era, however, global interconnect delay has become larger than gate delay and, consequently, interconnect has become the dominant factor determining speed.

Sakurai and Tamaru's equation (Equation 5.2) gives the propagation delay of a signal transmitted from an emitter gate to a receiver gate.

$$t_d = R_{out}\left(C_{out} + C_L\right) + R_{out} c l_W + 0{,}4\left[\left(crl_W^2\right)^{1{,}6} + \tau_{tof}^{1{,}6}\right]^{1/1{,}6} + 0{,}7 r l_w C_L \tag{5.1}$$

In this expression, R_{out} and C_{out} are, respectively, the output resistance and output capacitance of the emitter gate; C_L is the input capacitance of the receiver gate; *r*, *c*, and l_w are the lineic resistance, the lineic capacitance, and the length of the link between the emitter and the receiver; and τ_{tof} is the time of flight (i.e., the length of the line divided by the speed of the electromagnetic field).

In the case of local and intermediate interconnects, this equation reduces to:

$$t_d = R_{out}\left(C_{out} + C_L\right) + R_{out} c l_W \tag{5.2}$$

Equation 5.2 shows that the delay time is a combination of the gate output resistance, interconnect, and load capacitances. Gate sizing makes it possible to reduce delay by increasing gate strength, at the cost of increased area and power consumption.

In the case of global links, Sakurai's formula shows that the delay time in the line becomes predominant. To limit the delay time in the metallic line, global links are routed on the upper metal layers where it is possible to increase the width and the thickness of the line, and thus to reduce the lineic resistance. Reverse scaling (by reducing the thickness of the metal layer less than the scale factor) is commonplace, leading to high aspect ratios. Gate sizing makes it possible to minimize t_d, and it is possible to show that t_d varies with l_w^2. This increase of the delay time with the second power of the line length cannot be avoided. Repeater insertion makes it possible to make the delay vary with l_w, but this of course comes at the cost of a very large number of repeaters. In this scenario therefore, a relatively high percentage of silicon real estate and IC power consumption is devoted to interconnect instead of to data processing functions.

Subsequently, the problem facing us is that evolutionary solutions will not be sufficient to meet the performance roadmap. To tackle the issues developed previously, radically different interconnect approaches displaying a highly improved data-rate-to-power ratio must be developed. At present, the most prominent ideas are the use of integrated radio frequency or microwave interconnects [3], three-dimensional (3D) (nonplanar) integration [4], and optical interconnects [5]. This chapter focuses on the latter concept.

5.1.2 The Optical Alternative

A promising approach to the interconnect problem is the use of an optical interconnect layer. Such a layer could empower an enormous bandwidth increase, immunity to electromagnetic noise, a decrease in the power consumption, synchronous operation within the circuit and with other circuits, and reduced immunity to temperature variations. Important constraints when developing the optical interconnect layer are the fact that all fabrication steps have to be compatible with future IC technology and that the additional cost incurred remains affordable. Difficulties expected are obtaining a large enough optical-electrical conversion efficiency, reducing the optical transmission losses while allowing for a sufficient density of photonic waveguides on the circuit and reduction of the latency while operating above the 10-GHz mark. Sections 5.3, 5.4, and 5.5 describe, respectively, the issues involved in photonic waveguides, active devices, and optoelectronic conversion circuits.

5.1.3 Identified Applications

Optical links can be categorized into three broad domains, for which various analyses have been carried out and applications identified: single wavelength point-to-point (1-1 link); single and multiple wavelength broadcast (1-n link); multiple wavelength bus and switching (n-n link). The latter category is rather new and is discussed in Section 5.8.

5.1.3.1 Point-to-Point (1-1) Links

Today, complex chips typically need hundreds or thousands of global links [6]. The basic idea behind using point-to-point optical links consists of replacing electrical global links with optical ones. Research has been carried out on analyzing the benefits of introducing optical interconnect in critical data-intensive links, such as CPU-memory buses in processor architectures [7]. These analyses showed that point-to-point links do not present a sufficiently high performance gain to warrant their widespread use in future technologies. In essence, the bandwidth/power ratio for point-to-point optical links is higher than for electrical wires, but not high enough, when interface circuit power is taken into consideration. Instead, it is preferable to apply architectural modifications in order to enable bottlenecks to be overcome (in the given example application, the solution was to add more cache memory), even at the expense of greater silicon area and power. The benefits of optical interconnect in terms of physical cost in this situation are, in the long run, not viable for industrial manufacturers because the entire manufacturing process (from design to fabrication) would have to be changed: a very costly course of action. This proves that for optical interconnect to be accepted as a real alternative to metallic interconnect, performance gains of at least one order of magnitude must be demonstrated through circuit and device research advances, as well as through application targeting.

5.1.3.2 Broadcast (1-n) Links

Another and potentially more profitable application of optical interconnect technology is in clock distribution networks (CDN) [8]. To operate at high frequencies, CDNs require several hundreds of repeaters to drive the metallic tracks over the entire chip, resulting in using a high portion of overall IC power (up to 40–50%). This mode of operation also leads to stringent constraints on the design of the clock tree because an unbalanced tree will result in serious clock skew and, consequently, system failure. An electrical alternative is global clock distribution at a relatively low frequency and local clock multiplication to generate the required clock speed. Disadvantages of this approach include interzone synchronization and clock multiplication lock time. By replacing the electrical clock distribution tree by an optical one, the need for repeaters or clock multiplier circuits would be eliminated, thus reducing power consumption and clock skew; however, it would be illusory to believe that the optical clock signal could be routed down to the single-gate level: optoelectronic interface circuits are of course necessary and consume power. An example system realizing a clock distribution function, illustrated in Figure 5.1, requires a single photonic source coupled to a symmetrical waveguide structure routing to a number of optical receivers. At the receivers, the high-speed optical signal is converted to an electrical one and provided to local

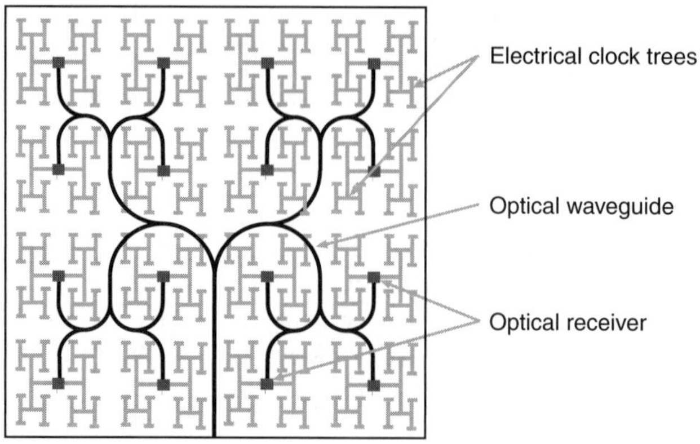

FIGURE 5.1 1-16 point optical clock distribution tree.

electrical networks. The number of clock distribution points is a particularly crucial parameter in the overall system.

5.2 Top-Down Link Design

5.2.1 Technology

Various technological solutions may be proposed for integrating an optical transport layer in a standard complementary metal oxide semiconductor (CMOS) system. The first choice is to specify where this optical layer has to be placed. Then one has to choose the different materials used for the active devices and the passive transport layer.

5.2.1.1 Materials

Materials have to be chosen with different constraints:

- Efficient light detection. Obviously, the active devices are of fundamental importance to the power budget of the link. Optics is suitable only if, for a given throughput, the global power consumption of the whole link is lower than the power consumption of classical metallic links. The quantum efficiency of the active devices is of prime importance in this context. In addition, particular attention has to be paid to the receiver: the signal to noise ratio determines the minimal optical power at the detector.
- Efficient signal transport. Attenuation and compactness are the main parameters for the choice of the passive waveguide. Technological compatibility with mature existing technologies (to ensure the required reproducibility and homogeneity of the device parameters).
- Technological compatibility with standard CMOS processes. An industrial solution is conceivable only if the optical process is completely separated from the CMOS process (the development cost of a new CMOS technology is so high that it seems very difficult to propose a solution which would require a fundamental rethink of IC fabrication processes).

Different materials are available for the realization of the optical passive guides but we focus here on silicon/silica waveguides. Silicon is an excellent material for wavelengths above 1.2 μm, and monomode waveguiding with attenuation as low as 0.8 dB/cm has been proven [9]. Moreover, the high refractive index difference between silicon and silica makes it possible to realize passive structures with dimensions compatible with DSM technologies (for example, it is possible to realize monomode waveguides less than 1 μm wide). The realization of silicon/silica waveguides is (at least in principle) compatible with a standard CMOS process. The choice of silicon waveguides leads to the use of wavelengths greater than 1.2 μm. To capitalize on the maturity of devices and concepts developed by the telecommunications industry, the choice of the wavelength is in practice limited to 1.3–1.55-μm windows.

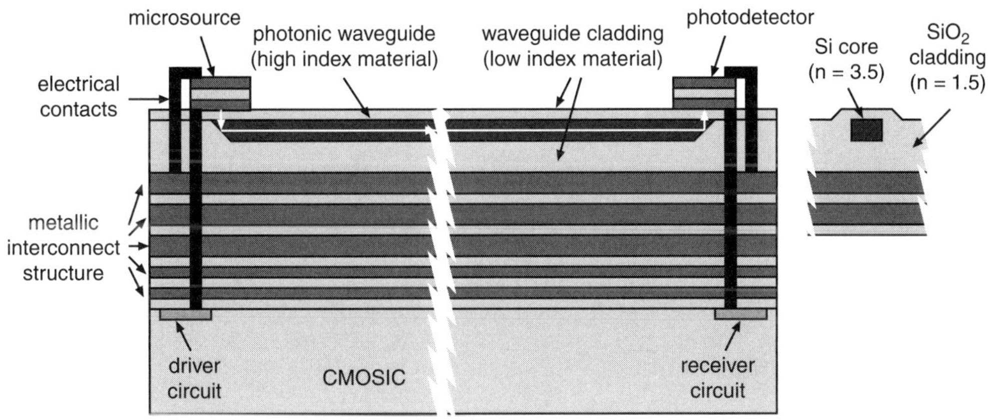

FIGURE 5.2 Cross section of hybridized interconnection structure.

5.2.1.2 Hybrid or Monolithic

The use of silicon waveguides makes it possible to imagine either monolithic (planar) or hybrid (3D) integration of the optical subsystem with CMOS systems. It is believed that the former solution is not realistic.

The integration of silicon waveguides at the front end of the CMOS process (i.e., before fabrication of the metallic interconnection layers) is certainly possible but other considerations have to be taken into account. At the transistor level, the routing of the waveguide is extremely difficult and requires routing space at the IC level. Further, the problem of the active devices remains: silicon-based sources cannot yet (and for the foreseeable future) be considered mature, while the growth of III-V devices on silicon faces strong technological barriers. The use of external sources and detectors bonded by flip-chip is unrealistic due to the high number of individual bonding operations required. In addition, silicon-based devices can only work at low wavelengths (850 nm), which translates to higher attenuation in the waveguides. This solution requires an extraordinary mutation in the CMOS process and, as such, is highly unattractive from an economic point of view.

Hybrid integration of the optical layer on top of a complete CMOS IC is much more practical and offers more scope for development. The source and detector devices are no longer bound to be realized in the host material. Figure 5.2 is a cross section of how a complete "above IC" photonic layer could be realized. The photonic source shown can be on- or off-chip: it seems likely that for some near-term applications, such as clock distribution, it is better to target off-chip signal generation for thermal reasons, even if it means higher assembly costs. It should be noted that this solution also applies to multichip module (MCM) technology. The optical process is completely independent from the CMOS process, which is appealing from an industrial point of view. Disadvantages of this approach include the more complex electrical link between the CMOS subcircuits (source drivers and detector amplifiers) and, inevitably, more advanced technological solutions for bonding.

In the system depicted in Figure 5.2, the microsource is coupled to the passive waveguide structure and provides a signal to an optical receiver (or possibly to several, as in the case of a broadcast function). At the receiver, the high-speed optical signal is converted to an electrical signal and, subsequently, distributed by a local electrical interconnect network.

To form a planar optical waveguide, silicon is used as the core and SiO_2 as the cladding material. Si/SiO_2 structures are compatible with conventional silicon technology and transparent for 1.3–1.55-μm wavelengths. Such waveguides with high relative refractive index difference $\Delta \approx (n_1^2-n_2^2)/2n_1^2$ between the core ($n_1 \approx 3.5$ for Si) and claddings ($n_2 \approx 1.5$ for SiO_2) allow the realization of a compact optical circuit, with bend radius of the order of a few μm [10]. To avoid modal dispersion, improve coupling efficiency, and reduce loss, single-mode conditions are applied to the waveguide dimensions. For a wavelength of 1.55 μm, this means a waveguide width of 0.3 μm.

The main criterion in evaluating the performance of digital transmission systems is the resulting bit error rate (BER), which may be defined as the rate of error occurrences. Typically, the BER figure required by Gigabit Ethernet and by Fiber Channel is 10^{-12} or better. For an on-chip interconnect network, a BER of 10^{-15} is acceptable. It should be noted here that BER is not commonly considered in IC design circles, and for good reason: metallic interconnects typically achieve BER figures better than 10^{-45}. Future operating frequencies are likely to change this, however, because the combination of necessarily faster rise and fall times, lower supply voltages, and higher cross talk increases the probability of wrongly interpreting the signal that was sent.

5.2.2 Design Requirements

To make a reasoned comparison between electrical and optical interconnect, a set of design requirements must be established. We have already mentioned BER; we must add to this the ubiquitous power/speed/area trinity found in any digital system. Power and speed can be compared directly, while area (in the 3D scenario) is more difficult to evaluate because we are essentially aiming at adding a photonic layer of the same size as the chip itself. What is important therefore is the average achievable area/bit ratio.

To evaluate and optimize link performance criteria correctly, predictive models and design methodologies are required. Concerning the power aspects, the aim is to establish the overall power dissipation for an optical link at a given data rate and BER. The receiver essentially conditions the calculation because the BER defines the lower limit for the received optical power. This lower limit can then be used to calculate the required power coupled into waveguides by optical sources, the required detector efficiency (including optical coupling), and acceptable transmission losses. Power can then be estimated from source bias current and photoreceiver front-end design methodologies.

For integration density aspects, source and detector sizes must be taken into account, while the width, pitch, and required bend radius of waveguides is fundamental to estimating the size of the photonic layer. On the circuit layer, the additional surface due to optical interconnect is in the driver and receiver circuits, as well as the contact and via stack to the photonic layer. The circuit layout problem is compounded by the necessity of using clean supply lines (i.e., separate from digital supplies) to reduce noise (for BER).

The data rate is essentially governed by the bandwidth of the photoreceiver: high modulation speed at the source is generally more easily attainable than similar detection speed at the receiver. This is essentially due to the photodiode parasitic capacitance at the input of the transimpedance amplifier.

Apart from these concerns, functional aspects also have to be considered. For example, using the same signal to drive two nodes is not trivial (as is the case in electrical interconnect) because the layout of a 1-2 splitter is crucial to the equal distribution of power to each node. More fundamentally, dividing the power has a direct influence on the power required at the source to achieve the lower power limit at the receiving nodes.

5.3 Passive Photonic Devices for Signal Routing

5.3.1 Waveguides

Optical system performance depends on the minimum optical power required by the receiver and on the efficiency of passive optical devices used in the system. The total loss in any optical link (represented in Figure 5.3) is the sum of losses (in decibels) of all optical components:

$$L_{total} = L_{CV} + L_B + L_Y + L_{CR} \tag{5.3}$$

where L_{CV} is the coupling coefficient between the photonic source and optical waveguide, L_W is the rectangular waveguide transmission loss, L_B is the bending loss, L_Y is the Y-coupler loss, and L_{CR} is the

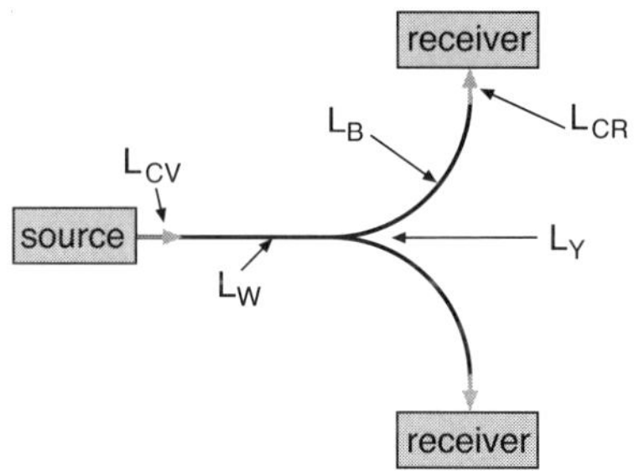

FIGURE 5.3 Losses in an optical link.

coupling loss from the waveguide to the optical receiver. To provide an unambiguous comparison in terms of dissipated power between optical and electrical on-chip interconnect networks, it is necessary to incorporate all of these quantities.

In the present technology, several methods are used to couple the beam emitted from the laser into the optical waveguide. In the proposed system, we assumed 50% coupling efficiency, L_{CV}, from the source to a single mode waveguide.

Transmission loss, L_W, describes the attenuation rate of the optical power, as light travels in the waveguide. Due to small waveguide dimensions and large index change at the core/cladding interface in the Si/SiO_2 waveguide the sidewall scattering is the dominant source of losses. To calculate the attenuation coefficient we used the Payne formula [11] associated with the Effective Index Method [12, 13]. For the waveguide fabricated by Lee et al. [9] with roughness of 2 nm, the calculated transmission loss is 1.3 dB/cm.

The bending loss L_B is highly dependent on the refractive index difference Δ between the core and cladding medium. For low Δ, the bending loss is very high, which prevents increasing the packing density. In Si/SiO_2 waveguides, Δ is relatively high and, therefore, due to this strong optical confinement, bend radii as small as a few μm may be realized. To assess the bending loss L_B we use the Marcuse method [14]. Figure 5.4 shows that the bending losses associated with a single mode strip waveguide are negligible if the radius of curvature is bigger than 2 μm.

The Y-junction loss L_Y depends on the reflection and scattering attenuation into the propagation path and surrounding medium. Different Y-branch structures have been analyzed by several methods [15,16]. For high index difference waveguides, the losses for the Y-branch are significantly smaller than for the low-Δ structures, and the simulated losses are less than 0.2 dB per split [17].

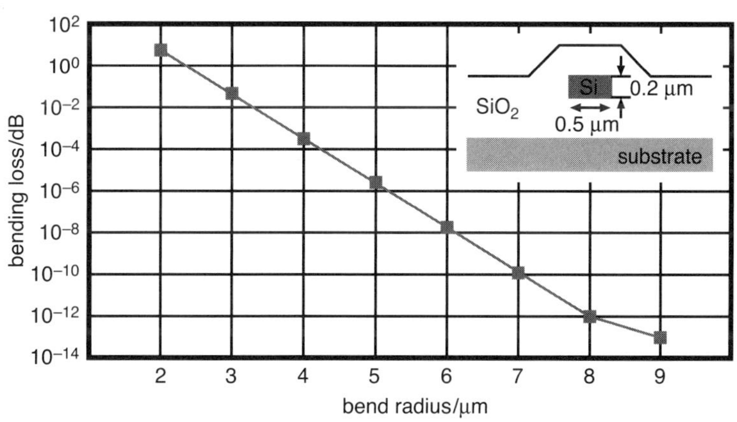

FIGURE 5.4 Simulated bending loss for Si/SiO_2 strip waveguide.

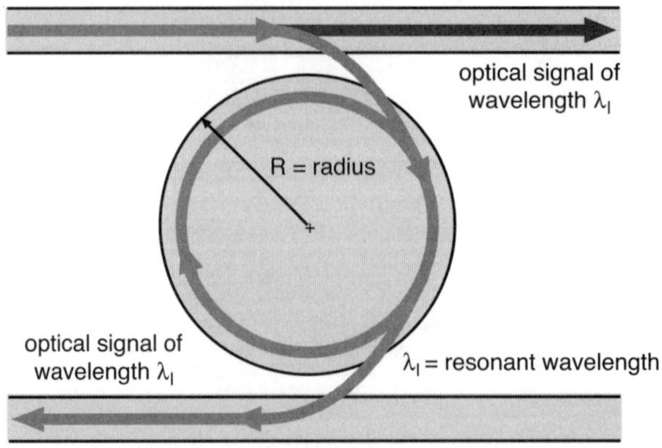

FIGURE 5.5 Micro-disk realization of an add-drop filter.

Using currently available materials and methods it is possible to achieve an almost 100% coupling efficiency from waveguide to optical receiver. In the proposed system, the coupling efficiency L_{CR} from the waveguide to the optical receiver is assumed to be 87% [18].

5.3.2 Resonators

Microdisks are resonating structures and are most commonly used in "add-drop" filters (so-called because of their capacity to add or subtract a signal from a waveguide based on its wavelength). The filter itself (Figure 5.5) is composed of one or more identical disks evanescently side-coupled to signal waveguides. The electromagnetic field is propagated within the structure only for modes corresponding to specific wavelengths, where these resonant wavelength values are determined by geometric and structural parameters (i.e., substrate and microdisk material index; thickness and radius of microdisk).

The basic function of a microresonator can be thought of as a wavelength-controlled switching function. If the wavelength of an optical signal passing through a waveguide in proximity to the resonator does not correspond to the resonant wavelength, then the electromagnetic field continues to propagate along the waveguide and not through the structure. If, however, the signal wavelength is close enough to the resonant wavelength (i.e., tolerance is of the order of a few nm, depending on the coupling strength between the disk and the waveguide), then the electromagnetic field propagates around the structure and then out along the second waveguide. Switching has occurred, based on the physical properties of the signal.

An obvious application of this device is in optical crossbar networks. More elaborate N × N switching networks have been devised [19], but experimental operation has yet to be proven. The advantages of such structures lie in the possibility of building highly complex, dense, and passive on-chip switching networks.

5.3.3 Photonic Crystals

Photonic crystals are nanostructures composed of, in the two-dimensional (2D) case, ultra-small cylinders periodically arranged in a background medium. 3D photonic crystals also exist but are much more difficult to fabricate from a technology point of view. Typically, for 2D photonic crystals, the cylinders are realized in a low-index material (such as SiO_2 or air), the background being a high-index material (such as Si). For light of certain wavelengths, such structures have a photonic band gap, leading to optical confinement. By introducing line defects (i.e., by removing one or more rows of cylinders), single-mode waveguides can be created. Other functions can be created using photonic crystals, such as couplers [20], multiplexers, demultiplexers, microresonators (using point defects instead of line defects), and even lasers. Photonic crystals are certainly good candidates for microscale optical integrated circuits due to their small size (a typical value for waveguide pitch is 0.5 μm) and massive fabrication potential; however, attenuation is an order of magnitude higher than that of planar waveguides (6 dB/mm [21]), although good progress has recently been made in this area.

5.4 Active Devices for Signal Conversion

5.4.1 III-V Sources

Fundamental requirements for integrated semiconductor lasers are small size, low threshold lasing operation, and single-mode operation (i.e., only one mode is allowed in the gain spectrum). From the viewpoint of mode field confinement and mirror reflection, two types of microcavity structures exist: multiple reflection (VCSELs and photonic crystals) and total internal reflection (microdisks). An overview of microcavity semiconductor lasers can be found in Baba [22].

5.4.1.1 Vertical Cavity Surface Emitting Lasers (VCSELs)

VCSELs are without doubt the most mature emitters for on-chip or chip-to-chip interconnections. As their name indicates, light is emitted vertically at the surface, by stimulated emission via a current above a few microamperes.

The active layer is formed by multiple quantum wells surrounded by III-V compound materials, and the whole forms the optical cavity of the desired wavelength. Above and below are Bragg reflectors, with deep proton implant to confine the current injected via the anode.

VCSELs are intrinsically single-mode due to their small cavity dimensions. They also have a very low threshold current and low divergence. Further, arrays of VCSELs are easy to fabricate; however, the internal cavity temperature can become quite high, and this is important because both wavelength and optical gain are dependent on the temperature.

Commercial VCSELs, when forward biased at a voltage well above 1.5 V, can emit optical power of the order of a few mW around 850 nm, with an efficiency of some 40%. Threshold currents are typically in the mA range. It is clear that effort is required from the research community if VCSELs are to compete in the on-chip optical interconnect arena, to increase wavelength, efficiency, and threshold current. Long wavelength and low-threshold VCSELs are only just beginning to emerge (e.g., a 1.5-μm, 2.5-Gb/s tuneable VCSEL [23] and an 850-nm, 70-μA threshold current, 2.6-μm diameter CMOS-compatible VCSEL [24] have been reported).

5.4.1.2 Microsources

Integrated microlasers differ from VCSELs in that light emission is in-plane to be able to inject light directly into a waveguide with minimum loss. Such devices, to be compatible with dense photonic integration, must satisfy the requirements of small volume, high optical confinement, low threshold current and emission in the 1.3–1.6-μm range. This wavelength implies the necessary use of indium phosphide (InP) and related materials, which leads to heterogeneous integration: bonding issues arise, which are covered in Section 5.6.

The structure of a microdisk laser is depicted in Figure 5.6 [25]. Upper and lower posts support the active region of the disk. Small cavity volume and strong optical confinement through semiconductor/air boundaries leads to low threshold currents. Current injection via the top contact causes carriers to diffuse to the disk edge and, consequently, produce optical gain. Lasing oscillation is generated by "whispering gallery" modes (so-called because of how the energy is distributed) rounding inside the disk edge. These modes, defined by the disk radius and representing the emission wavelengths, can be calculated using finite difference time domain (FDTD) simulations.

Photonic crystals can also be used as microsources. Although they are potentially smaller than microdisks and with better control of emission directivity and coupling, their mode behavior is complex and difficult to evaluate, and structures designed for lightwave frequencies are difficult to fabricate and to characterize. Research is ongoing in this area.

5.4.2 Detectors

Conventional positive-intrinsic-negative (PIN) photodiodes have relatively small area per unit capacitance values, meaning that the optical responsivity bandwidth product is low. This is a problem for high-speed operation in optical interconnect because transimpedance amplifier interface circuits cannot sup-

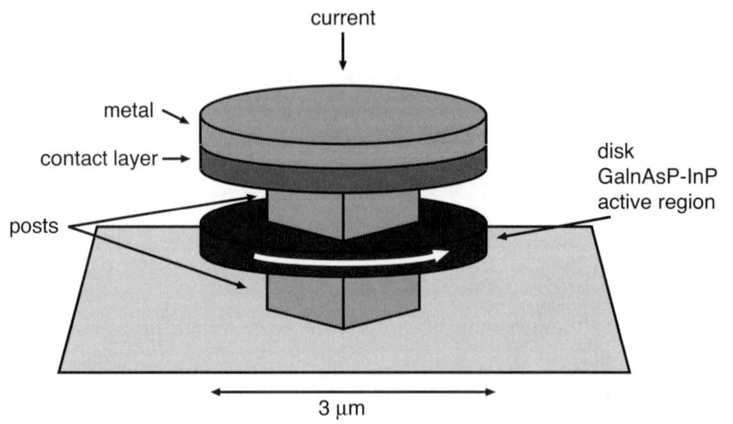

FIGURE 5.6 Structure of a microdisk laser.

port high photodetector capacitance values. Current research is focusing on thin-film, metal-semiconductor-metal (MSM) photodetectors, due to their improved area per unit capacitance [26].

5.5 Conversion Circuits

Between electronic data processing and photonic data transport lie crucial building blocks to the optical interconnect solution: high-speed optoelectronic interface circuits. On the emitter side, the power dissipated by the source driver is largely governed by the bias conditions required for the source itself. Advances in this area thus follow, to a large extent, improvements resulting from device research. On the receiver side however, things are rather different: most of the receiver power is due to the circuit. Only a small fraction is required for the photodetector device. The objective therefore is to attain the maximum speed/power ratio using dedicated circuit design methodologies.

5.5.1 Driver Circuits

The basic current modulation configuration of the source driver circuit is illustrated in Figure 5.7. The source is biased above its threshold current by M_2 to eliminate turn-on delays, and because the bias current value is the main contributing factor to emitter power, reducing the source threshold current is a primary device research objective. Figures of approximately 40 μA [27] have been reported. Device M_1 serves to modulate the current flowing through the source and, consequently, the output optical power injected into the waveguide. As with most current-mode circuits, high bandwidth can be achieved because the voltage over the source is held relatively constant and parasitic capacitances at this node have reduced influence on the speed.

5.5.2 Receiver Circuits

The classical structure for a receiver circuit is illustrated in Figure 5.8: a transimpedance amplifier (TIA) converts the photocurrent of a few μA into a voltage of a few mV; a comparator generates a rail-to-rail signal; and a data recovery circuit eliminates jitter from the restored signal.

Of these, the TIA is arguably the most critical component because it has to cope with a generally large photodiode capacitance situated at its input. Bandwidth/power ratio maximization can be achieved in several ways:

- Parametric optimization. For a given transimpedance structure, find the combination of component parameters necessary for maximum bandwidth.

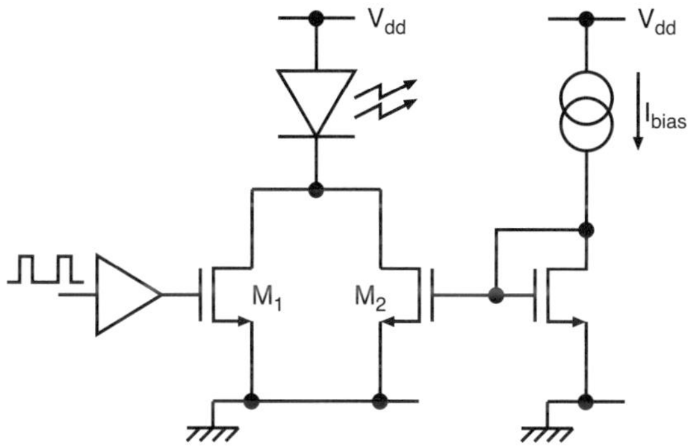

FIGURE 5.7 Basic current modulation source driver circuit.

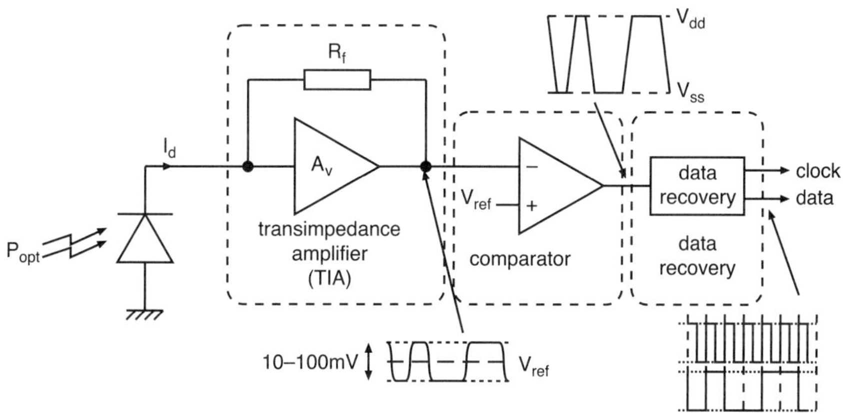

FIGURE 5.8 Typical photoreceiver circuit.

- Structural modification. For a given preamplifier architecture, make structural modifications, usually by adding elements, such as inductors for shunt peaking [28] or capacitors as artificial loads or feedback [29].
- Architectural exploration. Use complex architectures such as bootstrap or common-gate input stages [30].

The basic transimpedance amplifier structure in a typical configuration is depicted in Figure 5.9 [31]. The bandwidth/power ratio of this structure can be maximized by using small-signal analysis and mapping of the individual component values to a filter approximation of the Butterworth type, which gives analytical equations for the static transimpedance gain (Z_{g0}), the pole angular frequency (ω_0), and the pale quality factor (Q):

$$Z_{g0} = \frac{R_0 - R_f A_v}{1 + A_v} \tag{5.4}$$

$$\omega_0 = \frac{1}{R_0 C_y} \sqrt{\frac{1 + A_v}{M_f (M_x + M_m + M_x M_m)}} \tag{5.5}$$

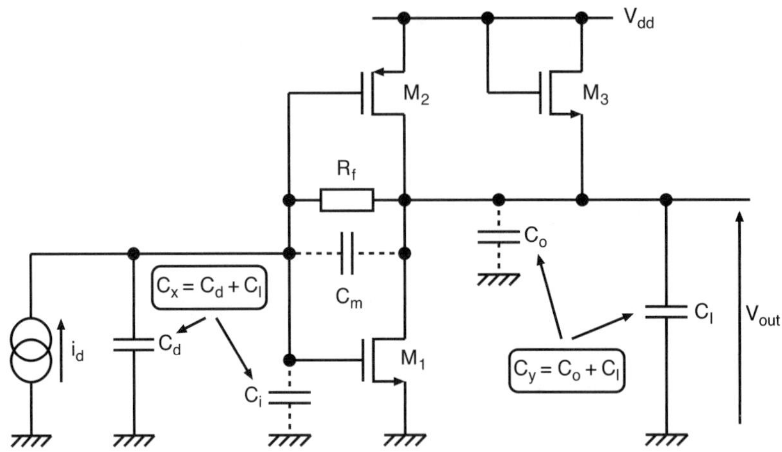

FIGURE 5.9 CMOS transimpedance amplifier structure.

$$Q = \frac{\sqrt{M_f\left(M_x + M_m\left(1+M_x\right)\right)\left(1+A_v\right)}}{1+M_x\left(1+M_f\right)+M_mM_f\left(1+A_v\right)} \tag{5.6}$$

where the multiplying factors $M_f = R_f/R_o$, $M_i = C_x/C_y$, and $M_m = C_m/C_y$ are introduced, normalizing all expressions to the time constant $\tau = R_oC_y$. By rearranging these equations, it is then possible to develop a synthesis procedure that, from desired transimpedance performance criteria (Z_{g0}, bandwidth and Q) and operating conditions (C_d, C_l), generates component values for the feedback resistance R_f and the voltage amplifier (A_v and R_o) (voltage gain A_v and output resistance R_o).

Taking into consideration the physical realization of the amplifier, those with requirements for low-gain and high-output resistance (high R_o/A_v ratio) are the easiest to build, and require the least quiescent current and area. Figure 5.10(a) shows a plot of this quantity against the TIA specifications (bandwidth and transimpedance gain) for $C_x = C_d = 500$ fF and $C_y = C_l = 100$ fF.

Approximate equations for the small-signal characteristics and bias conditions of the circuit allow a first-cut sizing of the amplifier. The solution can then be fine-tuned by numerical or manual optimization, using simulation for exact results [32].

Using this methodology and predictive BSIM3v3 models for technology nodes from 180 nm down to 70 nm [33], we generated design parameters for 1-THzΩ transimpedance amplifiers to evaluate the evolution in critical characteristics with technology node. Figure 5.10(b) shows the results of transistor level simulation of fully generated photoreceiver circuits at each technology node. According to traditional "shrink" predictions, which consider the effect of applying a unit-less scale factor of *1/s* to the geometry of metal-oxide semiconductor (MOS) transistors, the quiescent power and device area should decrease by a factor of $1/s^2$. Between the 180-nm and 70-nm technology nodes, $s^2 \approx 6.61$, which is verified through the sizing procedure. This methodology also allows us to find a particular specification to a given tolerance, as shown in Figure 5.10(c), which gives the active area and power of the generated TIA for bandwidths of 1GHz–5GHz (with $Z_{g0} = 1\text{k}\Omega$ and $Q = 1/\sqrt{2}$).

5.6 Bonding Issues

Connection of the optical interconnect network and the electronic IC is a nontrivial aspect to the whole optical interconnect concept. Probably the most effective and proven technique is flip-chip bonding [34]. This involves the depositing of gold solder bumps on either the electronic or photonic IC, then alignment and, finally, bonding, usually using thermocompression. At the wafer-scale bonding level, advanced machines are capable of precision alignment down to the order of 1 μm. In such cases, the solder bump

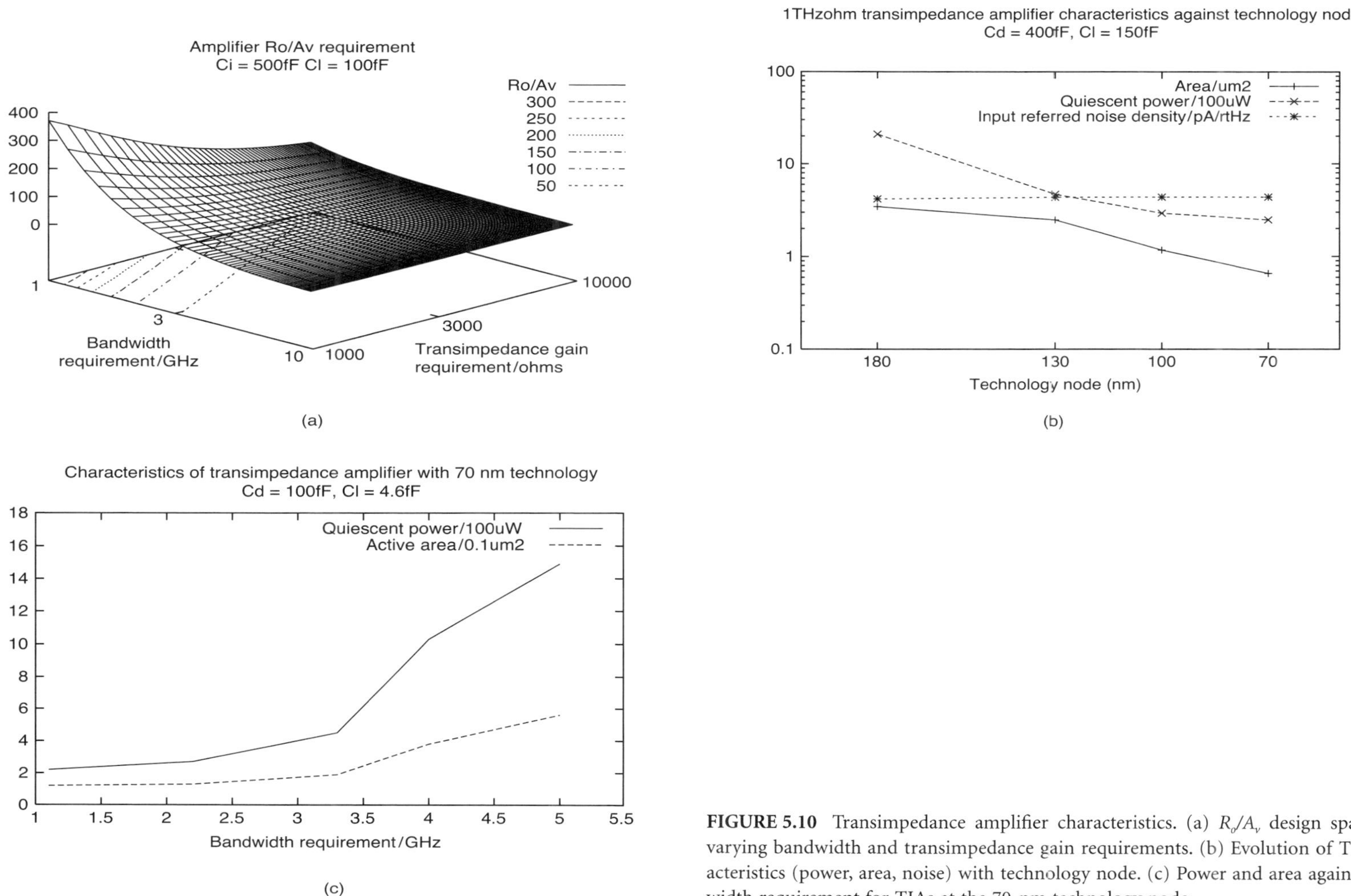

FIGURE 5.10 Transimpedance amplifier characteristics. (a) R_o/A_v design space with varying bandwidth and transimpedance gain requirements. (b) Evolution of TIA characteristics (power, area, noise) with technology node. (c) Power and area against bandwidth requirement for TIAs at the 70-nm technology node.

can be made small (under 10 μm in diameter) so that the total capacitance of the link, including pads, is of the order of a few tens of fF, compatible with high-speed interface circuit operation.

More futuristic ideas concern epitaxial integration, where the III-V material is grown directly onto the silicon substrate. This is possible, but the temperatures involved are usually rather high (800°C). However, recent research [35] has successfully demonstrated hydrophilic wafer bonding of an InP microdisk laser onto a silicon wafer, at room temperature, by means of SiO_2 layers on both InP and silicon substrates. The highest temperature in this process was 200°C (for annealing, to increase bonding energy), which is compatible with CMOS IC fabrication steps.

5.7 Link Performance (Comparison of Optical and Electrical Systems)

To provide a clear comparison in terms of dissipated power between the optical and electrical interconnect networks it is necessary to estimate the electrical power dissipated in both systems. As an example, the power dissipated in clock distribution networks was analyzed in both systems at the 70-nm technology node. Power dissipation figures for electrical and optical clock distribution networks (CDNs) were calculated based on the system performance summarized in Table 5.1 and Table 5.2. The power dissipated in the electrical system can be attributed to the charging and discharging of the wiring and load capacitance and to the static power dissipated by the buffers. To calculate the power, we used an internally developed simulator, which allows us to model and calculate the electrical parameters of clock networks for future technology nodes.

The first input to this program is the set of technology parameters for the process of interest, particularly the feature size, dielectric constant, and metal resistivity according to the ITRS roadmap. In the next step, the resistance, capacitance, and inductance values for a given metal layer are calculated, as well as the electrical parameters of minimum size inverters. Based on this information, it is then possible to determine the optimal number and size of buffers needed to drive the clock network. For such a system, the program creates the SPICE netlist where the interconnect is replaced by resistance-capacitance (RC) or resistance-inductance-capacitance (RLC) distributed lines coupled by buffers

TABLE 5.1 Electrical System Performance

Technology [μm]	0.07
V_{dd} [V]	0.9
T_{ox} [nm]	1.6
Chip size [mm^2]	400
Global wire width [μm]	1
Metal resistivity [μΩ-cm]	2.2
Dielectric constant	3
Optimal segment length [mm]	1.7
Optimal buffer size [μm]	90

TABLE 5.2 Optical System Performance

Wavelength λ [μm]	1.55
Waveguide core index (Si)	3.47
Waveguide cladding index (SiO_2)	1.44
Waveguide thickness [μm]	0.2
Waveguide width [μm]	0.5
Transmission loss [dB/cm]	1.3
Loss per Y-junction [dB]	0.2
Input coupling coefficient [%]	50
Photodiode capacitance [fF]	100
Photodiode responsivity [A/W]	0.95

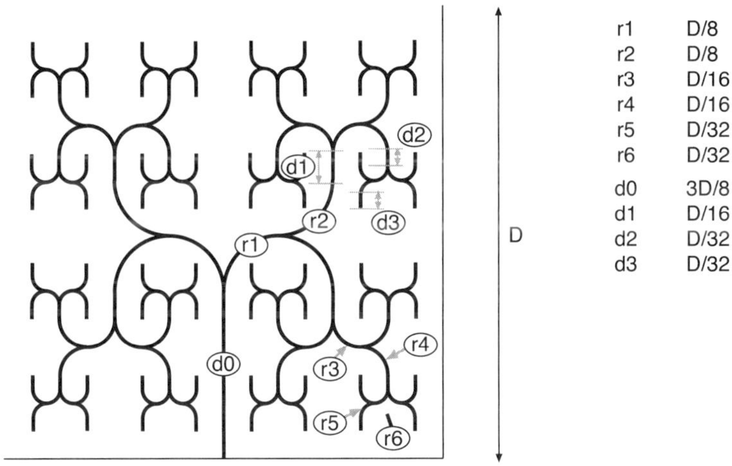

FIGURE 5.11 Optical H-tree network with 64 output nodes. $r_1..r_6$ are the bend radii and $d_0..d_3$ are the lengths of straight lines linked to the chip width D.

designed as CMOS inverters. Berkeley BSIM3v3 [33] parameters were used to model the transistors used in the inverters. The power dissipated in the system is extracted from transistor-level simulations.

Two main sources of electrical power dissipation exist in the optical clock distribution network (Figure 5.11):

1. Power dissipated by the optical receivers
2. Energy needed by the optical source to provide the required optical output power

To estimate the electrical power dissipated in the system, we used the methodology given in Figure 5.12. We assumed the use of an external VCSEL source, instead of an integrated microsource. The global optical H-tree was optimized to achieve minimal optical losses. The bend radii are designed to be as large as possible. For 20-mm die width and 64 output nodes in the H-tree at the 70-nm technology mode, the smallest radius of curvature (r_5, r_6 in Figure 5.11) is 625 μm, which leads to negligible pure bending loss.

First, based on the given photodiode parameters (C_d, R, I_{dark}), the method described in Section 5.5 is used for the design of the transimpedance amplifier. Next, for a given system performance (BER) and for the noise signal associated with the photodiode and transimpedance circuit, we calculate the minimum optical power required by the receiver to operate at the given error probability, using the Morikuni formula [36] in the preamplifier noise calculations.

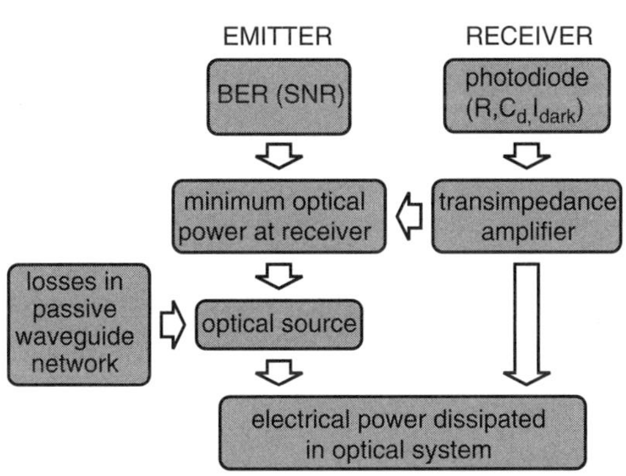

FIGURE 5.12 Methodology used to estimate the electrical power dissipation in an optical clock distribution network.

TABLE 5.3 Optical Power Budget for 20-mm Die Width at 3 GHz

Number of nodes in H-tree	4	8	16	32	64	128
Loss in straight lines [dB]	1.3	1.3	1.3	1.3	1.3	1.3
Loss in curved lines [dB]	1	1.31	1.53	1.66	1.78	1.85
Y-dividers [dB]	6	9	12	15	18	21
Loss in Y-couplers [dB]	0.4	0.6	0.8	1	1.2	1.4
Output coupling loss [dB]	0.6	0.6	0.6	0.6	0.6	0.6
Input coupling loss [dB]	3	3	3	3	3	3
Total optical loss [dB]	12.3	15.8	19.2	22.5	25.8	29.1
Min. receiver power [dBm]	−22.3	−22.3	−22.3	−22.3	−22.3	−22.3
Laser optical power [mW]	0.1	0.25	0.5	1.1	2.30	4.85

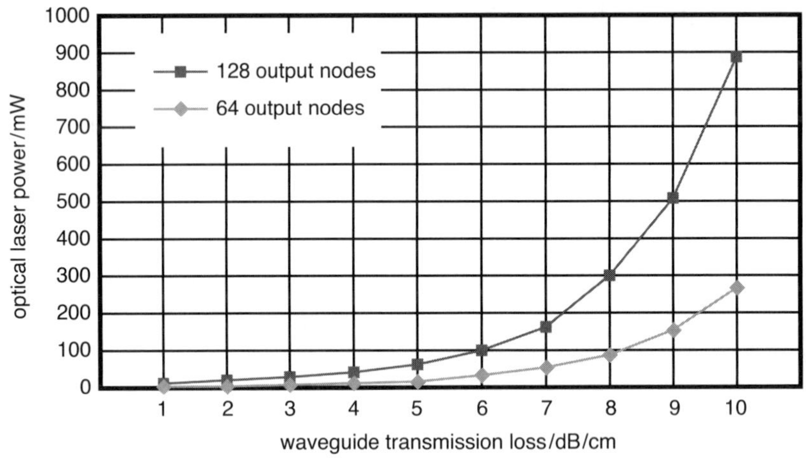

FIGURE 5.13 VCSEL optical output power required by the H-tree to provide a given BER for varying waveguide transmission loss.

To estimate the optical power emitted by the VCSEL, we took into account the previously calculated minimal required signal by the receiver and the losses incurred throughout passive optical waveguides. The electrical power dissipated in optical clock networks is the sum of the power dissipated by the number of optical receivers and the energy needed by the VCSEL to provide the required optical power. The electrical power dissipated by the receivers has been extracted from transistor-level simulations. To estimate the energy needed by the optical source, we use the laser light-current characteristics given by Amann et al. [37].

For a BER of $10^{-15,}$ the minimal power required by the receiver is −22.3 dBm (at 3 GHz). Losses incurred by passive components for various nodes in the H-tree are summarized in Table 5.3.

Figure 5.13 plots the optical power emitted by the VCSEL necessary to provide a given BER for various waveguide transmission losses. The comparison in terms of dissipated power between the optical and electrical global clock distribution networks is plotted in Figure 5.14. It can be seen that the power dissipated by the electrical system is highly dependent on the operating frequency. In the optical system, however, it remains almost the same. The difference between the power dissipated in both systems is clearly higher if we increase the frequency and number of nodes in H-trees. For a classical 64-node H-tree at 5GHz frequency, the power consumption in the optical CDN should be 5 times lower than in an electrical network.

5.8 Research Directions

Integrated optical interconnect is one potential technological solution to reduce the power required to move volumes of data between circuit blocks on integrated circuits, but it only makes sense to use this

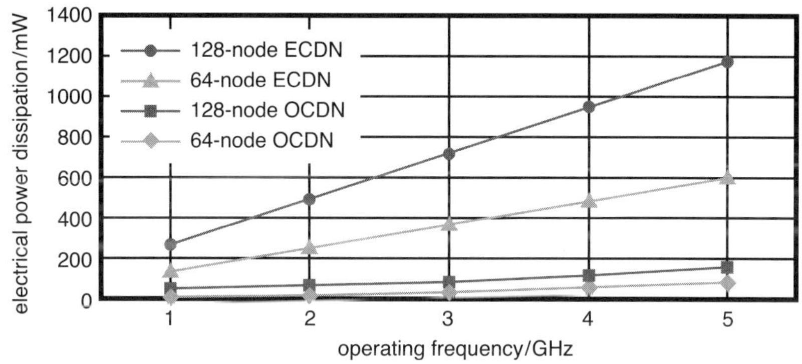

FIGURE 5.14 Electrical power dissipated by electrical (ECDN) and optical (OCDN) clock distribution networks for varying operating frequency.

technology for global high-speed data links. In addition, if the use of this technology implies a hard breach in terms of process technology and design methodologies, then architectural solutions may be an easier way to reduce power, by optimizing layout for the application such that the need for global high-speed data links is alleviated.

It is difficult to predict between these two scenarios. Future ICs will probably make use of advances in both areas. One parameter that is likely to make the difference in favor of optical interconnect is new research into the possibility of on-chip wavelength division multiplexing (WDM).

5.8.1 Network Links

Network, or n-n optical links, where wavelength routing may be used, would be targeted at (a) optical buses and possibly (b) reconfigurable networks. System architectures are moving rapidly toward platform-based designs, whereby every functional block on the chip (digital signal processors (DSP), analog/radio frequency [RF], video processors, memory, etc.) interfaces to an interconnect network architecture for data communication. Global system on chip (SOC) communication is around several tens of Gb/s. Commercial solutions for SoC development platforms are now based on bus architectures [38,39]: metallic interconnect architectures rely heavily on wide (64/128 bits) buses, as well as frequent use of switch boxes to dynamically define a communication route between two functional internet protocol (IP) blocks. Again, because the order of distance of communication is the chip die size, systematic use of repeaters (over the buses or within the actual switch boxes) is necessary and increases power consumption.

In the future, limitations (e.g., latency due to line delay, nonscalability, time sharing, and nonreconfigurability) of bus-based architectures will appear: future architectures of integrated systems will require new concepts for on-chip data exchange. The ever increasing number of transistors in a chip will lead to such complexity that IP reuse will be mandatory: a system is designed by integrating some hundreds of predesigned complex functional blocks, with the designer concentrating mainly on the organization of data transfer between these blocks.

A number of innovative interconnect architectures, often called networks on chip (NoC), have been recently proposed to overcome the limitations of bus-based platforms [40–42]. NoC architectures look much more like switching telecommunication networks than conventional bus-based architectures. Depending on the target application (multiprocessor SoCs or systems in which different functional blocks process heterogeneous signals), NoCs may have different structures, such as rings, meshes, hypercubes, or random networks. Latency, connectivity, global throughput, and reconfigurability constitute the main performance indicators of these networks.

Integrated optics may constitute an effective and attractive alternative for NoC. The superiority of optical interconnects in long distance links is established, and, possibly, some advantages of optical propagation may overcome the limitations of classical technologies for data exchanges at the integrated system scale. Some of the physical advantages of optical interconnects may be of a prime interest for NoC: flat frequency response (i.e., signal attenuation does not depend on frequency), limitation of cross talk, no repeaters, and power consumption. Above all, however, wavelength division multiplexing (WDM) may offer new and appealing solutions, such as optical buses and reconfigurable networks.

5.8.1.1 Optical Buses

As in telecommunications, WDM provides a route to very high data rates, even if the individual devices cannot be modulated much faster than electrical bus data rates. A single waveguide could be used to replace a 64-bit bus, for example, where each individual signal makes use of a distinct wavelength.

5.8.1.2 Reconfigurable Networks

By using a WDM approach, reconfigurable networks could be realized in the optical domain, leading to power reduction and higher integration density. Switch boxes, a key element for reconfigurable networks, could also be realized by using compact micro-resonators (about $10 \times 10\ \mu m^2$), capable of selecting and redirecting a signal based on its wavelength. Such networks would be entirely passive (i.e., no power would be required to transport the data, whatever the communication route necessary). Such a scheme, however, would imply a shift in the routing paradigm from a centralized arbiter acting on the switch boxes to one acting on the block interfaces to select the wavelength(s) to be used. In addition, tunable and thermally stable microlasers would be required.

5.9 Acknowledgments

The authors thank Dr. Xavier Letartre for his valuable discussions on active integrated devices for optical interconnect.

References

[1] Semiconductor Industry Association, *International Technology Roadmap for Semiconductors, 2001 ed.*, http://public.itrs.net, 2001.

[2] Sakurai, T. and Tamaru, K. Simple formulas for two- and three-dimensional capacitances, *IEEE Tran. Electron. Devices*, 30, 183, 1983.

[3] Chang, M.F. et al. RF/wireless interconnect for inter- and intra-chip communications, *Proc. IEEE*, 89, 456, 2001.

[4] Banerjee, K. et al. 3-D ICs: A novel chip design for improving deep-submicrometer interconnect performance and systems-on-chip integration, *Proc. IEEE*, 89, 602, 2001.

[5] Miller, D.A.B. Rationale and challenges for optical interconnects to electronic chips, *Proc. IEEE*, 88, 728, 2000.

[6] Davis, J.A., De, V.K., and Meindl, J.D. A stochastic wire-length distribution for gigascale integration (GSI) — Part I: derivation and validation, *IEEE Trans. Electron. Dev.*, 45, 580, 1998.

[7] Collet, J.H. et al. Architectural approach to the role of optics in mono- and multi-processor machines, *Applied Optics*, 39, 671, 2000.

[8] Friedman, E.G. Clock distribution networks in synchronous digital integrated circuits, *Proc. IEEE*, 89, 665, 2001.

[9] Lee, K.K. et al. Fabrication of ultralow-loss Si/SiO_2 waveguides by roughness reduction, *Optics Lett.*, 26, 1888, 2001.

[10] Sakai, A., Hara, G., and Baba, T. Propagation characteristics of ultrahigh-Δ optical waveguide on silicon-on-insulator substrate, *Japanese J. Appl. Phys. — Part 2*, 40, 383, 2001.

[11] Payne, F.P. and Lacey, J.P.R. A theoretical analysis of scattering loss from planar optical waveguides, *Optical Quantum Electron.*, 26, 977, 1994.

[12] Nishihara, H., Haruna, M., and Suhara, T. *Optical Integrated Circuits*, McGraw-Hill, New York, 1988.

[13] Kim, C.M., Jung, B.G., and Lee, C.W. Analysis of dielectric rectangular waveguide by modified effective-index method, *IEEE Electron. Lett.*, 22, 296, 1986.
[14] Marcuse, D. *Light Transmission Optics*, Van Nostrand Reinhold, New York, 1972.
[15] Chu, F.S. and Liu, P.L. Low-loss coherent-coupling Y branches, *Optics Lett.*, 16, 309, 1991.
[16] Rangaraj, M., Minakata, M. and Kawakami, S. Low-loss integrated optical Y-branch, *IEEE J. Lightwave Technol.*, 7, 753, 1989.
[17] Sakai, A., Fukazawa, T., and Baba, T. Low-loss ultra-small branches in a silicon photonic wire waveguide, *IEICE Trans. Electron.*, E85-C, 1033, 2002.
[18] Schultz, S.M., Glytsis, E.N., and Gaylord, T.K. Design, fabrication, and performance of preferential-order volume grating waveguide couplers, *Applied Optics-IP*, 39, 1223, 2000.
[19] Little B.E, Chu S.T., Pan W., and Kokubun Y. Microring resonator arrays for VLSI photonics, *IEEE Photonics Tech. Lett.*, 12, 323, 2000.
[20] Martinez, A., Cuesta, F., and Marti, J. Ultrashort 2-D photonic crystal directional couplers, *IEEE Photonics Tech. Lett.*, 15, 694, 2003.
[21] Notomi, M. et al. Structural tuning of guiding modes of line-defect waveguides of silicon-on-insulator photonic crystal slabs, *IEEE J. Quantum Electron.*, 38, 736, 2002.
[22] Baba, T. Photonic crystals and microdisk cavities based on GaInAsP-InP system, *IEEE J. Selected Topics in Quantum Electron.*, 3, 808, 1997.
[23] Filios, A. et al. Transmission performance of a 1.5-mm 2.5-Gb/s directly modulated tunable VCSEL, *IEEE Photonics Tech. Lett.*, 15, 599, 2003.
[24] Liu, J.J. et al. Ultralow-threshold sapphire substrate-bonded top-emitting 850-nm VCSEL array, *IEEE Photonics Lett.*, 14, 1234, 2002.
[25] Fujita, M., Sakai, A., and Baba, T. Ultrasmall and ultralow threshold GaInAsP-InP microdisk injection lasers: design, fabrication, lasing characteristics and spontaneous emission factor, *IEEE J. Selected Topics in Quantum Electron.*, 5, 673, 1999.
[26] Cho, S.Y. et al. Integrated detectors for embedded optical interconnections on electrical boards, modules and integrated circuits, *IEEE J. Selected Topics in Quantum Electron.*, 8, 1427, 2002.
[27] Fujita, M., Ushigome, R., and Baba, T. Continuous wave lasing in GaInAsP microdisk injection laser with threshold current of 40μA, *IEEE Electron. Lett.*, 36, 790, 2000.
[28] Mohan, S.S. et al. Bandwidth extension in CMOS with optimized chip inductors, *IEEE J. Solid-State Circuits*, 35, 3, March 2000.
[29] Kuo, C.W. et al. 2 Gbit/s transimpedance amplifier fabricated by 0.35μm CMOS technologies, *IEEE Electron. Lett.*, 37, 1158, 2001.
[30] Graeme, J. *Photodiode Amplifiers*, McGraw-Hill, New York, 1996, chap. 4.
[31] Ingels, M. and Steyaert, M.S.J. A 1-Gb/s, 0.7-μm CMOS optical receiver with full rail-to-rail output swing, *IEEE J. Solid-State Circuits*, 34, 971, 1999.
[32] O'Connor, I. et al. Exploration paramétrique d'amplificateurs de transimpédance CMOS à bande passante maximisée, *Proc. Traitement Analogique de l'Information, du Signal et ses Applications*, 73, Paris, September 12–13, 2002.
[33] Cao, Y. et al. New paradigm of predictive MOSFET and interconnect modeling for early circuit design, *Proc. Custom Integrated Circuit Conf.*, Orlando, FL, May 21–24, 2000.
[34] Krishnamoorthy, A.V. and Goossen, K.W. Optoelectronic-VLSI: Photonics integrated with VLSI circuits, *IEEE J. Selected Topics in Quantum Electron.*, 4, 899, 1998.
[35] Seassal, C. et al. InP microdisk lasers on silicon wafer: CW room temperature operation at 1.6 μm, *IEEE Electron. Lett.*, 37, 222, 2001.
[36] Morikuni, J.J. et al. Improvements to the standard theory for photoreceiver noise, *IEEE J. Lightwave Technol.*, 12, 1174, 1994.
[37] Amann, M.C., Ortsiefer, M. and Shau, R. Surface-emitting laser diodes for telecommunications, *Proc. Symp. Opto- and Microelectronic Devices and Circuits*, Stuttgart, March 10–16, 2002.
[38] IBM, The core connect bus architecture, http://www-306.ibm.com/products/coreconnect, 1999.

[39] VSI Alliance, http://www.vsi.org.

[40] Benini, L. and De Micheli, G. Networks on chip: a new SoC paradigm, *IEEE Computer*, 35, 70, 2002.

[41] Guerrier, P. and Greiner, A. A generic architecture for on-chip packet-switched interconnections, *Proc. Design, Automation and Test in Europe 2000*, 250, Paris, France, March 27–30, 2000.

[42] Dally, W.J. and Towles, B. Route packets, not wires: on-chip interconnection networks, *Proc. 38th Design Automation Conf.*, Las Vegas, June 18–22, 2001.

Part II

Low-Power Circuits

6

Modeling for Designing in Deep Submicron Technologies

Daniel Auvergne
Philippe Maurine
Nadine Azémard
LIRMM, University of Montpellier

6.1 Introduction

Much effort is being devoted to metal-oxide semiconductor (MOS) device modeling in deep submicron process [1–6]. Most of this effort addresses the modeling of short-channel and narrow-width effects to derive an accurate model of the threshold voltage and of the velocity saturation effect of the field dependent mobility. This has resulted in very complex modeling equations of the device performances. If these equations, validated on experimental silicon, can be used to accurately simulate analog designs with limited number of devices, they remain completely unpractical for digital design involving a huge number of components.

However, the performance of complementary metal-oxide semiconductor (CMOS) digital circuits depends on relatively few parameters, such as the threshold voltage and the drive current. In that case, a simple model of the average current available in any digital structure is sufficient to get an accurate representation of the delay and power performance of a design, if the model can easily be calibrated on the process. The objective of this chapter is to derive a design-oriented model of the delay and power consumption performances of digital structures. This model must give the explicit sensitivity of the performance parameters to the process, to the design and control conditions, as well as to the temperature and the value of the supply voltage. It is demonstrated here that using an engineering model [7] allows fulfillment of this objective, with a useful representation of the design space.

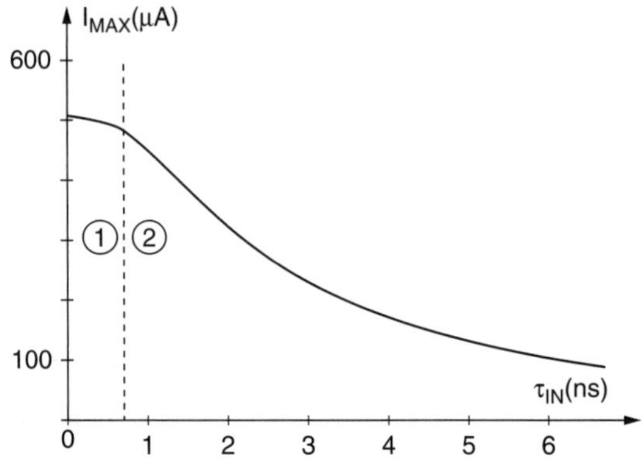

FIGURE 6.1 Sensitivity to the input ramp duration of the inverter maximum current value.

6.2 Current Modeling

Since the pioneering works of Mead and Rem [8] and Veendrick [9], it is well recognized that the speed and the total dynamic power consumption (including the short-circuit component) of any structure are directly related to the value of its switching current, which depends on the structure size and topology. As a result, the use of an accurate and design parameter explicit modeling of the switching current is the only way to define alternatives for low-power designing power-efficient circuits.

This design-oriented modeling can be obtained into two steps: by determining the maximum available current in the structure, and then considering the input slope effect, which defines the variation of this maximum value with the input control conditions.

6.2.1 Maximum Switching Current

An analytical direct current model to be used for hand calculations by designers has been presented in [7], where it is shown that considering transverse and high field effects on the carrier mobility, a drift velocity saturation current can be defined as:

$$I_{SAT} = K \cdot W \cdot (V_{GS} - V_T) \tag{6.1}$$

where $K = \kappa . v_{SAT} . C_{OX}$ is the product of the short-channel effect factor κ, by the carrier saturation speed and the oxide capacitance, W is the transistor width, and V_T the threshold voltage value. Note that Equation 6.1 corresponds to the Sakurai and Newton equation [10], for $\alpha = 1$.

Let us now evaluate the maximum value of this current on a simple CMOS structure, such as an inverter. The switching of an inverter is a dynamic process of which characteristics depend on the time duration of the voltage ramp applied to its input. Figure 6.1 represents the sensitivity to the input ramp duration of the maximum switching current of an inverter. In region 1, the current has a maximum value during all the discharge process, while in region 2, the maximum current value decreases when the input ramp duration increases.

The complete analysis of the input ramp effect, on both the speed and power performances, allows definition, as suggested in Maurine et al. [11], two switching ranges — the fast and slow input ranges in which a specific evaluation of the current must be performed.

6.2.2 Fast Input Range

The inverter maximum current value can easily be obtained from Equation 6.1 considering the maximum value of the controlling voltage, $V_{GSN} = V_{DD}$, which is instantaneously applied on the N transistor. In that case, the maximum current available in the N transistor is used to discharge the output node. We obtain:

TABLE 6.1 Relative Discrepancy between Simulated and Calculated Values of I_{MAX}

τ_{LN}/T_{HLS}	$F_o = 1$	$F_o = 3$	$F_o = 5$	$F_o = 10$	$F_o = 15$	$F_o = 20$
0.15	3%	2%	2%	1%	2%	1%
1	4%	3%	3%	3%	3%	2%
2	6%	5%	5%	5%	5%	4%
4	3%	2%	2%	0%	2%	1%
6	5%	3%	3%	2%	4%	1%
8	2%	1%	2%	3%	4%	3%
10	8%	2%	3%	5%	5%	5%
12	4%	3%	5%	6%	6%	6%
14	4%	5%	6%	7%	7%	7%
16	5%	6%	7%	8%	9%	8%
18	6%	7%	9%	9%	10%	9%
20	10%	9%	10%	10%	10%	10%

$$I_{MAX}^{Fast} = K_N \cdot W_N \cdot (V_{DD} - V_{TN}) \tag{6.2}$$

with an equivalent expression for the P transistor. In this expression K_N, is the conduction factor, previously defined, which can be directly calibrated on the process.

6.2.3 Slow Input Range

In this range, the transistor is still in saturation, but its driving voltage is lower than in the preceding part. Considering a nearly symmetric variation of the current around the maximum [11], its maximum value can be obtained as:

$$I_{MAX}^{Slow} = \sqrt{\frac{K_N \cdot W_N \cdot V_{DD}^2 \cdot C_L}{\tau_{IN}}} \tag{6.3}$$

which clearly exhibits the input ramp duration and load dependency of the current. This is illustrated by the nonlinear variation of I_{MAX}, observed in Figure 6.1. Table 6.1 gives an illustration of the accuracy obtained, with respect to Spice simulations, in determining the maximum current value (Equation 6.2 and Equation 6.3) in an inverter ($W_N = 1$ μm, $W_P = 2.2$ μm) implemented in a 0.18-μm process. As demonstrated in a large range of loading and controlling conditions, a very good agreement between simulated and calculated values is obtained.

6.2.4 Extension to Gates

For an *n* input gate the current available is input vector dependent, and the current can be evaluated as one or *m* times ($1 \le m \le n$) the current available in an inverter of identical size. It has been shown [12] that an array of series-connected transistors is equivalent to a current generator with a current capability reduced by the current reduction factor (DW), compared with an inverter of identical size:

$$DW = \frac{I_{Inverter}}{I_{Array}} \tag{6.4}$$

Detailed expressions of this coefficient have been given in Maurine et al. [11] for fast and slow input ranges and for controls applied on the top, bottom, or intermediate transistor of the array.

In Table 6.2, an example of validation obtained on a NAND2 (0.18 μm) illustrates the sensitivity of the reduction factor value to the input transition time.

TABLE 6.2 Sensitivity to the Input Transition Time of the Reduction Factor Value of a NAND2, for Top and Bottom Controls

	NAND2					
	Red_{Top}			Red_{Bot}		
τ_{IH}/t_{HLS}	Cal.	Sim.	Δ%	Cal.	Sim.	Δ%
1	1.54	1.54	0%	1.54	1.53	1%
2	1.54	1.50	3%	1.54	1.52	1%
3	1.54	1.50	3%	1.50	1.49	4%
8	1.24	1.28	3%	1.24	1.20	4%
13	1.24	1.24	0%	1.10	1.18	7%
18	1.24	1.23	1%	1.03	1.17	13%
20	1.24	1.22	2%	1.00	1.16	15%

TABLE 6.3 Comparison of the Reduction Factor Values in the Fast and Slow Input Range

CMOS 0.18 μm	Red^{Fast}	Red^{Slow}	Refs. [16, 22–25]
INV	1.00	1.00	1
NAND2	1.55	1.20	2
NAND3	2.10	1.48	3
NAND4	2.65	1.78	4
NOR2	1.93	1.39	2
NOR3	2.96	1.72	3
NOR4	3.76	1.94	4

Table 6.3 presents the variation of the DW reduction factor value with the gate complexity, for a top control of the series-connected transistors. For the process under study (0.18 μm), the values of DW^{Slow} obtained for NAND and NOR gates with 2, 3, and 4 inputs, respectively, are found quite different from the values obtained for fast input transition time conditions or from a direct reduction based on the number of serially connected transistors.

6.3 Definition of Metric for Performance

Using the preceding expressions for evaluating the maximum value of the switching current of CMOS structures, it is possible to obtain closed form expressions of the output transition time, the propagation delay, and the short-circuit power component.

6.3.1 Metric for the Transition Time

The output or input transition time (τ_{OUT}, τ_{IN}, respectively) is one of the fundamental performance parameters. It directly controls the value of the propagation delay and that of the short-circuit power component. It is defined as the time spent by the cell output (input) voltage to switch between the supply rail values. Its value is structure current capability (I_{MAX}) and output load dependent. Considering a linear variation of the output voltage, a simple first-order expression of the output transition time can be obtained from the charge conservation law initially introduced by Mead and Conway [13]:

$$\tau_{OUT} = \frac{C_L \cdot V_{DD}}{I_{MAX}} \tag{6.5}$$

where τ_{out} represents the time spent by the output voltage to swing over the full supply voltage value, V_{DD} is the variation of the output node voltage, C_L the output loading capacitance, and I_{MAX} the maximum value of the switching current. In this expression, currently used in roadmaps to predict the speed of extrapolated processes, the driving element is considered as a current generator supplying a constant current to the output loading capacitance. The output transition time value can directly be obtained by replacing I_{MAX}, from Equation 6.5, by one of the expressions (Equation 6.2 and Equation 6.3) previously introduced.

This gives for the fast input control range:

$$\tau_{OUT}(fall) = \frac{C_L \cdot V_{DD}}{I_{NMAX}} = C_L \cdot \frac{V_{DD}}{K_N \cdot W_N \cdot (V_{DD} - V_{TN})}$$

$$\tau_{OUT}(rise) = \frac{C_L \cdot V_{DD}}{I_{PMAX}} = C_L \cdot \frac{V_{DD}}{K_P \cdot W_P \cdot (V_{DD} - V_{TP})} \tag{6.6}$$

where the switching current has been obtained from Equation 6.2 considering a step input voltage.

To introduce a metric for the transition time, let us consider an ideal inverter (free from parasitic capacitance and I/O coupling), implemented with identically sized and minimum length, N and P transistors ($W_N = W_P$). Equation 6.6 becomes:

$$\tau_{OUTHL} = \frac{2Cox \cdot L_{MIN} \cdot V_{DD}}{K_N \cdot (V_{DD} - V_{TN})} = 2\tau_{ST}$$

$$\tau_{OUTLH} = 2\tau_{ST} \cdot R_\mu \tag{6.7a}$$

where R_μ is an indicator of the current imbalance between the N and P transistors.

$$R_\mu = \frac{K_N}{K_P} \cdot \frac{V_{DD} - V_{TN}}{V_{DD} - V_{TP}} \tag{6.7b}$$

For a general output load, we obtain:

$$\tau_{OUT}^{Fast}(fall) = \tau_{ST} \cdot \frac{C_L}{C_N} = \tau_{ST} \cdot (1+k) \cdot \frac{C_L}{C_{IN}}$$

$$\tau_{OUT}^{Fast}(rise) = R_\mu \cdot \tau_{ST} \cdot \frac{C_L}{C_P} = R_\mu \cdot \tau_{ST} \cdot \frac{(1+k)}{k} \cdot \frac{C_L}{C_{IN}}. \tag{6.8}$$

In these equations, k, C_{IN}, and C_L are, respectively, the configuration ratio, the input capacitance of the switching structure, and the total load evaluated at the output of the structure. As shown for a switching inverter controlled by a step-input voltage, the output transition time by unit load (C_{IN}) depends only on the configuration ratio of the inverter and on a technological factor τ_{ST}, which is a characteristic of the process speed [9]. This factor can be used, as well as a metric for the transition time.

For a slow input control, the reduction of the maximum current due to the input ramp effect has to be considered. From Equation 6.3 and Equation 6.5, we obtain:

$$\tau_{OUT}^{Slow}(Inv) = \sqrt{\frac{(V_{DD} - V_{TN})}{V_{DD}}} \cdot \sqrt{\tau_{OUT}^{Fast} \cdot \tau_{IN}} \tag{6.9}$$

Considering the full input transition range, the expression of the inverter output transition time is obtained from the maximum value of the Equation 6.8 and Equation 6.9:

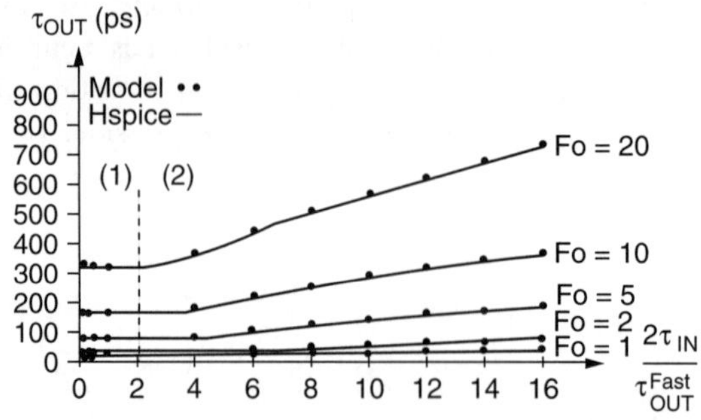

FIGURE 6.2 Inverter output transition time sensitivity to the load; Equation 6.1 and Equation 6.2 identify the fast and slow input ramp domains ($W_N = 1$ μm, $W_P = 2.5$ μm, $L_{MIN} = 0.18$ μm).

$$\tau_{OUT}(Inv) = MAX\left\{ \begin{array}{l} \tau_{OUT}^{Fast}, \\ \sqrt{\dfrac{(V_{DD} - V_{TN})}{V_{DD}}} \cdot \sqrt{\tau_{OUT}^{Fast} \cdot \tau_{IN}} \end{array} \right\} \tag{6.10}$$

In Figure 6.2 we compare the calculated and simulated input transition time sensitivity of an inverter output transition time.

In the same way, the transition time of gates is directly deduced from the inverter expressions corrected by the DW given in Equation 6.4:

$$\tau_{OUT}^{Fast}(Gate) = DW_{Top}^{Fast} \cdot \tau_{OUT}^{Fast}(Inv)$$

$$\tau_{OUT}^{Slow}(GateTop) = DW_{Top}^{Slow} \cdot \tau_{OUT}^{Slow}(INV) \tag{6.11}$$

As shown in Equation 6.11, the output (input) transition time expression can easily be obtained as the product of three terms:

1. A technological factor common to all the structures
2. A symmetrical factor (logical effort of Sutherland et al. [20]), characteristic of the implemented logical function and of the internal configuration ratio ($k = W_P/W_N$)
3. The ratio of load to input cell capacitance (electrical effort of Sutherland et al. [20]) that characterizes the cell environment

If the first factor characterizes the process and the second one is library specific, designing for low power requires a lot of effort to optimize at physical level the third factor.

6.3.2 Metric for the Process

As shown in Equation 6.7, in the fast input control domain, the output transition time of an ideal symmetrical inverter, loaded by an identical one, is a direct characteristic of the process and of the current difference between N and P transistors:

$$\frac{\tau_{OUT}(fall)}{2} = \frac{Cox \cdot L_{MIN} \cdot V_{DD}}{K_N \cdot (V_{DD} - V_{TN})} = \tau_{ST}$$

$$\frac{\tau_{OUT}(rise)}{\tau_{OUT}(fall)} = R_{\mu} \tag{6.12}$$

TABLE 6.4 τ_{ST} Value Evolution with the Process

L (μm)	Cox (fF/μm²)	V_{DD} (V)	τ_{ST}calc. (ps)	τ_{ST}sim (ps)	V_{TN} (V)
0.13	13.3	1.2	4.05	4.05	0.60
0.18	7.85	1.8	4.51	4.57	0.54
0.25	6.91	2.5	7.00	6.70	0.64
0.35	4.30	2.85	10.8	11.5	6.62
0.50	2.80	2.85	15.9	16.4	0.73
0.80	2.30	5.0	24.8	23.8	1.22
1.00	1.73	5.0	28.0	26.3	0.82
1.20	1.38	5.0	33.0	30.7	0.70

In Table 6.4 we compare, for different processes, the τ_{ST} values simulated or measured on ring oscillators to the values calculated from Equation 6.12. The calculated values have been obtained from the value determined on the 0.25-μm process, used as a reference, by updating the L_{MIN}, V_{DD}, and V_T values of the corresponding process. The good agreement between the calculated and the measured values gives evidence of the interest in using τ_{ST} as a metric for defining or predicting the process speed performance.

6.3.3 Supply Voltage and Temperature Sensitivity

As shown in Equation 6.12, τ_{ST} gives the explicit supply voltage sensitivity of the transition time.

The temperature sensitivity can easily be included in the model considering both the mobility and the threshold voltage variations described in Sze [14] and Power et al. [15]:

$$
\begin{aligned}
K_\theta &= K_{nom} \cdot \left(\frac{\theta_{nom}}{\theta}\right)^{XT} \\
V_T(\theta) &= V_{Tnom} - \delta(\theta - \theta_{nom})
\end{aligned}
\tag{6.13}
$$

where K and V_T are, respectively, the conductivity factor and the threshold voltage; θ_{nom} and θ represent, respectively, the reference and the targeted temperature; and XT and δ are the temperature coefficients of the mobility and of the threshold voltage.

Combining Equation 6.12 and Equation 6.13, a general expression for the τ_{ST} supply voltage and temperature sensitivity can be obtained:

$$
\frac{\tau_{ST}(V_{DD},\theta)}{\tau_{STnom}} = \left(\frac{\theta}{\theta_{nom}}\right)^{XT} \left(\frac{V_{DD}}{V_{DDnom}}\right) \cdot \frac{1}{\dfrac{V_{DD} - V_{Tnom} + \delta(\theta - \theta nom)}{V_{DDnom} - V_{Tnom}}}
\tag{6.14}
$$

In this expression, the different parameters X_T and δ can directly be determined from specific simulation conditions to be defined in the next section.

6.3.4 Metric for the Delay

A realistic delay model must be input slope dependent and must distinguish between falling and rising signals. As developed in Jeppson [16], considering the input-to-output coupling effect, the input slope effect can be introduced in the propagation delay as:

$$
\begin{aligned}
t_{fall}(i) &= \frac{v_{TN}}{2} \tau_{INrise}(i-1) + (1 + \frac{2C_M}{C_M + C_L}) \frac{\tau_{OUTfall}}{2}(i) \\
t_{rise}(i) &= \frac{v_{TP}}{2} \tau_{INfall}(i-1) + (1 + \frac{2C_M}{C_M + C_L}) \frac{\tau_{OUTrise}}{2}(i)
\end{aligned}
\tag{6.15}
$$

where $\tau_{IN(fall,rise)}$ is the duration time of the input signal, generated by the controlling gate. C_M is the coupling capacitance between the input and output nodes [17], that can be evaluated as one half the input capacitance of the P(N) transistor for input rising (falling) edge, respectively or directly calibrated from electrical simulations. Indexes (i) and (i-1) refer to the location of the cell in a gate array.

As shown in Equation 6.15, the considered input transition time is short enough to assume that the output switching of the gate still occurs under a constant value of the current. This assumption justifies the use of the transition times for evaluating the delay. Otherwise, in the Slow input control range, the Slow ramp expression of the internal cross talk [18] must be used in Equation 6.15, and consideration of the short-circuit current [19] must be given. This results in a more complex expression out of the scope of this part.

Considering Equation 6.8, Equation 6.10, and Equation 6.15, it can be concluded that both the transition time and the propagation delay can be accurately evaluated using few parameters, which can be obtained from the design specifications and calibrated on the process, as shown later.

6.3.4 Metric for the Short-Circuit Power Dissipation

The dynamic power dissipation of static CMOS structures contains two terms:

1. An external term, the switching component, required to charge or discharge the different capacitances involved in the design
2. An internal term, the short-circuit component, which appears when both the N and P array of transistors are conducting

This last component is directly proportional to the common part of the current flowing between the supply rails. In the previous sections, it was shown that from the modeling of the switching current of CMOS structures, it was possible to define metrics for evaluating the transition time and propagation delay values. Now, from the modeling of the short-circuit current, metrics for evaluating the short-circuit power component can be defined.

The typical waveforms of the switching and-short-circuit currents flowing in an inverter controlled respectively by a fast and slow input ramps are given in Figure 6.3. The analysis of these typical waveforms shows that:

- In the fast input range, the short-circuit current, and thus the short-circuit power, is negligible with respect to the switching current.
- In the slow input range, the amplitude and the duration of the short-circuit current have values comparable to that of the switching current.

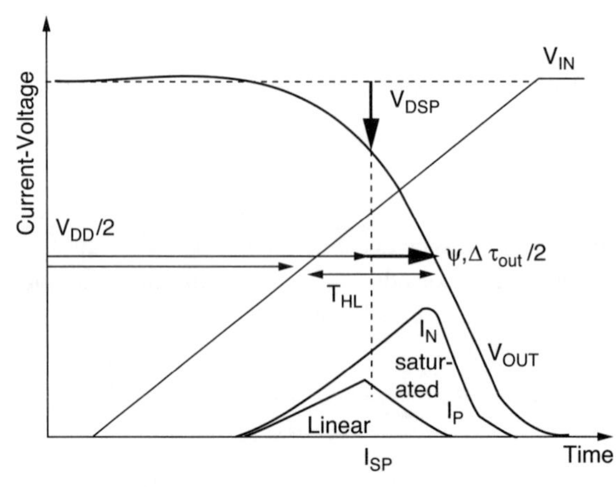

FIGURE 6.3 Illustration of the switching waveform of an inverter.

It is therefore obvious that, to reduce the short-circuit power dissipation, it is of prime importance to design CMOS circuit with a right control of the signal slopes along combinatorial paths. This is particularly true considering noncritical paths that are usually implemented with minimum drive gates.

Thus, to reduce, at gate level, the short-circuit power dissipated by noncritical paths, two steps have to be applied:

1. To develop a design-oriented model of the short-circuit power dissipation; such a model and the corresponding metric are introduced in the following
2. To deduce from this model a gate-sizing criterion to control effectively the signal slopes. This application of the model is given in Section 6.6

In his seminal work, Veendrick [9] first introduced a design-oriented model of the short-circuit power consumption. This model has been developed for micronic technology and unloaded structures, resulting in an upper bound of the short-circuit component evaluation. This model is extended here to submicronic process, considering realistically loaded structures; however, it is still assumed that:

- The short-circuit current waveform is symmetrical with respect to its maximum value, I_{SC}^{MAX}.
- The short-circuit current varies linearly between the times t_{OV} (end of the overshoot) and $t_{SP,N}$ (occurrence of I_{SC}^{MAX}, Figure 6.3).

Then evaluating, for an input rising edge, the maximum short-circuit current flowing between the supply rails [19] gives:

$$I_{SC}^{MAX} = \left\{ \psi_1 \cdot \frac{\tau_{IN}}{\tau_{OUT}} + \psi_2 \right\} . W_P \tag{6.16}$$

and the charge Q_{SC} supplied to the load by the P transistor as:

$$Q_{SC} = \frac{1}{2} \cdot (1 - v_{THP} - v_{TN}) \cdot \left\{ \psi_1 \cdot \frac{\tau_{IN}}{\tau_{OUT}} + \psi_2 \right\} . W_P \tag{6.17}$$

where v_{THP} (v_{TN}) is the normalized threshold voltage of the P (N) transistor working in linear (saturated) mode, and ψ_1 and ψ_2 are unique parameters to be calibrated on the process. As an example $\psi_1 = 10$ μa/μm and $\psi_2 = 2$ μa/μm, for a 0.18-μm process.

The energy short-circuit component can now be defined in the same way as the switching component:

$$E_{SC} = \frac{1}{2} \cdot C_{SC} \cdot V_{DD}^2 \tag{6.18}$$

Calculating the total charge transferred during the short-circuit period gives the following expression:

$$C_{SC} = \frac{Q_{SC}}{Vdd} = \frac{1}{2 \cdot Vdd} \cdot (1 - v_{THP} - v_{TN}) \cdot \left\{ \psi_1 \cdot \frac{\tau_{IN}}{\tau_{OUT}} + \psi_2 \right\} . W_P \tag{6.19}$$

with an identical expression for a falling input edge.

Equation 6.19 is of great interest for comparing the energy dissipated by the short-circuit process (E_{SC}) to the energy required to discharge (charge) the output load (E_{CL}). The following expression holds for an input rising edge:

$$R = \frac{E_{SC}}{E_{CL}} = \frac{\frac{1}{2} \cdot C_{SC} \cdot V_{DD}^2}{\frac{1}{2} \cdot C_L \cdot V_{DD}^2} = \frac{C_{SC}}{C_L} < 1 \tag{6.20}$$

In this expression, R represents the inefficiency of the gate in terms of energy consumption. Its value is necessarily lower than 1 because the short-circuit current waveform is included in the driving current waveform (Figure 6.3).

Note here (Equation 6.8, Equation 6.11, and Equation 6.19) that C_{SC} depends on the input and output transition time values, and C_L completely defines the output transition time. In this case, R appears as a good metric for evaluating the slope control on the design.

A value of R close to 1 indicates that the gate is badly controlled, while a value close to 0 indicates a good input slope control.

Considering Equation 6.19 and Equation 6.20, it clearly appears that R may reach significant values (empirically 0.3 or 0.4), if the designers do not control properly the signal slopes along combinatorial paths. Such values of R can be obtained in combinatorial noncritical paths where minimum size gates are used without proper control of the input slope. This may result in significant extra power dissipation on noncritical paths. Consequently, it could be of interest to develop a sizing criterion to properly control the signal slope along combinatorial paths in order to minimize this useless short-circuit power consumption.

6.4 Application to a Standard Cell Library

In a standard industrial approach, timing performance verification is obtained using a tabular method. The performance of each gate on a path, for each loading and control condition, is deduced from an interpolation between a set of predefined values. These values are determined from electrical simulations performed for a limited number of design conditions, such as load, input transition time, supply voltage value, and operating temperature. Characterizing each edge of the transition time and the propagation delay of each library cell, for typically five loading and input ramp conditions, involves 100 simulations. Then considering the process corners, defined for three supply voltage values (V_{max}, V_{nom}, V_{min}) and three temperature values (T_{max}, T_{nom}, T_{min}), the characterization of a logic function imposes 900 simulations by drive strength of this function. This huge number of simulations just allows representing the design space with five loading and controlling conditions. Intermediate conditions must then be interpolated using a linear characteristic equation (e.g., $f(\tau_{IN},C_L) = A\tau_{IN} + BC_L + C\tau_{IN}CL + D$). In submicron process, the transition time and the propagation delay exhibit a nonlinear variation with respect to the control and loading conditions that depends on each particular operating point imposed on the different combinatorial paths. This nonlinear range must clearly be located in the design space to adequately choose the simulation points to be inserted in the lookup table.

A continuous representation of the timing performance of a CMOS library will be introduced in the next paragraph to define the output transition time and propagation delay sensitivities of the cells to the design space parameters, such as the load, the input transition time, the supply voltage and the temperature values.

6.4.1 Continuous Representation of Standard Cell Performance

While considering Equation 6.10 to Equation 6.15, it clearly appears that in the fast input range, τ_{OUT}^{Fast} characterizes an inverter (gate) structure and its load. Considering the sensitivity of the different expressions to the input slope, τ_{OUT}^{Fast} can be used as an internal reference of the structure output transition time. In this condition, the following expression can be written:

$$\frac{\tau_{outfall}}{\tau_{outfall}^{Fast}}(Inv) = Max\left[\begin{array}{c} 1 \\ \sqrt{\frac{V_{DD}-V_{TN}}{V_{DD}}} \cdot \sqrt{\frac{\tau_{INrise}}{\tau_{outfall}^{Fast}}} \end{array} \right] \tag{6.21}$$

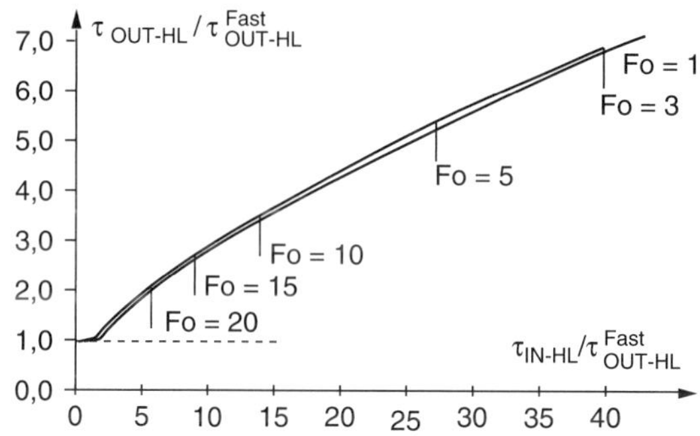

FIGURE 6.4 Full representation of the output transition time variation of the seven inverters of a 0.25-μm library.

Normalizing the output transition time with respect to $\tau_{outfall}^{Fast}$, used as a reference, the resulting expression only depends on the input transition time and is configuration ratio and load independent. Similar results can be obtained for gates and the representation of propagation delay.

This is illustrated in Figure 6.4 where the output transition time variation of the complete family of inverters of a 0.25-μm library is represented. As expected, all the curves pile up on the same one, representing the output transition time sensitivity to the input transition time. The final value for each specific cell is then directly obtained from the evaluation of τ_{OUT}^{Fast} given in Equation 6.11, which contains the structure and load dependency.

6.4.2 Calibration Procedure

From the preceding equations and considering the variation displayed in Figure 6.4, it appears that the output transition time and the propagation delay of all the gates of a library can be characterized with a reduced set of electrical simulations. The calibration of the parameters can be performed as follows.

1. The τ_{ST} value is obtained from the output transition τ_{HL} (falling edge) of a heavily loaded inverter (with a known configuration ratio k) controlled by a fast input ramp ($\tau_{IN} < \tau_{OUT}$).
2. R is obtained from the value of the ratio τ_{LH}/τ_{HL}.
3. For a small load, the variation of the apparent τ_{ST} value determines the value of C_{par} and C_M.
4. In the slow input range ($\tau_{IN} > \tau_{OUT}$) at constant load, varying τ_{IN} determines the input slope sensitivity (Equation 6.6).
5. Using the inverter as a reference, the gate parameters k and DW are directly determined from the ratio τ_{Gate}/τ_{Inv}.
6. Equation 6.9 completely determines the supply voltage sensitivity.
7. The temperature sensitivity parameters, XT and δ, are obtained from the preceding steps realized at different temperature values.

6.4.3 Validation

The validation of this representation has been done on a 0.13-μm library. The target is to get a continuous characterization of the timing performance with a robust identification of the design space (fast, slow input control range) including the temperature and supply voltage sensitivity. Only simple gates are considered, such as Inverter with seven different drives, NAND, and NOR gates with two and three inputs and five different drives. Initially the timing performance (transition time and propagation delay) of all these elements has been characterized from electrical simulations. They are available in tables (TLF, STF) that give, for each edge of the transition time and the propagation delay of each element and for three temperature and supply voltage values, the corresponding performance for five different values of the load and the input transition time.

Following the procedure described in Section 6.3, the values of the technology parameters are determined on the different tables, thus allowing to plot in Figure 6.5 to Figure 6.8 the variation of the transition time and the propagation delay of each logic family. As shown, the performance variation of all the elements of each family can be represented by one curve, as predicted by Equation 6.10.

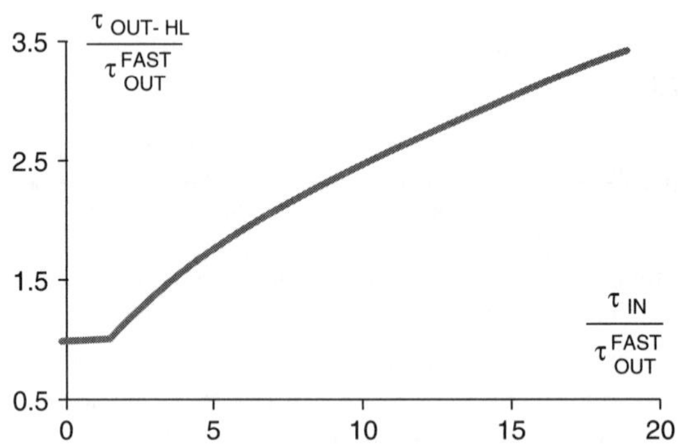

FIGURE 6.5 Output transition time representation of the seven inverters of a 0.13-μm process.

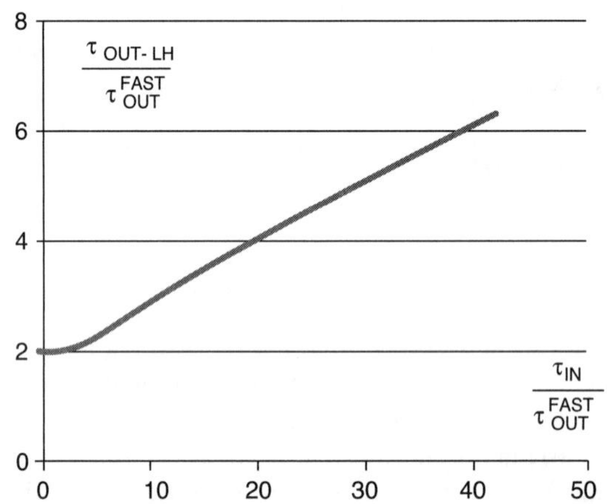

FIGURE 6.6 Output transition time representation of the five NAND2 of a 0.13-μm process.

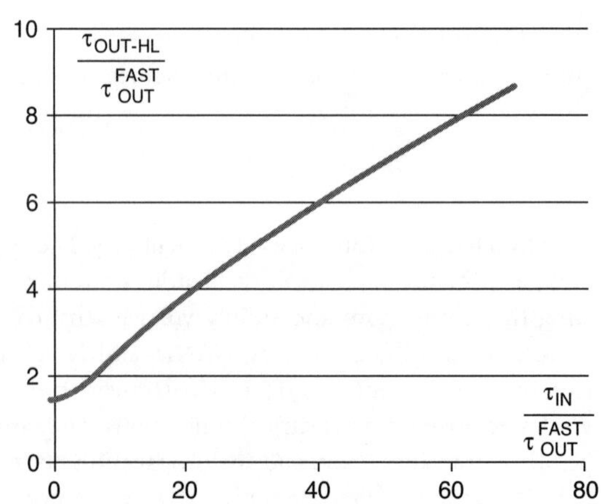

FIGURE 6.7 Output transition time representation of the five NOR2 of a 0.13-μm process.

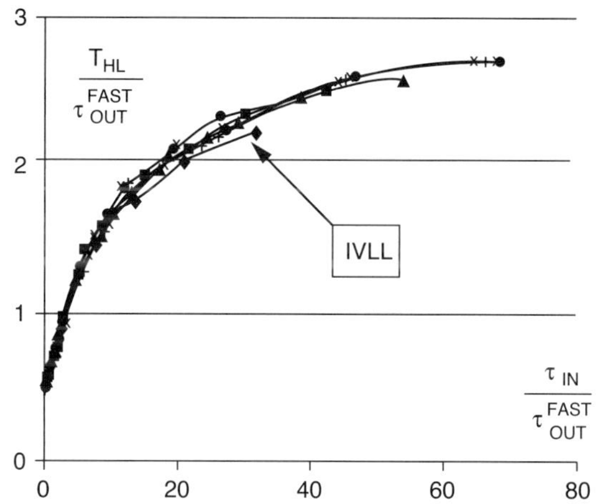

FIGURE 6.8 Propagation delay representation of the seven inverters of a 0.13-μm process.

TABLE 6.5 Voltage and Temperature Sensitivity of the τ_{ST} Parameter, XT=1.65, $\delta = 2.10^{-3}$

τ_{ST}		V_{PP} (V)					
		1.08		1.2		1.32	
		Model	Simul.	Model	Simul.	Model	Simul.
Temp. (°K)	233	4.26	3.92	3.86	3.56	3.02	3.3
	298	4.59	4.56	4.05	4.05	3.69	3.66
	398	5.16	5.45	4.85	4.93	4.62	4.58

Table 6.5 compares the τ_{ST} supply voltage and temperature sensitivity calculated from Equation 6.9 and Equation 6.11 with the value deduced from the simulated values on the lookup tables. As shown, an excellent agreement is obtained between calculated and simulated values for all the considered supply voltage and temperature range. These variations can then be completely represented by (11):

$$\frac{\tau_{ST}(V_{DD},\theta)}{\tau_{ST}} = \left(\frac{\theta}{298}\right)^{1.65} \cdot \left(\frac{V_{DD}}{1.2}\right) \cdot \frac{1}{\frac{V_{DD} - 0.62 + 2 \cdot 10^{-3}(\theta - 298)}{0.58}} \tag{6.22}$$

where the different coefficients have been directly determined, following the calibration procedure given in the preceding part.

6.5 Application to Low-Power Design

6.5.1 Rule for Slope Control

To minimize the short-circuit power consumption, this section presents a gate-sizing criterion for properly controlling the input slope. Let us consider the structure depicted in Figure 6.9, where C_A models a parasitic (including routing) capacitance. The main challenge here is to control the value of the input transition time value τ_{IN}, allowing the reduction of the short-circuit power dissipated by stage (i).

This can be accomplished by increasing the size of stage (i-1); however, this results in an increase of the energy required for its control ($1/2 \cdot C(i-1).V_{DD}\Sigma$). This means that, in the same way as for delay optimization, a trade-off must be defined between the reduction of the short-circuit energy consumption of stage (i) and the increase of the energy required to control stage (i-1). The optimal sizing of stage (i-1), for minimizing the total energy consumption E_{TOT} of the structure (Figure 6.9) can be obtained from:

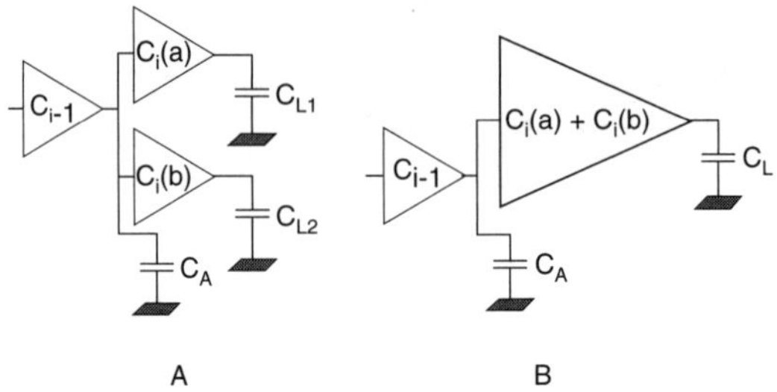

FIGURE 6.9 An example of divergence and its equivalent structure.

$$E_{TOT}=\frac{1}{2}C(i-1)\cdot V_{DD}^2+\frac{1}{2}\left(C(i)+C_{SC}(i-1)\right)\cdot V_{DD}^2+\frac{1}{2}\left(C_L+C_{SC}(i)\right)\cdot V_{DD}^2 \tag{6.23}$$

where C(k), C_{SC}(k) represent respectively the input capacitance and the short-circuit capacitance of stage k.

In this expression, the first term represents the energy required to control stage (i-1), the second term, the energy consumption of stage (i-1), and, finally, the last term is the energy consumption of stage (i).

Assuming that the stage (i-1) is controlled in such a way that its inefficiency R_{i-1} is minimized, Equation 6.23 becomes:

$$E_{TOT}=\frac{1}{2}C(i-1)\cdot V_{DD}^2+\frac{1}{2}C(i)\cdot V_{DD}^2+\frac{1}{2}\left(C_L+C_{SC}(i)\right)\cdot V_{DD}^2 \tag{6.24}$$

where only the first term can be reduced by a specific slope control. Searching analytically for the optimal value of C(i-1) gives:

$$C_{OPT}^{5/2}(i-1)\approx\frac{3}{2}\cdot A\cdot\frac{C^{3/2}(i).\left(C(i)+C_A\right)^{3/2}}{C_L^{1/2}} \tag{6.25}$$

where A is a process dependent parameter defined by:

$$A=\frac{\left(1-v_{THN}-v_{TP}\right).\left(\psi_1^{HL}+R_?.\psi_1^{LH}\right).\tau_{ST}}{2.V_{DD}.C_{OX}.L_{GEO}} \tag{6.26}$$

6.5.2 Application

The application of the sizing criterion (Equation 6.26) to an inverter tree is almost straightforward, processing backward from the output to the input of the tree; however, both the problems of divergence branches and of the output drivers have to be considered.

In minimizing the total power dissipated in an inverter tree, it appears that the optimal sizing of the output drivers depends strongly on the load content. For example, in optimizing the logic that drives a register or next gates, it can be considered that the output load is an active load or the sum of active and passive loads. Therefore, the sizing of the output driver has to be performed using Equation 6.25. If the output driver controls a passive load, however, no short-circuit power dissipation occurs in the load, and the driver must be sized at the minimum value satisfying the delay constraint.

The case of divergence branches presents a difficulty because the sizing criterion developed in the preceding section does not allow predicting the optimal sizing of the (i-1). The adopted solution is based on the fact that the power is an additive characteristic of the structure. To justify this approach, let us consider the structure represented in Figure 6.9.

The sizing criterion (Equation 6.26) supplies the optimal value of C_{i-1} only if $C_{L1} = C_{L2}$, in which case the two inverters can be lumped in a unique inverter with an input gate capacitance equal to $C_i(a) + C_i(b)$. In a general configuration, however, C_{L1} and C_{L2} have different values.

Nevertheless, as the short-circuit power dissipation is a decreasing function of C_L, the two inverters (a) and (b) are modeled by a unique inverter (Figure 6.9b) loaded by C_L=MAX(C_{L1},C_{L2}) to avoid any overestimation of the short-circuit power dissipated by (a) and (b).

6.5.3 Validation

This sizing heuristic, based on the sizing criterion defined by Equation 6.25, has been applied to an inverter tree represented in the Figure 6.10. The total power dissipated in the different implementations has been obtained from SPICE simulations.

Figure 6.11 illustrates the power gain and loss values obtained when comparing the proposed sizing solution to a minimal surface implementation. Different values of the parasitic routing capacitance P are considered to illustrate the sensitivity of the result to the parasitic content of the load.

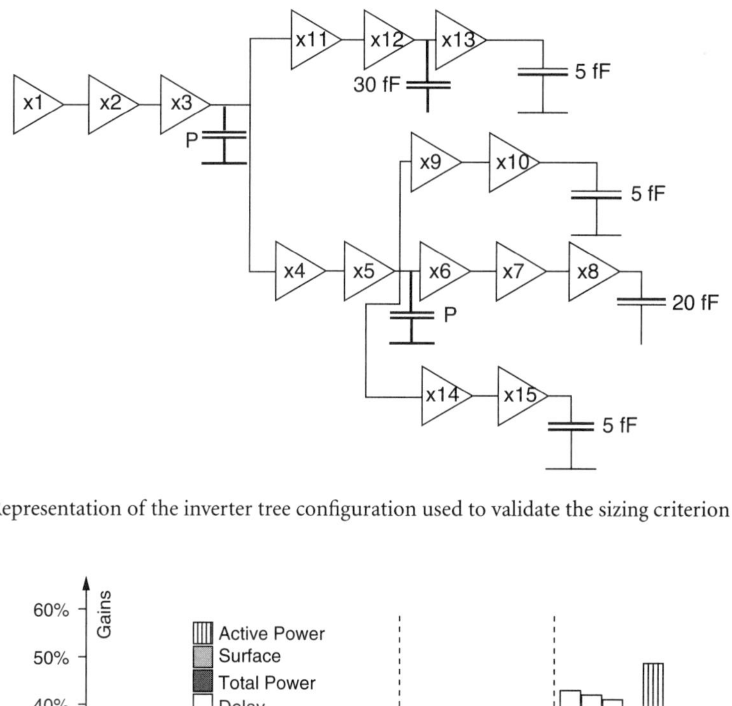

FIGURE 6.10 Representation of the inverter tree configuration used to validate the sizing criterion (Equation 6.25).

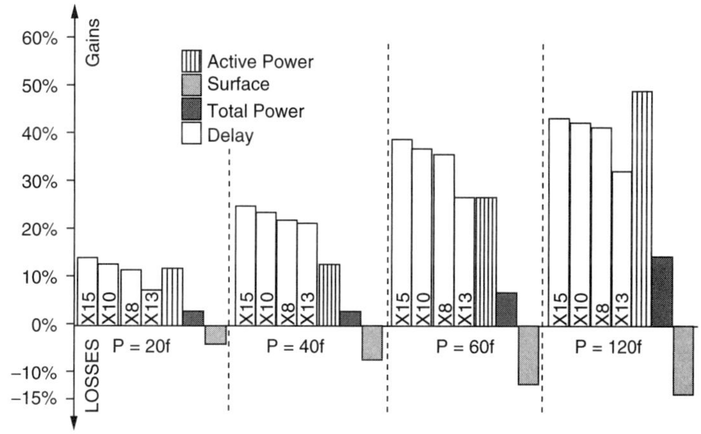

FIGURE 6.11 Gain and loss in delay, power, and area obtained on the inverter tree for different values of the parasitic capacitance $P = P_{3,4} = P_{5,6}$.

As shown, depending on the parasitic content of the inverter tree, the gain in power and speed is ranging from 3 to 15% and 13 to 45%, respectively.

The speed increase can be easily justified after a detailed analysis of the simulation results. For the considered example, the application of the sizing criterion increases the size of the stages X12, X5, and X3. This induces a reduction of the ramp duration applied at the input of the stage X14, X11, X9, X6, and X4 reducing their switching delays.

This ramp control allows to size noncritical paths at minimum dynamic power consumption. Moreover, this sizing method improves the speed, compared with a minimum size implementation. As a result, this solution can be recommended as an initial solution to be implemented before any critical path optimization.

6.6 Conclusion

Due to the fast evolution of the CMOS process, associated to the increasing complexity of the structures to be managed, it becomes necessary to define metric for performance allowing designers to use easy but robust indicators to evaluate alternatives at all the steps of the design flow. Using an analytical model to evaluate the maximum switching current value, a simple but accurate design oriented representation of the performance of CMOS logic was obtained. Simple and closed form formula for the output transition time, the propagation delay, and the short-circuit power component were derived. Metrics to characterize the speed of the CMOS process, as well as its sensitivity to the supply voltage and the temperature were defined.

Clear evidence was given that the transition time can be used as a simple and robust indicator for evaluating the cell performance and for defining the conditions of load and control satisfying the imposed constraints. The definition of the transition time through parameters, which are characteristics of the process, the structure and the load, gives to the designer opportunity to characterize a cell library in terms of load and critical transition time, and to improve lookup table centering in the useful design space.

A new way for a continuous representation of these performances was introduced, allowing modeling of the complete load and inputting ramp sensitivity by one curve. A method to calibrate the parameters of this representation was given, which that was completely validated on a 0.13-μm process for different temperature and supply voltage conditions.

Considering the power dissipation as a critical design parameter, a sizing criterion for minimizing the switching power dissipation component has been presented. The latter has been obtained by lowering the short-circuit component through a control of the gate input transition time. Using an analytical model of the short-circuit power dissipation and of the output transition time, it has been demonstrated that a sizing condition minimizing the short-circuit component, can be defined. Application has been given to general inverter configurations in various loading conditions. Gain in power and speed as large as 15 and 45% can be obtained, with respect to minimal size implementations.

These indicators also give facilities in controlling the load and input transition time distribution in combinatorial paths, which is, at the physical level, the most efficient way to manage the speed to power trade-off for circuit optimization.

References

[1] A. Chatterjee, C.F. Machala, and P. Yang, A submicron DC MOSFET model for simulation of analog circuits, *IEEE Trans. on CAD of Integrated Circuits and Syst.*, vol. 14, no. 10, pp. 1193–1207, 1995.

[2] A.F. Tasch, The challenges in achieving sub-100nm MOSFETS, *1997 Int. Conf. on Innovative Syst.*, pp. 53–60, Austin, TX.

[3] B.L. Austin, K.A. Bowman, X. Tang, and J.D. Meindl, A low-power transregional MOSFET model for complete power-delay analysis of CMOS gigascale integration, *11th Int. ASIC Conf*, pp. 125–129, Rochester, NY, 1998.

[4] S.H. Jen and B. Sheu, A compact unified MOS DC current model with highly continuous conductance for low-voltage ICs, *IEEE Trans. on CAD of Integrated Circuits and Syst.*, vol. 17, no. 2, pp. 169–172, 1998.

[5] Y. Cheng, K. Chen, K. Imai, and C. Hu, A unified MOSFET channel charge model for device modeling in circuit simulation, *IEEE Trans. on CAD of Integrated Circuits and Syst.*, vol. 17, no. 8, pp. 641–644, 1998.

[6] T. Skotnicki, Analysis of the silicon technology roadmap – how far can CMOS go? *C. R. Acad. Sci. Paris,* t.1, Série IV, pp. 885–909, 2000.

[7] K.-Y. Toh, P.-K. Ko, and R.G. Meyer, An engineering model for short-channel MOS devices, *IEEE J. Solid-State Circuits,* vol. 23, no. 4, pp. 959–958, 1988.

[8] C. Mead and M. Rem, Minimum propagation delays in VLSI, *IEEE J. Solid-State Circuits,* vol. SC17, no. 4, pp. 773–775, 1982.

[9] H.J.M. Veendrick, Short-circuit power dissipation estimation for CMOS logic gates, *IEEE J. Solid-State Circuits,* vol. 19, no. 4, pp. 468–473, 1984.

[10] T. Sakurai and A.R. Newton, A simple MOSFET model for circuit analysis, *IEEE Trans. on Electron. Devices,* vol. 38, no. 4, pp. 887–894, April 1991.

[11] P. Maurine, M. Rezzoug, N. Azemard, and D. Auvergne, Transition time modeling in deep submicron CMOS, *IEEE Trans. on CAD of Integrated Circuits and Syst.*, vol. 21, no. 11, pp. 1352–1361, Nov. 2002.

[12] P. Maurine, M. Rezzoug, and D. Auvergne, Output transition time modeling of CMOS structures, *ISCAS '01*, Sydney, Australia, May 2001, pp. V-363–V-366.

[13] C. Mead and L. Conway, Introduction to VLSI systems, in *Addison-Wesley Series in Computer Science, 2nd ed.*, Addison-Wesley, Reading, MA, 1980.

[14] S.M. Sze, *Physics of Semiconductor Devices,* John Wiley & Sons, New York, 1983.

[15] J.A. Power et al., An investigation of MOSFET statistical and temperature effects, *Proc. IEEE 1992 Int. Conf. on Microelectronic Test Structures,* vol. 5, March 1992.

[16] K.O. Jeppson, Modeling the influence of the transistor gain ratio and the input-to-output coupling capacitance on the CMOS inverter delay, *IEEE J. Solid-State Circuits,* vol. 29, pp. 646–654, 1994.

[17] J. Meyer, Semiconductor device modeling for CAD, chap. 5, G.K. Herskowitz and R.B. Schilling, Eds. Mc–Graw-Hill, New York, 1972.

[18] S. Turgis and D. Auvergne, A novel macro-model for power estimation in CMOS structures, *IEEE Trans. on CAD of Integrated Circuits and Syst.,* vol. 17, no. 11, pp. 1090–1098, Nov. 1998.

[19] P. Maurine, R. Poirier, N. Azemard, and D. Auvergne, Switching current modeling in CMOS inverter for speed and power estimation, *DCIS '01,* pp. 618–622, Porto, Portugal, November 2001.

[20] I. Sutherland, B. Sproull, and D. Harris, *Logical Effort: Designing Fast CMOS Circuits,* Morgan Kaufmann Publishers, Inc., San Francisco, CA, 1999.

7

Logic Circuits and Standard Cells

Christian Piguet

CSEM & LAP-EPFL

7.1 Introduction

Today, digital logic design is performed by using standard cell libraries and place and route computer-aided design (CAD) tools; however, many different logic styles have been and continue to be proposed for general-purpose and specialized standard cell libraries. Low power is even more important than speed and silicon area, but it is increasingly difficult to achieve in very deep submicron technologies as well as for specialized libraries for self-timed or cryptographic applications. This chapter summarizes logic design styles, stressing low power design issues. It also describes new and emerging logic styles for specialized libraries.

7.2 Logic Families

To achieve a very low dynamic power, a large number of logic families have been proposed and used in various designs. This section describes some of these logic families, which are assumably the most interesting regarding low-power designs. Various comparisons are proposed, but to be fair, some specific circuits are used for the comparison.

7.2.1 Static CMOS Logic

Static CMOS is the older and still most used logic family. It is still considered the most simple and robust logic style [1]. Each CMOS gate is constructed with two dual N-ch and P-ch networks, connected respectively between V_{ss} and V_{dd} and the gate output. Any logic Boolean function can be designed by

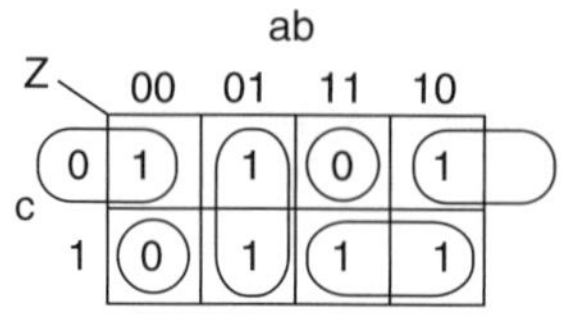

$$z = (\bar{a}\,\bar{b}c + ab\,\bar{c})\,\{0\} + (\bar{a}b + ac + \bar{b}\,\bar{c})\,\{1\}$$

$$zN = \bar{a}\,\bar{b}c + ab\bar{c}$$

$$zP = a\bar{b} + \bar{a}\,\bar{c} + bc$$

FIGURE 7.1 CMOS gate synthesized by the "separated simplification" method.

connecting, in series or parallel, CMOS transistors in the two N-ch and P-ch networks, if inputs are available in true and complemented forms.

The design of a CMOS gate is generally performed by synthesizing the N-ch network by taking the "0" cubes in the Karnaugh map of the Boolean function. The P-ch network is then derived as the dual network by connecting in series (parallel) the transistors that are in parallel (series) in the N-ch network. Figure 7.1 illustrates the synthesis of a Boolean function given by a Karnaugh map, in which the z symmetrical equation contains two terms, the first one with the "0" cubes indicated by {0} and the second term with the "1" cubes indicated with {1}. The metal-oxide semiconductor (MOS) structure of the N-ch and P-ch networks are then designed as zN and zP, by taking the first term {0} as such and by inverting each letter in the second term {1}, as P-ch transistors are conducting when they have a "0" on their gate. In the zN and zP expressions, AND operators mean a serial connection of transistors, while OR operators mean a parallel connection.

This method has been introduced in order to be capable of having the two N-ch and P-ch expressions as sums of products. As described in the next paragraph, it results in the so-called "branch-based" logic style that provides some advantages with respect to layout regularity, better performances in speed and power and a better testability [2,3].

7.2.2 Branch-Based Logic

In the "branch-based logic" [2], logic cells are designed exclusively with branches composed of transistors in series connected between a supply line and the gate output (Figure 7.2). The number of MOS in series is limited to three for speed performances. The main advantage of such an implementation is the layout density. For instance, the symbolic layout of the non-branch-based P-ch network (Figure 7.2) contains two supplementary contacts with two drain parasitic capacitances that can be removed in the more compact branch-based implementation. The symbolic layout of Figure 7.2 comes from the logical equation $S = (B + C)\ (A\bar{C} + \bar{A}\ D)$. If implemented as such, the P-ch network is presented at the top of Figure 7.2. If a Karnaugh map is designed, the minimum number of blocks of "1" that are necessary is three, resulting in the branch-based implementation with six transistors shown in the middle of Figure 7.2.

Figure 7.2 also describes a very regular geometrical branch-based layout consisting of three branches. It provides no diffusion interruption, common drain for two branches, a minimal number of contacts and few metal connections. This is not the case, for instance, for the implementation presented at the top of Figure 7.2, where a product of sum has to be implemented with a supplementary wire. This branch-based technique, first introduced to reduce parasitic capacitances for achieving low power [1], is also beneficial for high-speed logic such as fast adders in silicon-on-insulator (SOI) technology [3].

An obvious drawback of this technique is the possible increase of the number of transistors when realizing an MOS network of complex CMOS gates with a sum of products (a transistor controlled by the same input in two parallel branches is repeated); however, this problem is not so serious. A standard cell library does not contain a large number of complex gates. The most used cells are simple gates, flip-flops, latches, and multiplexers. In a 200-cell library, compared with non-branch-based logic, some cells contain one supplementary transistor (e.g., XOR, AOI, OAI, latch with reset, D-flip-flop [DFF],

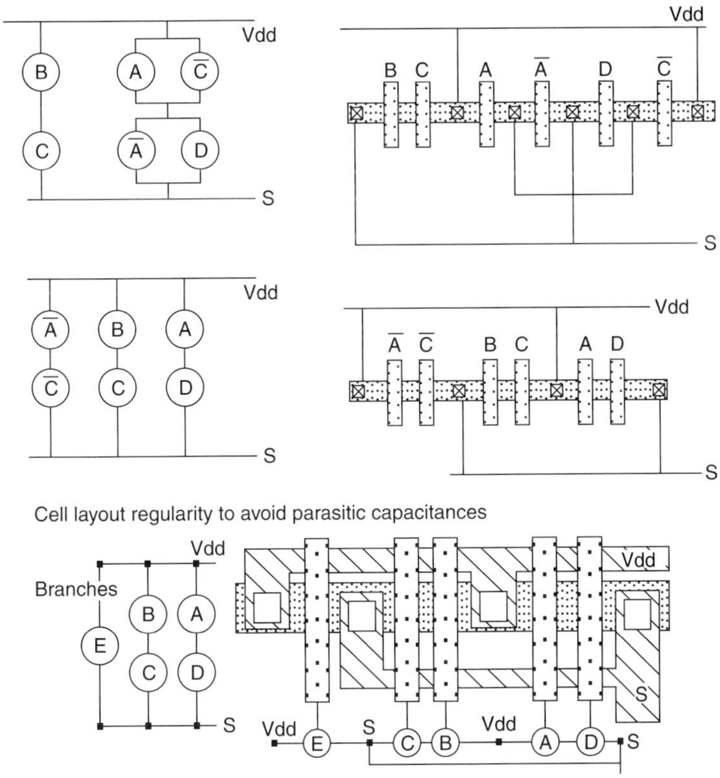

FIGURE 7.2 Branch-based layout.

and frequency dividers), and a few cells contain two supplementary transistors (e.g., latch and DFF with set/reset).

7.2.3 Transmission Gates

Transmission gate-based design has been largely used, ever since the CMOS technology introduction, because the MOS transistor is a very good switch. Transmission gates use two complementary transistors, as a single N-ch pass transistor, for instance, presenting a small gate-source V_{GS} when it has to conduct V_{dd} from input to output, reaches only V_{dd}-VT (threshold voltage). In the same situation, the complementary P-ch transistor has a full gate-source V_{GS} and provides V_{dd} at the output of the transmission gate. Rules of thumb are often used for the design of these transmission gate-based cells. This paragraph shows that the same basic methodology introduced for "branch-based" logic can be applied to transmission gates or pass-transistor circuits.

A transmission gate, controlled by an input variable, connects another input variable to the gate output. This means that some inputs are connected to transistor sources and not only to the gates of transistors. Compared to the branch-based style, for which sources of transistors or branches are always connected to V_{ss} {0} or to V_{dd} {1}, in transmission gate designs, sources of transistors or branches can also be connected to input variables. Thus, some cubes in the Karnaugh map are not only "0" cubes indicated by {0} or "1" cubes by {1}, but also some cubes identified by {input}. The content of the cubes is not "0" or "1," but is identical to a given input variable (or the complemented input). As a result, some cubes containing both "0" and "1" can be chosen, provided that the arrangement of "0" and "1" are identical to a given input.

Figure 7.3 provides an example for which one cube is a conventional cube with {0} and for which two other cubes are transmission cubes containing both "0" and "1." From Figure 7.3, it can be observed that

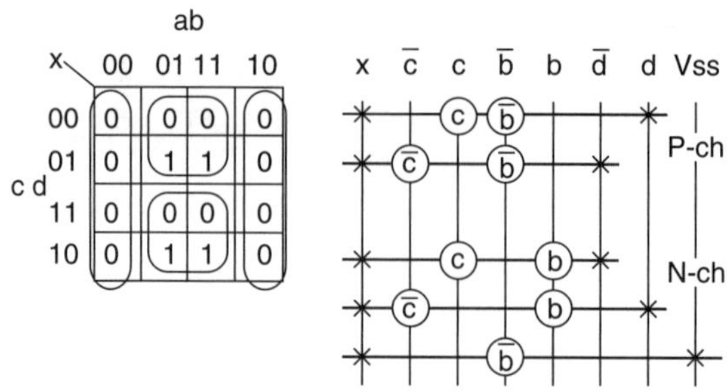

x = (b̄) {0} + (bc̄) {d} + (bc) {d̄}

xN = (b̄) {0} + (bc̄) {d} + (bc) {d̄}

xP = (b̄c) {d} + (b̄ c̄) {d̄}

FIGURE 7.3 Synthesis of a complex CMOS gate with transmission gates.

the top cube content is identical to the input variable d, while the other cube content is identical to the complement of the input variable d. As such, the designer can write the symmetrical equation of x by a first term representing the "0" cube {0} and by two other terms for which the only difference is that they refer to input variables (i.e., {d} and $\{\bar{d}\}$). The N-ch and P-ch networks are then derived using the same rules (i.e., the terms with {0} and {input} are selected without any change for the N-ch network xN, while the terms with {1} and {input} are selected for the P-ch network xP by inverting each letter in the expression of the cubes). The transmission cubes give a contribution in both N-ch and P-ch networks, as the designer wants to get a transmission gate-based circuit. If only N-pass transistors are required, only the contribution of the transmission cubes in the N-ch network is necessary. The example of Figure 7.3 demonstrates that both cube types {0} or {1} and {input} can be simultaneously synthesized. The resulting circuit will therefore contain some branches connected to V_{ss} (and V_{dd}) and other branches connected to input variables.

The symbolic schematic of the resulting circuit (Figure 7.3), designed in such a way that branches are highlighted, shows, for instance, that two branches are connected to input d (i.e., one P-ch branch controlled by c and $\bar{b}$ and one N-ch branch controlled by $\bar{c}$ and b). These two branches do implement two transmission gates connected in series.

Transmission gate-based circuits do have a smaller number of transistors compared to static CMOS logic. In terms of layout density, however, depending on the layout style, the cell area is often very similar. Some circuits are advantageously designed as transmission gate-based design, such as XOR gates and adders (based on XOR gates), but other basic cells in a library can be designed in static CMOS logic without any penalty.

7.2.4 N-Pass Logic

Before CMOS became the mainstream technology, N-MOS logic was extensively used with a depleted transistor as the load device. N-MOS pass transistor logic was also used for many cells resulting in a very low transistor count; however, in N-MOS logic, the gate output produces a V_{dd}-VT voltage. Although it was not a problem many years ago with V_{dd} at 5.0 V, it is a major drawback today with supply voltages close to or below 1.0 V.

To keep the transistor count low, N-pass logic can be used with an output keeper or a restoring transistor as illustrated in Figure 7.4 for an XOR gate. This logic style is called single pass transistor logic (SPL). The keeper device is used to force full V_{dd} at the output when the inverted output is "0." This logic style appears to be interesting only for multiplexers and XOR gates, explaining why it is generally benchmarked for adders and multipliers. This SPL logic could compete with static CMOS at high V_{dd}, but not at low V_{dd}, where SPL is slower and consumes much more than static CMOS.

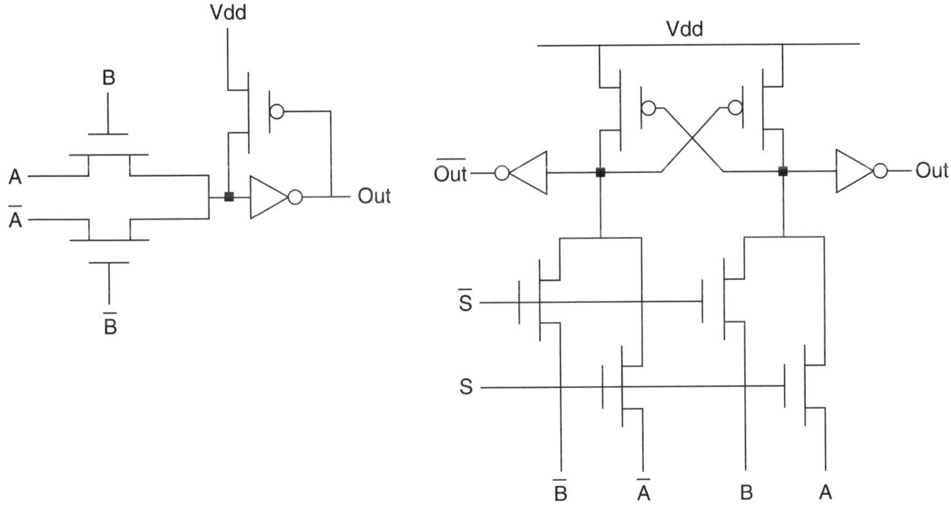

FIGURE 7.4 SPL XOR gate and CPL (dual-rail) 2:1 multiplexer.

TABLE 7.1 Comparison for Full Adder [6]

	Delay (ns)		Power (mW)		Power * Delay	
Logic Family	3.3 V	1.5 V	3.3 V	1.5 V	3.3 V	1.5 V
CMOS	1.89	7.88	32.9	6.4	1.00	1.00
CPL	1.39	8.33	34.1	6.0	0.76	0.99

Figure 7.4 presents another similar logic style called complementary pass transistor logic (CPL) and shows a 2:1 multiplexer. This CPL style is dual rail logic, as both true and complemented outputs are provided. The number of interconnect wires is therefore increased. A comparison (Table 7.1) with static CMOS shows some advantages in speed at high V_{dd}, but not better performances at low V_{dd} [4], the transistor count being similar to static CMOS. Many other logic styles are inspired from SPL and CPL, such as dual rail differential cascode voltage switch logic (DCVSL) [5], for which the two N-ch networks are not designed as N-pass logic but as conventional N-ch networks. Two cross-coupled P-ch MOS are used as load devices, similarly to two P-ch MOS of the CPL logic shown in Figure 7.4. It is a ratioed logic because the N-ch networks have to fight against the P-ch devices.

7.2.5 Dynamic Precharged Logic

Dynamic precharged logic is used to avoid the realization of P-ch networks to limit the transistor count in very complex logic gates. In single rail implementation, it is furthermore impossible to use resistive load devices if power consumption is an issue. This is why the P-ch network is replaced by a precharged P-ch transistor, which is used to precharge the gate output to V_{dd} in a first precharge phase. A second precharge N-ch transistor is used to cut off the N-ch network during the precharge phase. In the evaluation phase, the precharge N-ch MOS is conducting while the P-ch MOS is cut off. If the N-ch network is on, the gate output is switched to Vss, which is the right state for a logic gate with on N-ch network. If the N-ch network is off, the gate output keeps dynamically its precharged "1" state stored in the parasitic output capacitance.

Simple precharged logic gates cannot be connected in series, as the outputs of the first gates all precharged to "1" and connected to the inputs of the second gate result in a conducting N-ch network of the second gate. If the common precharge signal reaches the second gate a few nanoseconds before the first gate, the second gate can be discharged erroneously. Implementing a delayed precharge signal

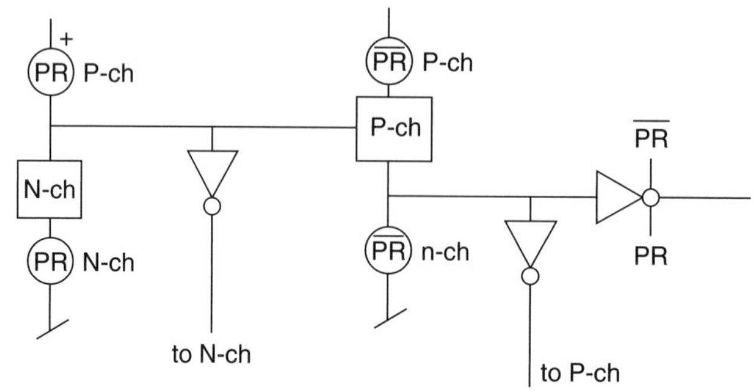

FIGURE 7.5 Precharged NORA gate.

for the second gate can solve this problem. Another solution is to implement a second gate with a P-ch network and precharged to "0" or to invert systematically the outputs of the precharged gates.

These solutions are used for dynamic DOMINO and NORA logic styles [7], as illustrated in Figure 7.5. A DOMINO gate is a conventional precharged gate with an inverted output, in such a way that DOMINO gates can be connected in series. The acronym "NORA" means "NO Race." This logic style combines DOMINO gates and precharged gates using alternate N-ch and P-ch networks (Figure 7.5). Furthermore, a NORA gate implements a dynamic latch (a simple tri-state gate) at its output. This latch is used for memorizing the output information and therefore to remove any hazard. Such logic has to work in pipeline due the output latch.

Dynamic logic is also used for precharged dual rail DCVSL logic [5]. By replacing the P-ch load devices with P-ch precharged transistors, the outputs in the evaluation phase switch necessarily to a complemented state (i.e., one output to "0" and the other to "1"). Consequently, for each computation, one output signal is always switched to "0," implying the same activity and the same power consumption for each computation. It is therefore more difficult to trace variation in power consumption during execution, which may be useful, for instance, for cryptographic applications attacked by differential power attack (DPA). Regarding power consumption, however, the activity of dual-rail gates is dramatically increased to 100%, as one of the outputs precharged to V_{dd} always switches to V_{ss} for each computation.

7.2.6 Memory Elements

The design of low-power flip-flops is crucial for the design of low-power circuits, as a digital block contains many memory elements. Besides power consumption issues, these elements have to be designed in such a way that they avoid any hazard for any gate delay. Master-slave structures, when true and complemented clocks respectively drive the master and the slave, present such a hazard [8]. Race-free flip-flops have therefore been designed to obtain speed-independent cells. The method [9], based on the fact that basic building blocks of CMOS are inverting or negative gates, result in structures that require inputs including the clock only in true forms, similar to the true single phase clock (TSPC).

For instance, using the method described in [10], a race-free DFF has been designed, containing only NAND gates and the single clock is connected to only four transistors to reduce the clock capacitance (Figure 7.6). Another interesting feature of this DFF is a weak sensitivity to the clock input slope. If a wire delay occurs on the clock fork connected to the two X and Y gates, this delay has to be shorter than three gate delays (e.g., gate delays Y, NM, and M) to guarantee a correct behavior.

7.2.7 Double-Edge Triggered Flip-Flops

Using the parallelization scheme proposed in [11], a flip-flop can be parallelized to obtain a double-edge triggered (DET) flip-flop clocked at half the master frequency [12]. A conventional DFF is implemented with two latches in series. Its parallelization results in two latches in parallel with an output multiplexer

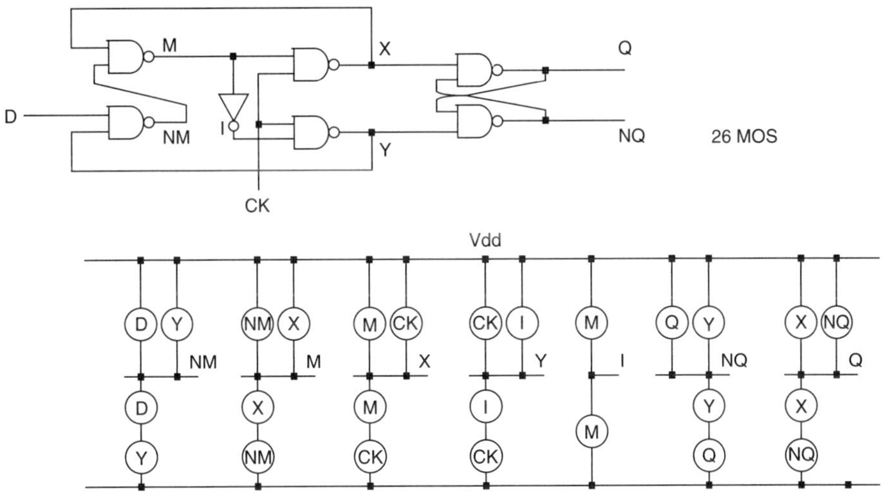

FIGURE 7.6 Race-free NAND-based DFF.

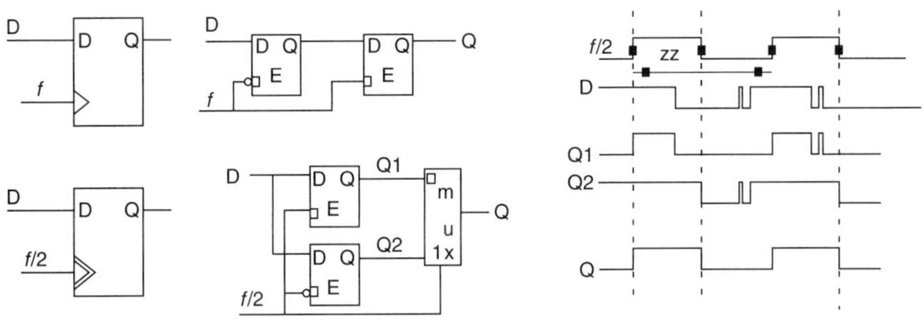

FIGURE 7.7 DFF parallelization.

(Figure 7.7). Such a flip-flop is sensitive to both edges of the input clock signal *f*/2 [13]. Its use in synchronous systems results in a master clock reduced of a factor 2. Any finite state machine may be implemented with double edge DFFs to reduce the input frequency by a factor 2 and the power consumption of the clock tree.

The choice of a logic family for a low-power design is generally not an issue. In systems on chip (SoC) design, a standard cell library is used. Its basic cells are generally designed in static CMOS. The controversy still holds about the benefits of transmission gates in some technologies, such as SOI [14]. For very fast microprocessors, some dynamic logic styles are used to achieve the huge speed required, but the layout is partially handcrafted. The design of fast and low-power flip-flops is still an issue for microprocessors [15]. For standard cell libraries, however, as the number of cells is increasingly reduced, conventional flip-flops instead of DET are used.

7.3 Low-Power and Standard Cell Libraries

The power consumption is today the major issue in the design of integrated circuits for portable devices. Design methodologies at different abstraction levels, such as systems, architectures, logic design, basic cells, as well as layout, must take into account the power consumption. The main goals of such design methods are V_{dd} reduction, activity reduction, as well as, reduction of parasitic capacitance [11,16]. It is well-known that most of the power can be saved at the highest levels; however, these choices are strongly application dependent. At the lowest levels, for instance, a low-power standard cell library, only a smaller factor in power reduction can be achieved, but the resulting library can be used for any design.

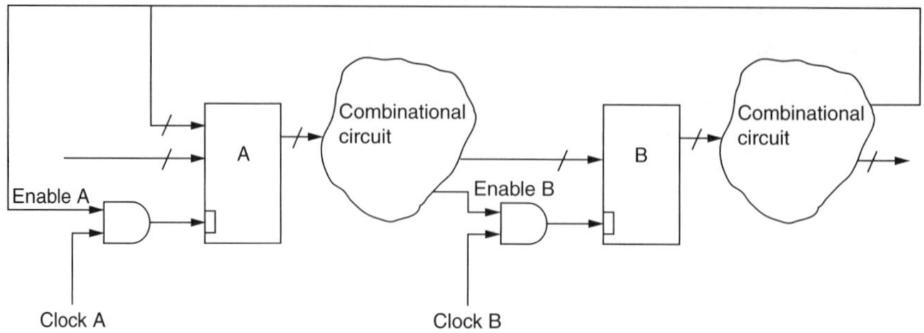

FIGURE 7.8 Latch-based design and gated clocks.

For standard cell libraries, besides the parasitic capacitance reduction for instance achieved by branch-based logic, V_{dd} reduction and gated clocks techniques are the most successful techniques used to reduce the power consumption.

7.3.1 Gated Clocks

The gated clock technique is extensively used in the design of low-power circuits [16,17]. It consists to gate the clock of sub-circuits that are in idle mode or that have just to keep their data as such. Arithmetic and logic units (ALU) of microcontrollers, for instance, have been designed with input and control registers that are loaded only when an ALU operation has to be executed. During the execution of another instruction (e.g., branch, load/store), these registers are not clocked avoiding any transition into the ALU. Gated clocks are also used to gate the clock of finite state machines [17] when the next state is identical to the present state.

Some logic synthesizers introduce gated clocks automatically; however, they could also gate clocks that have to be always active, which is useless. It is preferable to describe in very high speed hardware description language (VHDL) the necessary code to gate a clock and to introduce it only if it is useful. The most critical problem is to prevent the synthesizer from optimizing the clock gating "AND" gate with the rest of the combinational logic (Figure 7.8). This can be easily done manually by the designer by placing these AND gates in a separate level of hierarchy of his design or placing a "don't touch" attribute on them [18].

7.3.2 Latch-Based Designs

When designing a digital block using a standard cell library, the clocking scheme is extremely important for speed and power consumption. Generally, a single clock design is chosen, using master–slave flip-flops, well supported by the CAD tools; however, the clock tree synthesis is more and more difficult to achieve for avoiding a too large clock skew, resulting in large and power consumer buffers. Thirty percent of the total power could be in the clock circuits.

Latch-based designs with several nonoverlapping clocks have been proposed to solve this problem [18], and it has been demonstrated that they are more reliable at very low supply voltage. Conservative nonoverlapping clocks are used (i.e., an Ø1 clock pulse for the first period of the master clock CK (generated by an on-chip oscillator) and a second Ø2 pulse for the second period of the master clock). Therefore, the clock skew has to be shorter than half a period of the clock CK; however, such a scheme requires two clock cycles of the master clock CK to execute a single operation clocked by Ø1 and Ø2.

Latch-based designs provide several advantages over single clock master–slave flip-flop designs. In the design of a microcontroller, the power consumption can be reduced by about a factor of 2 [18]. The constraint with respect to the clock skew can be relaxed for both the Ø1 and Ø2 clock trees. This allows the synthesizer and router to use smaller clock buffers and to simplify the clock tree generation, which will reduce the power consumption of the clock tree.

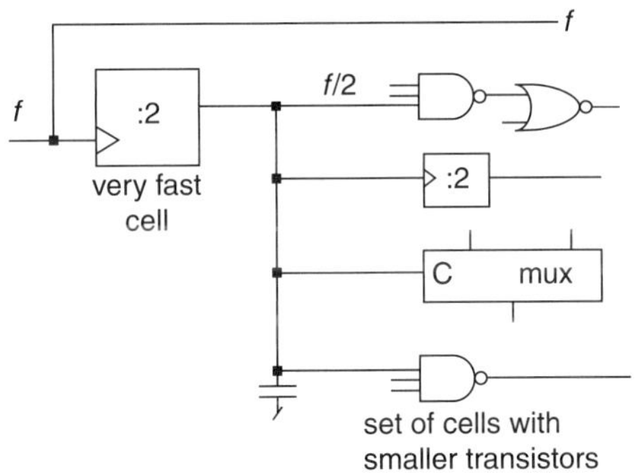

FIGURE 7.9 Several cell categories.

The latch design provides a smaller logic depth in combinational circuits and additional time barriers, which stop the transitions, avoid unneeded propagation of the signal, and thus reduce glitch power consumption. Using latches can also reduce the number of MOS of a design, for instance, in a register bank. With latches, the master part of the register bank can be common for all the registers, which gives a single master and many slaves, achieving a register bank area reduced by a factor of 2.

Using latches for pipeline structure is also very good for power consumption when using such a scheme in conjunction with clock gating. Figure 7.8 depicts a simple and safe way of generating enable signals for clock gating. This method gives glitch-free clock signals without the adding of memory elements, as it is needed with master–slave flip-flop clock gating [17]. Logic synthesizers very nicely handle the latch-based design methodology if the designer writes the description of the clock gating in his VHDL code.

7.3.3 Cell Drives

Standard cell libraries have to provide several drive versions of the same Boolean function (i.e., low-power, high-speed, and very high-speed cells as well as many buffer drives). This allows the designer or the logic synthesizer to place very high-speed cells on the critical path. However, if very high-speed cells are used outside the critical path, these cells will contain oversized transistors, which increase the power consumption and slow down the critical path. Figure 7.9 illustrates a simple example in which a first very fast cell is loaded with many other cells that are not on the critical path. If these cells contain oversized transistors, the load capacitance of the first very fast cell is increased, resulting in decreased speed. If the other cells are low-power cells with small transistors, however, the speed of the first cell will be higher and the power consumption reduced.

7.3.4 Complex Gate Decomposition

Complex gate decomposition is necessary if the number of transistors in series must be limited, as was the case in the proposed branch-based style [2]. As presented in Figure 7.10, the result is that simple gates with more than three inputs are decomposed into several simpler gates. This results in more transistors for the same Boolean function, but the total delay is reduced.

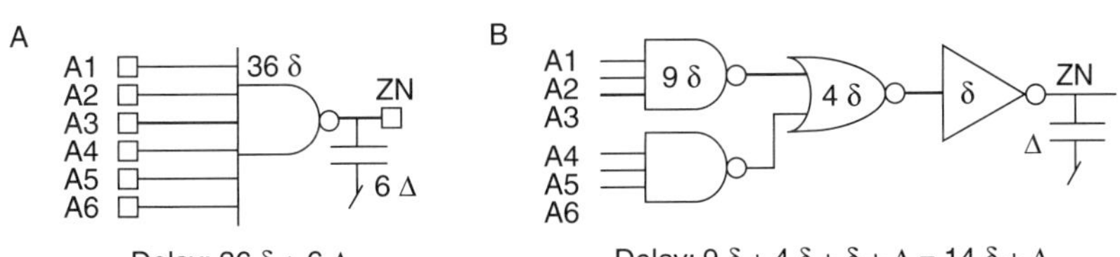

FIGURE 7.10 NAND6 decomposition.

TABLE 7.2 Gate Delay Comparison

Logic Gate	Delay
NAND6 (not decomposed)	0.70 ns
NAND6 (decomposed)	0.42 ns
NOR6 (not decomposed)	1.81 ns
NOR6 (decomposed: 2*NOR3 + NAND2)	0.65 ns
NOR6 (decomposed: 3*NOR2 + NAND3)	0.53 ns

According to a rough gate delay model [19], an N-input NAND gate contains a branch with N transistors in series, resulting in an increased internal resistance of $N^*\delta$. Furthermore, the internal parasitic capacitance is also increased roughly by a factor N (N drain capacitances). The internal delay of an N-input gate is therefore $N^{2*}\delta$. The load delay of an N-input gate is $N^*\Delta$ as the output capacitance must be charged or discharged by a branch with N transistors in series. Therefore, the total delay of an N-input gate is: delay = $N^{2*}\delta + N^*\Delta$. For a 6-input NAND gate, the total delay is $36\ \delta + 6\ \Delta$. If such a 6-input gate is decomposed as illustrated in Figure 7.10, the critical path of such a gate is made of three simple gates in series (e.g., a 3-input gate, a 2-input gate, and an inverter), resulting in a shorter total delay of $14\ \delta + \Delta$.

Table 7.2 gives some results in a 0.7-μm technology for both NAND6 and NOR6 gates with small-sized transistors (W = 2.2 μm). The load capacitance has been considered as six gate capacitances of similar logic gates. Obviously, the delay reduction for decomposed gates is better for NOR gates, which present P-ch transistors in series. At the same V_{dd} and for the same transistor sizes, a decomposed gate presents higher power consumption, as the simple gates could switch without an output transition.

7.3.5 Standard Cell Libraries

Standard cell libraries often provide a huge number of cells, up to 300 or even 500. A new approach is proposed, more or less similar to RISC vs. CISC processor architectures, which is based on a limited set of standard cells [20]. The number of functions for the new library has been reduced to 22 and the number of layouts to 92. It can be seen that the ratio between the number of layouts and the number of functions is larger (92/22 = 4.2 instead of 220/60 = 3.6 for the previous library). This means that the number of cell and buffer drives is larger. For speed and power optimization achieved by the logic synthesizer, the increased ratio of layouts to functions, as presented previously, is beneficial.

It appears obvious that the logic synthesizer could do a better job if the number of cells in the library is large. With a larger choice, it should be possible to provide a better solution, but this is not the case. Experiments in Table 7.3 and Table 7.4 demonstrate that the delay of some operators is significantly reduced with the new library resulting in a very small increase in silicon area (Table 7.3) and that the silicon area is reduced at the same speed with the new library (Table 7.4). These results show that the logic synthesizer is more efficient because it has a limited set of well-chosen cells and cell sizing adapted to the considered logic synthesizer. With significantly fewer cells than conventional libraries, the synthesizer is not lost in some optimization loops due to a too large choice of cells.

TABLE 7.3 Delay Comparison (synthesis for maximum speed, 0.5-μm process)

	Old Library		New Library	
	Delay [ns]	μm²	Delay [ns]	μm²
32-bit multiplier	16.4	907 K	12.1	999 K
Floating-point adder	27.7	510 K	21.1	548 K
CoolRISC ALU [18]	1.08	140 K	7.7	170 K

TABLE 7.4 Silicon Area Comparison (synthesis for a given delay, 0.5-µm process)

	Old Library		New Library	
	Delay [ns]	µm²	Delay [ns]	µm²
32-bit multiplier	17.1	868 K	17.0	830 K
Floating-point adder	28.1	484 K	28.0	472 K
CoolRISC ALU [18]	11.0	139 K	11.0	118 K

The design of the library has been based on keeping only very fast cells (i.e., to remove all the cells with 3 P-ch transistors in series and to have a very limited number of cells with 2 P-ch transistors in series). The number of cell layouts for the same function has been increased; however, it is not a simple increase from, for instance, sizing D1 (small transistors), D2, and D3 (medium-sized transistors) to D1, D2, D3, D4, and D5 (very large transistors). The cell sizing performed takes into account how the synthesizer uses the considered cells. The third consideration is based on buffer insertion (i.e., the combination of a given cell and of a buffer to replace complex gates).

Such a strategy must be checked through many experiments. The choice of the 22 functions was performed with a large number of experiments with and without a specific cell, and then the decision was made to either insert this cell in the library or not. Similar experiments were performed with various sizing and buffering of the cells. At the end, only 22 functions and 92 layouts were kept in the new library.

Furthermore, as the number of layouts is drastically reduced, it takes less time to design a new library for a more advanced process. Substantial time can also be saved for the library characterization, which is often the most time-consuming activity in library design. Reducing the number of layouts from 220 to 92 is a significant advantage. The reduction of the number of cells implies removal of complex gates from the library, forcing the logic synthesizer to decompose complex gates, which, as described, is beneficial in terms of speed.

It will also be a crucial point in future libraries for which more versions of the same function will be required while considering static power problems. The same function could be realized, for instance, with low or high VT for double VT technologies, or with several cells such as a generic cell with typical VT, a low-power cell with high VT and a fast cell with low VT.

7.3.6 Static Power

Static power is a dramatic issue for deep submicron technologies. Due to lower and lower supply voltages, the threshold voltages are also significantly reduced to keep some speed and the leakage is increased exponentially with the VT reduction [21].

Several circuit techniques have been proposed, which partially solve this issue. For standard cell libraries, three techniques directly affect the design of the library cells:

1. Multi-threshold CMOS (MTCMOS)
2. Stacked transistors
3. Dynamic threshold MOS (DTMOS)

The MTCMOS technique [22] is based on a technology offering two low and high VT for each MOS transistor. The low VT devices have to be used only on the critical path. Very fast cells of the library are therefore designed with low VT transistors on their critical path. Generally, only 10% of the transistors of a digital block are low VT devices, resulting in leakage reduction of about a factor of 10. Another method implements more stacked transistors [23], however, speed is impacted. This method implies to design logic cells of a library taking into account the trade-off between speed (the smaller number of MOS in series) and leakage (the larger number of MOS in series). The third technique, DTMOS [24], originally proposed for SOI technology, implements a connection between the transistor gate and the body of the transistor. It results in increased VT when the transistor is cut off.

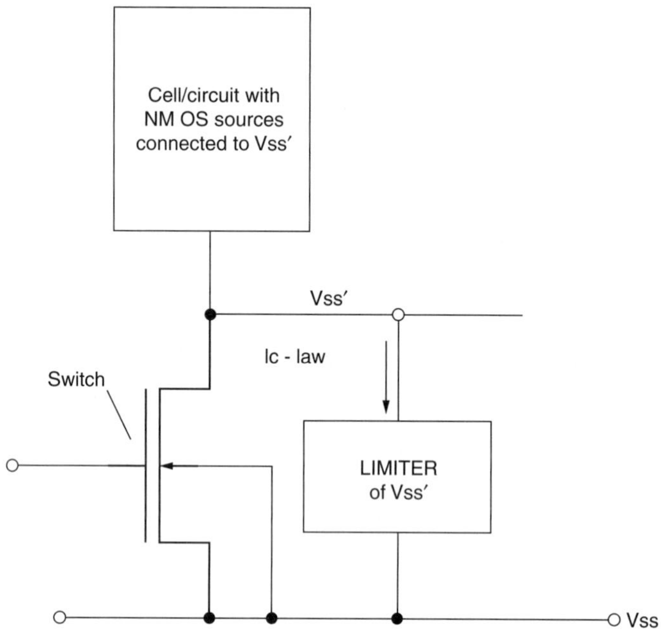

FIGURE 7.11 Switch in supply line with a limiter circuit.

Other well-known techniques, such as variables VT (variable threshold CMOS [VTCMOS] or self adjusted threshold scheme [SATS]) or switches in supply lines can be implemented at the fine grain (library cell) or large grain (block) levels. The former applies a well bias voltage to increase VT in idle or weak activity modes, while the second technique cut off the supply lines in idle modes. When cutting off transistors in idle modes, the resulting voltage of the considered cell or block may be so low that data in memory elements is lost. This is why some techniques, as illustrated in Figure 7.11, implement a limiter circuit in such a way that the resulting voltage is sufficient to keep the data in memory elements [25, 26]. Furthermore, when the switch is cut off, the N-MOS threshold voltages of the considered circuit are increased due to transistor source bias (V_{ss} is higher than the body of transistors).

Digital design is based on standard cell libraries, but it is also strongly impacted by the logic synthesis. As such, it is interesting to consider if synthesized digital architectures, such as pipeline, parallel, and asynchronous, may be better for reducing leakage. A very low activity factor does not provide a good ratio between dynamic and static power, as an idle circuit does present the same leakage than a very active circuit. To reduce the total power consumption, an attractive goal could be to have fewer, yet more active, transistors to perform the same logic function. If a given logic function is performed with 10,000 gates with an activity factor of 1%, this means that on average 100 gates are switching in a clock period. If the same logic function could be implemented with 1,000 gates, keeping the same number of switching gates (100), the activity will be 10%, with the same dynamic power but with a leakage reduced by a factor of 10 due to the reduction of the total number of gates.

Improved use of the gate switching is also dependent on the duty factor (Chapter 16) (i.e., the switching duration of a given logic gate over the clock period duration). The duty factor is defined as $\alpha = f^{*}Td$ (1/f: clock period; Td: switching duration). If there are many gates connected in series, as depicted in Figure 7.12, this duty factor is either equal to or less than $\alpha = 1/LD$ (LD: logical depth). Furthermore, as depicted in Figure 7.12 for five gates, a significant part of the clock period may be unused, without any switching, resulting in a smaller α than $\alpha = 1/LD$.

To compare digital architectures regarding leakage, the following design parameters can be used: activity factor *a*, duty factor α, the number N of logic gates, the capacitance C per gate, and the ratio of I_{OFF}/I_{ON}. The dynamic and static energies can be defined as:

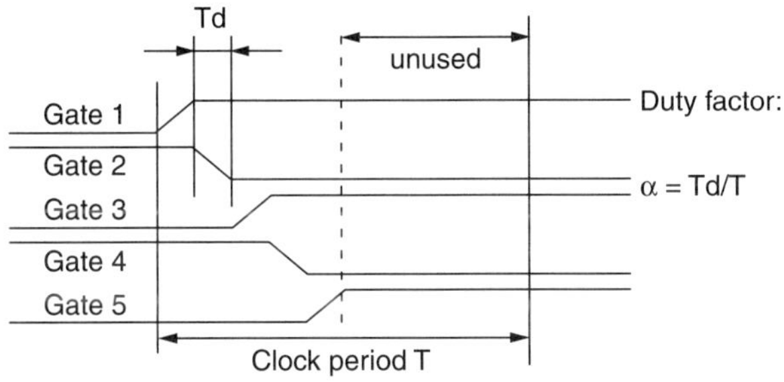

FIGURE 7.12 Duty factor and logic gates connected in series.

$$\text{Edyn} = a * \text{N} * \text{C} * \text{V}_{dd}^{2}$$

$$\text{Estat} = (1/\text{f}) * \text{N} * \text{V}_{dd} * \text{I}_{0FF}$$

Considering that the clock period is fully used, and introducing the fmax frequency (product of α by fmax of a logic gate) fmax $= \alpha * I_{ON}/(C*V_{dd})$ into the expression of the static energy:

$$\text{Estat} = (1/(\alpha * \text{I}_{ON})) * \text{N} * \text{C} * \text{V}_{dd}^{2} * \text{I}_{0FF}$$

Therefore, the total energy becomes:

$$\text{Etot} = (a + 1/\alpha * \text{I}_{0FF}/\text{I}_{ON}) * \text{N} * \text{C} * \text{V}_{dd}^{2}$$

$$\text{Etot} = (a + \text{LD} * \text{I}_{0FF}/\text{I}_{ON}) * \text{N} * \text{C} * \text{V}_{dd}^{2}$$

This last equation shows that the static energy is proportional to the number LD of gates connected in series. A large LD implies a small α and a smaller use of the available logic gates. It has been shown [33] that the optimum of the total power consumption is roughly at 50% of dynamic and 50% of static power (i.e., Edyn = Estat). This allows defining:

$$a = \text{LD} * \text{I}_{0FF}/\text{I}_{ON}$$

or

$$\text{I}_{0N}/\text{I}_{OFF} = \text{LD}/a = 1/(\alpha * a)$$

This expression indicates that relatively small values of I_{0N}/I_{OFF} are possible if LD is small and activity a relatively large. For $I_{0N}/I_{OFF} = 100$, LD = 10, and a = 10%. These values result in very small VT and V_{dd} as well as very low total power consumption at a reasonable speed [33]. It also means that LD=100 and a=1% will result in I_{0N}/I_{OFF} =10'000, value for which VT and V_{dd} could not be reduced significantly. It is therefore necessary for digital synthesis to achieve very low values of I_{0N}/I_{OFF}, and, consequently, small LD and high activity. Clearly, pipeline architectures (small LD) are better than nonpipelined [34], asynchronous architectures could also be interesting to avoid unused part of the clock period (Figure 7.12), but high activity architectures are more difficult to design, as low activity has been a goal in recent years to reduce dynamic power.

7.4 Logic Styles for Specific Applications

The design of cell libraries is largely considered independent of the applications (i.e., any library can be used for any application). This assumption no longer holds, as certain specific applications require special cell libraries. Some examples are described here, such as self-timed design, cryptographic applications, and fault-tolerant logic.

7.4.1 Library Cells for Self-Timed Design

Self-timed logic (i.e., digital circuits without any master clock) has been introduced to solve the problem of the clock tree synthesis, which has proven more difficult and power-consuming [27,28]. In SoC design, the master clock cannot be propagated through the chip without having clock skew larger than the clock period. Globally asynchronous locally synchronous (GALS) SoC architectures have been proposed to solve this problem, but another solution could be the complete removal of the clock by designing pure asynchronous logic blocks.

Several asynchronous techniques have been proposed, at the block and/or cell levels. They are based on handshaking (i.e., a local control of the data shifted in a pipeline). This control logic is largely based on C-Muller elements, generally not proposed by conventional libraries. For asynchronous design, it would be beneficial to have this C gate as a library cell. Another logic style that is used in self-timed architectures is dynamic DCVSL dual rail logic. During the precharged phase, both outputs are "1," an invalid state. After evaluation, the valid state is reached ("01" or "10"), indicating that the operation is completed. This signal is consequently used to start the next operation (request) and to acknowledge the previous pipeline stage (acknowledge). In a manner similar to global clocks, in which rising edges are used for synchronization, rising and falling edges of these "request" and "acknowledge" signals are also used in self-timed logic. A static logic family, called "event logic," can also be designed while using the signal edges for which only these edges are the events of interest. Two different protocols are used:

1. Two phases protocol, for which a rising edge has the same significance as a falling edge
2. Four phases protocol, for which only the rising edge is taken into account while the falling edge means only reset

Figure 7.13 depicts the four phases protocol, in which the rising edge of the "request" signal occurs when data is ready (valid data). It is important to note that the "request" signal is *not* generated by the control logic itself (in this case, a critical race can occur between the data and the "request" wire), but by the data, using a specific code, such as the dual rail code or other codes such as one hot, parity or Berger codes. In this way, no critical race exists between data and the "request." Such a scheme is called quasi delay insensitive (QDI). When the data has been used, the "acknowledge" rising edge occurs and the data can be resettled (empty data). Then the two signals are resettled and a new cycle can be started.

The basic logic gates of "event logic" are quite different from conventional logic. For instance, the AND gate is a C-element or Muller gate. This gate is switched on if the two inputs switch to "1" and switched off if the two inputs switch off. This means that this gate has a memory (i.e., if only one input goes to "0" after that the two inputs switch on, the gate keeps its output = 1, so it is not a combinational AND). A library of "Event Logic" standard cells can be designed using the dual rail approach. The code for "1" is Z0 = 0 and Z1 = 1, while a "0" is represented by Z0 = 1 and Z1 = 0. From a Karnaugh map, for instance, the "OR" function in Table 7.5, the two outputs Z0 and Z1 can be designed as the sum of the minterms of each cube. Figure 7.14 illustrates the logic implementation of the OR function using C-

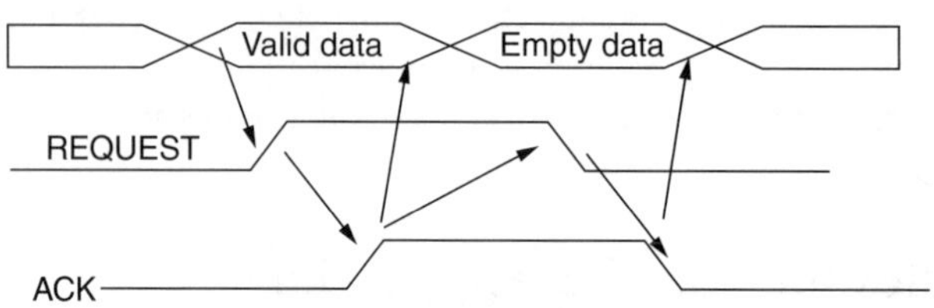

FIGURE 7.13 Four phases protocol.

TABLE 7.5 Karnaugh Map of the OR Function

Z	A = 0	A = 1
B = 0	0	1
B = 1	1	1

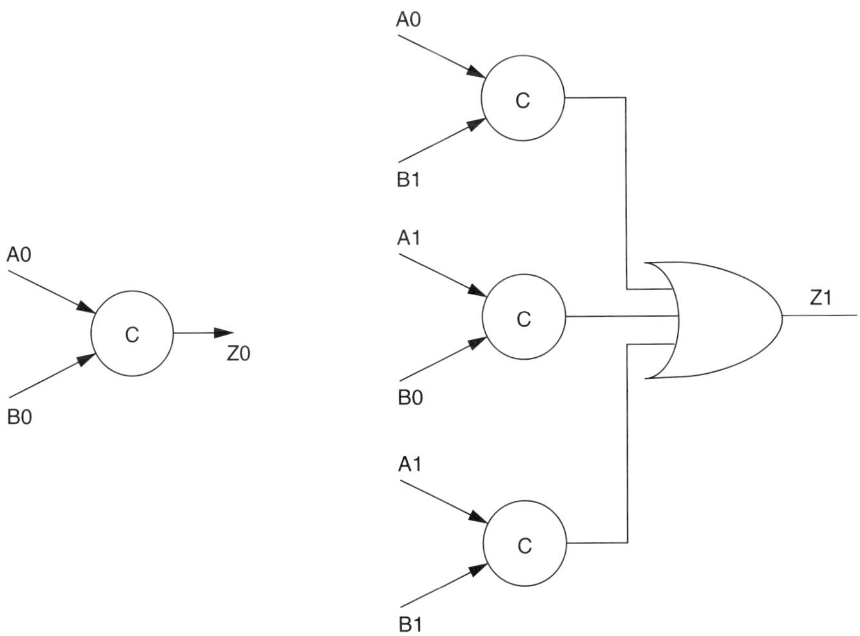

FIGURE 7.14 OR function in "Event" dual-rail logic.

elements. Any combinational circuit can be designed using this approach starting from its Karnaugh map. The cost in terms of number of transistors is significantly increased.

$$Z0 = A0^{*}B0$$

$$Z1 = A0^{*}B1 + A1^{*}B0 + A1^{*}B1$$

7.4.2 Library Cells for Cryptographic Applications

Application-specific integrated circuits (ASICs) for smart cards can be attacked by differential power attacks (DPA), which trace the power consumption and identify operations that are data-dependent after removal of the power consumption, which is data-independent. In this way, secret keys could be obtained. Consequently, DPA-resistant circuits will be of crucial importance in the future.

DPA was demonstrated in 1999 [29], thus conveying that it is possible to examine the power consumed by the circuit when processing data or executing instructions. By analyzing the variation in power consumption and the data processed, an attacker can discover the secure information being processed and the keys hidden in the circuit. The attacker can analyze a single power trace (SPA) or can perform a statistical analysis of many collected power traces (DPA). These power traces will provide an average power trace that represents the data-independent power trace. By having a given power trace with a given hidden key, the attacker can subtract the average power trace and determine the difference that represents only the data-dependent power. By comparing the difference to the simulated power traces, the attacker can quite easily deduce the secure key of the considered circuit.

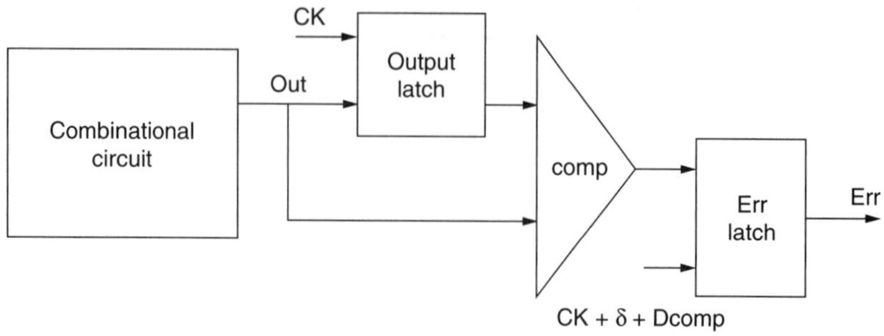

FIGURE 7.15 SEU fault-tolerant logic.

The DPA attack is because executed operations or instructions present power consumption that is data-dependent. A very general goal is to find circuit implementations that are not sensitive to operands regarding the power. Some circuit techniques, such as current steering logic or dual rail DCVSL logic, are known to be less data-dependent. Self-timed logic does not have a global clock, so there is not a global timing signal for use as a reference and the analysis and correlation of power traces of power consumption are more difficult [30]. It is difficult to determine when an operation or instruction starts and stops. Because power issues are crucial for smart cards, however, these logic families do not meet this requirement. Furthermore, recent research has revealed weaknesses in the basic DCVSL scheme and suggested improvements [31]. In dual-rail DCVSL logic, parasitic capacitances are different in the two N-ch networks implementing the dual Boolean functions and, therefore, the power consumption. A new sense amplifier based logic (SABL) logic family is introduced in [31], for which the output will charge the same capacitance for each clock, even if the output transition is 1-1 or 0-0. This SABL logic is based on the differential strong arm flip-flop (SAFF); it consumes, however, two times the power of a static CMOS. Balanced "Event Logic" can also be used, by adding into Figure 7.14, for instance, a dummy OR gate for z0.

7.4.3 SEU-Tolerant Logic

Another dramatic problem in deep submicron technologies is due to soft errors, which may arise due to particles that can discharge a logic node. Several years ago, only dynamic nodes were an issue, for instance, in dynamic RAM memories (DRAMs), and mainly for space applications. With technology scaling, however, such as V_{dd} and parasitic capacitance scaling, the charge at each node becomes smaller and smaller. It is therefore becoming easier for a particle to discharge such a node even on earth, resulting in a glitch that could affect the circuit behavior.

Techniques have been proposed to create a soft error (single event upset [SEU]) tolerant logic. It is based on duplication of the logic, while observing if the two results are the same. If not, the executed operation is repeated until the SEU has disappeared. Such logic duplication is very expensive in terms of silicon area and power consumption. This is why some of the most interesting techniques are based on timing redundancy [32] (i.e., the operation is executed in a single combinational block), but the result is stored in a first latch at the rising edge of the CK, and in an extra latch at CK + δ. If the SEU occurs after the CK edge, the two stored data will be different and the SEU detected. Figure 7.15 depicts an implementation in which even the extra latch is removed by performing the comparison at CK + δ. The latch contains the data memorized at the CK edge, while the other input of the comparator is the data at CK + δ. To obtain an SEU fault tolerant logic, it is possible to transform the VHDL description of any circuit into a VHDL description containing cells as depicted in Figure 7.15 [32].

7.5 Conclusion

For deep submicron technologies, robustness is the main challenge after low power consumption. Therefore, robust logic styles, such as the old but very simple static logic style, are the best candidates for future libraries [35]. Leakage reduction will be the most difficult problem to solve regarding low power, as well as leakage reduction in active modes. As more constraints must be satisfied for specific applications, specialized libraries will emerge, and, ultimately, these are expected to lead to application specific libraries.

References

[1] T.G. Noll, E. De Man, Pushing the performances limits due to power dissipation of future ULSI chips, *IEEE Int. Symp. on Circuits and Syst., ISCAS '92,* San Diego, CA, May 10–13, 1992, pp. 1652–1655.

[2] J-M. Masgonty et al. Technology- and power-supply-independent cell library, *IEEE CICC '91,* San Diego, CA, May 12–15, 1991, Conf. 25.5.

[3] A. Nève et al. Design of a branch-based 64-bit carry-select adder in 0.18 μm partially depleted SOI CMOS, *Proc. ISLPED '02,* Monterey, CA, August 12–14, 2002, pp. 108–111.

[4] S. Nikolaidis and A. Chatzigeorgiou, Circuit-level low-power design, Chapter 4 in *Designing CMOS Circuits for Low-Power,* D. Soudris, C. Piguet, and C. Goutis, Eds., Kluwer Academic Press, Dordrecht, 2002.

[5] L.G. Heller et al. Cascode voltage switch logic: a differential CMOS logic family, *Proc. ISSCC 1984,* San Francisco, CA, February 12–14, pp. 16–17.

[6] R. Zimmermann and W. Fichtner, Low-power logic styles: CMOS versus pass-transistor logic, *IEEE JSSC,* Vol. 37, No. 7, pp. 1079–1090, July 1997.

[7] N.F. Goncalvez and H.J. De Man NORA: a racefree dynamic CMOS technique for pipelined logic structure, *IEEE J. Solid-State Circuits,* SC-18, No. 3, June 1983, p. 261.

[8] C. Piguet, Supplementary condition for STG-designed speed-independent circuits, *Electron. Lett.,* Vol. 34, No. 7, April 2, 1998, pp. 620–622.

[9] C. Piguet and J. Zahnd, Signal-transition graphs-based design of speed-independent CMOS circuits, *ESSCIRC '98,* Den Haag, The Netherlands, September 21–24, 1998, pp. 432–435

[10] C. Piguet, Robustness of asynchronous sequential standard cells in a synchronous environment, *AINT '2000,* Delft, The Netherlands, July 1920, 2000.

[11] A.P. Chandrakasan, S. Sheng, and R.W. Brodersen, Low-power CMOS digital design, *IEEE J. of Solid-State Circuits,* Vol. 27, No. 4, April 1992, pp. 473–484.

[12] C. Piguet et al. Logic design for low-voltage/low-power CMOS circuits, *1995 Int. Symp. on Low-Power Design,* Dana Point, CA, April 24–26, 1995, pp. 117–122.

[13] R. Hossain et al. Low-power design using double edge triggered flip-flops, *IEEE Trans. on Very Large-Scale Integr. Syst.,* Vol. 2, No. 2, June 1994, p. 261.

[14] M. Belleville and O. Faynot, Low-power SOI design, *Proc. PATMOS 2001,* Yverdon, Switzerland, September 26–28, 2001, pp. 8.1.2–8.1.10.

[15] V. Stojanovic and V.G. Oklobdzija, Comparative analysis of master-slave latches and flip-flops for high-performance and low-power systems, *JSSC,* Vol. 34, No. 4, April 1999.

[16] C. Piguet, Low-power and low-voltage CMOS digital digital design, *Elsevier Microelectron. Eng.,* 39, 1997, pp. 179–208.

[17] L. Benini et al. Saving power by synthesizing gated clocks for sequential circuits, *IEEE Design and Test of Computers,* Vol. 11, No. 4, pp. 32–41, 1994.

[18] C. Arm et al. Double-latch clocking scheme for low-power I.P. cores, *Proc. PATMOS 2000,* Goettingen, Germany, September 13–15, 2000, pp. 217–224.

[19] C. Piguet et al. Low-power low-voltage digital CMOS cell design, *Proc. PATMOS '94,* Barcelona, Spain, Oct. 17–19, 1994, pp. 132–139.

[20] J-M. Masgonty et al. Low-power low-voltage standard library cells with a limited number of cells, *PATMOS 2001,* Yverdon, Switzerland, September 26–28, 2001, pp. 9.4.1–9.4.8.

[21] T. Sakurai, Perspectives on power-aware electronics. Plenary talk 1.2, *Proc. ISSCC 2003,* San Francisco, CA, Feb. 9–13, 2003, pp. 26–29.

[22] S.V. Kosonocky et al. Enhanced multi-threshold (MTCMOS) circuits using variable well bias, *Proc. ISLPED '01,* August 6–7, 2001, Huntington Beach, CA, pp. 165–169.

[23] S. Narendra et al. Scaling of stack effect and its application for leakage reduction, *Proc. ISLPED '01,* Huntington Beach, CA, August 6–7, 2001, pp. 195–200.

[24] F. Assaderaghi et al. A dynamic threshold voltage (DTMOS) for ultra-low voltage operation, *IEEE IEDM Tech. Dig., Conf. 33.1.1,* pp. 809–812, 1994.

[25] T. Enomoto, Y. Oka, H. Shikano, and T. Harada, A self-controllable-voltage-level (SVL) circuit for low-power high-speed CMOS circuits, *Proc. ESSCIRC 2002,* Forence, Italy, September 24–26, 2002, pp. 411–414.

[26] S. Cserveny, J-M. Masgonty, and C. Piguet, Stand-by power reduction for storage circuits, *Proc. PATMOS 2003,* Torino, Italy, September 10–12, 2003, pp. 229–238.

[27] Asynchronous circuits and systems, *Special Issue of IEEE Proc.,* Vol. 87, No. 2, February 1999.

[28] J. Sparso and S. Furber, *Principles of Asynchronous Circuit Design, A Systems Perspective,* Kluwer Academic Publishers, Dordrecht, 2001.

[29] P. Kocher, Differential power analysis, Advanced in Cryptology–Crypto 99, Springer LNCS, Vol. 1666, pp. 388–397.

[30] M. Renaudin and C. Piguet, Asynchronous and locally synchronous low-power SoCs, *DATE 2001,* Münich, Germany, March 13–16, 2001, pp. 490–491.

[31] C. Tiri et al. A dynamic and differential CMOS logic with signal independent power consumption to withstand differential power analysis on smart cards, *Proc. ESSCIRC 2002,* Florence, Italy, September 2002, pp. 403–406.

[32] L. Anghel and M. Nicolaidis, Cost reduction and evaluation of a temporary faults detecting technique, *Proc. DATE 2000,* Paris, France, March 27–30, 2000, pp. 591–598.

[33] C. Heer et al. Designing low-power circuits: an industrial point of view, *PATMOS 2001,* Yverdon, September 26–28, 2001.

[34] C. Piguet et al. Techniques de circuits et méthodes de conception pour réduire la consommation statique dans les technologies profondément submicroniques, *Proc. FTFC '03,* Paris, France, May 15–16, 2003, pp. 21–29.

[35] M. Allam et al. Effect of Technology Scaling on Digital CMOS Logic Styles, *Proc. IEEE CICC 2000,* Conf. 19.1, Orlando, Florida, May 21–24, 2000, pp. 401–408.

8

Low-Power Very Fast Dynamic Logic Circuits

Jiren Yuan
Lund University

8.1 Introduction

To achieve a high throughput usually means more power consumption in a given CMOS technology because the dynamic power consumption is proportional to the activity ratio. It implies that at a low activity ratio static logic circuits may consume less power compared with clocked dynamic logic circuits. For a given logic function, however, high-speed or a short-propagation delay does not necessarily mean high-power consumption, if highly efficient dynamic logic circuits with low power-delay products are used. Such low-power and very fast dynamic circuits are introduced in this chapter. A simple way to distinguish dynamic logic from static logic is to see whether the logic states are still correctly maintained, as for the static circuits, or destroyed, as for the dynamic circuits when the clock is turned off. This is because dynamic logic circuits need to be regularly refreshed for the charge stored on the logic nodes while static logic circuits need not.

Although clocking is used in all synchronous circuits, it is used only as synchronization for static logic circuits while as both synchronization and refreshment for dynamic logic circuits. Note that both complementary logic circuits and precharged logic circuits can be either static or dynamic, depending on whether the logic states are locked, for example by a cross-coupled loop, or not.

Section 8.2 focuses on the basic synchronizing components, single-clock latches and flip-flops, with comparisons in power and delay. Section 8.3 presents high throughput logic styles based on these components. Section 8.4 demonstrates examples of very fast dynamic CMOS functional circuits. Section 8.5 discusses the future of dynamic logic when the leakage current becomes a serious problem in deep submicron technologies. The conclusion is given in Section 8.6.

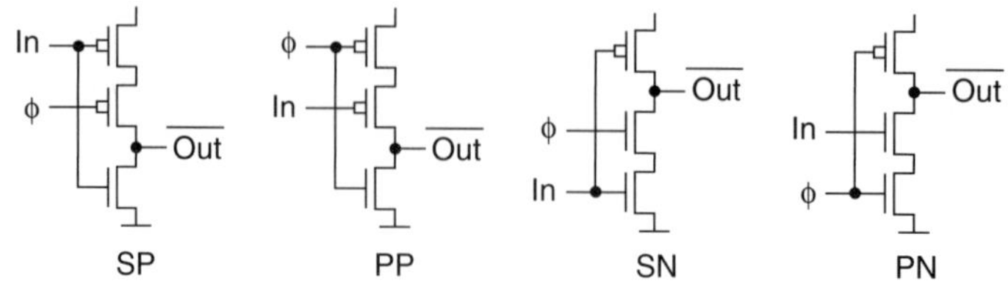

FIGURE 8.1 Basic stages in TSPC. (© 2004 IEEE.)

8.2 Single-Clock Latches and Flip-Flops

Latches and flip-flops controlled by clock(s) are the fundamental blocks of a synchronous system. It is well-known that $P_D = C_L V_{DD}^2 f_C A$, where P_D is the dynamic (usually the dominating) power consumption, C_L the load capacitance, V_{DD} the power supply voltage, f_C the clock frequency, and A the activity ratio. Clock is considered a fully active signal with a reference activity ratio of 1.0. Clocking strategy and the types of latches and flip-flops used for this system thus have a significant impact on its power consumption. Regarding the dynamic power consumption, a smaller number of clock wires and a smaller number of clocked devices will likely result in lower power dissipation. Based on this principle, we prefer to have as few clocked devices as possible and to use a single clock if it does not mean more clocked devices.

8.2.1 TSPC Latches and Flip-Flops

The true single phase clock (TSPC) circuit technique [1,2] uses only a single clock and two to three clocked transistors in each latch without local inversion of the clock as such an inversion requires more clocked devices. The basic stages SP, PP, SN, and PN in TSPC are depicted in Figure 8.1, where the first letter represents the logic style (S for nonprecharged and P for precharged), and the second represents the type of clocked devices (P for p-type and N for n-type). Stages SP and PP are identical except the exchange of data and clock inputs, the same for stages SN and PN. Two cascaded SP stages (nontransparent when clock is high) or SN stages (nontransparent when clock is low) become a p-type or n-type nonprecharged TSPC latch, respectively (see Figure 8.2(a)). A PP stage followed by an SP stage or a PN stage followed by an SN stage become a p-type or n-type precharged TSPC latch, respectively (see Figure

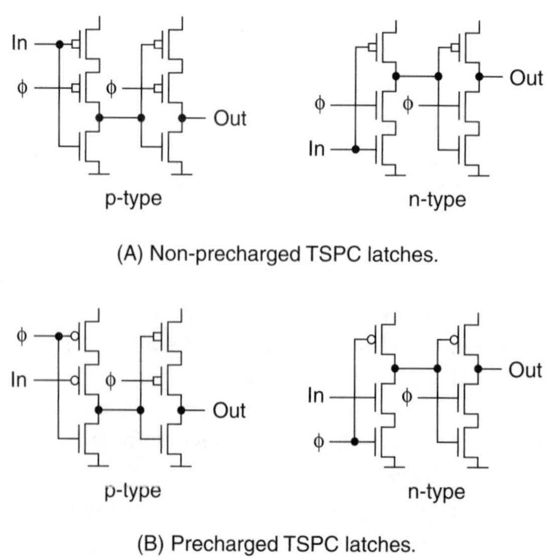

FIGURE 8.2 TSPC latches. (© 2004 IEEE.)

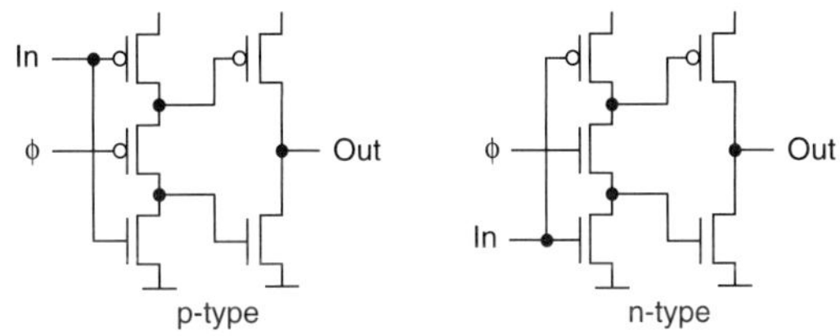

FIGURE 8.3 TSPC split-output latches. (© 2004 IEEE.)

8.2(b)). A TSPC nonprecharged flip-flop consists of two cascaded nonprecharged latches, a p-type and an n-type, and it becomes positive edge-triggered when the p-type is before the n-type, or negative edge-triggered otherwise. A TSPC precharged flip-flop is formed by a nonprecharged TSPC latch followed by a precharged TSPC latch in an opposite type, and it becomes positive edge-triggered when the nonprecharged TSPC latch is a p-type, or negative edge-triggered otherwise. The nonprecharged TSPC latches and flip-flops are superior in low-power performance [3].

To reduce power consumption, it is possible to use only a single clocked transistor for each latch. Figure 8.3 depicts such latches in p-type and n-type. They are so-called TSPC split-output latches in which the output of the first stage is split [2]. Edge triggered flip-flops can be built by cascading the split-output latches. As the number of clocked devices is at its minimum, the power spent on the clocked node is minimized. Because the clocked transistor propagates both high state and low state, however, one of the two states will not have a full swing. The clocked transistor should be properly sized when the supply voltage is low. In submicron technologies, the TSPC split-output latches can still be used due to reduced threshold voltages.

A very efficient and fast TSPC flip-flop using only nine transistors, based on a nonclassic flip-flop concept, is depicted in Figure 8.4(a), which gives an inverted data output [2]. The nonclassic flip-flop concept is illustrated in Figure 8.5(b), in comparison with the classic flip-flop concept depicted in Figure 8.5(a). A flip-flop transfers the input to the output only when the right clock edge comes and must be nontransparent otherwise. In a classic flip-flop the master and slave are completely nontransparent in its latching phase regardless the input logic states. In a nonclassic flip-flop, the master may be transparent in its latching phase for either a high or a low input, but the slave (which is in its nonlatching phase) must be nontransparent for the output of the master. In the example given in Figure 8.5(b), the master is transparent for a high input in its latching phase but the slave is nontransparent for the low output of the master although the slave is in its nonlatching phase. In such a way, the flip-flop is still nontransparent, which is exactly the case for the nine-transistor TSPC flip-flop depicted in Figure 8.4(a). The master is

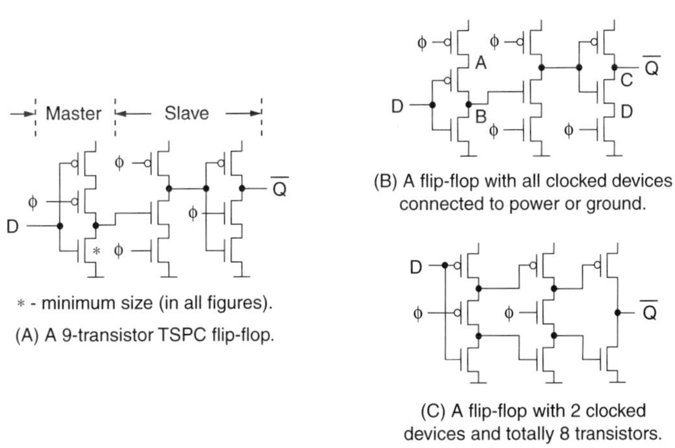

FIGURE 8.4 Efficient nonclassic single-clock flip-flops. (© 2004 IEEE.)

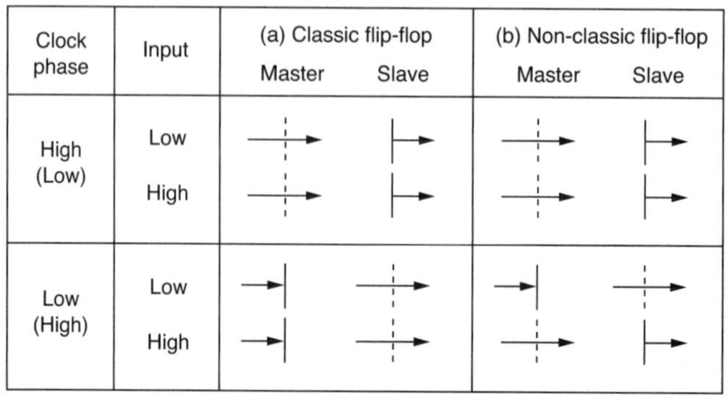

FIGURE 8.5 Illustration of classic and nonclassic flip-flop concepts.

a p-type half-latch with only an SP-stage and transparent for a high input. When the clock is high (latching phase), it gives a low output. The slave, an n-type precharged latch, however, is nontransparent for the low input after the PN stage finishes its evaluation. The condition of finishing evaluation increases the required hold time but just slightly because the evaluation of PN stage takes very short time. The speed increase, however, is significant because the p-type master latch with two SP stages is slower than the n-type precharged slave latch, and the removal of one of the two SP stages balances the delays of both latches. In the same time, the power consumption is reduced due to the removal of three transistors, especially the clocked p-transistor. A similar circuit but in a different transistor stacking order, illustrated in Figure 8.4(b), was published in 1973 [4] and its advantage in speed optimization was addressed in Huang and Rogenmoser [5].

Because all clocked devices are connected to power and ground rails, they can be sized without excessive loading to improve speed, though care must be taken for the charge sharing between nodes A and B and between nodes C and D. In another example, a nonclassic flip-flop can be built from a split-output SP stage and an n-type split-output latch (or an SN stage and a p-type split-output latch), reducing the number of total transistors to eight and the number of clocked devices to two, which is shown in Figure 8.4(c). For a high clock, the SP stage is half-transparent, but the split outputs respectively to the p- and n-transistors in the following latch stage can never make them transparent, although the latch is in its nonlatching phase. Another method using flow tables and signal transition graphs (STG) has been presented in Piguet [6] and Piguet and Zahnd [7] to design similar circuits including dynamic flip-flops, aiming at race-free (or, today, speed-independent [SI]) circuits.

To reduce the hold time of the flip-flop in Figure 8.4(a), a 10-transistor TSPC flip-flop illustrated in Figure 8.6(a) can be used [8]. Only the hold time for a low input needs to be reduced. The added nMOS transistor controlled by the precharged node signal will firstly increase the delay for a high input to a low output and secondly make the single stage master completely nontransparent (i.e., a full latch) without any additional clock or clocked device. A similar counterpart but with the single stage full latch

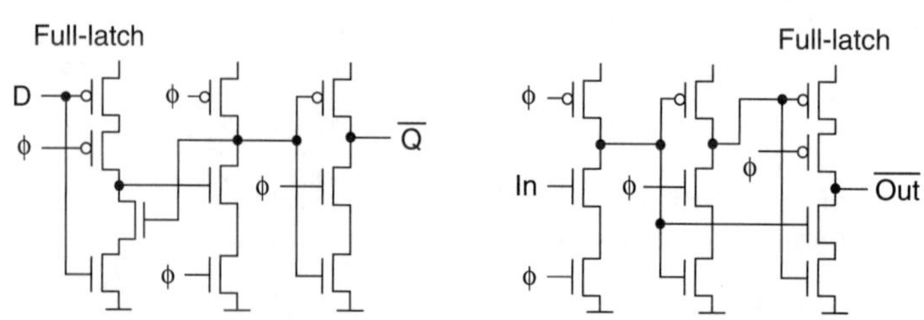

(A) A flip-flop with reduced hold-time. (B) A full-latch in a pipeline.

FIGURE 8.6 Single-stage full latches using the precharged node signal. (© 2004 IEEE.)

at the output is shown in Figure 8.6(b). The two circuits can be used for a TSPC double pipeline [9,10] to improve its robustness, which will be discussed later.

8.2.2 Differential Single-Clock Latches and Flip-Flops

It is necessary to avoid precharge for low-power applications, especially for a circuit with a low activity ratio. The aforementioned nonprecharged latches and flip-flops are therefore preferred in this case. To obtain differential outputs, however, an inverter has to be added for all single-ended latches and flip-flops, which consumes additional power. Differential latches and flip-flops can produce complementary outputs without an additional inverter. One example is the CVSL dynamic latches [11] illustrated in Figure 8.7. The problem with a cross-coupled differential latch is that it is sensitive to the ratio between p- and n-transistors, especially for the p-type latch (see the chart in Figure 8.7). This problem can be avoided by using dynamic ratio insensitive (DRIS) differential latches in which there is no fighting between p- and n-transistors [8]. The p-type latch of this kind is shown in Figure 8.8 along with the comparison between CVSL and DRIS. All latches and flip-flops introduced so far are dynamic (i.e., they have to be refreshed above a minimum clock rate). Static latches and flip-flops can accept a zero clock rate (clock off for low power) without losing data. The fully static counterpart to CVSL and DRIS latches are the random-access memory (RAM)-type [11] and SRIS latches [8], respectively, depicted in Figure 8.9. SRIS p-latch is faster than its RAM-type counterpart due to the absence of fighting.

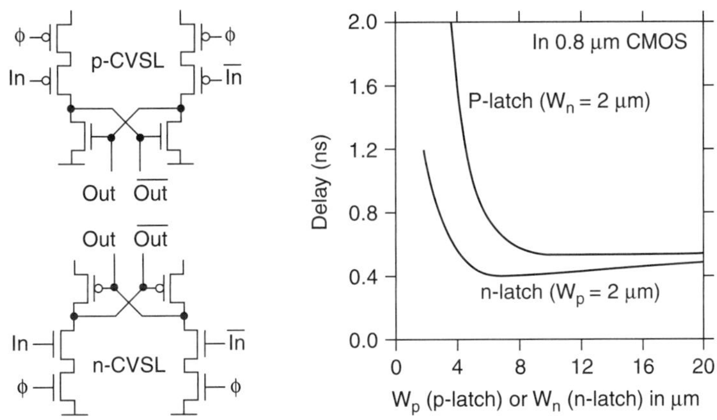

FIGURE 8.7 CVSL latches and their ratio sensitivities. (© 2004 IEEE.)

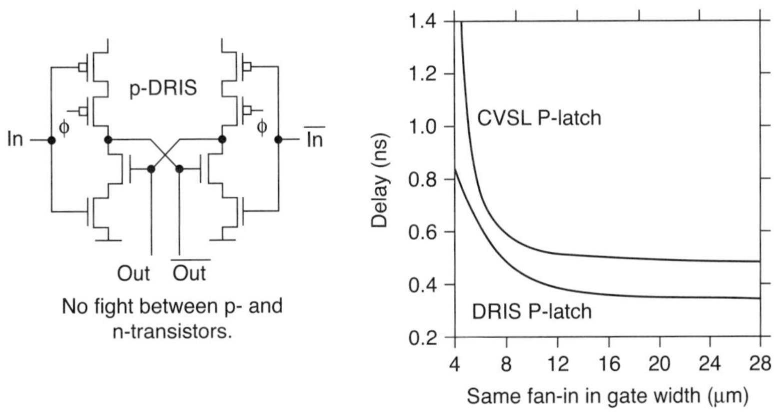

FIGURE 8.8 DRIS p-latch. (© 2004 IEEE.)

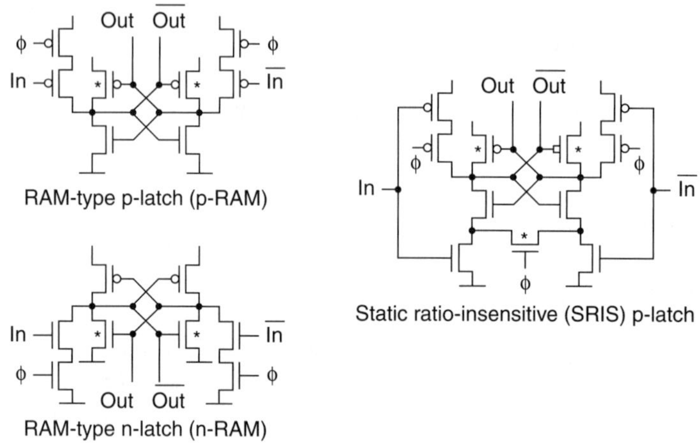

FIGURE 8.9 Static RAM-type latches and static ratio-insensitive p-latch. (© 2004 IEEE.)

It is also possible to use only a single clocked transistor for a differential latch to reduce power consumption, as the p-type and n-type dynamic single-transistor clocked latches (version 1) depicted in Figure 8.10, named p-DSTC1 and n-DSTC1 latches [8]. They have the advantages of minimized clock load, minimized input load, and available differential outputs. Caution is needed for using DSTC lathes. There is a risk of charge sharing between the two output nodes when the two inputs have overlapped periods or glitches during which both input transistors are conductive. The same as the p-CVSL latch, p-DSTC1 latch is severely transistor ratio sensitive and slow, which will be handled later by its second version. To handle the charge sharing, static single-transistor clocked (SSTC) latches can be used [8]. Its n-type (n-SSTC) is depicted in Figure 8.10, which was developed from the RAM-type latches but with only a single clocked device. The two added n-transistors compared to n-DSTC1 need only the minimum size for reducing the load. The full output differential swing, if degraded by the charge sharing through two input transistors, could be effectively recovered by the cross-coupled inverter pair, which makes the SSTC latch highly robust.

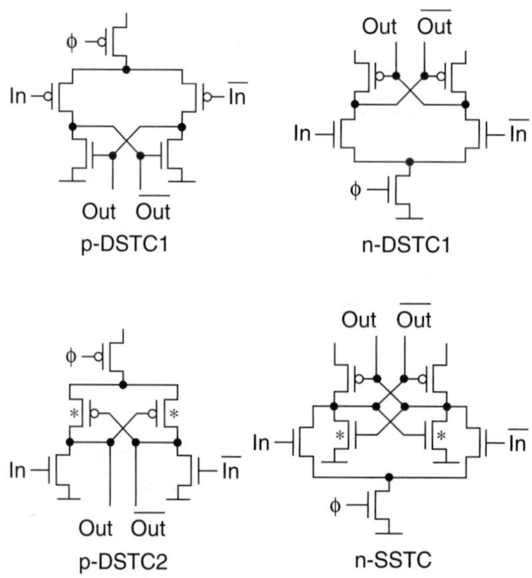

FIGURE 8.10 Single transistor clocked differential latches. (© 2004 IEEE.)

To break the speed bottleneck, the p-DSTC1 latch can be replaced by the p-DSTC2 latch [8], shown in Figure 8.10. The p-DSTC2 latch looks similar to the n-DSTC1 latch but with the clocked transistor being p-type and at the top of the latch. The p-DSTC2 latch is not a full latch but half-transparent when the clock is high, and a low-to-high input transition will result in a high-to-low output transition. If the p-DSTC2 latch (as the master) is followed by an n-type differential latch (as the slave) (e.g., an n-DCST1 or n-SSTC latch) it becomes a positive edge-triggered nonclassic flip-flop. The high-to-low output transition from the master will not propagate through the slave because the slave's input n-transistor will not respond to a high-to-low transition as long as the data propagation in the slave finishes. The advantages are obvious. First, the input transistors of both latches are n-type, resulting in low fan-in and high speed due to all logic in nMOS. Second, the p-DSTC2 latch is less sensitive to transistor ratio. Third, the two flip-flops are highly robust because the p-DSTC2 latch produces nonoverlap and glitch-free signals to the n-DSCT1 or n-SSTC latch. Finally, the flip-flop with the n-SSTC latch is semi-static, allowing the clock to standby at low state.

8.2.3 Power-Delay Comparison

The worst delay and power consumption of the aforementioned flip-flops are compared with classic and conventional solutions. For this purpose, the dynamic flip-flop with p-classic and n-classic latches, the dynamic flip-flop with p-C²MOS and n-C²MOS latches, and the static flip-flop with classic latches are presented in Figure 8.11(a), Figure 8.11(b), and Figure 8.11(c), respectively. The comparison is done by using the parameters of a 0.8 μm CMOS process through SPICE simulations to extract the worst delay and power consumption values under different activity ratios. It is not easy to fairly compare different circuits. To make the comparison as fair as possible, the following conditions are applied. The minimum length, 0.8 μm, is used for all transistors. For the widths, Wp = 6 μm, Wn = 3 μm, and Wmin = 2 μm (the transistor marked with *). For ratio sensitive n-type latches, such as n-CVSL, n-RAM and n-DSTC1 and n-SSTC latches, Wp = 4 μm, Wn = 6 μm, and Wmin = 2 μm. For ratio sensitive p-type latches, such as p-CVSL and p-RAM latches, Wp = 12 μm, Wn = 3 μm, and Wmin = 2 μm. The load to each flip-flop is the input capacitance of two inverters with Wp = 6 μm and Wn = 3 μm. In the simulation, both the power for driving the inputs of each flip-flop and the power for driving the output load are included in the power consumption. At an activity ratio of 0.25, the values of power consumption and worst delay are listed in Table 8.1. The power-delay products for all flip-flops with activity ratios from 0.0 to 0.5 are plotted in Figure 8.12. The minimum delay is given by the flip-flop with p-DSTC2 and n-DSTC1. At an

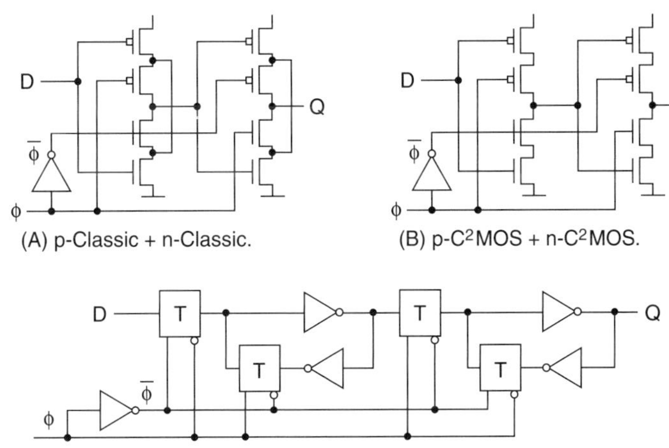

FIGURE 8.11 Three flip-flops for comparison.

TABLE 8.1 Comparison of Flip-Flops

Type	No.	Master + Slave	Power (μW)	Delay (ns)
Dynamic	1	p-CVSL + n-CVSL	699.4	0.691
	2	p-C²MOS + n-C²MOS	491.8	0.950
	3	p-Classic + n-Classic	512.4	0.776
	4	(SP + SP) + (PN + SN)	404.3	0.835
	5	SP + (PN + SN)	331.6	0.832
	6	(SP + SP) + (SN + SN)	317.6	0.802
	7	p-DSTC2 + n-DSTC1	313.1	0.717
Static	8	Classic + Classic	668.8	1.008
	9	p-RAM + n-RAM	685.4	0.673
	10	p-DSTC2 + n-SSTC	393.5	0.705

Note: Activity ratio = 0.25 and load = two inverters, in 0.8 μm CMOS.

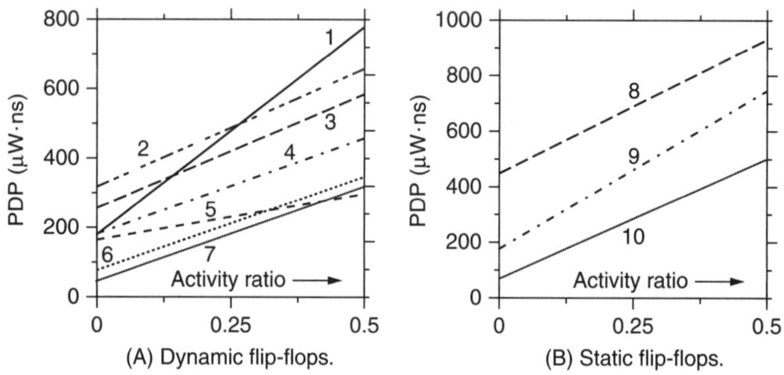

FIGURE 8.12 Comparison of power-delay products.

activity ratio of 0.25, the minimum power consumption is given by the flip-flop with p-RAM and n-RAM. As far as power-delay products are concerned, the best dynamic flip-flop is made of p-DSTC2 and n-DSTC1, and the best static (including semi-static) flip-flop is made of p-DSTC2 and n-SSTC.

8.3 High-Throughput CMOS Circuit Techniques

8.3.1 TSPC Pipeline

TSPC flip-flops can be used as edge-triggered elements in a synchronous pipeline. Its short setup-time, hold-time, and propagation delay contribute to high speed. Complementary logic stages can be placed between two TSPC latches in the pipeline. More efficiently, the logic gates can be embedded within TSPC latches [8], as depicted in Figure 8.13(a) and Figure 8.13(b). The previously mentioned pipeline can be divided into p-blocks and n-blocks, as depicted in Figure 8.13(c). A p-block consists of a p-type latch, which may embed logic, associated with the complementary logic stages before and after the p-latch, and it is the same for an n-block but with an n-type latch instead. The blocks must be connected with p-type and n-type latches alternately. Feedback is allowed but must also follow the rule from p-type to n-type or vice versa. In such a pipeline, p-blocks are the speed bottlenecks, especially when logic gates are included in the p-blocks or embedded in the p-latch with many stacked p-transistors. Therefore, in order to achieve a high throughput, logic gates are preferably placed in the n-blocks, leaving the p-block as a passing stage or with very simple logic. The nonclassic concept may be used to simplify the p-block to just a single SP stage if directly (or indirectly after an even number of complementary stages) followed by an n-type precharged latch. An all n-logic true-single-phase dynamic CMOS circuit technique was proposed in Gu and Elmasry [12] to speed up the p-block, in which the logic embedded in a p-type precharged latch uses n-transistors instead of p-transistors.

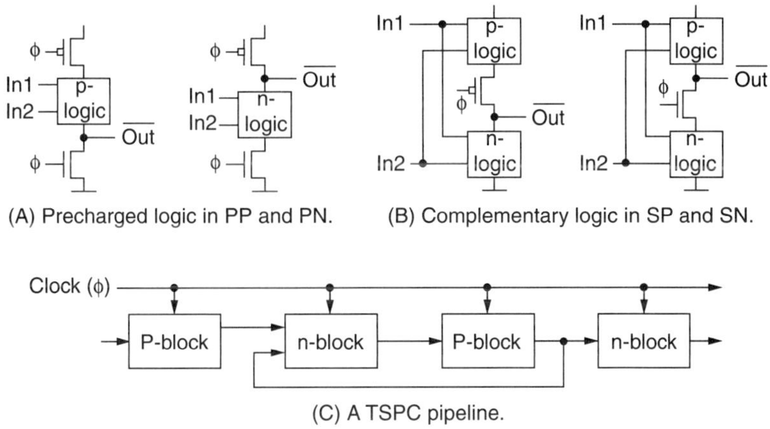

FIGURE 8.13 Logic embedded in latch stages in a TSPC pipeline.

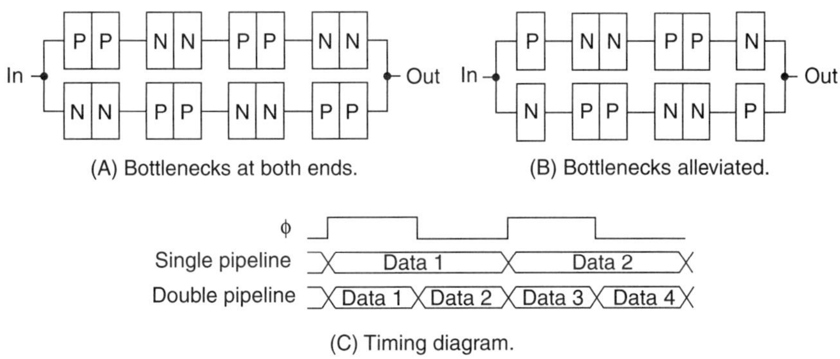

FIGURE 8.14 TSPC double pipelines.

8.3.2 TSPC Double Pipeline

Synchronous elements in a pipeline are usually triggered by a single clock edge. In a double pipeline, however, both edges of a clock are utilized for achieving high throughput and efficiency [9]. The data rate at the input and output of a double pipeline is at twice the clock rate. Internally, each pipeline works as normal, and data can be cross-connected between the two lines as long as following the n-to-p or p-to-n rule. As illustrated in Figure 8.14(a), two TSPC pipelines starting and finishing with opposite types of blocks can be such a double pipeline, and the two input-connected input blocks and the two output-connected output blocks become a multiplexer and a demultiplexer respectively. Because the input and output blocks have to work at a double data rate, they are the speed bottlenecks. It is therefore preferred to have the double pipeline configured as shown in Figure 8.14(b) (i.e., single-stage latches at both ends). This can be done by using the single-stage full latches [10], depicted in Figure 8.6(a) and Figure 8.6(b), which narrows the forbidden windows of low-to-high and high-to-low data transition by almost half, increasing speed and robustness. To reduce power consumption at a given data rate, a low-swing clock double-edge triggered flip-flop was proposed in Kim and Kang [13], in which both edges of a low swing clock are used to trigger a single flip-flop to reduce overall clock rate and associated power consumption.

8.3.3 Clock-and-Data Precharged Circuit Technique

All the preceding circuits are aiming at a high throughput regardless of the latency or the number of operating clock cycles for a final output. In many applications, however, the decision has to be made in one clock cycle. A technique named clock-and-data precharged dynamic (CDPD) circuit technique may

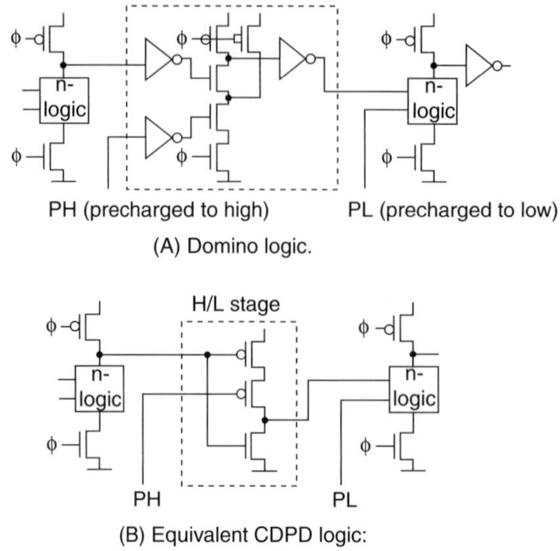

FIGURE 8.15 Domino logic and its equivalent CDPD logic.

offer an alternative for a fast one-clock-cycle decision and in the same time reduce the power consumption [14]. Domino logic is often used for logic calculations with a large depth as the logic parts can be distributed along the domino chain and are all in nMOS. As illustrated in Figure 8.15(a), however, an inverter has to be placed between two precharged stages to prevent an erroneous high-output to the next stage at the beginning of evaluation. Moreover, charge sharing may occur between the output node and the intermediate nodes so extra precharging transistors have to be used. As illustrated in Figure 8.15(b), all contents in the dashed line box can be replaced by only three transistors in CDPD technique, and no clocked transistor is contained in it. This CDPD block is named an H/L (high-to-low) stage in which the output is precharged to low by a high data input, and the NOR function is simply fulfilled by the two p-transistors. An H/L stage can be followed by an L/H (low-to-high) stage in which the output is precharged to high by a low data input. An n-type CDPD chain can be formed by the original domino precharged stages along with the H/L and L/H stages in between, as illustrated in Figure 8.16(a). It needs

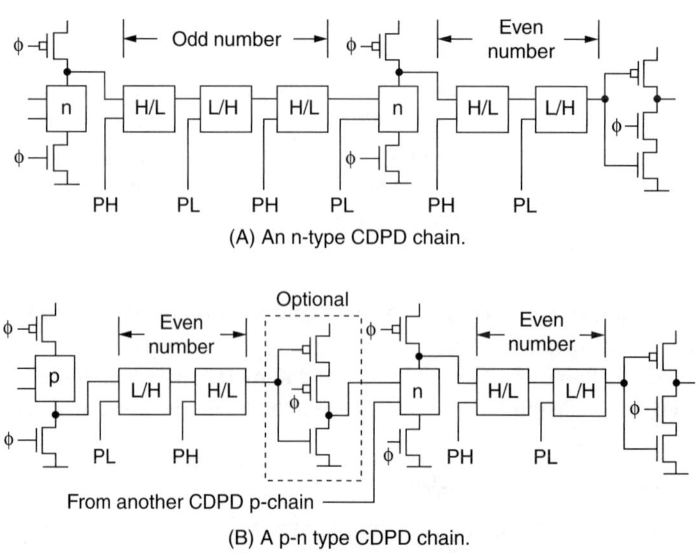

FIGURE 8.16 Two types of CDPD chains each ended with an SN latching stage.

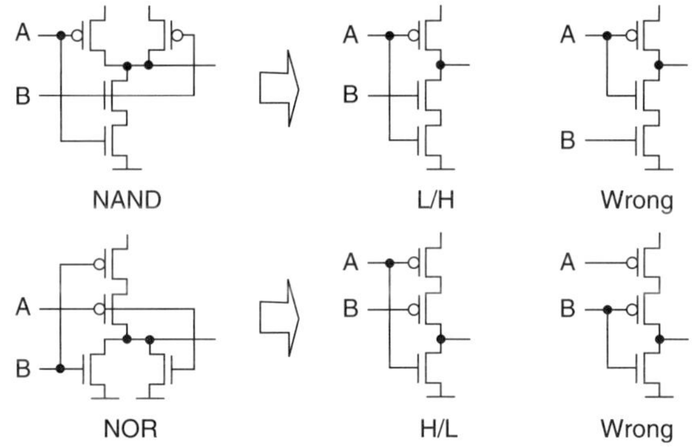

FIGURE 8.17 NAND and NOR gates transferred into L/H or H/L stages.

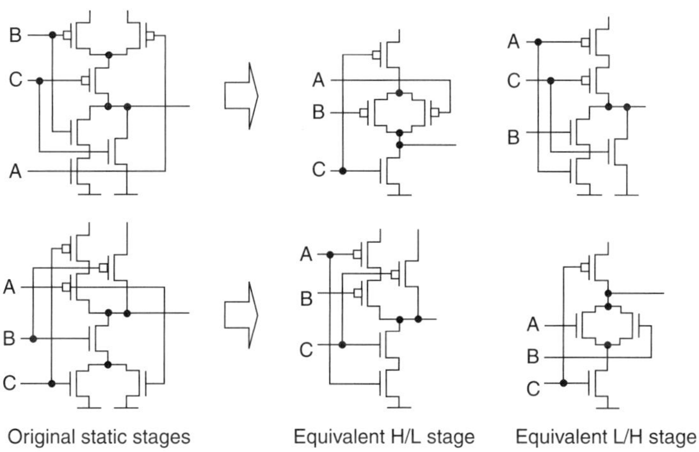

FIGURE 8.18 Logic gates transferred into either L/H or H/L stages.

an odd number of CDPD stages between two domino precharged stages, and an even number of CDPD stages between a domino stage and an output latch.

A number of advantages can be cited. First and second, all domino inverters are removed, and the number of clocked devices is minimized, reducing unnecessary power consumption. Third, the skewed precharging of CDPD stages effectively reduces the peak current. A p-n type CDPD chain is presented in Figure 8.16(b), and the rules can be found in the figure.

The p-n type CDPD chain has additional advantages. First, the logic operations are completed in both high and low clock periods so each duty cycle of the clock is fully utilized. Second, not only the number of clocked devices but also the number of latch stages is reduced. As indicated in Figure 8.16(b), the latch before the n-type precharged stage is optional, only depending on the need of inversion. Cares and skills are needed for designing CDPD stages to avoid erroneous results, as illustrated in Figure 8.17. A "NAND" function can be simplified in an L/H stage but is directly used for an H/L stage, while a "NOR" function can be simplified in an H/L stage but is directly used for an L/H stage. The wrong connections, which will result in charge sharing, should be avoided. Generally, complementary gates are simplified differently in an H/L or an L/H stage, (see Figure 8.18). In the worst-case scenario, complementary gates can be directly used for either an L/H or an H/L stage.

8.3.4 United Connection Rules of TSPC and CDPD Stages

It is important to follow the connection rules for constructing TSPC and CDPD circuits, and computer-aided design (CAD) tools should be able to check the correctness of the circuit connection according to

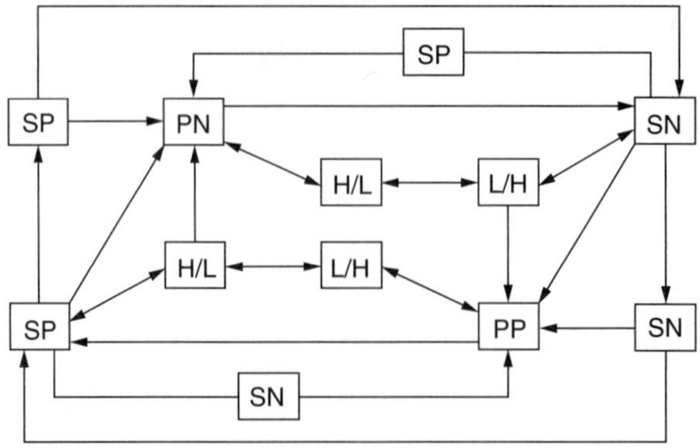

FIGURE 8.19 Unified connection rules of TSPC and CDPD stages.

the rules. If connections are correct, the circuit will undoubtedly work but the target function and speed have to be checked by simulation. The unified connection rules of TSPC and CDPD stages are illustrated in Figure 8.19. For example, SP→SP→SN→SN represents a TSPC nonprecharged flip-flop, and SP→SP→PN→SN becomes a TSPC precharged flip-flop. Nonclassic flip-flops are represented by the connections of SP→PN→SN (positive edge-triggered) and SN→PP→SP (negative edge-triggered). The connection rules between CDPD and TSPC stages are also clearly included.

8.4 Fast and Efficient CMOS Functional Circuits

The CMOS functional circuits introduced in this section are featured with high efficiency and high speed. High efficiency leads to a small number of both clocked and logic-operating devices, resulting in low-power consumption, and high speed offers a large delay margin that can be used for trading power at a lower supply voltage.

8.4.1 Dividers and Ripple Counters

A very fast divider can be constructed simply by connecting the output and the input of the nonclassic nine-transistor TSPC flip-flop depicted in Figure 8.4(a). Its dynamic version is shown in Figure 8.20 while its semi-static version is presented in Figure 8.21, respectively. The transistor widths in 2-μm CMOS, given in Figure 8.20, are optimized for speed. An 8-bit ripple counter built from the dynamic divide-by-two stage reached an input frequency of 750 MHz [15]. The semi-static divider in Figure 8.21 can be used in a very long ripple counter where the frequencies get very low in later stages. Note that the narrow pulse signal (out 1) should be fed to the next stage so that the condition for a dynamic circuit will be always satisfied for all stages in the ripple chain, while the 50% duty-cycle signal (out 2) used for bit-output.

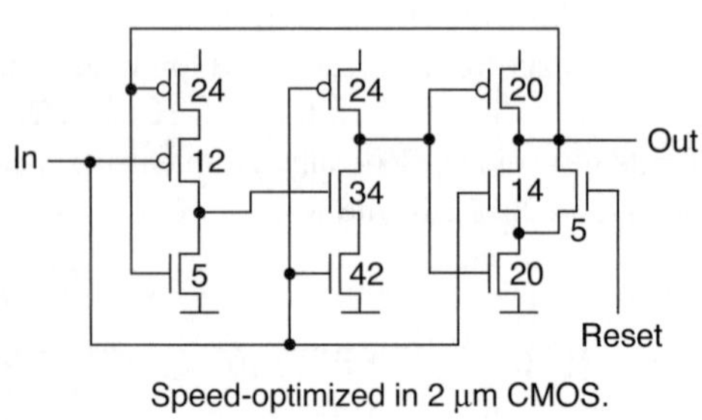

FIGURE 8.20 A dynamic divide-by-two circuit (D-1/2).

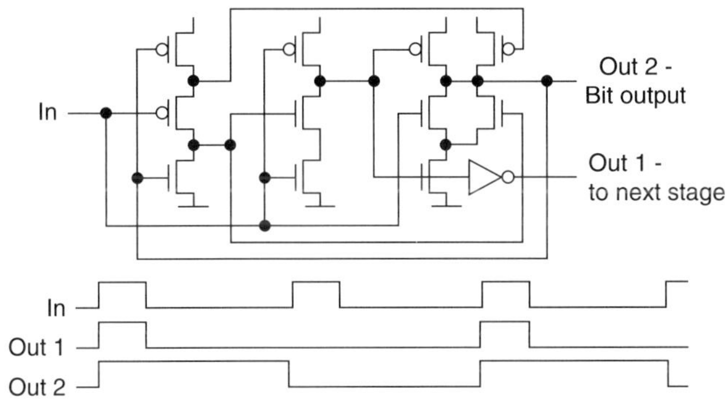

FIGURE 8.21 A semi-static divide-by-two circuit.

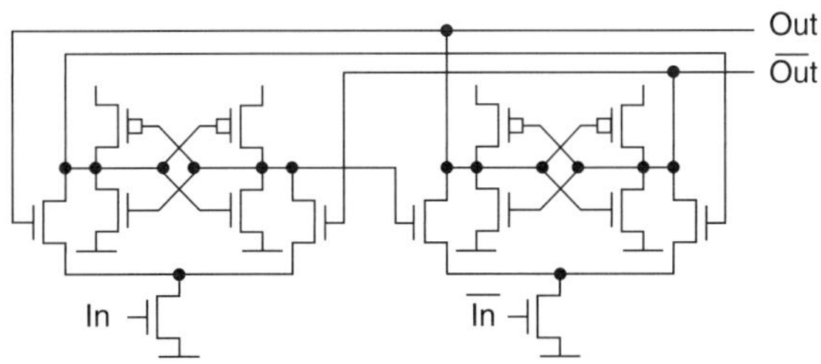

(A) A static differential divide-by-two circuit (S-1/2).

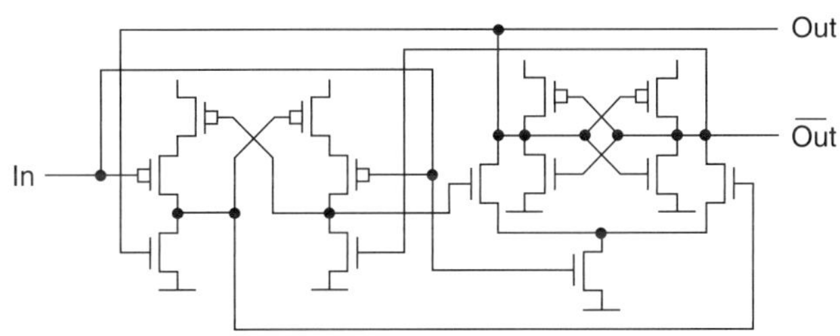

(B) A single-in double-out semi-static divide-by-two circuit (HS-1/2).

FIGURE 8.22 Differential divide-by-two circuits.

A differential divider offers differential outputs, which is sometimes a quite useful feature. In favor of speed, two n-SSTC latches can be cascaded to form a divider instead of using a p-SSTC and an n-SSTC, by using differential output signals available from the previous stage, named an S-1/2 stage and presented in Figure 8.22(a). For a single ended input, a semi-static differential divider may be used, named an HS-1/2 stage and presented in Figure 8.22(b). Two divider chains, one constructed by four D-1/2 stages with a buffer between stage 1 and stage 2 and another one constructed by an HS-1/2 stage followed by three S-1/2 stages, were constructed in IBM's partially scaled 0.1-μm CMOS process [16]. The measured input frequencies achieved 16.6 GHz for the dynamic divider chain and 12.5 GHz for the static divider chain [17].

8.4.2 Synchronous Counter

A TSPC synchronous counter is depicted in Figure 8.23 as an example showing how the carry-logic can be arranged in a p-block in favor of speed while using the dynamic divider as the toggle stage with the carry control function embedded [15]. The transistor widths optimized for speed are valid in a 3-μm technology. An 8-bit synchronous counter of this kind in the 3-μm technology was measured to reach a clock rate of 200 MHz. For a very long counter, however, the carry propagation becomes a serious speed bottleneck even with the parallel carry-logic depicted in Figure 8.23. A so-called backward carry propagation topology, in contrast to the conventional forward carry propagation, can be used to break the limit [18]. The principle block diagram of a backward carry propagation synchronous counter is presented in Figure 8.24. In a conventional counter, the worst-case scenario happens at output 0111 ... 111, and the 0→1 flip of LSB has to be propagated through the whole chain of "AND" gates to MSB to enable the next output 1000 ... 000. In Figure 8.24, however, when the out is 0111 ... 110, the carry propagation is almost finished, and when the 0→1 flip of LSB comes all bits are ready for next output simultaneously. A more practical architecture is presented in Figure 8.25, mixing backward with forward carry propagation, at a lowest area and power penalty. The interface between the two propagation strategies depends on the counter length. Generally, a few bits using the backward carry propagation are enough.

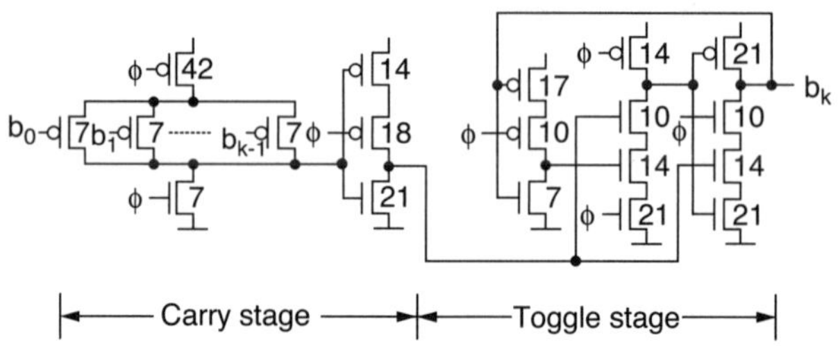

FIGURE 8.23 A bit-slice of a synchronous counter with parallel carry-logic in TSPC.

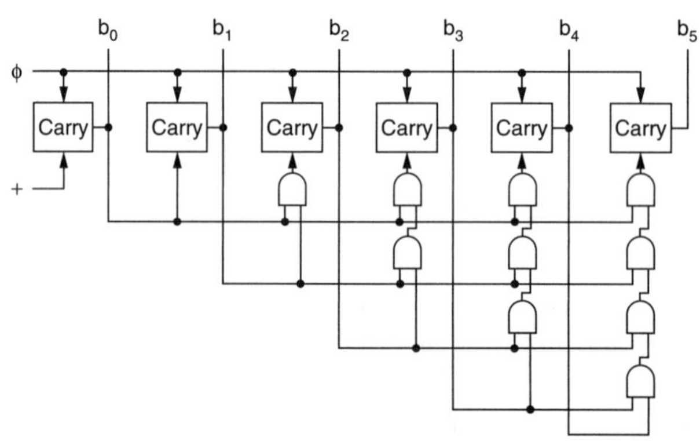

FIGURE 8.24 A synchronous counter with fully backward carry propagation.

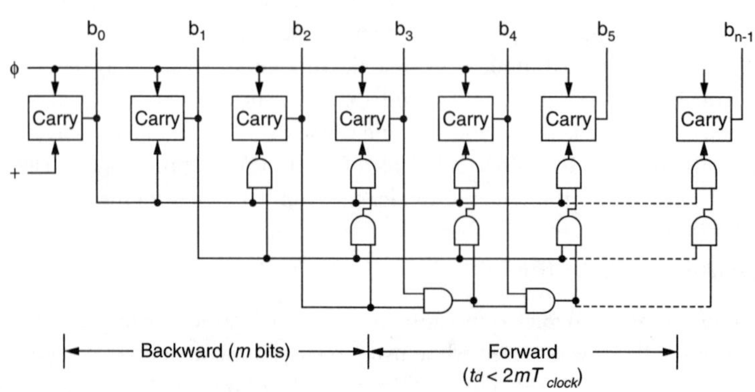

FIGURE 8.25 A synchronous counter mixing backward and forward carry propagations.

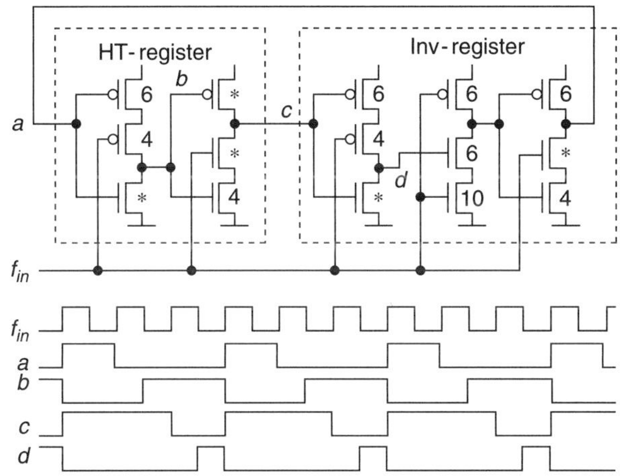

FIGURE 8.26 A dynamic nonbinary divider (1/3).

8.4.3 Nonbinary Divider and Prescaler

A nonbinary divider is usually constructed by a synchronous counter plus a decoding logic (i.e., when the output code reaches the target dividing ratio, the counter is reset). Such a topology can offer any dividing ratio; however, a synchronous counter is slow and the decoding logic adds additional delay. It was found that an SP stage followed by an SN stage in TSPC becomes a half-transparent register (HT-register) (i.e., registering a low-input [imposing a clock cycle delay] but no-register function for a high-input [transparent]). This feature can be utilized for constructing a nonbinary divider [19]. A divide-by-three circuit is depicted in Figure 8.26 along with the waveforms at different nodes. At node *b*, a symmetric waveform is obtained at a frequency of $f_{in}/3$, assuming the input clock is symmetric. If (n-2) HT-registers are used, it becomes a divide-by-n circuit. Because no decoding and no carry propagation exist, this circuit can work at the same speed as a 1/2 divider. The long propagation delay due to many cascaded transparent stages can be solved by a few speed-up transistors, see Yuan and Svensson [19]. Note that in a nonbinary divider the output is still edge-triggered i.e., there is no skew between input and output as could be a problem for a ripple counter. Although a single nonbinary divider can offer any dividing ratio, it is more efficient to cascade two or more than two nonbinary dividers to achieve a high dividing ratio. For example, a 1/3 divider cascaded by a 1/7 divider becomes a 1/21 divider, and so on. The nonbinary divider is extremely useful for prescalers as the needed operating speed is often very high and the dynamic feature is usually not a problem. A dual-modulus prescaler, divided by either n or (n-1), is presented in Figure 8.27 for the purpose of frequency synthesis, where the "Inv-register"

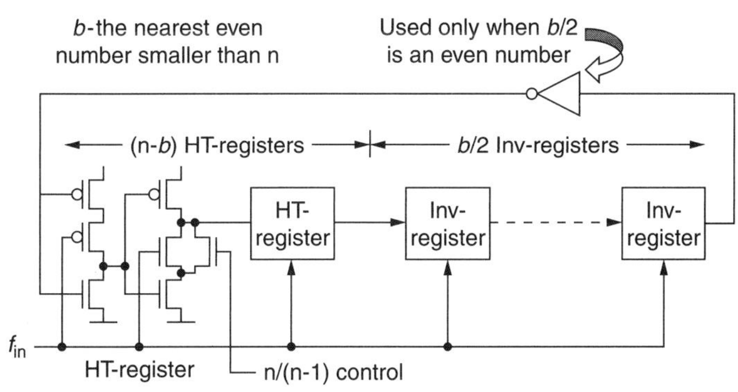

FIGURE 8.27 A divide-by-n/(n-1) prescaler.

represents the 9-transistor TSPC flip-flop. The control of divide-by-n and divide-by-(n-1) is extremely simple, only a single n-transistor in one of the (n-b) HT-registers, making this circuit highly attractive. When the input of the transistor is high, this HT-register becomes fully transparent. Other techniques in dual-modulus prescalers based on the modification of the nine-transistor TSPC flip-flop can be found in Chang et al. [20] and Yang et al. [21].

8.4.4 Adder and Accumulator

The core part of an adder is the "XOR" logic. A highly efficient pipelined XOR gate in TSPC is shown in Figure 8.28. The basic topology is to implement the XOR function in two steps respectively in a p-block and an n-block and to embed logic into latches [22]. The logic diagram is given at the upper part while the circuit diagram is given at the lower part of the figure. The NAND, OR, and AND functions are respectively embedded in an SN stage, an n-type precharged latch and a p-type precharged latch. The connection exactly follows the rule mentioned previously. The efficiency comes from two facts. First, the nonclassic principle is applied to the pair of SN and precharged p-type latch, so a single SN stage is used to embed the NAND function in favor of speed. Second, both the OR function in the n-type precharged latch and the AND function in the p-type precharged latch use parallel transistors which is also in favor of speed. The pipelined XOR gate can be directly cascaded to deliver the sum output for a full adder, and the sum can be fed back to its own input for accumulation. An accumulator can be therefore configured efficiently by using the pipelined XOR gate, and one of the bit-slices is given in Figure 8.29. A 24-bit pipelined accumulator in 1.2-μm CMOS for a numerically controlled oscillator based on the topology achieved a clock rate of 700 MHz [23].

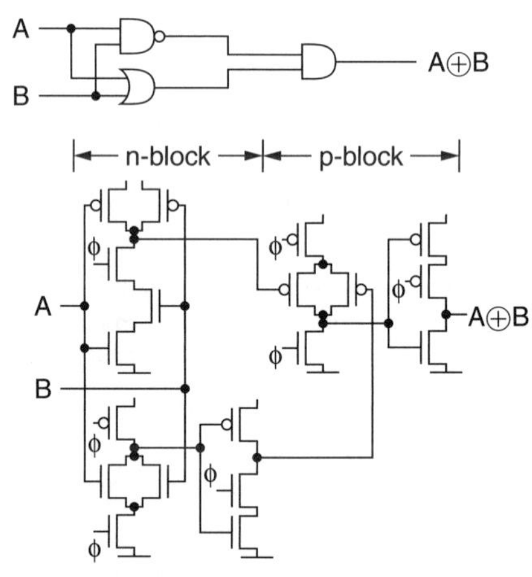

FIGURE 8.28 A pipelined XOR gate in TSPC.

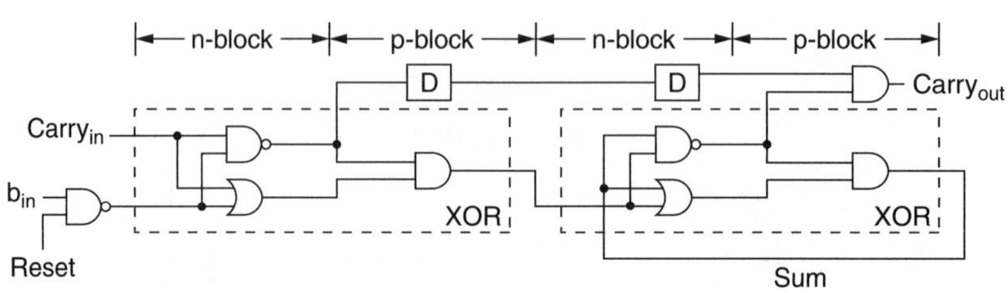

FIGURE 8.29 A bit-slice of an accumulator using the pipelined XOR in TSPC.

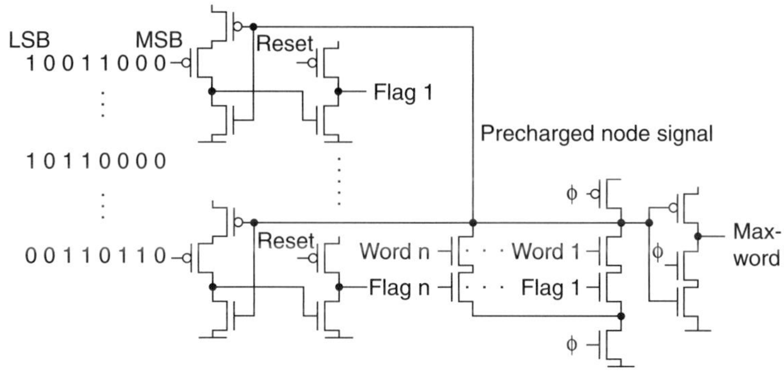

FIGURE 8.30 A bit-serial and word-parallel max/min selector.

8.4.5 Bit-Serial Comparator and Sorter

Dynamic logic may greatly simplify a complex circuit function and minimize the number of devices, resulting in low power without sacrificing speed or even with improved speed. One example is the bit-serial word-parallel maximum/minimum selector [24] in Figure 8.30. The main part of the selector is an n-type TSPC precharged latch embedding the selecting logic with AND functions (two n-transistors in stack) in parallel. The purpose to show this circuit is to emphasize that the precharged node signal can be used to effectively simplify the configuration. This signal is used as a second "clock" for a number of parallel PP stages. Each PP stage receives an input word. The precharged node signal of each PP stage is again used for the nMOS input of the flag logic stage, which consists of only an n- and a p-transistor. The p-transistor sets the flag high before the new input words start, and therefore all flags are high from the beginning. All word inputs start with MSB and are compared digits by digits. If all digits are the same, no matter zero or one, all flags are kept high. If partial digits become zero, the input with a zero digit will make the output of the PP stage high and the flag low to disable this input. In the end, only the maximum input is left. During comparison and selection, the output never stops. A minimum selector can be easily completed by inverting the input words and, of course, inverting the output again.

A bit-serial and word-parallel compare-and-swap cell [24] is presented in Figure 8.31. The maximum selector is used along with a minimum selector, which uses an inverted flag and the logic opposite to the maximum selector. When the two digits are equal, both go to the outputs, and when the two digits are different, the upper output will be "one" and the lower output will be "zero." In the same time, the smaller input will be disabled in the maximum selector, while the larger input will be disabled in the minimum

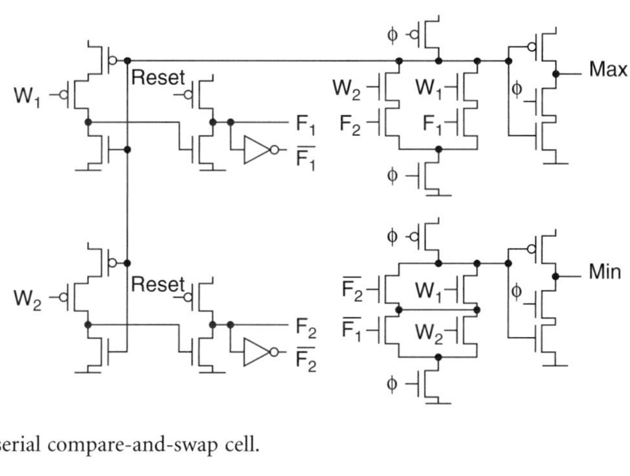

FIGURE 8.31 A bit-serial compare-and-swap cell.

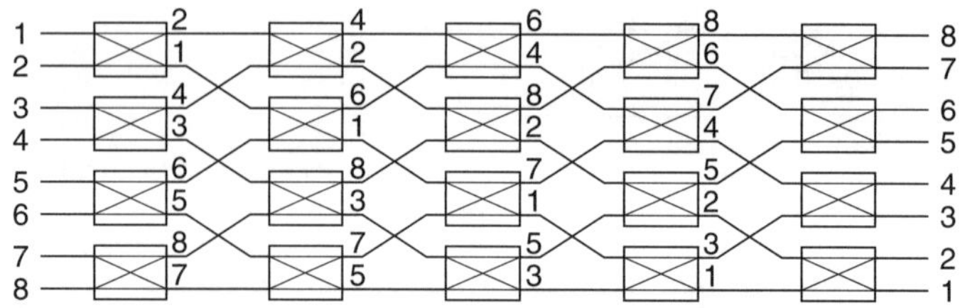

FIGURE 8.32 A bit-serial pipelined sorter using the compare-and-swap cell.

selector. It must be noted that the complete cell is an n-block in TSPC, and to cascade two such cells, a p-type latch must be used in between. An 8-input bit-serial and word-parallel sorter is depicted in Figure 8.32, where each box represents the compare-and-swap cell plus a p-type latch. This pipelined sorter can achieve a very high data throughput.

8.5 The Future of Dynamic Logic

It is well-known that dynamic CMOS logic circuits suffer from the leakage current, which puts a limit on the minimum possible clock frequency. This became a serious problem in deep submicron technologies. When V_{DD} is scaled down, V_{TH}, the threshold voltage, has to be scaled down to not lose the advantage of scaling, which is associated with the exponential increase in subthreshold leakage current [25]. It not only limits the use of dynamic logic but also dominates the overall power consumption. Advanced leakage control methods will become indispensable for future technologies [26].

Fortunately, some techniques have already emerged. One technique uses the self-reverse-biasing effect of stacked transistors, called stacking-effect, to reduce the standby leakage current [27]. This effect was successfully used in a low-leakage, gated-ground cache in which two off transistors connected in series reducing the leakage current by orders of magnitude [26]. It means that the topology and design methods have to change to meet the new challenge. Of course, to use static logic may avoid the reliability problem due to the leakage current; however, the reason to use dynamic logic has been its efficiency and high speed over its static counterpart under the same or a lower power budget. Therefore, dynamic logic may still find its position in deep submicron technologies if new techniques reducing the subthreshold leakage current are discovered by active researches in this field.

8.6 Conclusion

The comparisons in Table 8.1 and Figure 8.12 demonstrate that carefully designed dynamic CMOS flip-flops and latches — nonprecharged or precharged, single-ended or differential — present obvious short delays and small power-delay products over traditional dynamic and static flip-flops at different activity ratios. The strategies of single clock, low count of clocked devices, fewer transistors, nonclassic concept, and simple configurations lead to the results. The latches and flip-flops can be used in a pipeline or a double pipeline, resulting in a very high data throughput. When logic is embedded in an n-block, the number of devices and the overall capacitance is significantly less than what is necessary for a complementary logic, gaining high speed and low-power advantages. Advanced circuit topologies presented in the examples of functional circuits make the dynamic logic circuits highly attractive; however, precautions have to be taken. The robustness of dynamic (especially precharged) logic against power and ground noises is not as good as complementary logic. The node leakage current is another problem that has to be handled in deep submicron technology, and new leakage control techniques have to be found to make the dynamic logic circuits continuously attractive.

References

[1] Y. Ji-ren, I. Karlsson, and C. Svensson, A true single-phase-clock dynamic CMOS circuit technique, *IEEE J. Solid-State Circuits,* vol. SC-22, pp. 899–901, October 1987.

[2] J. Yuan and C. Svensson, High-speed CMOS circuit technique, *IEEE J. Solid-State Circuits,* vol. 24, pp. 62–70, February 1989.

[3] C. Svensson and D. Liu, Low-power circuit techniques, *Low-Power Design Methodol.,* J.M. Rabaey and M. Pedram, Eds., Kluwer Academic Publishers, Dordrecht, 1996, chap. 3.

[4] H. Oguey and E. Vittoz, CODYMOS frequency dividers achieve low-power consumption and high frequency, *Electron. Lett.,* pp. 386–387, August 23, 1973.

[5] Q. Huang and R. Rogenmoser, Speed optimization of edge-triggered CMOS circuits for Gigahertz single-phase clocks, *IEEE J. Solid-State Circuits,* vol. 31, pp. 456–465, March 1996.

[6] C. Piguet, Logic synthesis of race-free asynchronous CMOS circuits, *IEEE J. Solid-State Circuits,* vol. 26, pp. 271–380, March 1991.

[7] C. Piguet and J. Zahnd, Electrical design of dynamic and static speed-independent CMOS circuits from signal transition graphs, *ACiD Workshop,* Newcastle, U.K., January 18–19, 1999.

[8] J. Yuan and C. Svensson, New single clock CMOS latches and flip-flops with improved speed and power savings, *IEEE J. Solid-State Circuits,* vol. 32, pp. 62–69, January 1997.

[9] M. Afghahi and J. Yuan, Double edge-trigged flip-flop for high-speed CMOS, *IEEE J. Solid-State Circuits,* vol. 26, pp. 1168–1170, 1991.

[10] J. Yuan and C. Svensson, Fast and robust CMOS double pipeline using new TSPC multiplexer and demultiplexer, *Proc. 2nd Int. Conf. on ASIC,* pp. 271–274, October 1996.

[11] N. Weste and K. Eshraghian, *Principles of CMOS VLSI Design, 2nd ed.,* Addison-Wesley, Reading, MA, 1993, chap. 5.

[12] R.X. Gu and M.I. Elmasry, All-N-logic high-speed true-single-phase dynamic CMOS logic, *IEEE J. Solid-State Circuits,* vol. 31, pp. 221–229, February 1997.

[13] C. Kim and S-M. Kang, A low-swing clock double-edge triggered flip-flop, *IEEE J. Solid-State Circuits,* vol. 37, pp. 648–652, May 2002.

[14] J. Yuan, C. Svensson, and P. Larsson, New domino logic precharged by clock and data, *Electron. Lett.,* vol. 29, pp. 2188–2189, December 1993.

[15] J. Yuan, Efficient CMOS counter circuits, *Electron. Lett.,* vol. 24, pp. 1311–1313, October 1988.

[16] Y. Taur et al. CMOS scaling into the 21st century: 0.1 μm and beyond, *IBM J. Res. Dev.,* vol. 39, pp. 245–260, January/March 1995.

[17] J. Yuan and C. Svensson, Multigigahertz TSPC circuits in deep submicron CMOS, *Physica Scripta,* vol. T-79, pp. 283–286, 1999.

[18] P. Larsson and J. Yuan, Novel carry propagation in high-speed synchronous counters and dividers, *Electron. Lett.,* vol. 29, pp. 1457–1458, August 1993.

[19] J. Yuan and C. Svensson, Fast CMOS nonbinary divider and counter, *Electron. Lett.,* vol. 29, pp. 1222–1223, June 1993.

[20] B. Chang, J. Park, and W. Kim, A 1.2 GHz dual-modulus prescaler using new dynamic D-type flip-flops, *IEEE J. Solid-State Circuits,* vol. 31, pp. 749–752, May 1996.

[21] C-Y. Yang et al. New dynamic flipflops for high-speed dual-modulus prescaler, *IEEE J. Solid-State Circuits,* vol. 33, pp. 1568–1571, October 1998.

[22] J. Yuan, C. Svensson, F. Lu, and H. Samueli, A high speed pipelined CMOS accumulator for implementing numerically controlled oscillators, *Proc. ISCAS '90,* vol. 1, pp. 113–116, May 1990.

[23] F. Lu, H. Samueli, J. Yuan, and C. Svensson, A 700-MHz 24-bit pipelined accumulator in 1.2-um CMOS for application as a numerically controlled oscillator, *IEEE J. Solid-State Circuits,* vol. 28, pp. 878–886, August 1993.

[24] J. Yuan and K. Chen, Bit-serial realization of maximum and minimum filters, *Electron. Lett.,* vol. 24, pp. 485–486, April 1988.

[25] E. Vittoz and J. Fellrath, CMOS analog integrated circuits based on weak inversion operation, *IEEE J. Solid-State Circuits,* vol. SC-12, pp. 224–231, June 1977.

[26] A. Agarwal, H. Li, and K. Roy, A single V_t low-leakage gated-ground cache for deep submicron, *IEEE J. Solid-State Circuits,* vol. 38, pp. 319–328, February 2003.

[27] Y. Ye, S. Borkar, and V. De, A new technique for standby leakage reduction in high-performance circuits, *Symp. on VLSI Circuits, Dig. Tech. Papers,* pp. 40–41, 1998.

9 Low-Power Arithmetic Operators

Arnaud Tisserand
INRIA LIP Arénaire

9.1 Introduction

Arithmetic operators are among the most used basic blocks in digital integrated circuits. They are the core of functional units such as arithmetic and logic units (ALUs), integer multipliers, floating-point units (FPUs), or multimedia units. They also play a part in memory address generation units and in some controllers. In a complex system, such as a system on chip (SoC), the power consumption of those operators is often smaller than the power consumption of memories and buses. Nevertheless, some significant energy savings can be achieved at the arithmetic level without performance penalty. Furthermore, it is important to reduce power wastage wherever possible.

This chapter presents some low-power consumption aspects of the main arithmetic operators and representations of numbers used in high-performance circuits. A basic knowledge on computer arithmetic is assumed. More details on arithmetic algorithms and number systems can be found in reference computer arithmetic books such as Ercegovac and Lang [1] and Koren [2].

The optimization of a circuit can be done at different levels: system, algorithm, architecture, circuit, and technology. Arithmetic operators have no concern with the system and technology levels. Algorithm, architecture, and circuit levels widely impact the design of arithmetic operators and vice versa. More precisely, there is a complex trade-off among:

- The number system(s) used to represent the data (i.e., width and number coding)
- The algorithms used to compute the mathematical operations (i.e., evaluation methods, speed/area trade-offs, and fused operations)

- The characteristics of the data (i.e., accuracy, signal activity, and space/time correlations, etc.)
- Some circuit constraints (i.e., specific cells in the standard library and logic style, etc.)

This chapter focuses on standard-cell complementary metal oxide semiconductor (CMOS) circuits. As discussed in the previous chapters of this book, the power consumption of CMOS circuits can be decomposed into three main parts:

$$P = P_{switching} + P_{short-circuit} + P_{leakage}$$

The switching power $P_{switching}$ is due to the charge and discharge of the capacitors driven by the circuit. The short-circuit power $P_{short\text{-}circuit}$ is caused by the simultaneous conductance of P and N transistor networks. The leakage power $P_{leakage}$ is due to the leakage current that flows in the circuit such as subthreshold or reverse-biased PN junction leakages for instance. The sum of the switching and the short-circuit powers is called the dynamic power, while the leakage power is called the static power.

The activity into an operator is caused by two kinds of signal transitions. The useful activity due to the input transitions that produce the internal transitions required to perform the computation. The redundant activity (also called glitching activity) is caused by different delays from the inputs to the same output and circuit defects. Specific circuit styles or careful placement and routing of the gates can significantly reduce the glitches, but important energy savings can also be achieved on the useful activity by using specific number systems or optimized evaluation algorithms. Some methods for reducing both kinds of activity are presented.

This chapter focuses on the dynamic power and mainly on the switching contribution, as it is the largest one for arithmetic operators. The static power has a smaller contribution. The methods presented in other chapters for reducing the dynamic power or the static power can be used on arithmetic operators; however, the goal of this chapter is to present some energy savings that can be simply achieved at the arithmetic level.

The two's complement representation of integers and fixed-point numbers is the most used number system in digital circuits. Unless it is explicitly mentioned, we will assume two's complement representation of numbers in this chapter. Suppose $X = x_{n-1}x_{n-2}\ldots x_1x_0$ is an n-bit two's complement number, then x_{n-1} denotes the sign of the number and the magnitude of X is given by $-x_{n-1}2^{n-1} + \sum_{i=0}^{n-2} x_i 2^i$.

The chapter is organized as follows. Section 9.2 and Section 9.3 focus on the two main integer operations: addition and multiplication. Finally, Section 9.4 briefly presents other operations, number systems, and possible evolutions of low-power arithmetic.

9.2 Addition

Addition is the most-used arithmetic operation in microprocessors, digital signal processors (DSPs), and many digital circuits. It is the core operation in some computation units such as ALUs, but it also plays a part in multipliers, dividers, FPUs, and address generation units.

Assuming two's complement representation of numbers, the addition and the subtraction operations are very close. In the following, we denote addition both operations (i.e., addition and subtraction).

9.2.1 1-Bit Addition Cells

Most of current digital circuits are designed using standard cell libraries and place and route tools. The type of the basic cells (complexity, drive strength, etc.) and their characteristics are of first importance. Besides the various logic gates, flip-flops, and latches, a few cells are dedicated to 1-bit addition: the half-adder (HA) and the full-adder (FA) cells. The purpose of these cells, also called counters, is to *count* the number of "1" at the inputs and to output their sum using a standard binary representation (i.e., $X = \Sigma_{i=0}^{n-1} x_i 2^i$).

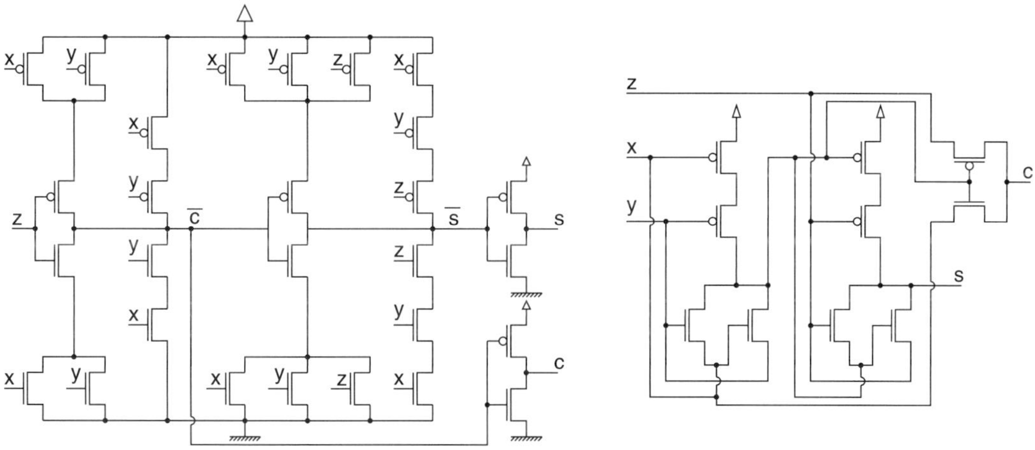

FIGURE 9.1 28 and 10-transistor implementations of the FA cell.

The half-adder is a (2,2) counter. The sum of the two input bits x and y is represented by the two output bits s (sum) and c (carry) such as $s + 2c = x + y$. A logic implementation of the HA cell gives $s = a \oplus b$ and $c = ab$. The full-adder is a (3,2) counter. The sum of the three input bits x, y, and z is represented by the two output bits s (sum) and c (carry) such as $s + 2c = x + y + z$. A logic implementation of the FA cell gives $s = a \oplus b \oplus c$ and $c = ab + ac + bc$. In an FA cell, the third input z is also called the carry-in c_{in} bit and the output c the carry-out c_{out}. Other counter cells are sometimes in libraries, such as the "4-to-2" cell used for reduction trees in multipliers (see Section 9.3).

Several possible circuit styles are used for the implementation of the HA and FA cells: pure CMOS, transmission gate, and complementary pass-transistor logic (CPL). The circuit implementation style of the gates is not an arithmetic parameter but it widely impacts the efficiency of larger operators. Depending on the circuit style, the transistor count of the FA cell may vary between 10 and 28. Figure 9.1 presents two implementations of the FA cell at the transistor level with different circuit characteristics.

In Shams and Bayoumi [3], a 16-transistor FA cell is presented. Energy savings up to 30% are achieved for a 16-bit carry select adder based on this cell compared with a version based on a previous 14-transistor FA cell without any speed penalty. Therefore, it shows that the power reduction is not just a problem of transistor number. A recent article presents a complete circuit style comparison for FA cell design in submicron technologies and many details about transistor sizing of those cells [4].

9.2.2 Sequential Adder

The simplest addition architecture is based on a linear array of FA cells as depicted in Figure 9.2. It is known as sequential adder or ripple carry adder (RCA). This adder is the slowest useful adder, but it is also the smallest.

Due to the very simple structure of the sequential adder, its power consumption has been the subject of many studies. For instance, Guyor and Abou-Samra [5] present a formal model of the sequential adder activity (and also for some other simple adders). In accordance with experimental results, this model shows that the average activity overhead (glitches) is about 50% for a sequential adder.

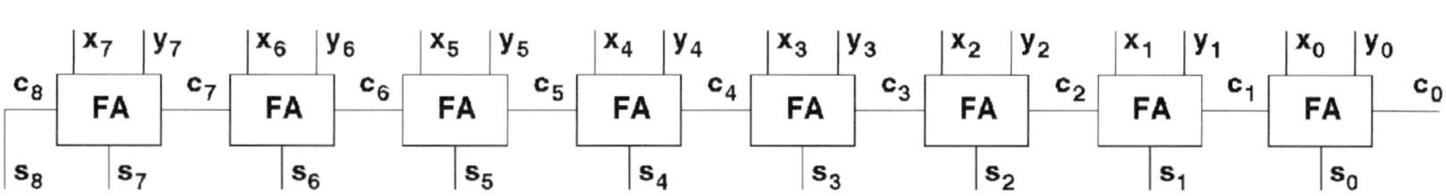

FIGURE 9.2 8-bit RCA or sequential adder.

TABLE 9.1 Bit-level Propagation and Generation of Carries

x	y	c_{out}	s	Carry Transfer
0	0	0	c_{in}	generate 0
0	1	c_{in}	$\overline{c_{in}}$	propagate
1	0	c_{in}	$\overline{c_{in}}$	propagate
1	1	1	c_{in}	generate 1
$x = y$		x	c_{in}	generate
$x \neq y$		c_{in}	$\overline{c_{in}}$	propagate

9.2.3 Propagate and Generate Mechanisms

One key point in the addition process is the computation of the carry-in bit for each rank (i.e., the $c_{in,i}$ values for all i in $\{0,1,\ldots n-1\}$). Once $c_{in,i}$ is known, the sum bit at rank i can be easily computed using $s_i = a_i \oplus b_i \oplus c_{in,i}$. As illustrated in Table 9.1, the carry-out bit c_{out} can be computed without need of the carry-in bit *cin* for some values of x and y. In this table, if x and y are different; the carry-in bit is propagated to the carry-out bit. If the input bits x and y are equal, the carry-out bit is generated without any need on the value of the carry-in. Sometimes, the generation of a carry-out bit equal to 0 is called a kill or absorption. Generation is then reserved for the case $c_{out} = 1$.

Three bits are defined to compute the carry-out bit at each rank. Those bits are: the propagate bit p, the generate bit g, and the kill bit k. Some of those bits are used in adders, such as carry lookahead or Manchester chain adders. A possible logic implementation of those bits is given by equation. Notice that the bits p and g can be computed using an HA gate ($p = c$ and $g = s$).

$$p = x \oplus y, \quad g = xy, \quad k = \bar{x}.\bar{y} = \overline{x + y} \tag{9.1}$$

9.2.4 Carry Select Adder

The main idea in a carry select adder (CSeA) is to split a sequential adder into two parts and performing the computation of the most significant bit (MSB) part with the two possible carry-in bits in parallel and selecting the right one using the carry-out bit of the least significant bit (LSB) part. Recursively applied, this method leads to a logarithmic time adder at the algorithmic level. As an example, Figure 9.3 depicts a 4-bit CseA, but CSeAs are not so fast in practice because of high fan out problems. This type of adder is used in combination with a faster scheme in some multiple size operators.

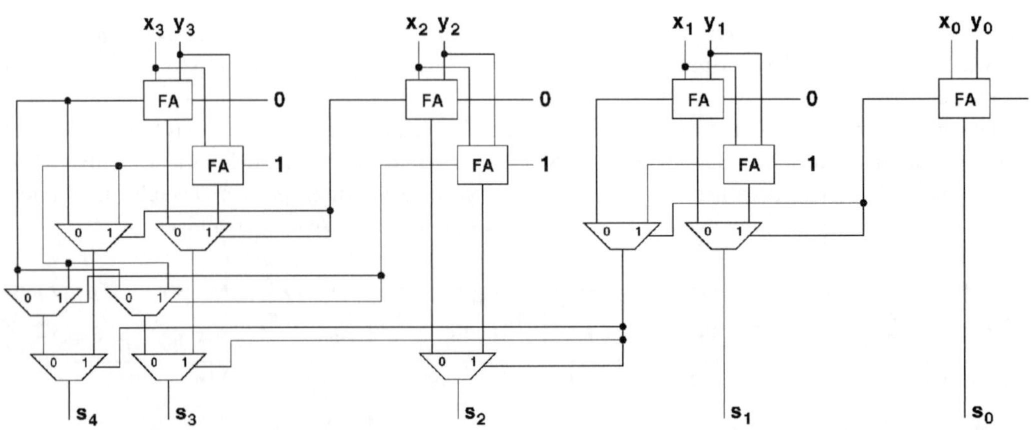

FIGURE 9.3 4-bit CSeA.

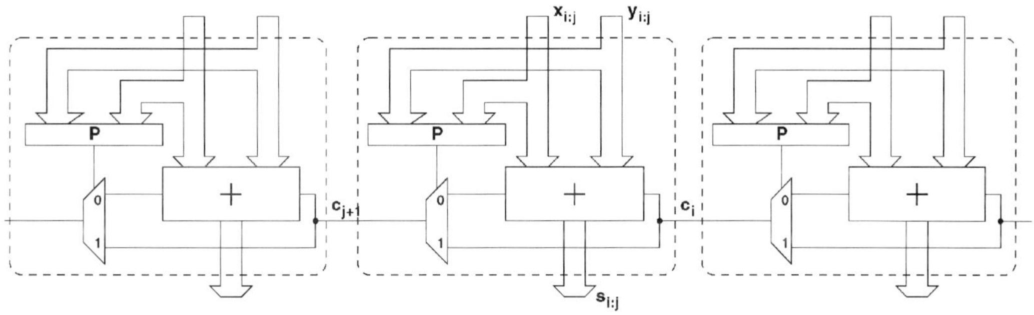

FIGURE 9.4 Principle of 1-level CSkAs.

9.2.5 Carry Skip Adder

The carry skip adder (CSkA) is based on a linear structure of blocks of sequential adders and additional logic used to skip blocks when all ranks propagate the carry inside the block. There is a skip over a block when $\prod_{i \in block} p_i = 1$. They are constant or variable block widths CSkAs. The principle of carry skip adders is depicted in Figure 9.4. The speed of those adders is $O(\sqrt{n})$ for n-bit operands. Although this adder has not the highest theoretical speed, it is widely used for fast but not critical additions because of its small area and regular layout. A hierarchical application of the carry-skip scheme can be used. This leads to multiple-level carry skip adders. In practice, a simple 1-level solution with variable block widths leads to simple and quite efficient adders.

9.2.6 Logarithmic Adders

Several kinds of logarithmic adders are available, such as carry-lookahead adders (CLAs) or parallel-prefix adders. The CLA is one special case of parallel prefix. This type of adder is widely used in fast circuits because of its high speed. The principle of CLAs is to compute the values $c_{in,i}$ using propagate/generate trees in parallel for all ranks instead of trying to propagate them as fast as possible. At rank i, a carry-in equal to 1 occurs in the following cases:

- Rank i-1 generates a carry-out equal to 1 (i.e., $g_{i-1} = 1$)
- Or rank i-1 propagates a carry generated at rank i-2 (i.e., $p_{i-1} = g_{i-2} = 1$)
- Or ranks i-1 and i-2 propagate a carry generated at rank i-3 (i.e., $p_{i-1} = p_{i-2} = g_{i-3} = 1$)

  ⋮
- Or ranks i-1 to 0 propagate the adder carry-in $c0$ equal to 1 (i.e., $p_{i-1} = p_{i-2} = \ldots = p_1 = p_0 = c_0 = 1$)

Therefore, all the carry-in bits can be computed using the relation:

$$c_i = g_{i-1} + p_{i-1}g_{i-2} + p_{i-1}p_{i-2}g_{i-3} + \ldots + p_{i-1} \cdots p_1 g_0 + p_{i-1} \cdots p_0 c_0 \tag{9.2}$$

The architecture of CLAs is based on the three following steps for all rank i:

1. Parallel computation of (g_i, p_i)
2. Parallel computation of c_i using Equation 9.2
3. Parallel computation of $s_i = a_i \oplus b_i \oplus c_i = p_i \oplus c_i$

The example of the computation of the carries in a 4-bit CLA is depicted Figure 9.5. Only small CLAs can be built because of their fan-in and fan-out limitations. Small CLAs, such as 4-bit block, are used to build larger adders. Various circuit styles have been evaluated on 32-bit CLAs in Ko et al. [6]. Very efficient implementations of such a small CLA can be achieved using AND-OR complex gates.

Parallel prefix adders are based on the same kind of structure: performing the computation of the carry-in bits for all ranks using partially shared generate/propagate trees. The sharing of the different

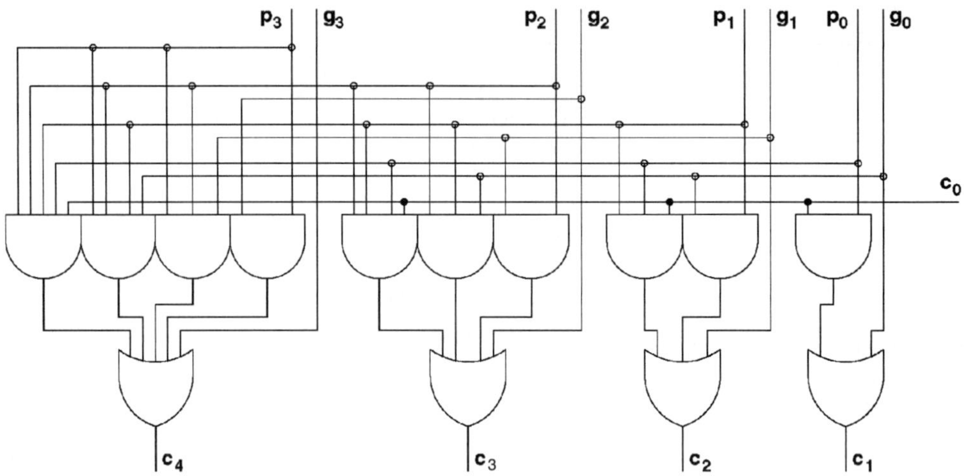

FIGURE 9.5 Computation of the carries in a 4-bit CLA.

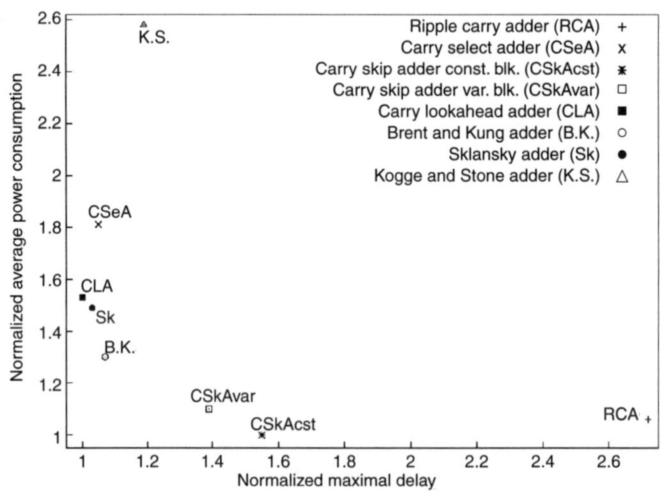

FIGURE 9.6 Power-delay comparison of standard 32-bit adders.

trees can be done using different schemes. This leads to the various types of parallel-prefix adders. The most well-known parallel prefix adders are: Brent and Kung, Sklansky, and Kogge and Stone. For a complete discussion on parallel prefix adders design, we refer the reader to Zimmerman [7].

9.2.7 Power/Delay Comparison

There are only a few papers on accurate SPICE-level power/delay comparison of the previous adders. For instance, Nagendra et al. [8] discusses numerous details about former technologies. Figure 9.6 presents a synthesis of accurate comparisons of various adders.

9.2.8 Redundant Adders

The use of redundant number systems allows constant time addition. The carry-save (CS) number system is widely used. This is a radix-2 representation in which the digits belong to {0,1,2}. The representation is considered redundant because some numbers have several representations. For instance, the integer 6 can be represented in CS using $(0110)_{cs}$ or $(0102)_{cs}$. In CS, each digit x of {0,1,2} is represented by the

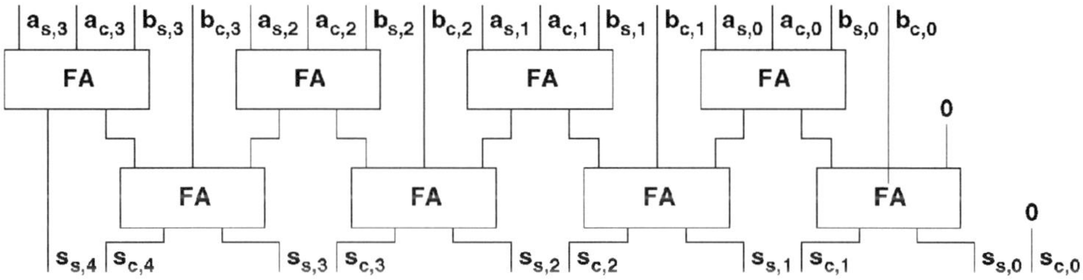

FIGURE 9.7 A 4-digit CSA.

sum of two bits: $x = x_c + x_s$ where x_c and x_s are in $\{0,1\}$. As an example, a four-digit carry-save adder (CSA) is depicted in Figure 9.7. We refer the reader to the books cited in the introduction for a comprehensive presentation of redundant number systems.

The conversion from a value represented using a redundant number system into a value represented using a nonredundant number system can be done using a nonredundant addition. In the case of CS numbers for instance, the conversion can be done using a standard fast adder. The sum vector and the carry vector of the CS representation are considered as the two nonredundant inputs in the adder.

Although a redundant number system allows to compute the addition of two arbitrary large numbers in a constant time, it has some drawbacks. First, more bits are required to represent the values than using a nonredundant number system. Second, due to the possible multiple different representations of a value, the comparison operation is quite complex. Therefore, redundant addition is mainly used as an internal representation. One frequent use of redundant adders is to perform the sum of three or more values. For instance, the sum of k values can be done using a $\log(k)$ -level tree of CSAs and a fast, nonredundant adder at the end. The adder tree produces the sum of k values represented in CS. The conversion into a nonredundant number system is done using a standard adder.

Nagendra et al. [8] gives the power/delay comparison of CSAs and CLAs. For 32-digit operands, the CSA is three times faster than the CLA and its average power consumption is 20% lower compared to the CLA just for the arithmetic part.

The comparison of redundant adders, such as CSAs and standard adders, sometimes called carry-propagate adder (CPA), is not straightforward. Indeed, redundant number systems require more bits than nonredundant ones for the same interval of representable values. This leads to storage or additional bus resources. For instance, in the case of a n-digit CSA, $2n$ bits are required. One good point about redundant adders is the fact that the glitching power of those operators is very limited due to their structure in comparison to CPAs, but this gain should be very small in front of the additional power consumption due to the storage or communication bus overhead.

9.3 Multiplication

Multiplication is the most area consuming arithmetic operation in high-performance circuits. This large area is required to implement high-speed multipliers. We refer the reader to Flynn and Oberman [9] for a comprehensive presentation of speed optimizations of multipliers. These large and complex operators also have the highest power consumption among all the arithmetic operators. Therefore, there are many research works on low-power design of high-speed multipliers.

Multiplication involves two basic operations: the generation of the partial products and their sum. Therefore, there are two possible ways to accelerate the multiplication: reduce the number of partial products or speed up their sum. Both solutions can be applied simultaneously. Reducing the number of partial products also reduces the time required to perform their sum.

High-speed multiplication have no concern with sequential or array multipliers, which are too slow. In this section, we only present tree based parallel multipliers. In practice, most of those high-performance multipliers are based on the three following steps:

- Parallel generation of the partial products
- Tree reduction of the partial products into a redundant sum (e.g., carry-save tree)
- Conversion of the redundant sum into a nonredundant representation using a fast adder

Several power optimization methods can be applied to multiplier design: circuit styles, improved computation algorithms, specific internal number systems. This section presents such optimizations for each basic step of high-speed parallel multipliers.

We denote X and Y the two n-bit operands of the multiplication where X is the multiplier and Y the multiplicand. We assume that the operands and the result are represented using the standard unsigned binary number system (i.e., $X = \Sigma_{i=0}^{n-1} x_i 2^i$). Unless it is explicitly mentioned, we look at the $2n$-bit full-width product $P = X \times Y$. In some applications, only the most significant bits of P are required, then we talk about truncated multiplication.

The modifications required to handle signed numbers in two's complement representation are small, and they do not affect significantly the power consumption. Therefore, we only focus on unsigned multiplication. More details on signed multiplication can be found in the books mentioned in the introduction.

9.3.1 Partial Products Generation

The generation of the partial products consists in the computation of all the $p_{i,j} = x_i \times y_j$ for all i and j in $\{0,1,\ldots,n-1\}$. The basic architecture is based on n^2 AND gates array. This generation appears to be straightforward, but the algorithmic description hides some load problems. Indeed, in an $n \times n$-bit multiplier, each input bit x_i or y_j is used as an input in n different gates. For instance x_i will be used to produce the n partial products $p_{i,0}, p_{i,1},\ldots, p_{i,n-1}$. Therefore, the load on those signals can be very high in large multipliers. A careful buffering and gate sizing is then required.

Using a standard radix-2 coding of both n-bit operands, there are $n \times n$ partial product bits or n partial product n-bit words. For each bit of the multiplier X, two possible partial product words are available: 0 and Y. The reduction of the number of the partial products can be done using a high-radix recoding of one operand. Usually the multiplier is the recoded operand. Using a radix-4 recoding of the multiplier, the number of partial products reduces to $n/2$. For a radix-2^k recoding of the multiplier, only n/k partial products are required.

The recoding is performed by examining several bits of the multiplier X for each partial product. Using a naive radix-4 recoding, an examination of two bits of the multiplier X leads to four possible partial products: 0, Y, $2Y$, and $3Y$. Of course, this recoding is not efficient in practice because of the generation of the value $3Y$, which requires a preliminary addition. A better radix-4 recoding uses the partial products -$2Y$, -Y, 0, Y, and $2Y$. Multiplication by two is done by shifting one bit left. Negation is done by complementing and adding a correction +1 in the LSB.

The modified Booth's recoding is a radix-4 recoding that generates $n/2$ partial products for n-bit multipliers. It examines three bits of the multiplier X with an overlap of one bit. Booth's recoding is based on a careful application of the identity: $2^{i+k} + 2^{i+k-1} + 2^{i+k-2} + \cdots + 2^i = 2^{i+k+1} - 2^i$. Table 9.2 presents the truth table of the modified Booth's recoding used for all odd values of i, namely $= 1,3,5,\ldots$, to produce the recoded multiplier X'. Only one nonnull value of the recoded multiplier X' exists for each line. This ensures that only $n/2$ partial products are generated.

Based on recoding schemes, the complete generation of the partial products requires two types of cells: recoding cells and partial product generator cells. In some rich standard cell libraries, there are specific gates to implement those functions. Indeed, those functions can be easily implemented using AOI complex gates. In the case of the radix-4 modified Booth's recoding applied to an $n \times n$-bit multiplier, there are:

- $n/2$ recoding cells that generate the operation signals, such as -$2Y$, -Y, 0, +Y, and +$2Y$, from three bits of the multiplier X

TABLE 9.2 Modified Booth's Recoding Transformations

x_i	x_{i-1}	x_{i-2}	x'_i	x'_{i-1}	Comment	Operation
0	0	0	0	0	string of 0's	+0
0	0	1	0	1	end of 1's	$+Y$
0	1	0	0	1	a single 1	$+Y$
0	1	1	1	0	end of 1's	$+2Y$
1	0	0	$\bar{1}$	0	beginning of 1's	$-2Y$
1	0	1	1	$\bar{1}$	a single 0	$-Y$
1	1	0	0	$\bar{1}$	beginning of 1's	$-Y$
1	1	1	0	0	string of 1's	+0

- $n/2 \times n$ partial product generator cells that actually generate the partial products using the y_j's and the operation signals

Higher radices can be used to recode the multiplier, but this leads to very complex recoders. In practice, it seems that a radix-4 modified Booth recoding offers a good trade-off between circuit complexity and number of partial products reduction.

If the basic functions used for the generation of the partial products are quite simple, their implementation require to be careful at the layout level (see Abu-Khater et al. [10], for instance). Indeed, many signals in the generation of the partial products have to be generated at the "right" time in order to avoid spurious transitions. For instance, recoding cells drive large loads and glitches on the corresponding wires would lead to higher power consumption and slower circuits. This is a challenging problem for standard place and route tools.

9.3.2 Reduction Trees

Once all the partial products have been generated, the second basic operation to perform in the multiplication is the sum of those partial products. In high-speed multipliers, this is performed using a tree of redundant additions. As described in the previous section, a fast way to sum up several numbers is the use of a tree of redundant additions such as a CS tree.

An h-level CS tree can reduce $n(h)$ nonredundant inputs to a CS sum. The "CS function" $n(h)$ is defined by $n(h) = \lfloor 3n(h-1)/2 \rfloor$ and $n(0) = 2$. Some useful values of this function are presented in Table 9.3. This means that to reduce 24 partial products for instance, a 6-level CS tree should be used.

CS trees and Wallace's trees are very close. In the multiplier architecture proposed by Wallace, a recursive decomposition of the computation is performed but the reduction of the different terms is done using CS trees. Some other full-adder based reduction trees can be used such as Dadda's trees or those generated using fast reduction algorithms (see Stelling et al. [11], for instance).

To improve the reduction step performance, higher order counters can be used. In Goldovsky et al. [12], a reduction tree based on (3,2), (5,3), and (7,4) counters is presented. Several logic and circuit level optimizations can be used when considering such counters instead of simple FA cells. The (5,3) counter is also called 4-to-2 compressor. Possible implementations of the 4-to-2 compressor are presented in Figure 9.8. Using such a cell leads to faster reduction trees. The power dissipation is also reduced because of the more regular layout and the smaller number of signal transitions in the tree. Typical power reduction is around 30%. One of the main problems that remain is the efficient placement of complex and irregular structures such as trees using standard electronic design automation (EDA) tools.

TABLE 9.3 Some Useful Values of the CS Function for Multiplier Design

h	1	2	3	4	5	6	7	8	9	10	11
$n(h)$	3	4	6	9	13	19	28	42	63	94	141

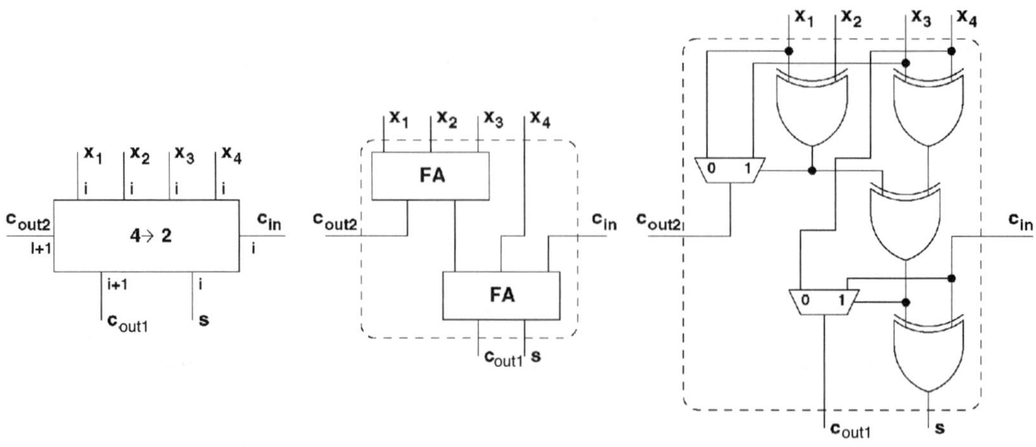

FIGURE 9.8 Some implementations of a 4-to-2 compressor.

9.3.3 Final Addition

The last step in the high-speed multiplication process is the conversion of the redundant sum produced by the reduction tree into a nonredundant representation. This step is also called assimilation of the carries. As described in Section 9.2, this conversion can be performed using a standard nonredundant adder.

Most adders are built assuming a uniform arrival time profile for their inputs (i.e., all the input signals are stable at the beginning of the computation). Nevertheless, the arrival times of the reduction tree outputs are nonuniform. A typical arrival time profile of the reduction tree outputs is presented Figure 9.9. This profile is expressed in XOR gate delays in the case of a 53 × 53-bit multiplier.

Based on these timing characteristics, the final adder structure can be optimized. The addition in region 1 can be done using a slow adder, such as an RCA. In region 2, a fast adder is required because the bits of this region arrive late. A CLA is often used in this region. The carry-out of region 1 can be used to select between the output of the adder of region 2 and this value plus one. High-speed adders such as CLA or parallel prefix adders can be modified to compute $a + b$ and $a + b + 1$ in the same time with a very small area overhead. Finally, the addition in region 3 can be done using a fast but simple adder such as a carry select adder. Using a CSeA for this last region allows to simply integrate the carry-out of region 2.

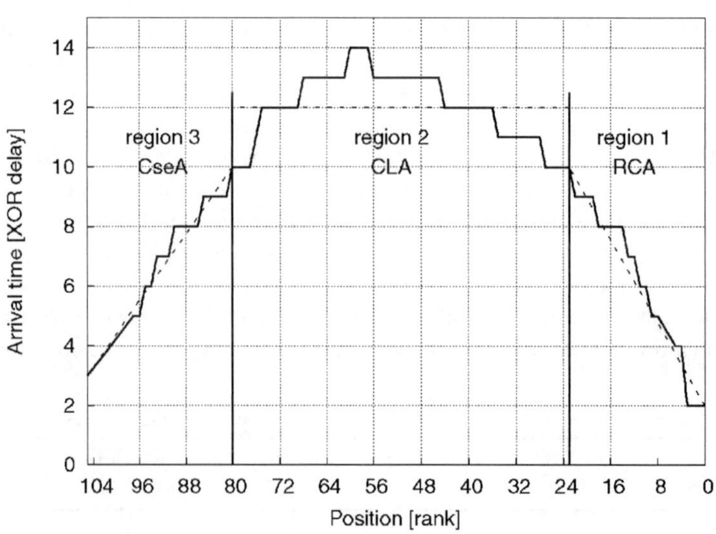

FIGURE 9.9 Typical arrival time of reduction tree output of a 53 × 53-bit multiplier.

The optimization of the final adder in a high-speed multiplier has two main advantages. First, high-speed adders require large area, and then by using slower and smaller adders for region 1 and 3, the area of the whole multiplier is improved. Second, the use of a structure that computes as late as possible avoids some spurious transitions. A power reduction around 20% is achieved using an optimized final adder instead of a standard high-speed one.

9.3.4 Fused Multiply and Add

The fused multiply and add operation $P = X \times Y + Z$ was introduced in the 1980s in DSPs to accelerate some signal processing algorithms. In many algorithms, such as filters, vector products, convolutions, or fast Fourier transforms, a multiplication is often followed by an addition in which one operand is the previously computed product. Therefore, a lot of modern general processors or DSPs have built-in hardware multiply-add and/or multiply-accumulate capabilities.

The basic idea in fast multiply-add units is to consider the third operand Z as an additional line of partial products. This leads to significant speed improvements compared to a simple concatenation of the multiplier and the adder. Indeed, for some values k, a reduction tree with h levels can reduce k or $k + 1$ partial products at the same speed. For instance, in the case of a 4-level CS tree, 9 up to 12 partial products can be reduced to a CS value in the same delay (see Table 9.3).

Although the fusion of the multiplication and the addition into a single operation leads to significant speed improvements, similar power consumption reductions can be achieved. The multiply-add operation has a smaller circuit and no intermediate storage of the product $X \times Y$ is required. Up to 40% power dissipation reduction can be achieved compared with two separate operations. The only drawback of the fused multiply-and-add unit is the need for fetching a third operand. In practice, the third operand is the output of a local accumulator to avoid heavy modifications of the processor register file.

9.3.5 Truncated Multiplication

In some applications, the $2n$-bit product produced by the multiplier is rounded to the n most significant bits to avoid growth in word size. Truncated multiplication is a method used to produce an accurate rounded evaluation of the n MSB of the product without computing most of the least significant bits. Several methods have been proposed to minimize the error produced by omitting some of the LSBs. In Schulte and Stine [13], post-layout simulations demonstrate that truncated 32-bit multipliers dissipate 40% less power than standard high-speed multipliers with a ±1 LSB error on 32-bit result.

9.3.6 Square

Some algorithms such as norm computation in image processing applications or Viterbi decoders involve a lot of square functions. Any standard multiplier can be used for computing those squares, but dedicated operators lead to significantly improved performances.

The optimization of a squaring unit is based on some simple identities applied to the generation of the partial products. As an example, we look for the square $P = X^2$ of an unsigned n-bit integer X represented using the standard binary format. The generation of the partial products leads to the values $x_i x_j$ for all i and j in $\{0,1,\ldots,n-1\}$. Several identities can be applied to these partial products:

- The term $x_i x_i$ reduces to x_i
- The sum of the two terms $x_i x_j + x_j x_i$ in a given column can be replaced by $2x_i x_j$ in the next column (of higher weight)
- Another simplification is possible using $x_i x_{i-1} + x_i = 2x_i x_{i-1} + x_i \bar{x}_{i-1}$ (i.e., two partial products in a column are replaced a by one partial product in the same column and another one in the next column [of higher weight])

The first simplification suppresses n partial products. The second one halves the number of partial products of the form $x_i x_j$, where $i \neq j$ (among the n^2 possible partial products, there are $n^2 - n$ such products). Finally, the last simplification decreases by one the height of highest column of partial products. It also reduces the length of the final adder.

Based on these simplifications, a dedicated square operator is significantly faster, smaller, and dissipates less power than a standard multiplier. Up to 60% power reduction can be achieved by using such an optimized squarer instead of a standard high-performance multiplier. Similar optimized operators can be designed for the cube function $P = X^3$.

9.4 Other Operations, Number Systems, and Constraints

Integer addition and multiplication are the two main blocks in arithmetic circuits. Most of the other operations or representations involve those two basic blocks. In this last section, we briefly present some low-power design aspects of other operations, number systems and some technology evolutions and constraints that may impact arithmetic operators.

9.4.1 Division and Square Root

Because of the lack of efficient division unit in former processors, some numerical and signal processing algorithms have been modified to avoid the use of division. Today, most of current processors have more or less complex hardware support for division. Some processors integrate a complete dedicated division unit while some other processors use functional iteration methods based on multiplication and table look-up. Even DSPs have dedicated instructions to speed-up basic division algorithms. There are a lot of division algorithms. A complete survey on division algorithms and implementations can be found in Oberman and Flynn [14]. Two main methods are used for high-speed division: digit recurrence and functional iteration.

The class of digit recurrence division algorithms is the simplest and most widely implemented. In those algorithms, a fixed number of quotient digits is produced in every iteration. The basic version of this algorithm is the "paper-and-pencil" algorithm taught at school, with just small modifications to use a radix-2 representation instead of a radix-10 one. A very efficient form of this algorithm is the SRT division (for Sweeney, Robertson, and Tocher). A comprehensive presentation of digit recurrence algorithms and implementations can be found in Ercegovac and Lang [15].

The SRT iteration is based on the following residual recurrence:

$$w[j+1] = rw[j] - q_{j+1}d$$

where j is the iteration number, r the radix, x the dividend, d the divisor, and q_{j+1} the new digit of the quotient, with an initial residual $w[0] = x$. To simplify the product $r\,w[j]$ the radix is chosen as a power of two. At each iteration, a new digit of the quotient is produced by a table lookup addressed by a few most significant digits of $w[j]$ and d. A redundant representation of the residual allows constant time subtraction and simplifies the selection of the new quotient digit. A complete presentation of the complex parameter space of the SRT division algorithms is out of the scope of this chapter. We refer the reader to the references given above.

One complete reference is available on low-power design of SRT dividers: Nannarelli and Lang [16]. Many optimizations are used in this chapter: retiming and path equalization to reduce the spurious transitions, modification of the internal redundant representation to reduce the number of flip-flops, gates with low drive capability, dual voltage supplies for the gates that are not in the critical path, clock gating, and "switch off" the power supply of inactive blocks. Based on all these optimizations, an impressive up to 60% power reduction can be achieved for a 53-bit radix-4 divider without speed penalty.

The other widely used class of division algorithms is based on functional iteration. In these algorithms, the multiplication is the fundamental operation. The most well-known algorithm is the Newton–Raphson method. It is based on the following iteration:

$$x_{i+1} = x_i(2 - x_i d)$$

where x_0 is an approximation of $1/d$ produced by a lookup in a small table. This iteration converges quadratically toward $1/d$ under some assumptions. The multiplication of the dividend by the reciprocal of the divisor finishes the division $q = x \times 1/d$. The cost of one iteration is two multiplications and one addition (performed in one's complement). The functional iteration has two main advantages over digit recurrence. First, due to the quadratic convergence of the functional iteration method, the number of digits of the quotient doubles at each iteration. This leads to very fast divisions. Second, the multiplier can be used as a shared unit. This avoids the need of dedicated division unit. The only hardware requirement is a small lookup table for the initial approximation of the divisor reciprocal (usually addressed by 6 to 10 MSB of the divisor).

Unfortunately, to our knowledge, there is no accurate comparison of digit recurrence and functional iteration algorithms for low power aspects. This task is very complex in practice. Indeed, as the functional iteration solution does not have a stand-alone unit, a processor model is required to perform the comparison. It appears, however, that the functional iteration method may have higher power consumption because of the use of the heavily loaded wires for the data transfers between the multiplier unit and the register file.

Both digit recurrence and functional iteration algorithms can be modified to efficiently compute square-roots.

9.4.2 Elementary Functions Evaluation

The evaluation of the elementary functions (e.g., sine, cosine, exponential, logarithm, and arctangent) requires quite complex algorithms. A complete presentation of the various algorithms used for evaluating these functions is given in Muller [17].

Three main classes of algorithms are available for the evaluation of the elementary functions: polynomial or rational approximations, shift-and-add methods, and table-based methods. The algorithms based on polynomial or rational approximations do not require specific hardware support. The shift and add algorithms such as the famous CORDIC are similar to SRT division algorithms, but their implementations are quite complex in practice. The last class is based on table look-up and addition, see Dinechin and Tisserand [18] for efficient implementations of this method.

It appears that the last class could lead to power efficient implementations because of their simple architecture for moderate precision (up to 32 bits). But for higher precision, the choice of the best method is still open. To our knowledge, there is no general and accurate comparison of these algorithms with respect to low-power considerations. This task is very complex, but it seems to be a motivating challenge for the research on computer arithmetic in the next years.

9.4.3 Floating-Point Arithmetic

Besides the fixed-point notation, the other widely used representation of real numbers is the floating-point number system and especially the IEEE 754 standard. In this system, a real number X is represented using the following coding:

$$X = (-1)^{s_x} \times m_x \times 2^{e_x}$$

where s_x is the sign bit, m_x the fixed-point mantissa, and e_x the integer exponent. The mantissa is a value in the interval [1,2). The exponent is a biased integer value (i.e., the stored exponent is $e_x + b$ where b is a given integer constant). All the characteristics of the IEEE 754 floating-point representations can be found in the books mentioned in the introduction.

The algorithms used to perform the arithmetic operations in the floating-point number system are more complex than in the fixed-point system. Indeed, they require some shifts, comparisons, and additive corrections to perform the alignment of the mantissa, the normalization, and the rounding. Usually, several prediction schemes are used to accelerate the computations. This leads to large and complex operators.

To reduce the power consumption of the floating-point operators, two ways have been investigated: optimization of the algorithms and reduction of the operands width. Most of the solutions presented for the integer operators can be used for floating-point operators, but additional improvements can be achieved on the shifts, the comparisons, and the corrections. The optimization of the length of the mantissa and the exponent leads to significant power reduction, but it requires a complex evaluation of the specifications of the algorithms.

9.4.4 Logarithmic Number System

In the logarithmic number system (LNS), the real numbers are represented using a sign bit and a fixed-point approximation of the logarithm of their absolute value. The value zero requires a specific representation, usually this is done by a dedicated bit. In the radix-2 LNS, the main operations can be performed using:

$$\log_2(a \times b) = \log_2 a + \log_2 b$$

$$\log_2(a \div b) = \log_2 a - \log_2 b$$

$$\log_2(a + b) = \log_2 a + \log_2(1 + 2^{\log_2 b - \log_2 a})$$

$$\log_2(a - b) = \log_2 a + \log_2(1 + 2^{\log_2 b - \log_2 a})$$

$$\log_2(a^2) = 2 \times \log_2 a$$

$$\log_2(\sqrt{a}) = \frac{\log_2 a}{2}$$

where the functions $\log_2(1+2^x)$ and $\log_2(1-2^x)$ are usually tabulated or evaluated by specific hardware operators. Based on these equations, it is clear that the LNS can lead to significant performance improvements in applications that involve a lot of multiplications, divisions, squares, or square roots. The main drawback in the implementation of the LNS is the cost of the huge tables used for the addition and subtraction functions.

In some signal processing applications, up to 30% power reduction is achieved using an LNS DSP instead of a fixed-point one. In Paliouras and Stouraitis [19], an interesting comparison between LNS and fixed-point circuit activity is investigated. It shows that to up to 50% power reduction can be achieved using the logarithmic number system.

9.4.5 Technology Evolution

The static power used to be very low and neglected in the past. In current technologies, the leakage power contribution increases significantly. The only thing that can be done at the arithmetic level is to use smaller operators, but this chapter has emphasized that large areas are often required to implement high-speed operators. For higher static to dynamic power ratios, we may have to change the algorithms used for the evaluation of arithmetic operations. For instance, sequential algorithms may be preferable to wide parallel ones.

References

[1] M.D. Ercegovac and T. Lang. *Digital Arithmetic.* Morgan Kaufman Publishers, San Francisco, CA, 2003.

[2] I. Koren. *Computer Arithmetic Algorithms, 2nd ed.* A.K. Peters Ltd., Natick, MA, 2001.

[3] A.M. Shams and M.A. Bayoumi. A novel high-performance CMOS 1-bit full-adder cell. *IEEE Trans. on Circuits and Syst. — II: Analog and Digital Signal Processing,* 47(5):478–481, May 2000.

[4] M. Alioto and G. Palumbo. Analysis and comparison on full adder block in submicron technology. *IEEE Trans. on Very Large Scale Integration (VLSI) Syst.,* 10(6):806–823, December 2002.

[5] A. Guyot and S. Abou-Samra. Modeling power consumption in arithmetic operators. *Microelectron. Eng.,* 39:245–253, 1997.

[6] U. Ko, P.T. Balsara, and W. Lee. Low-power design techniques for high-performance CMOS adders. *IEEE Trans. on Very Large Scale Integration (VLSI) Syst.,* 3(2):327–333, June 1995.

[7] R. Zimmerman. Binary adder architectures for cell-based VLSI and their synthesis. Ph.D. thesis, Swiss Federal Institute of Technology (ETH), Zurich, Hartung-Gorre Verlag, 1998.

[8] C. Nagendra, M.J. Irwin, and R.M. Owens. Area-time-power trade-offs in parallel adders. *IEEE Trans. on Circuits and Systems — II: Analog and Digital Signal Processing,* 43(10):689–702, October 1996.

[9] M.J. Flynn and S.F. Oberman. *Advanced Computer Arithmetic Design.* Wiley Interscience, New York, 2001.

[10] I.S. Abu-Khater, A. Bellaouar, and M.I. Elmasry. Circuit techniques for CMOS low-power high-performance multipliers. *IEEE J. Solid-State Circuits,* 31(10):1535–1546, October 1996.

[11] P.F. Stelling, C.U. Martel, V.G. Oklobdzija, and R. Ravi. Optimal circuits for parallel multipliers. IEEE Trans. on Computers, 47(3):273–285, March 1998.

[12] A. Goldovsky, B. Patel, M. Schulte, R. Kolagotla, H. Srinivas, and G. Burns. Design and implementation of a 16 by 16 low-power two's complement multiplier. *IEEE Int. Symp. on Circuits and Syst.,* pp. 345–348, 2000.

[13] M.J. Schulte and J.E. Stine. Reduced power dissipation through truncated multiplication. *IEEE Alessandro Volta Memorial Workshop on Low-Power Design,* pp. 61–69, 1999.

[14] S.F. Oberman and M.J. Flynn. Division algorithms and implementations. *IEEE Trans. on Computers,* 46(8):833–854, August 1997.

[15] M.D. Ercegovac and T. Lang. *Division and Square-Root Algorithms: Digit-Recurrence Algorithms and Implementations.* Kluwer Academic, Dordrecht, 1994.

[16] A. Nannarelli and T. Lang. Low-power divider. *IEEE Trans. on Computers,* 48(1):2–14, January 1999.

[17] J.-M. Muller. *Elementary Functions: Algorithms and Implementations.* Birkhauser, Boston, MA, 1997.

[18] F. de Dinechin and A. Tisserand. Some improvements on multipartite tables methods. In N. Burgess and L. Ciminiera, Eds., *IEEE 15th Int. Symp. on Computer Arithmetic ARITH15,* pp. 128–135, Vail, CO, June 2001.

[19] V. Paliouras and T. Stouraitis. Low-power properties of the logarithmic number system. In N. Burgess and L. Ciminiera, Eds., *IEEE 15th Int. Symp. on Computer Arithmetic ARITH15,* pp. 229–236, Vail, CO, June 2001.

10

Circuits Techniques for Dynamic Power Reduction

Dimitrios Soudris
Democritus University

10.1 Introduction

Power consumption has emerged as a very significant design parameter, which should be taken into consideration by the designer. Market-driven aggressive demands and technology-related limitations have steered researchers to try to invent new design techniques and methodologies to confront the power requirements. In particular, the field of personal computing devices (i.e., laptops, palmtops, as well as video- and audio-based multimedia products), wireless communication systems (i.e., personal digital assistants and mobile phones), home entertainment (i.e., consumer set-top boxes and video games) and wearable computers are some from the plethora of the market products, which are becoming increasingly popular. On the other hand, physical and technology issues related to chip packaging, cooling, signal integrity, threshold-voltage fluctuations, and variable supply voltages are some challenging problems, which should be studied and solved.

10.2 Dynamic Power Consumption Component

The dynamic power dissipation, P_{dyn}, is caused by the charging and discharging of parasitic capacitances in the circuit. We illustrate the computation of the dynamic dissipation through the example of a CMOS

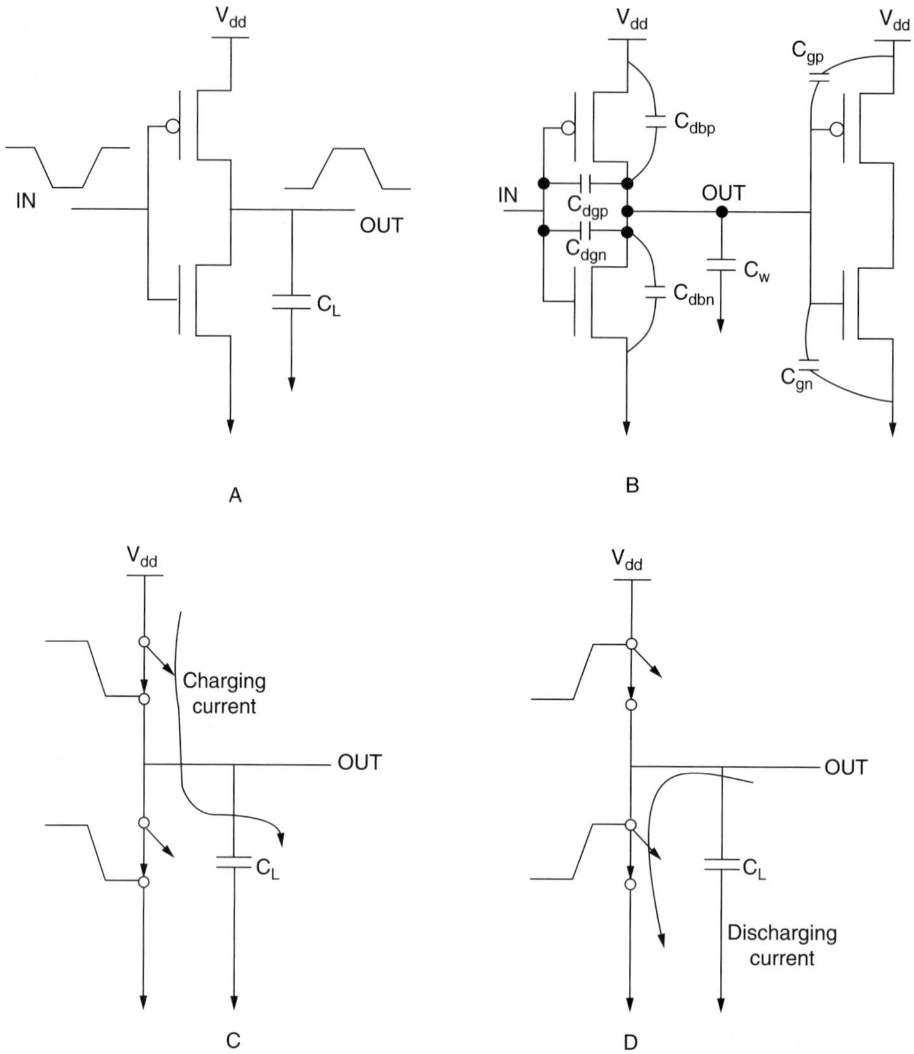

FIGURE 10.1 The function of CMOS inverter: (a) CMOS inverter, (b) capacitor C_L elements, (c) charging phase, and (d) discharging phase.

inverter driving a load capacitor C_L, as it is shown in Figure 10.1(a). The load capacitance C_L depicted in Figure 10.1(b) consists of the gate capacitance of subsequent inputs attached to the inverter output C_{gp} and C_{gn}, interconnect wire capacitance C_W, and the diffusion capacitance on drains of the inverter transistors C_{dgn}, C_{dgp}, C_{dbn}, and C_{dbp}. The total P_{dyn} comprises the sum of two power components: the first one occurs during low-to-high the output transition (i.e., charging phase), while the second one during high-to-low transition (i.e., discharging phase). More specifically, for every low-to-high output transition in a digital CMOS gate, the capacitance C_L on the output node incurs a voltage change ΔV, drawing energy of $C_L \langle \Delta V \langle V_{dd}$ joules from the supply voltage [8]. During this process, one-half of the energy is stored in the capacitor, whereas the second half is dissipated in the PMOS and interconnect wire. In the case of simple inverter, it holds $\Delta V = V_{dd}$, and thus, the power consumption is given by:

$$P_{dyn} = C_L \cdot V_{dd}^2 \cdot a_{0\rightarrow 1} \cdot f \tag{10.1}$$

where $a_{0\to1}$ is an activity factor that represents the average fraction of clock cycles in which a low-to-high transition occurs, and f is the clock frequency. Similarly, a high-to-low transition dissipates the energy stored on the capacitor C_L in NMOS transistor, pulling the output low. Consequently, the total dynamic power consumption is given by the golden formula:

$$P_{dyn} = C_L \cdot V_{dd}^2 \cdot a \cdot f \tag{10.2}$$

Here, we focus on the circuit level techniques for reducing the P_{dyn} power component. The remaining two power components are analyzed and appropriate techniques are discussed in other chapters. It should be stressed that the circuit techniques described for reducing dynamic power consumption may have impacts on the performance and silicon areas as well as on the remaining power components.

10.2.1 Power Reduction Approaches

Equation (10.3) calculates the dynamic power consumption, P_{dyn}, of CMOS logic gates. It can be easily inferred that P_{dyn} is proportional to the load capacitance, C_L, the square of V_{dd}, the switching activity, a, and the clock frequency, f. Consequently, power consumption reduction can be achieved by:

- Reducing of output capacitance, C_L
- Reducing of supply voltage, V_{dd}
- Reducing of switching activity, a
- Reducing of clock frequency, f

Thus, a designer should devise new techniques aiming at the decrease of each above-mentioned parameter or any combination of them. A very popular low strategy concerns the reduction of the *switched capacitance* or *effective capacitance*, C_{eff}, which is defined as the product of output capacitance times switching activity (i.e., $a \cdot C_L$).

Generally, the two main low-power reduction strategies concern the reduction of supply voltage and the switched capacitance. The reason is that we consider the throughput rate of a low powered-designed circuit remains the same with an existing circuit. In particular, the reduction of power supply voltage has the major impact on the power consumption due to the quadratic dependence of V_{dd}. Although such reduction is usually very effective, the circuit delay increases and system throughput degrades. In addition, the shift of industry from a supply voltage to a smaller one is quite expensive and slow due to, for instance, the compatibility issues of input/output signals with the peripheral circuits. In contrast, the reduction of the switching activity or the capacitance for a certain technology depends mainly on the designer's creativity. Thus, someone can reuse an existing silicon technology achieving satisfactorily level of power consumption without the need for purchasing new technology libraries, which may lead to design cost reduction. In other words, a designer may proceed to a more advance silicon technology only if he or she has explored all the possibilities for realizing a circuit with an existing technology considering the design cost and time-to-market constraints. The reduction of switching activity requires among others a detailed analysis of signal transition probabilities, careful redesign of circuit nodes with high activity, balanced paths, and selection of appropriate logic style. The capacitance load can be reduced by, for instance, technology scaling, transistor resizing, and logic family selection.

10.3 Circuit Parallelization

Circuit parallelization has been proposed to maintain, at a reduced V_{dd}, the throughput of logic modules that are placed on the critical path [8,17,23,24]. It can be achieved with M parallel units clocked at f/M. Results are provided at the nominal frequency f through an output multiplexer controlled at f (Figure 10.2(a)). Each unit can compute its result in a time slot M times longer (Figure 10.2(b)), and can therefore be supplied at a reduced supply voltage. If the units are datapaths or processors [23], the latter have to

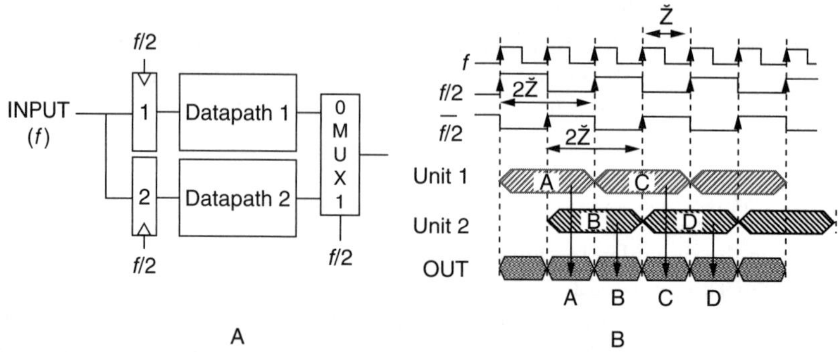

FIGURE 10.2 (a) Datapath parallelization concept, and (b) timing diagram.

TABLE 10.1 8-bit Adder Power Simulation With the CoolChip Library

2 μm Technology	F	V_{dd} (V)	Power	%
8-bit adder	f = 7 MHz	4.5	540 μW	100
2-// 8-bit adder	f/2 = 3.5 MHz	4.5	760 μW	140
2-// 8-bit adder	f/2 = 3.5 MHz	3.0	339 μW	63
2-// 8-bit adder	f/2 = 3.5 MHz	2.5	235 μW	44

be duplicated, resulting in an M times area and switched capacitance increase. Applying the well-known power formula, one can write:

$$P = M \cdot C_{eff} \cdot f / M \cdot V_{dd}^2 = C_{eff} \cdot f \cdot V_{dd}^2 \tag{10.3}$$

Table 10.1 presents the reduction of power of an 8-bit adder. One could deduce that power is saved only if V_{dd} is reduced. As operating frequency is reduced, however, the use of cells with smaller or unsized transistors results in a power reduction. Furthermore, some parallelized logic modules do not require M-unit duplication. It is the case, for instance, for memories [25], in which each unit contains $1/M$ data or instructions, resulting in the same total area to store the information and in the same C_{eff} or smaller C'_{eff} total switched capacitance, if cells with unsized transistors are used (Figure 10.3). In such a case, the power is the following:

$$P = C'_{eff} \cdot f / M \cdot V_{dd}^2 \tag{10.4}$$

At first order, power could be saved even if V_{dd} is not reduced; however, some overhead has to be considered, such as the address registers duplication and the output multiplexer (Figure 10.3). If this overhead is not too expensive, such a parallelization scheme has to be considered for logic modules that are not on the critical path. At a low V_{dd}, the latter are working without parallelization. At the same low V_{dd}, power could be saved if they are parallelized at the cost of a small overhead. Memories, shift registers, and serial-parallel converters provide interesting examples.

10.3.1 Memory Parallelization

In a parallelized module, operations of the execution units or data accesses in memories are performed in an overlapped or interleaved fashion (Figure 10.2(b)). Therefore, the result is provided with an M-1 latency delay compared to a nonparallel architecture. One can see on the timing diagram of Figure 10.2(b)

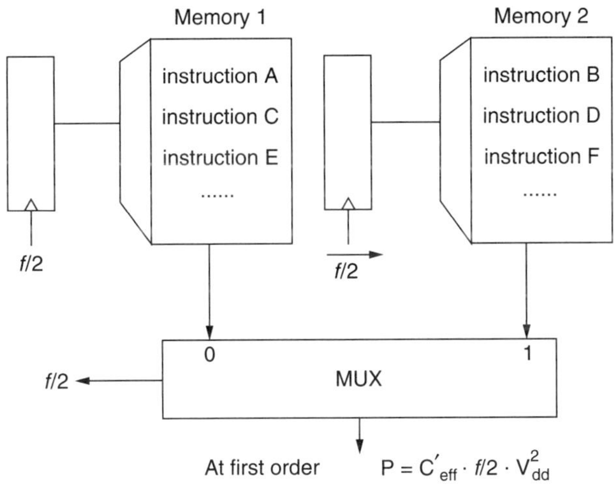

FIGURE 10.3 Memory parallelization.

that the output multiplexer can be controlled at $f/2$. The operation or access of a Unit 2 is started before the completion of the operation of Unit 1. Therefore, M successive computations do not have to be dependent on each other.

Controllers with a fixed sequence of commands without any branch instruction, or specialized processors for special linear computation, or random-access memories (RAMs) used to store coefficients for programmable finite impulse response (FIR) filters, can be parallelized according to the structure of Figure 10.3. It can be used, for instance, for transcoders in which several lookup tables (i.e., read-only memory [ROM]) are connected in parallel; however, parallel memories are difficult to use if branch instructions are used. Interleaved or parallelized memories (Figure 10.4) with branch instructions were used in the 1960s for computers [14]. With, for instance, 32 memory modules and an access time of 10 cycles ($f/10$), the probability to insert a branch delay is reduced as 10 successive instructions are, most of the time, stored in different modules.

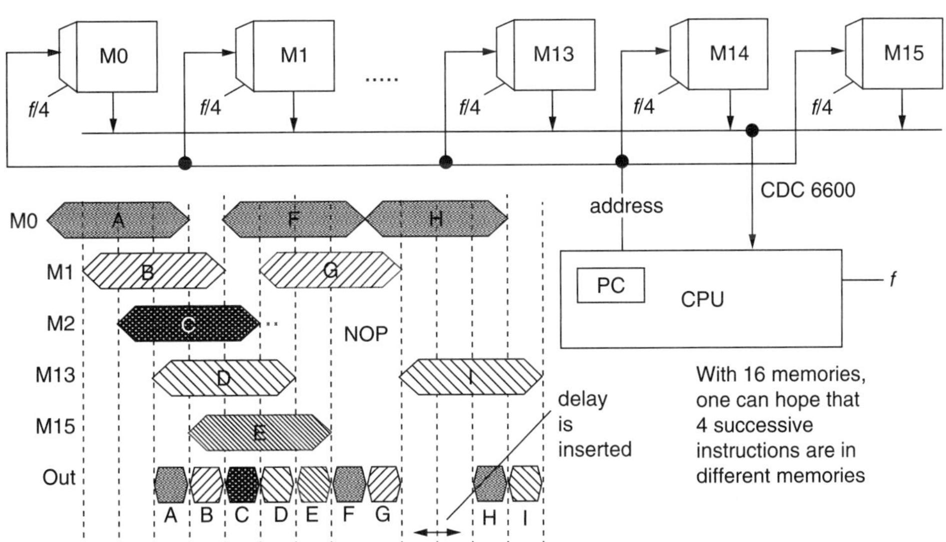

FIGURE 10.4 Memory parallelization in computers.

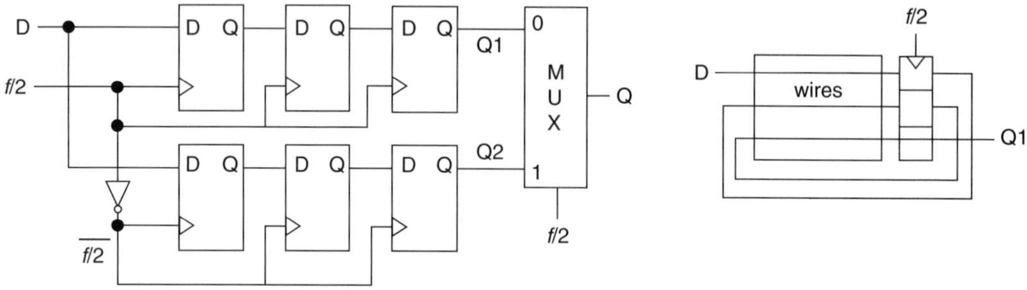

FIGURE 10.5 Parallelized shift register.

TABLE 10.2 Power Simulation with the CoolChip Library for: (i) Nonparallelized 16-bit Shift Register and (ii) 2- and 4-Parallelized 16-bit Shift Registers

2 μm Technology	f (MHz)	V_{dd} (V)	Power (μW)	%
16-bit SR	f = 33	4.5	1535	100
2-// 16 bit SR	f/2 = 16.5	4.5	887	58
4-// 16 bit SR	f/4 = 8.25	4.5	738	48
16-bit SR	f = 33	3.2	797	100
2-// 16 bit SR	f/2 = 16.5	3.2	448	56
4-// 16 bit SR	f/4 = 8.25	4.0	585	83

10.3.2 Parallelized Shift Register

Figure 10.5 depicts a parallelized shift register. Such a concept has been proposed for CCD serial memories [14,21]. The input is successively provided to the upper or to the lower half shift register at a reduced frequency, while the output multiplexer restores the output at the frequency f. No latency exists because the combinatorial circuit of the state machine "shift register" is implemented by simple wires, resulting in no associated delay. The total number of D-flip-flops (DFFs) is the same as in the nonparallelized shift register [24,25].

For the nonparallelized shift register, the maximum frequency is limited by the delays of the latches of the DFF. For the parallelized shift register, the maximum frequency is limited by one latch delay and the output multiplexer delay. Thus, the maximum frequency of the parallelized structure is the same as the classic structure (an f_{max} = 100 MHz classic shift register can be replaced by an $f/2$ = 50 MHz parallelized shift register, but it is impossible to increase $f/2 > 50$ MHz). Such a parallelization does not provide faster shift registers. It is therefore impossible to reduce V_{dd} if the shift register is on the critical path. For shift registers, which are not on the critical path, one can reduce both f and V_{dd}.

Table 10.2 presents the power consumption of nonparallelized and parallelized shift registers, depending on the degree of parallelism. Such a comparison is only valid for shift registers, which are not at their frequency limits, however, because an 8- or 4-parallelized cannot provide the same throughput as a nonparallelized shift register.

10.3.3 Serial-Parallel Converter

Figure 10.6 depicts a parallelized structure of a 16-bit serial-parallel converter in which the 1-bit input is successively loaded in four 4-bit shift registers clocked at $f/4$. Power consumption is reduced by a factor of four with the same throughput. Because no output multiplexer exists, the maximum frequency of such a structure can be much higher than the nonparallelized serial-parallel converter.

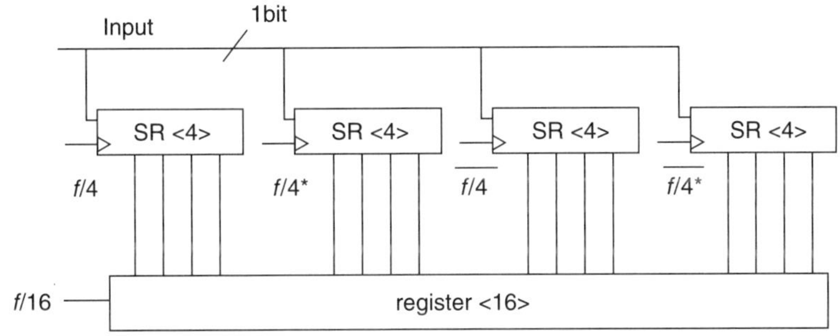

FIGURE 10.6 Parallel-serial converter.

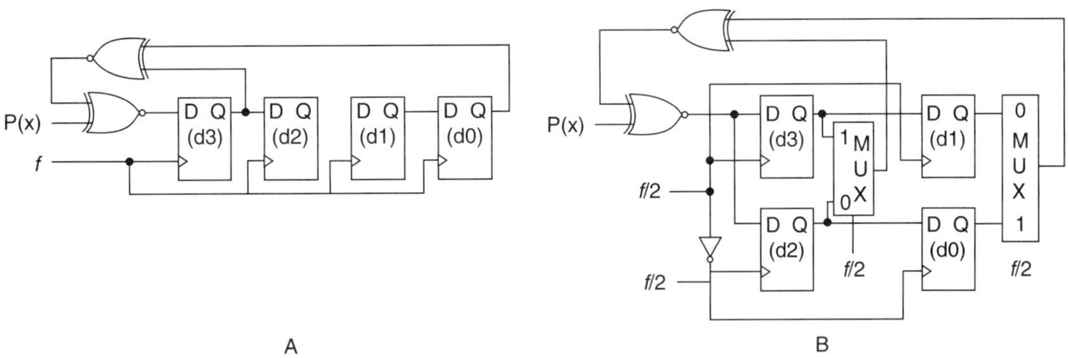

FIGURE 10.7 Linear feed-back shift register.

10.3.4 Linear Feed-Back Shift Registers

Shift register parallelization can be used for linear feed-back shift registers [19] with as many output multiplexers as the number of inputs of the XOR tree. Figure 10.7 gives an example with two output multiplexers. The example in [19] is an *M*-parallelized M-bit shift register with *M*-input simplified multiplexers. In addition, a parallelized LFSR architecture was used for the development of a division algorithm [16] and for the implementation of steam ciphers in cryptography [11].

10.3.5 Double-Edge Triggered Flip-Flop

Figure 10.8(a) and Figure 10.8(b) as well as Figure 10.8(c) to Figure 10.8(e) show the schematic and various circuit designs of single-edge triggered flip-flop (SET-FF) and double-edge triggered flip-flop (DET-FF), respectively. A classic SET-FF is implemented with two latches in series, while its parallelization results in two latches in parallel with an output multiplexer (i.e., derivation). A DET-FF is triggered on both rising and falling edge of a clock pulse. Using DET-FF the clock frequency, *f*, can be halved for the same throughput rate, thus reducing the power dissipation on the clock distribution network. Although many alternative DET-FF designs have been proposed, they have not been used extensively, due to the increased silicon area (i.e., increased input capacitance and number of transistors). This implies a larger number of internal nodes, which is strongly dependent on the input signal transition probability, *a*. It was proved [26] that if the switching activity *a* is low, significant power savings may be achieved, while high activity *a* may lead to increased power consumption. In addition, DET-FFs exhibit increased glitching activity compared with SET-FFs. SPICE simulation results show power savings around 10% using DET-FFs at the expense of a reduction of 10% in performance.

In Chung et al. [9], a detailed comparative study of five existing DET-FFs in terms of performance (i.e., latency^{-1}), total power consumption and power × delay product (PDP) is given in Table 10.3. It is

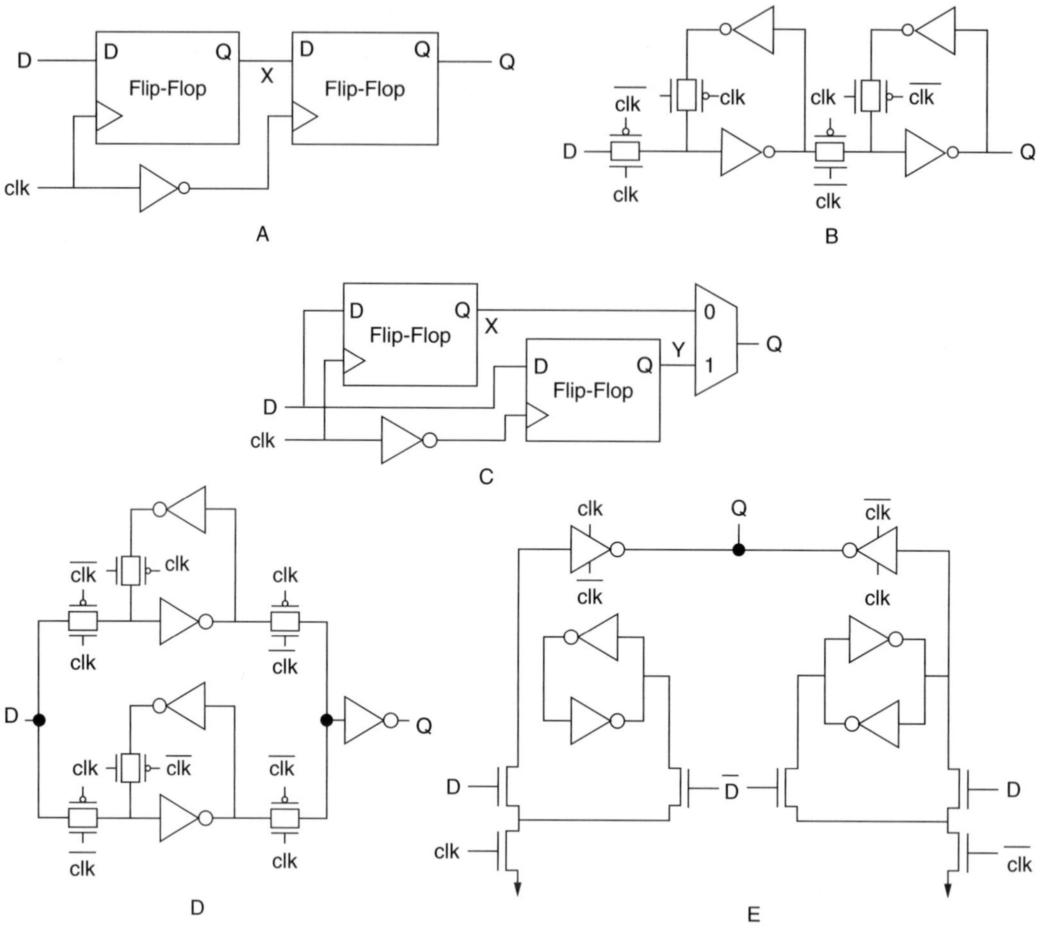

FIGURE 10.8 Flip-Flops: (a) block diagram of a single-edge triggered flip-flop (SET-FF), (b) circuit design of a SET-FF, (c) block diagram of a double-edge triggered flip-flop (DET-FF), (d) circuit design of a DET-FF [18], and (e) circuit design of a DET-FF [9].

TABLE 10.3 Comparison Results of DET-FF in Terms of Power Consumption, Latency, and Power-Delay Product

Type DET-FF	Clock Power (μW)	Data Power (μW)	Internal Power (μW)	Total Power (μW)	Latency (ps)	PDP (fJ)
[22]	17.6	65.6	241.7	324.9	245.4	79.6
[18]	17.0	4.6	153.4	175.0	312.3	54.7
[11]	23.2	11.6	131.4	166.2	262.2	43.6
[26]	30.0	13.4	194.5	237.8	235.3	56.0
[9]	18.1	10.9	189.4	218.4	230.5	50.3

assumed 0.18-μm technology and supply voltage of 1.8 volts. Notice that the total power consumption consists of three components:

1. Internal power dissipation
2. Data power
3. Local clock power, where the contribution of the internal power is over the 70% of total power consumption

Specifically, the first component concerns the power consumed inside a DET-FF including the power consumed for driving C_L. Thus, the power optimization techniques should concern the careful design of DET topology reducing capacitance or switching activity.

10.4 Voltage Scaling-Based Circuit Techniques

Because dynamic power is proportional to V_{dd}^2, even a small reduction in supply voltage causes a quadratic decrease in power consumption. However, a supply voltage reduction influences circuit's delay negatively. To preserve a constant system throughput using lower supply voltages, there exist three main approaches:

1. To redesign the circuits exploiting the principles of parallelism and pipelining
2. To reduce the threshold voltage, V_{th}, in order to compensate V_{dd} reduction
3. To assign lower V_{dd} to noncritical paths

10.4.1 Multiple Voltages Techniques

To preserve performance, while also reducing power consumption, a dual-V_{dd} approach can be used. The main concept is to assign the high V_{dd}, V_{ddH} to the gates that belong to the critical path, while the low V_{dd}, V_{ddL} is assigned to off-critical path remaining gates; however, the designer should be very careful to avoid the creation of static current. More specifically, the output of V_{ddL} gates cannot be fed directly to V_{ddH} gates because the output of a V_{ddL} gate can never be raised higher than V_{ddL}. Therefore, if connected to a V_{ddH} circuit, static current flows due to the pMOS in the V_{ddH} circuit are never being completely cut-off (Figure 10.9) [29].

To remove the static current, one possible solution is the use of level converters placed between the V_{ddL}- and V_{ddH}-supplied gates, which may increase area and power. To alleviate level converters' power and area overhead, one approach is to insert the level shifting function of a flip-flop circuit (FFLC) depicted in Figure 10.10. More specifically, the master latch is the same as a conventional flip-flop, while the slave latch also realizes the level-conversion function. This results in the power of the flip-flop being less than that of V_{ddH} flip-flop, while increasing delay slightly [30].

Layout is another important issue when dealing with multiple supply voltages. V_{ddL} and V_{ddH} cells should be separated because they have different n-well voltages. Generally, a row-by-row separation, depicted in Figure 10.11 is used due to high performance and applicability to both standard-cell and gate-arrays. Novel algorithms have been developed for optimal assignment of cells to the layout rows with V_{ddH} and V_{ddL} supply voltages [29].

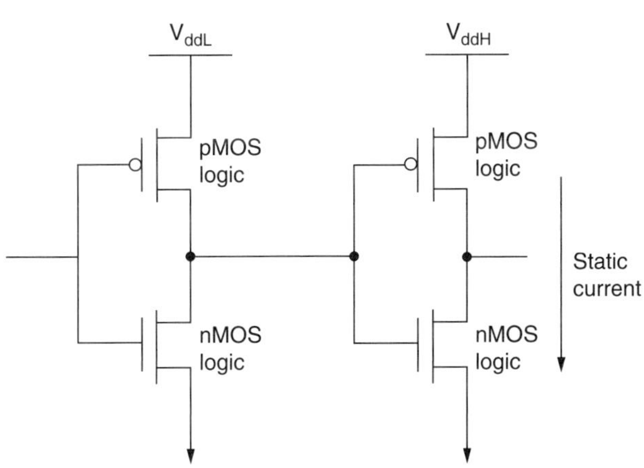

FIGURE 10.9 Dual supply voltages assignment concept.

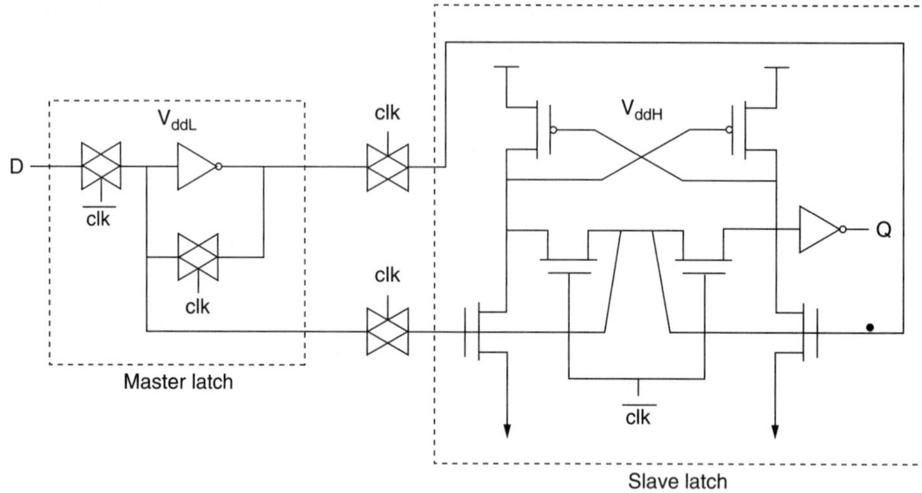

FIGURE 10.10 Circuit design of flip-flop with level-conversion function (FFLC) [30].

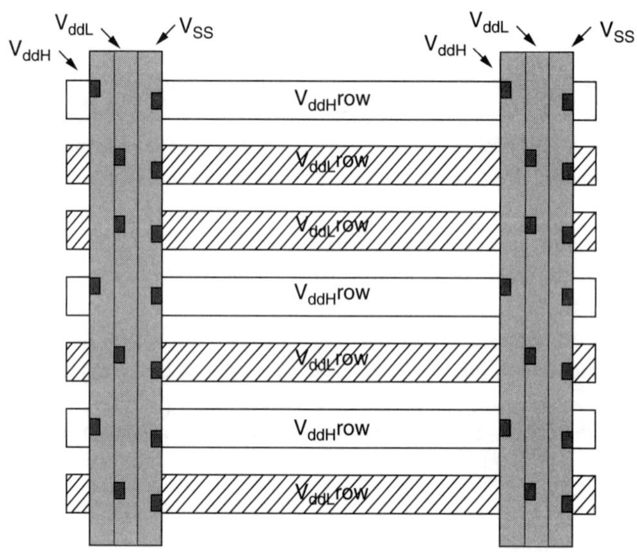

FIGURE 10.11 Placement of V_{ddH} and V_{ddL} supplied logic in a dual-V_{dd} layout.

Clustered voltage scaling (CVS) is one structure proposed to implement dual-V_{dd} design. With the CVS technique, all cells should be placed with a certain order starting from the primary inputs to primary outputs through V_{ddH}-supplied gates, V_{ddH}-supplied gates and level converters, as shown in Figure 10.12. This structure leads to clusters of V_{ddH} cells and V_{ddL} cells. By introducing level converters only at the end of a path, FFLC can be used and a minimal number of them is needed. To assign the V_{ddL} cells, a depth-first-search algorithm is used from the primary outputs to the primary inputs. As each cell is visited, an attempt is made to replace it with a V_{ddL} cell. If it can be replaced, the algorithm continues, otherwise the traversal is stopped. Dealing with multiple fan-outs can be tricky. To replace a cell, all of the cells in the fan-out of that cell should be replaced with V_{ddL} cells.

An extended CVS technique was developed [30], where the main difference compared with CVS is the fact it allows placement of a level converter even between logic cells. This can be useful in the case where a gate has multiple inputs and only one critical path. Before a level converter is inserted between logic cells, the insertion is checked to see if it does indeed reduce power consumption.

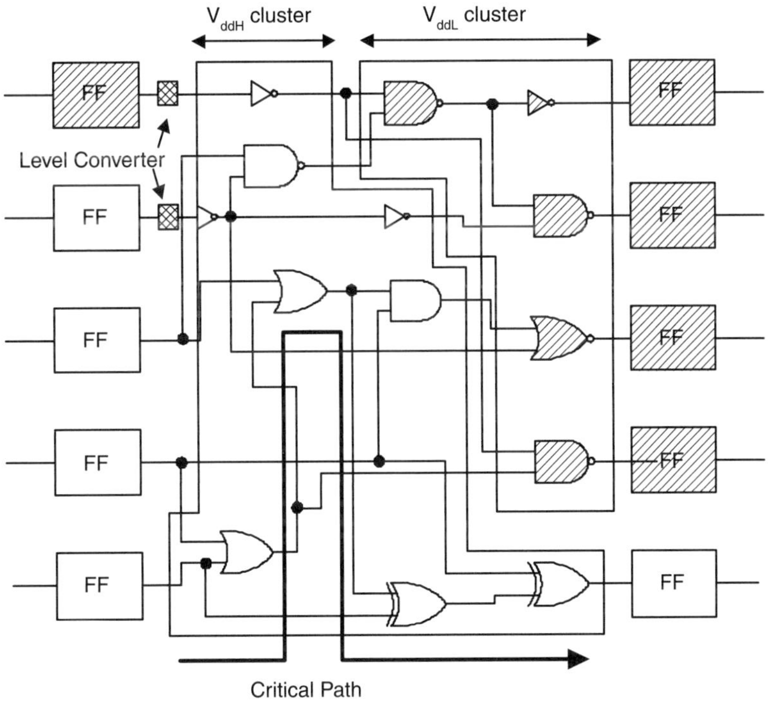

FIGURE 10.12 The concept of CVS technique.

In the case of multiple power supplies and multiple threshold voltages, theoretical models have been developed by Kuroda [15], which are used to determine "rules of thumb" for optimal multiple V_{dd} and V_{th}. These rules can be summarized as follows:

Multiple Voltages

For $\{V_{dd1}, V_{dd2}\}$:

$$\frac{V_{dd2}}{V_{dd1}} = 0.5 + 0.5\frac{V_{th}}{V_{dd1}}$$

For $\{V_{dd1}, V_{dd2}, V_{dd3}\}$:

$$\frac{V_{dd2}}{V_{dd1}} = \frac{V_{dd3}}{V_{dd2}} = 0.6 + 0.4\frac{V_{th}}{V_1}$$

For $\{V_{dd1}, V_{dd2}, V_{dd3}, V_{dd4}\}$:

$$\frac{V_{dd2}}{V_{dd1}} = \frac{V_{dd3}}{V_{dd2}} = \frac{V_{dd4}}{V_{dd3}} 0.7 + 0.3\frac{V_{th}}{V_{dd1}}$$

Multiple V_{th}

For $\{V_{th1}, V_{th2}\}$:

$$V_{th2} = 0.1 + V_{th1}$$

For $\{V_{th1}, V_{th2}, V_{th3}\}$:

TABLE 10.4 Comparison Results for Multiple V_{dd} and V_{th}

Approach	Technique	Power Reduction
[15][1,2]	Multiple V_{dd}	50%
[15][1]	Multiple V_{dd}, low- V_{th} device	30%
[15][1]	Multiple V_{th}, high- V_{th} device	15%
[29]	Dual- V_{dd}	10–20%

[1] Theoretical results. No real circuits were used
[2] Level converters power consumption not included

$$V_{th2} = 0.06\,V_{dd} + V_{th1}\,,\quad V_{th3} = 0.07\,V_{dd} + V_{th2}$$

For $\{V_{th1}, V_{th2}, V_{th3}, V_{th4}\}$:

$$V_{th2} = 0.04\,V_{dd} + V_{th1}\,,\quad V_{th3} = 0.05\,V_{dd} + V_{th2}\,,\quad V_{th4} = 0.05\,V_{dd} + V_{th2}$$

Table 10.4 gives some comparison results between different techniques based on multiple supply voltages. Despite the strong dependence of power consumption on supply voltage, Table 10.4 indicates that the power savings arising from the adoption of multiple supply voltages technique may be insignificant, due to the use of additional level converters.

Although the reduction of V_{dd} has a large impact on power consumption due to its quadratic dependence, leakage power may be the best candidate for reducing power especially for systems with nonuniform load and many standby periods. Consequently, a designer should be very careful when he or she attempts to reduce power consumption.

10.4.2 Low Voltage Swing

The low-swing voltage design technique aims at the power reduction on a long interconnect wire (i.e., large capacitance) through the use of reduced voltage swing on the wire. Given the fact that we are working in a specific design process (i.e., capacitance, frequency clock, and supply voltage remain unchanged) it is proved that on-chip lower supply voltages can be achieved [7,37], using specially designed circuits or DC-DC converters. We will discuss the concept of reduced voltage swing and the associated circuit implementations as well as its impact on P_{dyn}.

A typical form of signaling is the classical two inverters-based configuration scheme with rail-to-rail signal swing, as shown in Figure 10.13(a). The two CMOS inverters correspond to the driver and receiver circuit of the signal. In addition, intermediate repeaters/buffers are frequently used to improve signal characteristics [1]; however, the rail-to-rail swing (i.e., full swing, is an inefficient energy design approach). A possible solution is based on the reduction of the signal swing at the output driver and over the interconnect wire. Generally, the reduced voltage swing may increase not only the circuit performance, but also the major gain coming from the reduced dynamic power consumption, which can be significant in the case of large load capacitances (i.e., long wire).

Figure 10.13(b) depicts a typical circuit structure where a low swing design technique can be applied. In particular, having a long bus (or interconnection) (i.e., a large capacitive load, specially designed circuits for converting the normal swing signal to a low swing voltage and vice versa are placed at the interconnection ends. In this scheme, power savings can be achieved by two ways:

1. The charge needed for charge/discharge of C_L is smaller.
2. The driver size can be reduced because the current to be delivered by the driver to charge/discharge C_L in a certain time is smaller than the full-swing case.

The design of an efficient low-swing scheme has become a difficult problem with the deep-submicron process technology, due to very small supply voltage, V_{dd}, and threshold voltage, V_{th}.

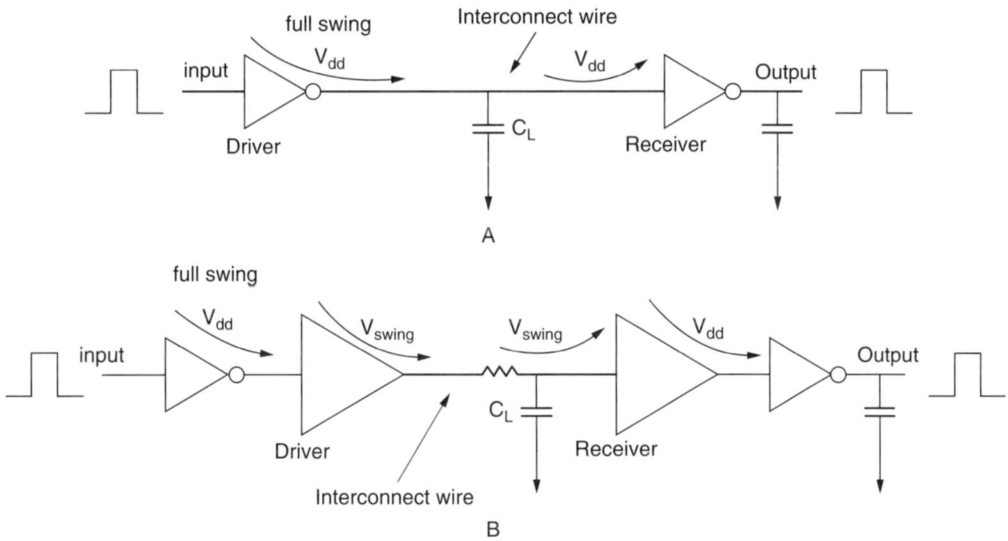

FIGURE 10.13 (a) Two-inverters-based signaling configuration scheme, and (b) typical reduced-swing circuit with a driver and receiver, and the long interconnect wire.

To design efficient low-swing-like circuits, the following parameters should be taken into account:

Energy. The dynamic energy consumption, E_{dyn}, of a interconnect wire in one cycle is given by:

$$E_{dyn} = a \cdot C_L \cdot V_{dd} \cdot V_{swing} \tag{10.5}$$

where a is the switching activity of the signal to be transmitted, and V_{swing} is the voltage across the wire.

It has been proven [32] that the selection of static drivers is preferable due to lower signal switching activity, and the supply voltage of the chosen driver should be as low as possible. The key challenge is how to detect a "one" signal at the receiver end. Novel low-swing designs using threshold voltage drops reduce the energy consumption by half order of magnitude. Additional power savings up to 4 to 6 times order of magnitude can be achieved by using very small supply voltages (from on-chip DC-DC converters).

Design Complexity. To meet certain design constraints, the designer should pay attention to how easy or complex the driver/receiver circuit design is and how much silicon area requires a new design. In addition, among other design issues related to the use of extra DC-DC converters for realizing different supply voltages and the use of only single-ended signaling schemes should be considered.

Delay. The use of voltage swing $V_{swing} < V_{dd}$ increases the propagation delay through a long interconnect wire. However, if a designer with careful architectural design can hide the increased bus delay and place a latch at the receiver, the low-swing transceiver can provide significant reduction in energy consumption.

A plethora of efficient low-swing-based techniques was reported [32]. The existing reduced voltage swing techniques can be classified as dynamic and static depending on the existence of a precharge phase (e.g., use of clock signal) during logic operation or not, respectively. Furthermore, depending on the chosen signaling approach, a low-swing circuit can implement either a single-ended or a differential signaling technique. Specifically, using the former signaling technique, a receiver detects an absolute change in a voltage in a single wire, while with the latter technique, a receiver should detect a relative difference in voltage between two wires.

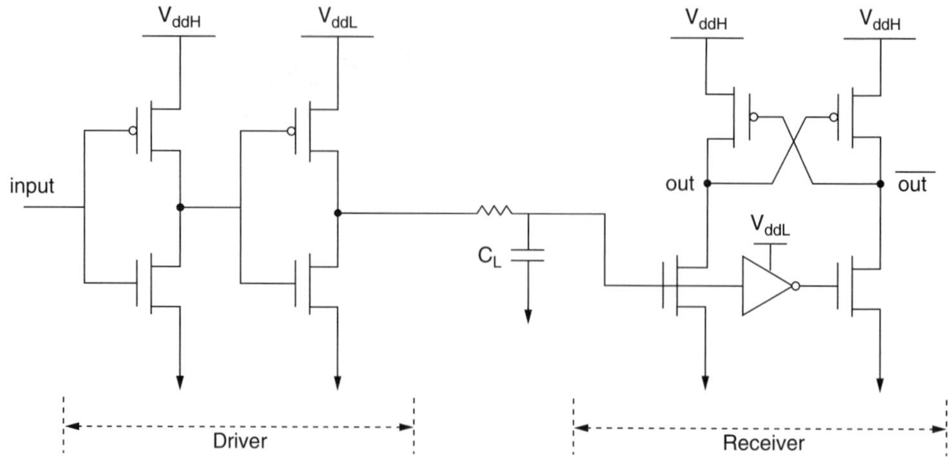

FIGURE 10.14 The architecture of the conventional level converter.

Several low-swing interfaces have been proposed in the literature; conventional level converter (CLC), differential interconnect (DIFF) [7], pulsed-controlled driver (PCD) [10], charge intershared bus (CISB) [12], charge-recycling bus (CRB) [31], capacitive-coupled level converter (CCLC) [32], level-converting register (LCR) [32], and pseudo-differential interconnect (PDIFF) [32]. It is beyond the scope of this chapter to describe in detail the architecture, the operation, and the advantages and disadvantages of each low-swing scheme; however, some comments about a typical static and differential swing scheme are provided for CLC and DIFF [7]. In particular, the CLC scheme shown in Figure 10.14 represents the conventional way for converting a full-swing signal to a low-swing one and vice versa. Moreover, the driver's circuit uses an additional supply voltage V_{ddL} to drive the load capacitance (i.e., interconnect). This voltage value is the voltage swing on the interconnect wire. From Table 10.5 it can be deduced that the noise margin remains in an acceptable level, because the receiver exhibits a differential behavior, while the circuit delay increases. Figure 10.15 illustrates the circuit of DIFF [7]. Differential signaling is an attractive choice due to its high common-mode noise rejection, leading to a very small signal swing.

Detailed comparison results of existing low-swing circuit techniques in terms of energy consumption, delay, energy × delay product, voltage swing, and SNR value are given in Table 10.5. The CMOS scheme of the first row represents the conventional full swing scheme, and it is assumed $V_{dd} = 2$ volts, $C_L = 1$pF. All swing schemes can achieve energy savings starting from 50% to a factor of seven. The PDIFF scheme provides the optimal solution with a very small energy consumption, acceptable performance, and perfect noise immunity level, employing a very low swing voltage of 0.5 V. Furthermore, a qualitative comparison of the plethora of low-swing techniques is depicted in Table 10.6.

TABLE 10.5 Low-Swing Techniques [32]

Approach	Energy (pJ)	Delay (ns)	Energy–Delay Product	Voltage Swing (V)	SNR
CMOS	11.6	2.1	24.5	2	1.52
CLC	4.4	3.1	13.6	1.1	1.24
DIFF [7]	3.0	2.7	8.1	0.25	1.64
PCD [10]	3.5	2.0	7.0	0.5	0.70
CISB [12]	3.5	4.4	15.4	0.25	0.66
CRB [13]	3.1	3.1	10.9	0.25	0.74
CCLC [32]	2.67	2.6	6.94	0.8	0.81
LCR [32]	2.44	2.59	6.32	0.8	1.23
PDIFF [32]	1.92	2.4	4.6	0.5	1.92

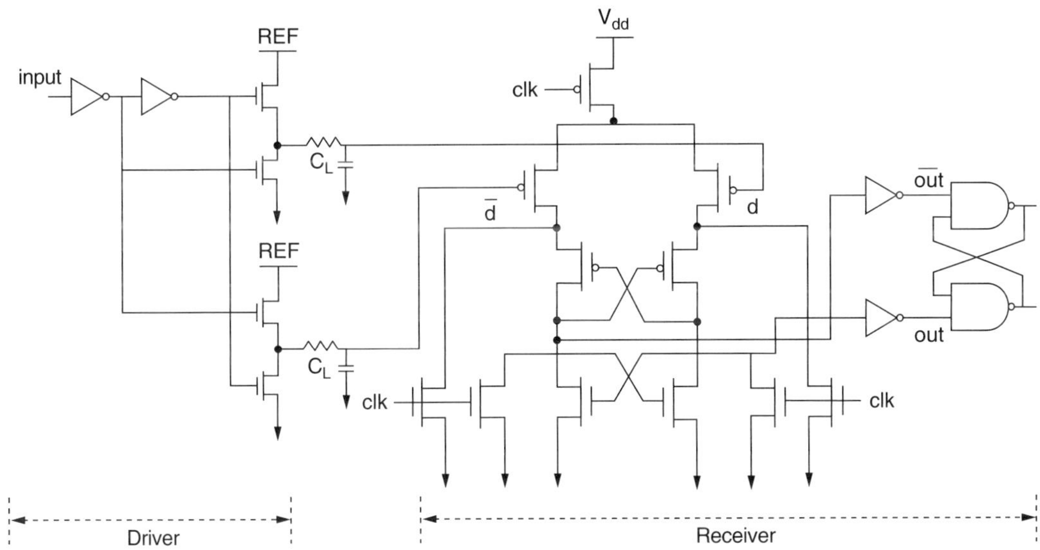

FIGURE 10.15 The architecture of differential low-swing scheme [7].

TABLE 10.6 Qualitative Comparison of Low-Swing Techniques

	CMOS	CLC	DIFF	PDC	CISB	CRB	CCLC	LCR	PDIFF
Extra power supplies	√				√	√			√
Reference voltages	√					√			
Multiple V_{th}	√		√	√	√	√	√	√	√
Voltage scaling	√	√	√	√	√	√	√	√	√
Low power		√	√	√	√	√	√	√	√
Low delay		√	√	√			√	√	√
Good SNR	√	√	√				√	√	√
Area penalty	√	√		√			√	√	√
Interconnect	Single	Single	Double	Single	Single	Double	Single	Single	Single

10.5 Circuit Technology-Independent Power Reduction

During a design process, it is possible to have only the behavioral of circuit (e.g., Boolean expression, state equations) with a few gates or flip-flops of a circuit. Although sometimes the actual implementation technology is not selected, it is possible to employ design techniques, which target circuit capacitance and switching activity reduction. Such low-power techniques may also reduce the total design time cost because a designer may have a good power consumption estimate during the early design phases.

10.5.1 Precomputation

This optimization technique is based on selectively precomputing the output logic values of a circuit one clock cycle before they are required, and then use the precomputed values to reduce the internal switching activity of the combinational logic in the successive clock cycle [2].

A simple example of this idea-based structure is shown in Figure 10.16, where the inputs of sequential block *A* have been partitioned into two sets, corresponding to registers R_1 and R_2, and the output of block *A* is the input of register R_3. The Boolean functions g_1 and g_2 serve as the predictor functions according to the following equations:

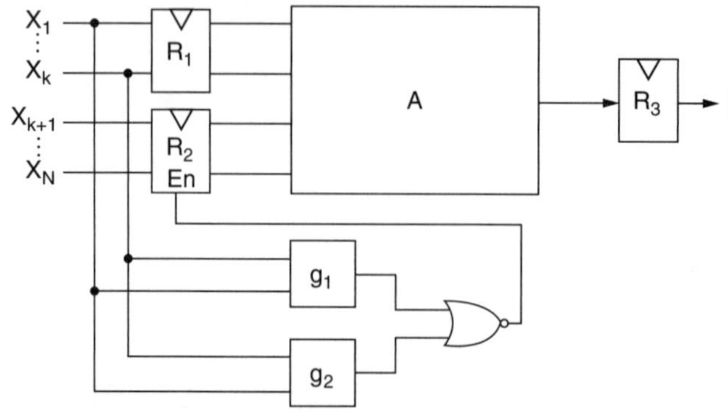

FIGURE 10.16 Precomputation structure for sequential circuits.

$$g_1 = 1 \Rightarrow f = 1$$
$$g_2 = 1 \Rightarrow f = 0 \quad (10.6)$$

The physical meaning of Equation (10.6) is that if function g_1 or g_2 equals 1, the value of the output function, *f*, is fully determined; however, the Boolean variables of functions g_1 and g_2 are a subset of the input signals of block A. Thus, the remaining signals, $(X_{k+1},\ldots, X_N)$, can be frozen. It must be stressed that it is not allowed for both g_1 and g_2 to be evaluated to 1.

Consequently, if the logic level of g_1 or g_2 is high, during clock cycle *T*, then the enable signal of register R_2 is low. Thus, the outputs of R_2, during clock cycle $T + 1$, are not changed. Because the output of R_1 is updated, the function *f* is evaluated correctly. As a subset of the inputs of block *A* changes, the switching activity of this block is reduced, implying a remarkable power saving. Because functions g_1 and g_2 occupy extra area and consume additional power, attention should be paid to the construction of g_1 and g_2 and an appropriate trade-off analysis should be performed.

10.5.2 Retiming

A novel method for reducing the power consumed in pipelined sequential circuits has been proposed in [20]. The method is based on retiming, which is a technique that repositions the flip-flops of the circuit resulting in the minimization of either the area or the delay of the circuit.

Because a flip-flop output makes at most one transition when the clock is asserted, the idea is to place a flip-flop in a circuit node with high glitching activity and high load capacitance. Thus, glitches are not propagated to the transitive fan-out of the node resulting in a reduction of the total switching activity, as shown in Figure 10.17.

However, attention should be paid because the switching activity of some nodes of the circuit may be changed due to retiming, which may result in an increase of the power consumption. In addition, the number of the used registers should be minimized because the power dissipation of the registers and clock line are not negligible. Finally, attention should also be paid to preserve the timing behavior of the circuit when retiming for low power is performed.

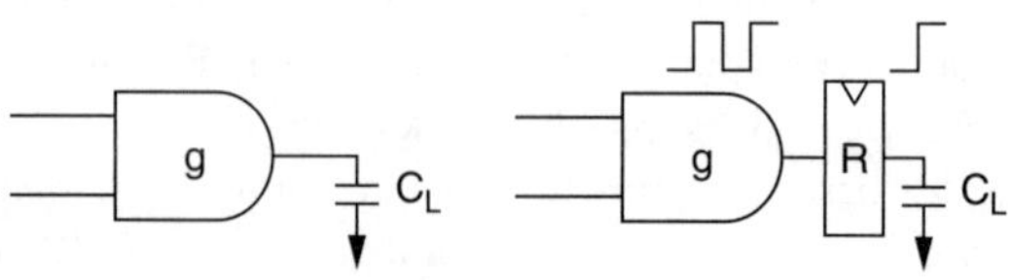

FIGURE 10.17 Retiming for low-power.

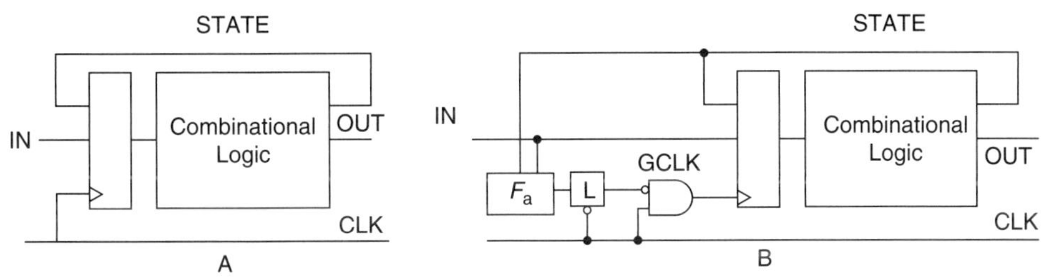

FIGURE 10.18 (a) Single clock, flip-flop based FSM, and (b) gated-clock version.

10.5.3 Synthesis of FSMs with Gated Clocks

This technique, presented in Benini et al. [3–6], refers to the power reduction that can be achieved in finite-state machine (FSM) circuits by using gated clocks. The key idea is that during the operation of an FSM there are conditions where the state and the output of the FSM do not change. Thus, clocking the circuit in the corresponding time intervals wastes power both in the combinational logic and in registers. Thus if we can detect the idle conditions of the FSM, we can also stop the clock during the corresponding time intervals. The benefit of using a gated-clock is twofold: first, when the clock is stopped, no power is consumed by the combinational logic because its inputs remain unchanged. Second, no power is consumed by the flip-flops and gated-clock line.

The flip-flop-based architecture is modified by setting a new activation signal, *GCLK*, as illustrated in Figure 10.18. The purpose of this signal is to selectively stop the local clock for the FSM when the machine is idle. The combinational circuit, F_a, which provides the activation signal, uses as its inputs the primary inputs and the state lines of the machine. It has been found that the application of this technique to such circuits provides power savings ranging between 10 and 30%. Because the function F_a consumes power, it is recommended to select a subset of all idle conditions such that this subset takes place with high probability during the circuit operation.

10.6 Circuit Technology-Dependent Power Reduction

The physical implementation of circuit behavior (e.g., Boolean function) may be differentiated by the chosen technology. In other words, the realized circuit topology and the chosen circuit components (e.g., from a library) may result in circuit designs with different hardware features (e.g., chain topology vs. tree topology), which affect the circuit capacitance and switching activity. The next paragraphs describe a series of low-power techniques, which achieve power savings through the reduction of a or C_L (Equation 10.2).

10.6.1 Path Balancing

The way the gates of a logic circuit are interconnected can strongly affect the overall switching activity, and hence the power dissipation. For example, timing skew between signals in a circuit can cause spurious transitions (glitches) resulting in extra power. To reduce the possible spurious activity in a circuit, delay of all true paths that converge at each gate must be balanced, as depicted in Figure 10.19, where the logic

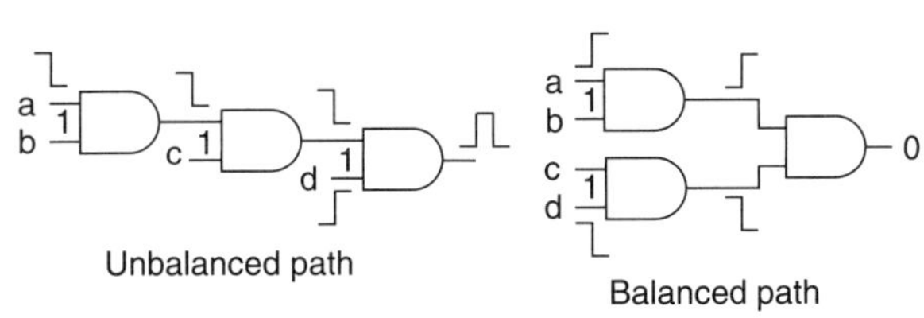

FIGURE 10.19 Path balancing for glitching reduction.

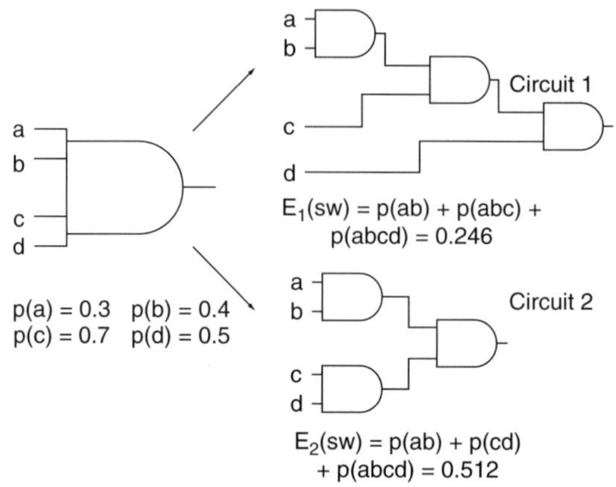

FIGURE 10.20 Technology decomposition for minimizing switching activity.

function $f = abcd$ is implemented in two alternative ways (i.e., chain structure and tree structure). In addition, notice that the tree implementation of function f provides glitches elimination, thus reducing effectively the total power dissipation.

Path balancing can be achieved before technology mapping by selective collapsing and logic decomposition or after technology mapping by delay insertion and pin reordering. The advantage of this technique is that by selectively collapsing the fan-ins of a node, the arrival time at the output of the node can be changed. Logic decomposition and extraction can be performed to minimize the level difference between the inputs of the nodes that are driving high capacitive nodes. Additionally, by inserting variable-delay buffers in a circuit, the delays of all paths in the circuit can be made equal. The issue in delay insertion is to use the minimum number of delay elements to achieve the maximum reduction in glitching activity. Path delays may sometimes be balanced by an appropriate signal to the pin assignment. This is possible, because the delay characteristics of CMOS gates vary as a function of the input pin that is causing a transition at the output.

10.6.2 Technology Decomposition

The next step during logic synthesis of a network is to convert the network to another, which only contains two-input AND/NAND and inverter gates. This step, named technology decomposition, is very useful for network synthesis and is carried out before the mapping of the network, according to the current cell library, takes place. Therefore, a decomposition scheme that minimizes the total switching activities of the network is a good starting point for power-efficient technology mapping.

Given the switching activity at each input of a node, Tsui et al. [28] suggested a technique for AND decomposition of this node, which reduces the total switching activity in the resulting two-input AND structure under a zero-delay model. The idea is to inject the high switching activity inputs into the decomposition model as late as possible, as shown in Figure 10.20, where two different decomposition structures for the four-input AND gate are depicted.

Note that signal *d*, which has the highest switching activity, is injected last in configuration A, thus implying better power performance for this configuration. This technique has been found as being optimal for dynamic CMOS circuits, but also produces very good results for static CMOS circuits. In general, the low-power technology decomposition procedure reduces the total switching activity in the circuits by 5% over the conventional balanced tree decomposition method.

10.6.3 Technology Mapping

Technology mapping refers to the process of binding a given Boolean network to the gates included in a target cell library. In Lin and Man [17], Tiwari et al. [27], and Tsui et al. [28], some design techniques

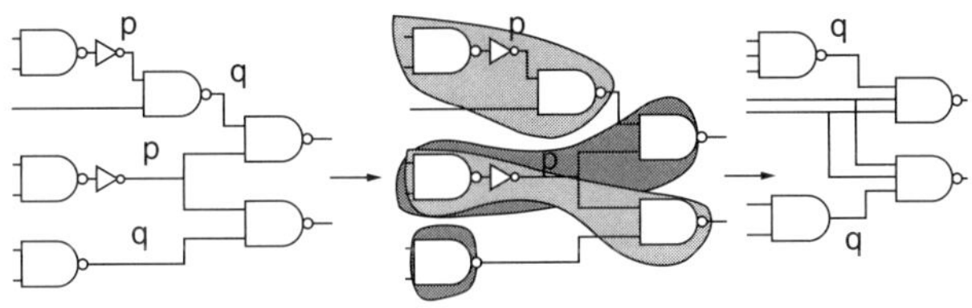

FIGURE 10.21 Technology mapping for minimizing switching activity.

for low-power consumption during technology mapping have been proposed. The main concept is to hide nodes with high switching activity inside the gates, thus they can drive smaller load capacitance, as presented in Figure 10.21.

According to Tsui et al. [28], the whole process consists of two steps. The first step requires the computation of power-delay curves (i.e., power consumption vs. arrival time) of all nodes in the network. The second step produces the mapping solution according to the previous curves and the required times at the primary inputs. This method has been proven to imply an 18% power savings at the expense of a 16% increase in area, without any penalty in network performance. In other words, we can say that the power-delay mapper reduces the number of high switching activity sub-networks at the expense of increasing the number of them having low switching activity. In addition, it reduces the network average load.

Although the approach mentioned previously refers to mapping for zero-delay circuits, an extension to a real-delay model is considered in Tsui et al. [28], resulting in optimum power solutions. According to [28], every point on the power-delay curve of a specific node uniquely defines a mapped subnet from the circuit inputs up to the node. The principle is to compute each such point with the probability waveform for the node in the corresponding mapped subnet. Thus, the total power cost, owing to steady-state transitions and hazards, of a candidate match can be calculated from the computed power-delay curves at the inputs of the gate and the power-delay characteristics of the gate itself.

10.7 Conclusions

The dynamic power consumption is a dominant power component for the current and future design technologies. Dynamic power substantially increases in nanometer technologies because of increased number of on-chip functions as well as a prolonging trend on getting higher clock frequencies. A multi-objective approach for reducing dynamic power consumption should combine multiple supply and threshold voltages with flexible gates from suitable cell libraries and efficient signaling schemes. Two design strategies can be adopted to reduce dynamic power. The first strategy concerns the supply voltage reduction, where substantial power savings can be achieved due to its quadratic dependence (i.e., $P \propto V_{dd}^2$). The second strategy concerns the capacitance or switching activity reduction, which is very useful when the design process is fixed. Four different sets of low-power design techniques were presented. More specifically, circuit techniques based on the principle of parallelism, techniques that use multiple supply voltages and low on-chip voltage swing, and techniques that are circuit technology-dependent and technology-independent. The key challenges to using multiple voltage supplies on a chip are minimizing area cost, placing logic cells under appropriate clustering constraints, as well as using dual power rails and efficient cell libraries that are capable of assigning the appropriate threshold voltage to each cell.

References

[1] V. Adler and E. Friedman, Repeater design to reduce delay and power in resistive interconnect, *Trans. Circuits and Systems — II*, Vol. 45, May 1998, pp. 607–616.

[2] M. Alidima, J. Monteiro, S. Devadas, A. Ghosh, and M. Papaefthimiou, Precomputation-based sequential logic optimization for low power, *IEEE Trans. on VLSI*, Vol. 2, No. 4, pp. 426–435, Dec. 1994.

[3] L. Benini, G. De Micheli, E. Macii, M. Poncino, and R. Scrasi, Symbolic synthesis of clock-gating logic for power optimization of synchronous controllers, *ACM Trans. on Design Automation of Electron. Syst.*, Vol. 4, No. 4, pp. 351–375, Oct. 1999.

[4] L. Benini, G. De Micheli, A. Lioy, E. Macii, G. Odasso, and M. Poncino, Synthesis of power-managed sequential components based on computation kernel extraction, *IEEE Trans. on CAD,* Vol. 20, No. 9, pp. 1118–1131, Sept. 2001.

[5] L. Benini, P. Siegel, and G. De Micheli, Saving power by synthesizing gated clocks for sequential circuits, *IEEE Design Test of Comput.*, pp. 32–41, Winter 1994.

[6] L. Benini and G. De Micheli, Automatic Synthesis of low-power gated-clock finite-state machines, *IEEE Trans. on CAD,* Vol. 15, No. 6, pp. 630–643, June 1996.

[7] T. Burd and R.W. Brodersen, *Energy Efficient Microprocessor Design,* Kluwer Academic Publishers, Boston, 2002.

[8] A.P. Chandrakasan and R.W. Brodersen, *Low-Power Digital CMOS Design,* Kluwer Academic Publishers, Boston, 1995.

[9] W. Chung, T. Lo, and M. Sachdev, A comparative analysis of low-power low-voltage dual-edge triggered flip-flops, *Trans. on VLSI Syst.*, Vol. 10, No. 6, Dec. 2002, pp. 913–918.

[10] R. Colshan and B. Jaroun, A novel reduced swing CMOS BUS interface circuit for high-speed low-power VLSI systems, *Proc. of Int. Symp. on Circuits and Syst. (ISCAS),* 30 May 1994, London, UK, Vol. IV, pp. 351–354.

[11] J. Goodman and A.P. Chandrakasan, Low-power scalable encryption for wireless systems, *Wireless Networks,* 4, 1998, pp. 55–70.

[12] M. Hiraki et al., Data-dependent logic swing internal bus architecture for ultra low-power LSI's, *IEEE J. Solid-State Circuits,* Vol. 30, Apr. 1995, pp. 397–402.

[13] R. Hossain, L. Wronski, and A. Albicki, Low-power design using double edge triggered flip-flops, *Trans. on VLSI Syst.*, Vol. 2, No. 2, June 1994, pp. 261–265.

[14] J.P. Hayes, *Computer Architecture and Organization,* McGraw-Hill, New York, 1978, p. 382.

[15] T. Kuroda, Low-power CMOS design challenges, *IEICE Trans. on Electron.*, Vol. E84-C, Aug. 2001, pp. 1021–1028.

[16] H.-J. Kwon and K. Lee, A new division algorithm based on lookahead of partial-remainder (LAPR) for high-speed/low-power coding applications, *IEEE Trans. of CAS-II,* Vol. 46, No. 2, Feb. 1999, pp. 202–209.

[17] B. Lin and H. De Man, Low-power driven technology mapping under timing constraints, *Proc. ICCAD,* 1993, pp. 421–427.

[18] R.P. Llopis and M. Sachdev, Low-power, testable dual-edge triggered flip-flops, *Proc. Int. Symp. Low-Power Electronics and Design,* 1996, pp. 341–345.

[19] M. Lowy, Low-power spread spectrum code generator based on parallel shift registers, *1994 IEEE Symp. on Low-Power Electron.*, San Diego, CA, Oct. 10–12, 1994, pp. 22–23.

[20] J. Monteiro, S. Devadas, and A. Ghosh, Retiming sequential circuits for low power, *Proc. ICCAD,* Nov. 7–11, Santa Clara, CA, pp. 398–402, 1993.

[21] G. Panigrahi, The implications of electronic serial memories, *Computer,* July 1977, pp.18–25.

[22] M. Pedram, Q. Wu, and X. Wu, A new design of double-edge triggered flip-flops, *Proc. ASP-DAC '98 Asian and South Pacific Design Automation Conf.*, Feb. 10–13, 1998, Yokohama, Japan, pp. 417–421.

[23] C. Piguet, J.-M. Masgonty, V. von Kaenel, and T. Schneider, Logic design for low-voltage/low-power CMOS circuits, *1995 Int. Symp. on Low-Power Design,* Dana Point, CA, Apr. 23–26, 1995, pp. 117–122.

[24] C. Piguet, Logic design for low-power CMOS circuits. Invited talk at TENCON '95, Hong-Kong, Nov. 7–10, 1995, pp. 299–302.

[25] T. Schneider, V. von Kaenel, and C. Piguet, Low-voltage/low-power parallelized logic modules, *Proc. PATMOS '95,* Paper S4.2, Oldenburg, Germany, Oct. 4–6, 1995, pp. 147–160.

[26] A. Antonio, G.M. Strollo, E. Napoli, and C. Cimino, Analysis of power dissipation in double edge-triggered flip-flops, *Trans. on VLSI Syst.,* Vol. 8., No. 5, Oct. 2000, pp. 624–629.
[27] V. Tiwari, P. Ashar, and S. Malik, Technology mapping for low-power in logic synthesis, *Integration, the VLSI J.,* July 1996.
[28] C.-Y. Tsui, M. Pedram, and A. Despain, Power-efficient technology decomposition and mapping under extended power consumption model, *IEEE Trans. on CAD,* Vol. 13, No. 9, Sept. 1994.
[29] K. Usami and M. Horowitz, Clustered voltage scaling technique for low-power design, *Proc. Int. Symp. on Low-Power Design,* Apr. 1995, pp. 3–8.
[30] K. Usami and M. Igarashi, Low-power design methodology and applications utilizing dual supply voltages, *Proc. Asia and South Pacific Design Automation Conf.,* Jan. 25–28, 2000, Yokohama, Japan, pp. 123–128.
[31] H. Yamauchi et al., An asymptotically zero power charge–recycling bus architecture for battery-operated ultra-high data rate ULSIs *IEEE J. Solid-State Circuits,* Vol. 30, Apr. 1995, pp. 423–431.
[32] H. Zhang, G. Varghese, and J. Rabaey, Low-swing on-chip signaling techniques: effectiveness and robustness, *Trans. on VLSI Syst.,* Vol. 8, No. 3, June 2000, pp. 264–272.

11

VHDL for Low Power

Amara Amara
ISEP

Philippe Royannez
Texas Instruments

11.1 Introduction

The purpose of this chapter is to provide front-end designers with guidelines and good design practices for writing efficient register transfer logic (RTL) code from a low-power standpoint. It is suitable for engineers that are already familiar with RTL coding for synthesis, but are not necessarily aware of low-power techniques.

RTL-level techniques are very efficient because hardware description language (HDL) programmers are knowledgeable about the circuit architecture and functionality. They can ask and answer questions such as: Why should we clock a register if the input data has not changed? Why should we update a data in a register or on a heavily loaded bus if this data is not used by anybody? A lot of power can be saved based on this information, but obviously, this can not be observed at the standard cell library level or at the technology level. It must be done at the RTL level.

After a brief reminder of basic coding rules and techniques, we address different types of blocks like operative parts or control logic. For each type of block, we introduce the appropriate techniques including clock gating, finite-state machine (FSM) state assignment, bus encoding, and conditional computing to optimize RTL code and to obtain a significant power consumption reduction after synthesis. In this chapter, we only consider the case of RTL synthesis for silicon complementary metal oxide semiconductor (CMOS) digital circuits. All examples are coded in VHDL, and synthesis script examples use synopsys DC.

11.2 Basics

11.2.1 Power Consumption

RTL synthesis has been a major improvement in the field of digital integrated circuit (IC) design. Besides the higher reusability and the better verification methodology, RTL-based design has definitely changed the way designers consider a digital circuit. Indeed, most of them do not see a circuit as a netlist of electronic components anymore but instead as a high-level software functional description. In that sense, RTL coding has improved the productivity, but, on the other hand, it has introduced a major disconnect between the front-end design and the real electronic devices where the power is burnt. For many years, however, this has not been a major issue. RTL synthesis was mostly timing driven with several iterations to optimize area and fix timing violations. Power was not a real concern, but with CMOS process scaling and ever increasing switching speeds, the power density can reach tens of W/cm^2 in today's digital ICs. Moreover, the exploding market of the portable electronic devices is also driving very strongly the need for low-power solutions. Therefore, the problem must also be tackled at the RTL stage, and RTL programmers cannot ignore power anymore. In particular, they need to bear in mind the underlying circuitry where it is consumed. Thus, let us briefly summarize the sources of power consumption.

Power is either static or dynamic. The static power consumption is due to MOS sub-threshold leakage and to a lesser degree extends to gate-induced drain leakage (GIDL), gate leakage, and diode leakage. The static current consumption used to be in the range of μA. Without leakage-reduction techniques, it is now in the range of mA and, in some cases, can account for more than 50% of the total power consumption. Temperature makes the picture even worse.

The dynamic power is due to the switching of CMOS gates. Ideally, this power is the well-known $P_D = \alpha CV^2 f$, where α is the switching activity. This includes the clock distribution network consumption and the parasitic power due to glitches. Often neglected, however, the latter can account for up to 15%.

Metal-oxide semiconductor (MOS) transistors do not switch instantaneously, and signals do not have zero transition time. There is always a short amount of time where both the pull-up and pull-down paths of a CMOS gate are on simultaneously, thus creating a parasitic current that is wasted. This additional power consumption sometimes called "short-circuit" power depends on the input and output transition times, on the output capacitance load and on the transistor characteristics. P_{Short} can account for 10% of the total dynamic power. Unlike the P_D term, this consumption decreases with the output capacitance load.

11.2.2 RTL Coding Applicability to Power Reduction

Static power-reduction techniques include multiple V_T, multiple V_{DD}, back biasing, and power supply scaling or switching, among others. These techniques affect the MOS process technology or the global system architecture. They are not really implemented at the RTL level. Thus, static power reduction is out of the scope of this chapter.

The dynamic power $P_D = \alpha CV^2 f$ can be optimized by reducing each factor. The power supply, V, as well as the operating frequency, f, is not handled at the RTL level, and voltage and frequency scaling are covered by other chapters of this book. The capacitance factor, C, is mostly dependent on the process technology and the standard cell library. The fan-out can be controlled by the dc_shell command set_max_fanout, but RTL code has a limited impact on the C factor. Thus, the RTL techniques will be used mainly to reduce the switching activity α and, to a lesser extent, the C factor.

As far as P_{Short} is concerned, this wasted power can be minimized by controlling the rise and fall times. For instance, we can use the set_max_transition directive for synthesis and check for post-synthesis reports as well as post-backend report. This constraint is not specific to power reduction. Slow nodes affect signal integrity, reliability, and timing. They should be avoided anyway. RTL coding has very little impact on this part of the power consumption.

To summarize, RTL level techniques are not applicable to reduce every type of power consumption. Therefore, the techniques presented in this chapter focus on dynamic power consumption reduction. This reduction will be mainly due to a better management of the switching activity.

The various strategies to reduce this switching activity often use the same types of basic techniques that are reviewed briefly in the next subsections.

11.2.3 Latch Inference

Latch insertion is a common practice to suppress or reduce unnecessary switching; however, latches, as memory elements, can cause race condition, logic hazard, or metastability. It is mandatory that the latch-enable command signals are spike-free, and have the appropriate setup and hold margins with regard to the data inputs. It is also recommended that these latches be initialized with the common hardware reset used by the other sequential elements of the block.

To infer a latch, it is recommended that either an incomplete IF clause or an explicit instantiation is used (see the VHDL example next).

```
EN_T <= EN or scan_mode;
LT: process (RST, EN_T)
begin
if (RST='1') then
      Q<='0';
elsif (EN_T='1') then
      Q<=D;
end if;
end process LT;
```

RTL designers should keep in mind that latch insertion can affect the testability, the static timing analysis (STA), and the equivalence checking. To avoid unwanted latches, check that your sensitivity lists are complete, that IF and CASE statements have default clauses, and that the variable contents are always initialized before being used. In addition, check the inference reports and the postsynthesis results against the report_reference and the all_registers options.

11.2.4 Direct Component Instantiation

Design for low power might require a very predictable and reproducible synthesis mapping. It is therefore very common to directly instantiate in the RTL code-specific cells, such as metal programmable delay buffers and clock-gating cells of clock tree buffers; however, this is at variance with the RTL reusability principle. To avoid this problem, it is a good practice to keep these direct inferences in some separate wrapper around the reusable RTL code. A generic component name should be used in the architecture part of the VHDL code. Then, for each target library, the explicit component can be defined in the specific configuration part of the VHDL description.

```
configuration my_block_cfg of my_block is
 for my_block_arch
 for INST0: generic_special_cell_name
 use configuration WORK.target_lib_special_cell_name_cfg
 end for;
 end for;
end my_block_cfg;
```

11.2.5 Explicit-State Encoding

In the RTL approach, the designer focuses on the high-level description and relies on the synthesis tool to implement the functionality. Most often, the FSM states or any symbols declared as enumerated types are encoded automatically. For low-power optimization, however, the designer might need to precisely control this encoding. This can be done either at the synthesis scrip or at the RTL level. With synopsys DC, the encoding is controlled by the set_fsm_state_vector and set_fsm_encoding commands.

To define the encoding at the RTL level, various synthesis tools support several attributes, but because there is no standardization, we recommend avoiding the enumeration type, such as that in the following portable code:

```
type state_type is std_ulogic_vector(1 downto 0);
constant S0: state_type :="01";
constant S1: state_type:="10";
signal curr_st, next_state : state_type;
```

11.3 Glitch Reduction

Glitches are due to converging combinatorial paths with different propagation delays. Let us consider the simple example depicted in Figure 11.1. A 32-bit adder is followed by an XOR-tree that counts the number of "1" in the sum and gives the parity. In the case of the -1 +1 addition the parity bit will oscillate many times before stabilizing to the valid state "0." Because the final and initial results are the same, all activity on the output node brings no information yet consumes both dynamic and short circuit power. This oscillation can also propagate to other combinatorial blocks and generate activity that is even more spurious. This propagation will stop either with a sequential element or by pulse swallowing once the transition times will become shorter than the intrinsic gate propagation delay. Glitches are not an issue for power consumption only. Because of the parasitic capacitive coupling, glitches also affect the signal integrity and the timing closure with effects like dynamic cross talk of driver weakening.

11.3.1 Gate-Level Control

To interrupt the propagation of glitches, a first idea is to pipeline the design, but this very efficient method comes at the expense of additional registers, latency, control logic, and clock tree distribution. The clock tree and registers will consume both static and dynamic power, and the designer must find the best compromise. Moreover, pipelining is not always possible because it delays the output data delivery by one or more clock cycles. In some cases, the architecture change requirements can go up to the compiler and the real time operating system (RTOS), which is often not possible. An intermediate solution is to subdivide a clock cycle into two or more phases. The multiple phase clocks can be used to mask the datapath signals with simple AND gates or with latches. A common implementation is the well-known two-phase master–slave latch logic; however, the overhead in terms of clock generation and distribution, static timing analysis, and design complexity must be carefully evaluated before using such complex clocking schemes.

Another approach consists in balancing the delay between different combinatorial paths. Delay cells are directly instantiated in the RTL and fine-tuned at the place and route step. Because delays vary with process variation and temperature, this method is difficult to implement. With CMOS device scaling and advanced technologies, those techniques are not recommended except for some full-custom, high-performance blocks.

Glitches activity can also be reduced by using sum of products style of Boolean equations. It is even more convenient to handle it at the synthesis step using the flattening options of the synthesis tool with, for instance, set_flatten true. These options prioritize the speed (i.e., logic depth) and thus reduce glitches; however, it is, in most cases, at the expense of area and dynamic power consumption.

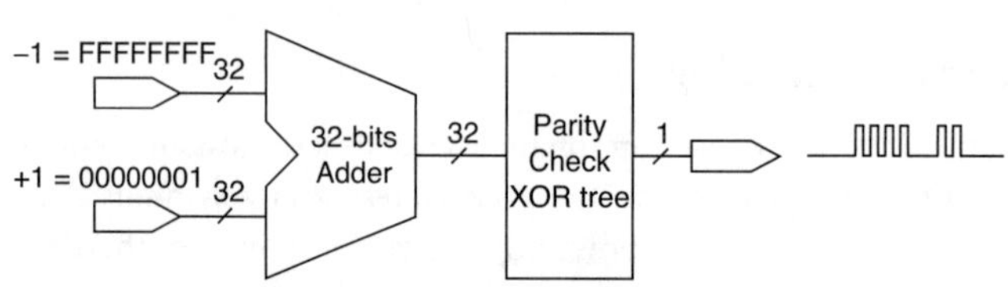

FIGURE 11.1 Example of glitches generation.

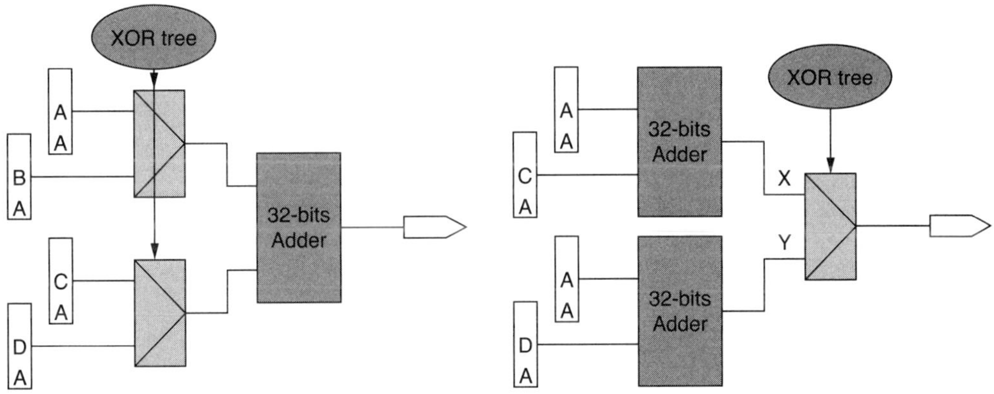

FIGURE 11.2 Glitch reduction by block reordering.

In case of a high-speed logic block, we can also use a monotonic design like domino logic. This type of logic is glitch-free, but requires a dedicated library of cells and an additional clock signal. To map the RTL code, we can again use direct instances or synthesis scripts to control the inferences (e.g., set_dont_use and set_use_only).

11.3.2 Block-Level Control

The last technique presented consists of rearranging the logic structure. To illustrate this approach, let us consider the block diagram in Figure 11.2. We use the previous XOR tree to select either A or B as the first operand of a 32-bit adder and C or D as the second operand. Because A, B, C, and D come from registers, they are stable data; but if the control signal of the multiplexers is oscillating, then the operands of the adder are unstable and propagate glitches which consume power. If we use two adders to compute X and Y sums first and then multiplex them, then adders see stable inputs and have much less power due to glitches. This reduction comes at the expense of one additional 32-bit adder block. In addition, note that synthesis tools are able to detect the two adders and, after a resource allocation step, could move back to the single-adder structure. To prevent this, the set_dont_touch attribute on net X and Y might be useful.

11.4 Clock Gating

Clock gating, which is probably one of the most well-known low-power techniques, is very effective in reducing the power consumption in digital circuits. The goal of this technique is to disable or suppress transitions from propagating to parts of the clock path (i.e., flip-flops, clock network, and logic) under a certain condition computed by clock-gating circuits. The savings are mainly due to the switching capacitance reduction in the clock network and the switching activity in the logic fed by the storage elements because unnecessary transitions are not loaded when the clock is not active.

Clock gating (CG) is illustrated in Figure 11.3. A block CG, which inhibits the clock signal when the idle condition is true, is associated with each sequential functional unit.. The clock signal is computed by function F_{cg}. CLK is the system clock and CLKG the gated clock of the functional unit. Clock-gating techniques have been successfully implemented in many microprocessors [1,2].

11.4.1 Flip-Flop-Based Design

Many implementations have been proposed for function CG. The simplest one uses an AND or an OR gate (Figure 11.4(a) and Figure 11.4(b)), but is not efficient because of the possible spikes at the output of the gate. An alternative and better solution is given in Figure 11.5 and is based on a latch L transparent

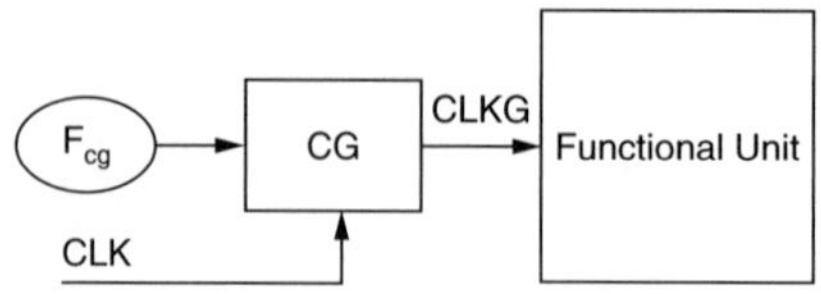

FIGURE 11.3 Clock-gating principle.

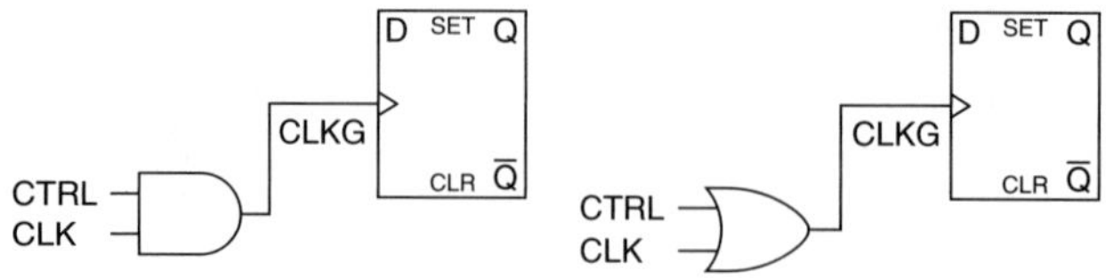

FIGURE 11.4 AND/OR CG block implementation.

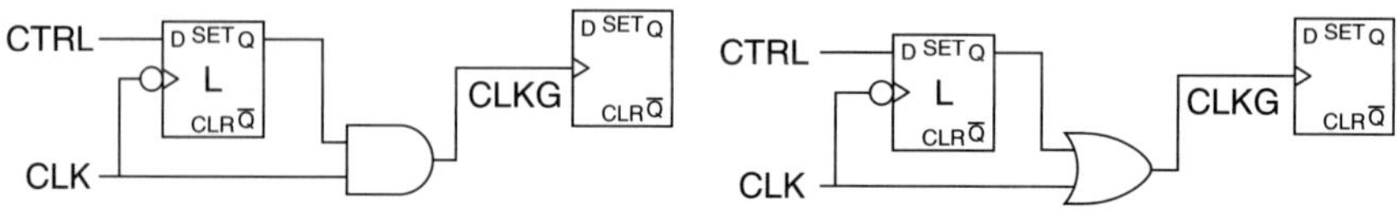

FIGURE 11.5 LATCH/AND/OR CG block implementation.

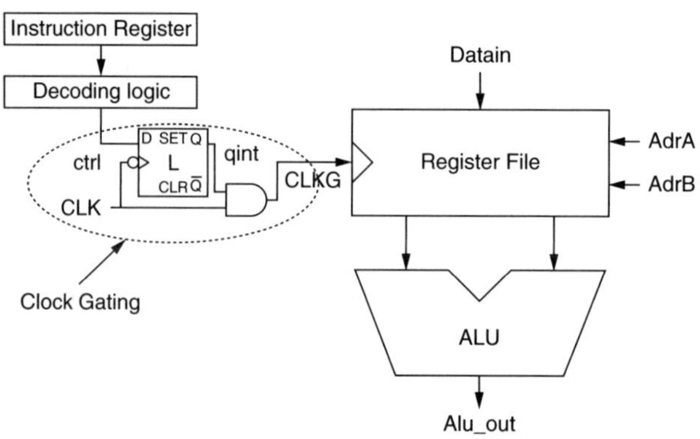

FIGURE 11.6 Clock gating example.

(when the clock is low) and an AND gate. With this configuration, the spurious transitions generated by function F_{cg} are filtered.

As an example, the circuit given in Figure 11.6 illustrates the clock gating of a datapath register file.

The clock-gating file and the register file must be physically close to reduce the impact on the skew and to prevent unwanted optimizations during the synthesis phase. They can be modeled by two separate processes in the same hierarchical block, synthesized, and then inserted into the parent hierarchy with a "don't touch" attribute. The following VHDL code describes the file register and its clock-gating circuit:

```
library ieee;
use ieee.std_logic_1164.ALL;
entity CG_RF_e is
port(
clk    : in std_logic;
adrA   : in std_logic_vector(4 downto 0);
adrB   : in std_logic_vector(4 downto 0);
datain: in std_logic_vector(7 downto 0);
```

```
wr    : in std_logic;
ctrl  : in std_logic;
A,B   : out std_logic_vector(7 downto 0));
end CG_RF_e;
architecture CG_RF_a of CG_RF_e is
signal clkg : std_logic;
type ram is array (0 to 31) of std_logic_vector(7 downto 0);
signal RF : ram;
begin
CG : Process (CLK, CTRL)
variable qint : std_logic;
begin
if clk = '0' then qint := ctrl; end if;
     CLKG <= (not Qint) and CLK;
end process;
process (CLKG)
begin
if CLKG = '1' then
          if WR ='1' then
               RF(conv_integer(adrA)) <= datain;
          else
               A <= RF(conv_integer(adrA));
               B <= RF(conv_integer(adrB));
          end if;
end if;
end process;
end;
```

In some applications, conditionally executed parts of the VHDL code can be identified and separated. Clock gating can be applied for each part. This technique has been proposed by Raghavan et al. [3]. The following VHDL code gives an example illustrating this technique. The initial process (P0) in the architecture listing2_1_a has been transformed into three processes (i.e., P1, P2, P3) as depicted in architecture listing2_2_a. A glitch-free load signal c_load is generated and combined to the clock, clk, to generate the gated clock, clkg, to be used by process P2.

```
Library ieee;
use ieee.std_logic_1164.ALL;
entity listing2_e is
port(
clk   : in std_logic;
load  : in std_logic;
A, B, C, E: in std_logic_vector(7 downto 0);
X, D, Z : out std_logic_vector(7 downto 0));
end listing2_e;
architecture listing2_1_a of listing2_e is
begin
P0 : process (clk)
begin
if (clk'event and clk='1')then
          X <= A + B;
          D <= E;
          if (load='1') then
```

```
                Z <= C;
            end if;
end if;
end process;
end;
architecture listing2_2_a of listing2_e is
signal Gclk : std_logic;
begin
P1 : process (clk)
begin
    if clk'event and clk='1' then
        X <= A + B;
        D <= E;
    end if;
end process;
P2 : process (Gclk)
begin
    if Gclk'event and Gclk='1' then
        if (load='1') then
            Z <= C;
        end if;
    end if;
end process;
P3: process (clk, load)
Variable c_load: std_logic;
begin
if clk = '0' then
        c_load <= load;
end if;
Gclk <= clk and c_load;
end process;
end;
```

In some designs, enabled flip-flops are used as shown in Figure 11.7(a). It is well-known that this kind of flip-flops are area and power-consuming, but their advantage compared with gated-clock-based design is that testability can be easily implemented and clock skew is more manageable. This kind of structure can be easily transformed into a gated clock structure (see Figure 11.7(b)). It is noteworthy that this transformation leads to important savings in area and power consumption. The following VHDL code gives the description of enabled flip-flop and its corresponding gated clock version:

```
library ieee;
use ieee.std_logic_1164.ALL;
entity EDFF_e is
```

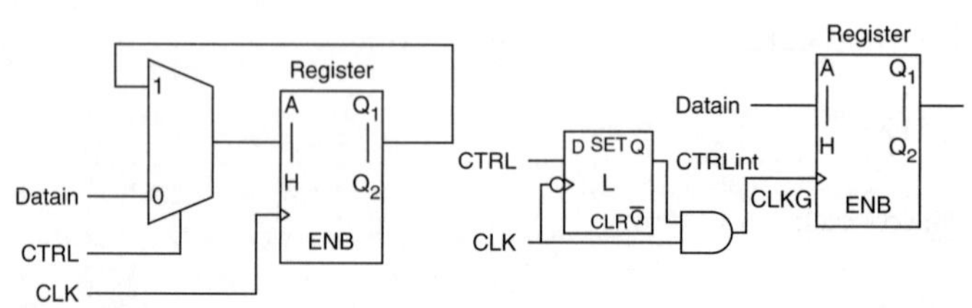

FIGURE 11.7 Enabled (a) to gated clock transformation (b).

```
port(
clk   : in std_logic;
ctrl  : in std_logic;
datain: in std_logic_vector(7 downto 0);
Q     : out std_logic_vector(7 downto 0));
end EDFF_e;
architecture EDFF_a of EDFF_e is
begin
process (clk)
begin
if clk'event and clk = '1' then
            if (ctrl='1') then
                  Q <= datain;
            end if;
end if;
end process;
end;
library ieee;
use ieee.std_logic_1164.ALL;
entity CG_DFF_e is
port(
clk   : in std_logic;
ctrl  : in std_logic;
datain: in std_logic;
Q     : out std_logic);
end CG_DFF_e;
architecture CG_DFF_a of CG_DFF_e is
signal clkg : std_logic;
begin
process (clkg)
begin
if clkg'event and clkg = '1' then
            Q <= datain;
end if;
end process;
process (clk, ctrl)
variable ctrl_int: std_logic;
begin
if clk = '0' then
            ctrl_int := ctrl;
end if;
clkg <= clk and ctrl_int;
end process;
end;
```

11.4.2 Issues in Clock Gating of DFF-Based Design

11.4.2.1 Timing Issues

The clock gate (i.e., AND or OR) must not alter the waveform of the clock other than turning the clock on or off. Unfortunately, introducing clock gating may result in setup time or hold time violations. Moreover, in most power design flows [4,5], the clock gating is inserted before the clock tree synthesis.

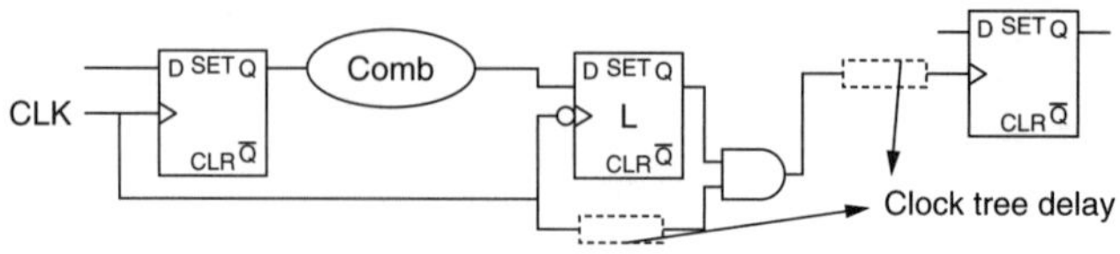

FIGURE 11.8 Timing issues in clock gating.

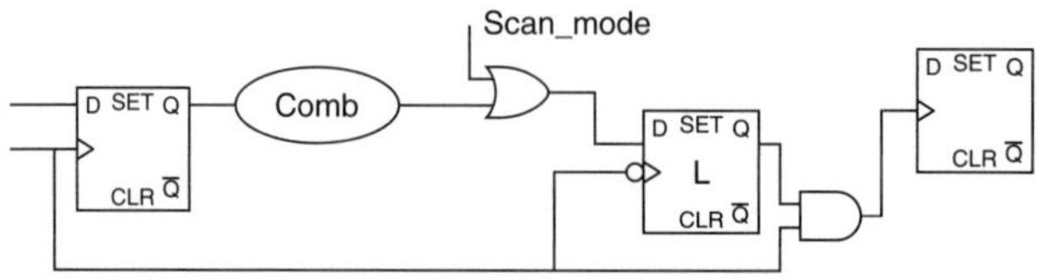

FIGURE 11.9 Testability issues in clock gating.

The existing synthesis tools by setting some variables allow the designer to specify these critical times before synthesis. When choosing these times, the designer has to estimate the delay impact of the clock tree from the clock gate to the gated register as depicted in Figure 11.8.

11.4.2.2 Testability Issues

Clock gating introduces multiple clock domains in the design, and this will affect the testability of the circuit. One way to improve the testability of the design is to insert a control point, which is an OR gate as indicated in Figure 11.9, controlled by an additional signal scan_mode. Its task is to eliminate the function of the clock gate during the test phase and thus restores the controllability of the clock signal.

11.4.2.3 Computer-Aided Design (CAD) Issues

Determining which flip-flops should be grouped for clock gating is an issue. Two techniques have been proposed:

1. Hold condition detection [6]. Flip-flops that share the same hold condition are detected and grouped to share the clock-gating circuitry. This method is not applicable to enabled flip-flops.
2. Redundant-clocking detection [7]. The method is simulation-based. Flip-flops are grouped with regard to the simulation traces to share the clock-gating circuitry. It is obvious that this method cannot be automated.

11.4.3 Latch-Based Design

In some applications, latch-based designs are preferred to D Flip Flop (DFF)–based designs. The basic concept is that a DFF can be split into two latches, and each one is clocked with an independent clock signal. The two clocks are nonoverlapping clocks as presented in Figure 11.10. Combinational network is usually inserted between the two latches to build a pipelined datapath (Figure 11.11). The main advantage is that this kind of design supports greater clock skew before failing than a similar DFF-based design. The second advantage is that time borrowing is achieved naturally in the pipelined datapath.

The clock gating is easy to implement. Figure 11.11 depicts a simple and robust way to do it [8]. A simple AND gate is used to generate the gated clock. This configuration is glitch-free because the control

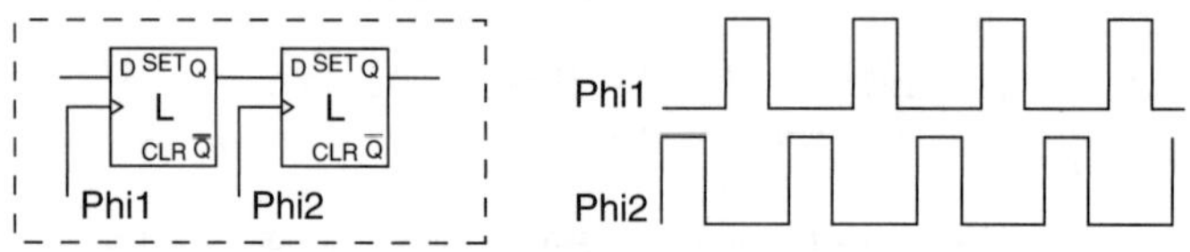

FIGURE 11.10 Master-slave latch and nonoverlapping clock concepts.

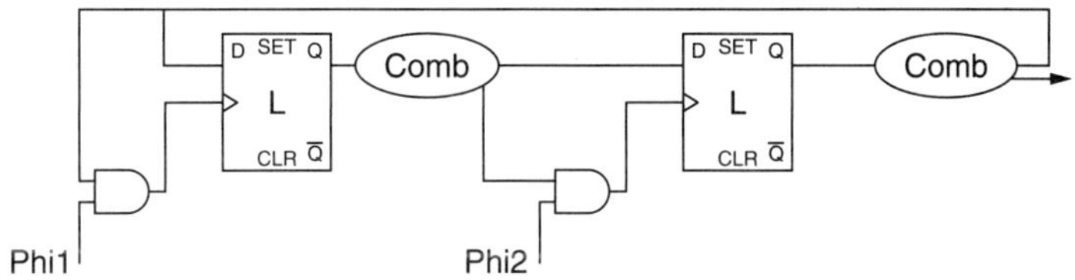

FIGURE 11.11 Clock gating of latch-based design.

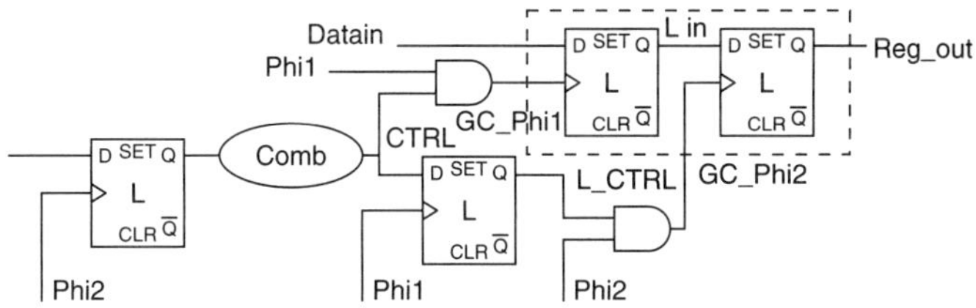

FIGURE 11.12 Clock gating in latch-based datapath.

signal, generated when Phi1 is high, is stable and remains stable when Phi2 goes high. In the case of registers, given the fact that the control signal is coming either from a latch clocked by Phi2 or a combinational function in which inputs are clocked by Phi2, it is necessary to add a latch clocked by Phi1 to delay the control signal as indicated in Figure 11.12 [9]. Notice that the AND gate must be handled carefully.

To prevent any optimization by the synthesis tool, the gate must be placed in a separate hierarchy level and assigned a "don't touch" attribute, for example. The code that follows this paragraph gives a VHDL description of a latch-based 32-bit register bank with the clock gating of Figure 11.12. Block CG contains the clock gating circuits (i.e., latch, 2 AND) in a separate hierarchy. This block can be assigned the "don't touch" attribute. Block RB contains the register bank, and, finally, GRB includes the structural description of the gated clock register bank.

```
Library ieee;
use ieee.std_logic_1164.ALL;
Entity RB is
Port (Phi1, Phi2 : in std_logic;
datain: in std_logic_vector(31 downto 0);
Reg_out: out std_logic_vector(31 downto 0));
End RB;
architecture Register_Bank of RB is
signal L_In: std_logic_vector(31 downto 0);
begin
process (GC_Phi1, datain)
begin
if (GC_Phi1 = '1') then
L_In <= datain;
end if;
end process;
process (GC_Phi2, L_In)
begin
if (GC_Phi2 = '1') then
Reg_out <= L_In;
end if;
end process;
```

```
end Register_Bank;
library ieee;
use ieee.std_logic_1164.ALL;
entity GC is
port(
CTRL, Phi1, Phi2: in std_logic;
GC_Phi1, GC_Phi2: out std_logic);
end GC;
architecture Gated_Clock of GC is
signal L_CTRL: std_logic;
begin
process (Phi1, CTRL)
begin
     If (Phi1 = '1') then
          L_CTRL <= CTRL;
end if;
end process;
GC_Phi1 <= CTRL AND Phi1;
GC_Phi2 <= L_CTRL AND Phi2;
end Gated_Clock;
library ieee;
use ieee.std_logic_1164.ALL;
entity GRB is
Port (
Phi1, Phi2 : in std_logic;
ctrl  : in std_logic;
datain: in std_logic_vector(31 downto 0);
Reg_out: out std_logic_vector(31 downto 0));
end GRB;
architecture Gated_Clock_Register_Bank of GRB is
component RB
port (Phi1, Phi2: in std_logic;
     datain: in std_logic_vector(31 downto 0);
     Reg_out: out std_logic_vector(31 downto 0));
end Component;
component GC
port (CTRL, Phi1, Phi2: in std_logic;
     CG_Phi1, CG_Phi2: out std_logic);
end Component;
signal GC_Phi1, GC_Phi2: std_logic;
begin
RB_instance: RB port map
     (GC_Phi1, GC_Phi2, datain, Reg_out);
GC_instance: GC port map
     (CTRL, Phi1, Phi2, GC_Phi1, GC_Phi2);
end Gated_Clock_Register_Bank;
```

11.4.4 Issues in Latch-Based Design

One of the design issues related to latch-based clock gating has been reported in Arm et al. [8]. In fact, the synthesis tool finds timing loops going through the control paths. As we can see from Figure 11.11,

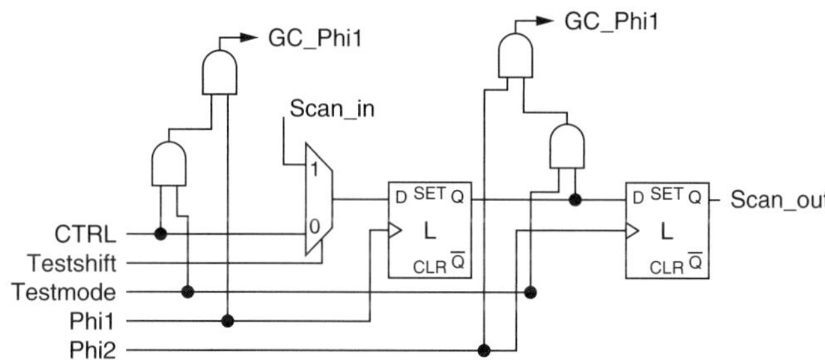

FIGURE 11.13 Clock-gating block with testability improvement.

the clock enable signal of each latch depends on the value computed by the logic fed by the other latches. As mentioned in Arm et al. [8], at least one of these paths can be cut using tool-dependent attributes (i.e., set-disable-timing for Synopsys design compiler).

The second design issue is related to testability. Figure 11.13 [9] illustrates the modification of clock-gating block to improve the testability of latch-based design.

11.5 Finite-State Machines

Finite-state machines (FSMs) are very common parts of digital systems. They are intensively used to generate signal sequences, to check an input signal sequence or to control datapath parts. The basic structure of an FSM is a state register and two logic blocks. The input (or "next-state") logic block computes the next state as a function of the current state and of the new set of inputs. The output logic block generates the outputs as a function of the current state (for a Moore FSM). The power can be burnt either in the logic blocks or in the clock distribution to the flip-flops of the state register. We present here various techniques to minimize this power consumption, using explicit state encoding and clock gating. The RTL coding of these techniques have been presented in the previous sections.

11.5.1 Gated-Clock FSM

The basic idea of gated-clock FSM is that it is not useful to have switching activity in the next-state logic or to distribute the clock if the state register will sample the same vector [20]. Let us take a simple, yet very common, example.

Figure 11.14 depicts a state machine that interacts with a timer-counter to implement a very long delay of thousands of clock cycles before executing a complex but very short operation (in the DO_IT state). We can use the clock-gating techniques to freeze the clock and the input signals as long as the ZERO flag from the time-out counter is not raised. This idea is efficient because this FSM spends most of the time in the WAIT state. It can be even more efficient if we assume that the FSM is used to control a very large datapath which outputs will not be used in the WAIT state. We can gate the clock or mask the inputs of this datapath and, therefore, avoid dynamic power consumption during all the countdown phases.

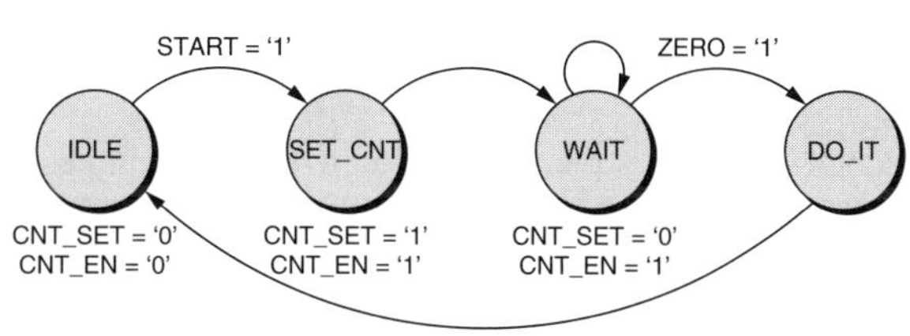

FIGURE 11.14 Example of FSM where the clock gating is easy and efficient.

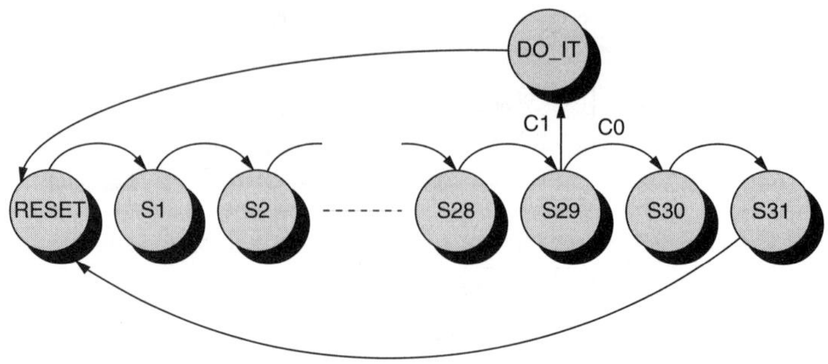

FIGURE 11.15 Example of FSM with gray encoding.

Software programmers are used to the 90/10 rule that states that 90% of the time a program loops over only 10% of the code. Actually, this rule is often valid for the FSM as well. It is the RTL designer's task to try to extract these small subparts of the FSM, isolate them, and then freeze the rest of the logic that is large and that most of the time does not achieve any useful computation.

11.5.2 State Encoding

This kind of technique uses the state encoding to reduce the logic activity in the input and output combinatorial blocks. One simple idea is to use an encoding that minimizes changes from one state to another if this transition is very likely to happen. In other words, we should minimize the hamming distance of the transition with high probability. This requires tools that propagate transition probabilities on the FSM inputs and calculate the probability of each transition. Again, we can make the parallel with software because this is similar to the branch prediction techniques. Although this probability estimation can be difficult, very common cases exist where this can be applied easily.

In the example depicted in Figure 11.15, states from RESET to S29 are chained sequentially with 100% probability of transition. Therefore, a gray encoding is the best choice. If we assume that condition C0 has a much lower probability than C1, the gray encoding should be not be incremented from S29 to S30 and S31.

However, what we gain in the next-state logic might be lost in the output logic activity. The designer has to find the best trade-off. If we consider now the power reduction on the output logic, we can also choose a judicious state encoding. A very common choice is the "one hot" encoding to optimize speed, area, and power for the output logic [19]. This approach is only valid for a small FSM (i.e., less than 8 to 10 states) because of the large state register and the increasing complexity of the next-state logic. For a larger FSM, a case-by-case analysis is needed. A good practice is to group states that generate the same outputs and assign them codes with minimum hamming distance.

A simple example is given in Figure 11.16, where an FSM is used to recognize the sequence "BEEFBEEF" on the input I and generates a flag Y = 1 if the sequence is complete. The encoding proposed achieves both a minimum "next-state logic" activity due to the "gray-like" encoding as well as no power consumption at all in the output logic because the orthogonal encoding defines the most significant bit of the state register as the flag Y itself [22].

11.5.3 FSM Partitioning

Often, FSM can be partitioned into smaller pieces. The idea here is to decompose a large FSM into several simpler FSMs with smaller state registers and combinatorial logic blocks. Only the active FSM receives clock and switching inputs. The others are static and do not consume any dynamic power [21]. Let us illustrate this technique with a simple example.

We consider a large FSM that includes a small subroutine, which is used very often in a real application scenario. We can easily partition the big FSM into two parts and isolate the subroutine loop. We add a

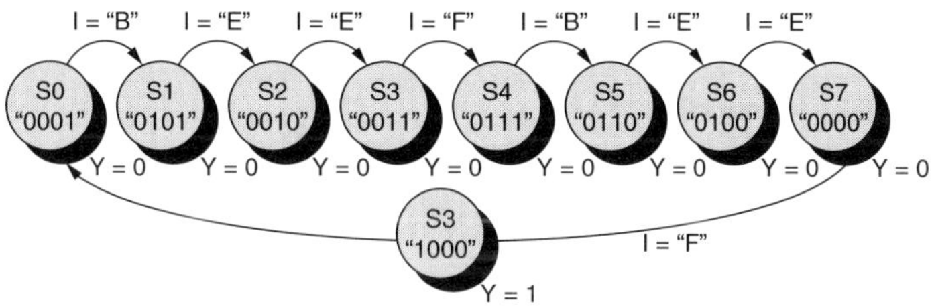

FIGURE 11.16 Example of FSM with zero-output logic.

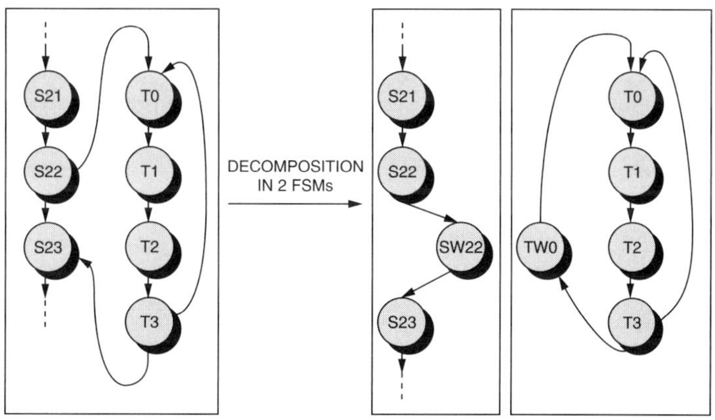

FIGURE 11.17 Example of FSM decomposition.

wait state, SW22 and TW0, between the entry and exit points of the subroutine in both FSMs. The new FSMs are mutually exclusive; when one is operating, the other one remains in its wait state. In such a state, clock and inputs can be gated to prevent any dynamic power (power supply could even be switched off to save leakage). The power savings is even higher if we can isolate very small subsets of states where the initial FSM remains most of the time.

11.6 Datapaths

An important amount of energy may be wasted in the datapath due to switching activity that does not contribute to the functionality of the circuit. Different techniques have been proposed to suppress or reduce dynamically this energy. Among these techniques, precomputation logic [10], guarded evaluation [11], and control-signal gating [12] techniques are widely used by low-power circuit designers. These techniques can be used early in the design flow (i.e., at the RTL level).

11.6.1 Precomputation Design Techniques

The principle of precomputation is to identify a logic condition on some inputs of a combinational circuit for which the output does not vary. Figure 11.18(a) gives a generic example of such a circuit. The inputs of the combinational logic f(X) are partitioned into precomputed inputs and gated inputs. If the output Y is independent of the gated inputs, then the function g generates a control signal for the register R2 that freezes its outputs. Yeap [13] describes a systematic method to derive the function g, but unfortunately, for given inputs, partitioning the solution is not unique, and the designer has to find the one that gives the best power-performance-area trade-off. Many implementation alternatives of precomputation logic are given in Yeap [13]. Figure 11.18(b) is a simple and realistic example of precomputation logic. It is a binary comparator that computes A > B (see the VHDL code that follows this paragraph).

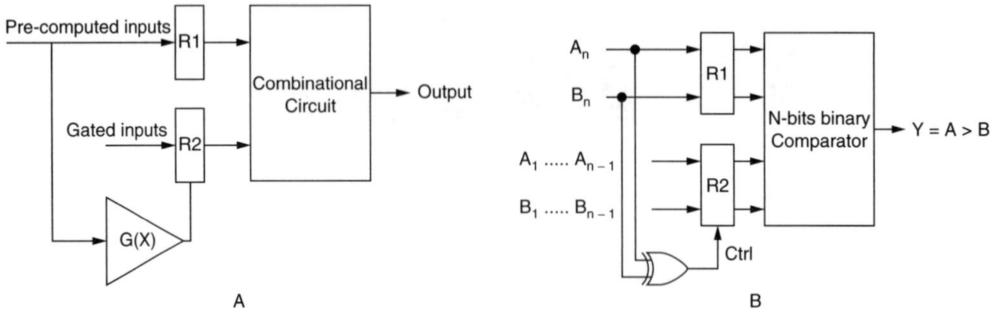

FIGURE 11.18 (a) Precomputation logic, and (b) application to a comparator.

In this case, the precomputation condition is very easy to derive. The precomputed inputs are A_n and B_n, and the most significant bits and the remaining bits are the gated inputs, thus the precomputation is a simple XOR gate. It is obvious that the designer needs to have some knowledge on the input statistics to apply efficiently the precomputation techniques. In practice, the selection of R1, R2, and the precomputation function depends heavily on the designer's experience.

```
Library ieee;
use ieee.std_logic_1164.ALL;
entity PC_Comp is
port (
A, B: in Std_Logic_Vector(31 downto 0);
Clk : in Std_Logic;
Y : out Std_Logic);
end PC_Comp;
architecture b32Comp of PC_comp is
signal Ctrl : std_logic;
signal A_R1, B_R1 : Std_Logic;
signal PC_A_R2, PC_B_R2 : Std_Logic_Vector(30 downto 0);
begin
Ctrl <= A(31) Xor B(31);
Y <= '1' when ((A_R1 & PC_A_R2) > (B_R1 & PC_B_R2)) else '0';
R1 : process (Clk)
begin
      if (Clk'event and Clk ='1') then
            A_R1 <= A(31);
            B_R1 <= B(31);
      end if;
end process;
R2 : process (Clk)
begin
      if (Clk'Event and Clk ='1') then
            if (Ctrl = '0') then
                  PC_A_R2 <= A(30 downto 0);
                  PC_B_R2 <= B(30 downto 0);
            end if;
      end if;
end process;
end b32Comp;
```

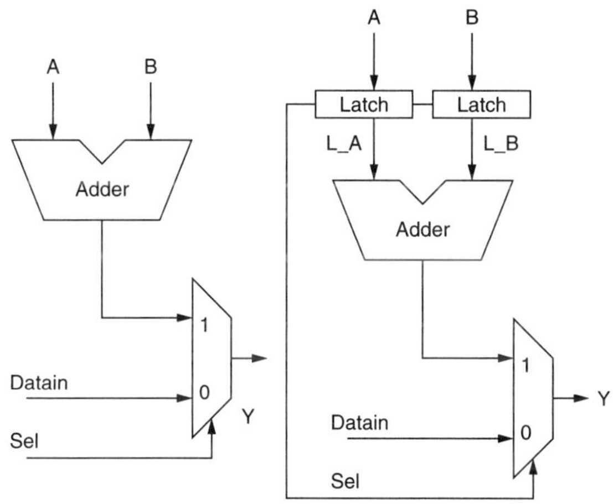

FIGURE 11.19 (a) Original circuit, and (b) its guarded evaluation version.

11.6.2 Guarded Evaluation Design Techniques

The technique is applicable to embedded combinational blocks from which outputs are in idle condition. Transparent latches are inserted at all the inputs of the embedded block. Control circuitry is added to determine the idle condition, which is then used to disable the latches. Figure 11.19 is a simple example that illustrates this technique. The arithemetic and logic unit (ALU) output may or may not be used depending on the condition selection of the multiplexer. If it is not used, the latches preserve the previous output values of the ALU. It is obvious that for wide buses, the area and power dissipation overhead should be nonnegligible. Following is the VHDL description of the Figure 11.19(b) circuit:

```
library ieee;
use ieee.std_logic_1164.ALL;
entity GE_Alu is
port(
A, B, Datain : in Std_Logic_Vector(31 downto 0);
Y : out Std_Logic_Vector(31 downto 0);
Sel : in Std_Logic);
end GE_Alu;
architecture Garded_Evaluation_Alu of GE_Alu is
signal L_A, L_B : Std_Logic_Vector(31 downto 0);
begin
process (Sel, A, B)
begin
    if Sel ='1' then
        L_A <= A;
        L_B <= B;
    end if;
end process;
Y <= (L_A + L_B) when (Sel ='1') else Datain;
end Garded_Evaluation_Alu;
```

11.6.3 Control-Signal Gating Design Techniques

The techniques we presented previously all aim to reduce the switching activity in a datapath module. The control-signal technique takes advantage of a fine granularity analysis to reduce the switching activity

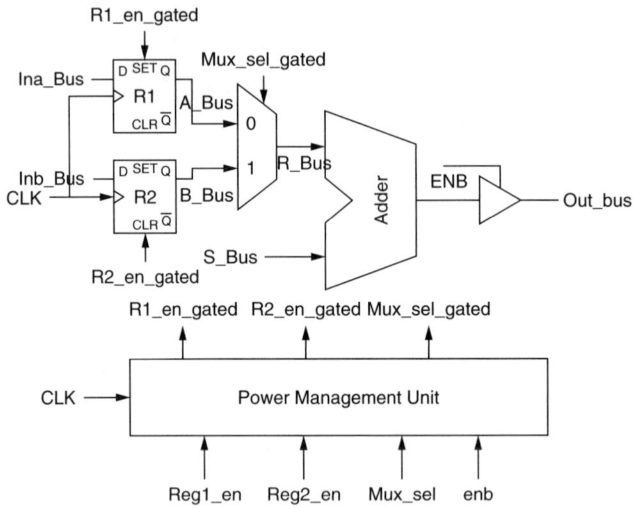

FIGURE 11.20 Control-signal gating technique.

in the datapath buses. The method is based on the observability don't care concept (ODC) [14] to detect when a bus is not used and to stop the propagation of the switching activity through the module(s) driving the bus. We can find in Kapadia et al. [12] a good formulation of the concept and its application to gating the datapath control signals. Figure 11.20 illustrates a datapath example. When enb is not active, mux_sel, reg1_en, and reg2_en can be gated, leading to a 100% switching activity reduction in R_Bus, A_bus, and B_Bus. When mux-sel is active, either reg1_en and reg2_en can be gated depending on the value of mux-sel. The gating conditions for the datapath of Figure 11.20 have been derived in Kapadia et al. [12] and are given next.

```
R1_en_gated = reg1_en AND (not(mux_sel OR (not enb)))@(T+1)
R2_en_gated = reg2_en AND (not(not mux_sel OR not enb))@(T+1)
(mux_sel_gated) @T = (mux_sel_gated) @(T-1) if ((not enb) @(T+1) = = True)
```

The suffix @T means the value of a variable or a function at the current clock cycle, @T-1 is the value one clock cycle before, and, finally, @T+1 is the value at the next clock cycle.

These equations can be implemented in a power management unit as depicted in Figure 11.20. The power management unit (PMU) generates all the gated control-signals for the datapath. The VHDL description of the PMU is given next.

```
library ieee;
use ieee.std_logic_1164.ALL;
entity PMU is
port (
Reg1_en, Reg2_en, Mux_sel, Enb, Clk : in Std_Logic;
R1_en_gated, R2_en_gated, Mux_sel_gated : out Std_Logic);
end PMU;
architecture Power_Management_Unit of PMU is
signal Enb_int, R1_en_tmp, R2_en_tmp : Std_Logic;
begin
R1EG : process (Clk)
begin
if (Clk'Event and Clk='1') then
R1_en_tmp <= NOT(mux_sel OR (NOT Enb));
end if;
```

```
R1_en_gated <= R1_en_tmp AND reg1_en;
end process;
R2EG : process (Clk)
begin
if (Clk'Event and Clk='1') then
R2_en_tmp<= NOT(NOT(mux_sel) OR (NOT Enb));
end if;
R2_en_gated <= R2_en_tmp AND reg2_en;
end process;
MSG : process (Clk)
begin
if (Clk'Event and Clk='1') then
Enb_int <= NOT Enb;
          if (Enb_int = '0') then
mux_sel_gated <= mux_sel;
          end if;
end if;
end process;
end Power_Management_Unit;
```

11.7 Bus Encoding

Advanced systems are typically characterized by wide and long buses, which consume a large amount of power mainly due to large capacitance and a significant switching activity. Many techniques have been proposed to deal with this issue at different design levels: low-swing bus [15], charge recycling bus [16], bus pipelining [17], bus multiplexing, and bus encoding techniques. The latter are more suitable for VHDL coding for low power. We will focus on one of them and briefly present other bus encoding techniques.

11.7.1 Bus Invert Encoding

Bus invert encoding is suitable for a set of parallel and synchronous signals such as internal buses in modern system on chip (SoC) architectures [18]. The idea behind is very simple: Before sending the data, the emitter compares its current value with the previous one and decides whether to send it or to send its inverted value along with a polarity signal. A bank of XOR gates at the sending and receiving ends inverts the bus data if necessary. Figure 11.21 depicts the bus encoding architecture, and the corresponding VHDL code is given next.

```
library ieee;
use ieee.std_logic_1164.ALL;
```

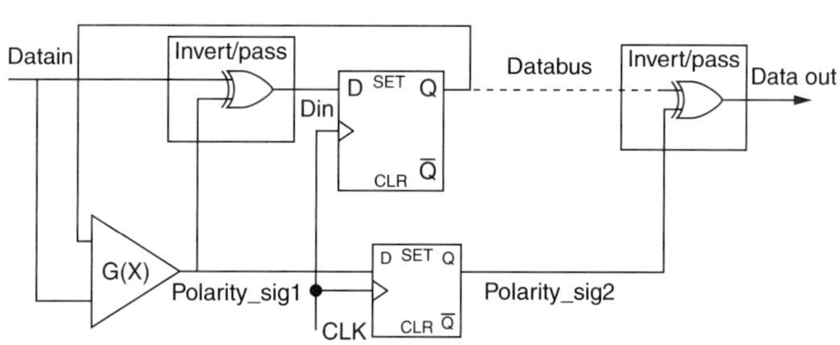

FIGURE 11.21 Bus invert encoding scheme.

```
entity BIE is
port(
Datain : in Std_Logic_Vector(31 downto 0);
Dataout : out Std_Logic_Vector(31 downto 0);
CLK : in Std_Logic);
end BIE;
architecture Bus_Invert_Encoding of BIE is
signal Din, Databus: Std_Logic_Vector(31 downto 0);
signal polarty_sig1, polarity_sig2: Boolean;
begin
Polarity_sig1 <= Polarity_Gen(datain, databus);
– Polarity_Gen is a procedure that returns True
– if the bus value has to be inverted
process (CLK)
begin
      if CLK'Event and CLK = '1' then
            if Polarity_sig1(datain,databus) == True; then
Databus <= not Datain;
else
Databus <= Datain;
            end if;
            polarity_sig2 <= polarity_sig1;
      end if;
end process;
Dataout <= polarity_sig2 xor Databus;
end Bus_Invert_Encoding;
```

11.7.2 Other Bus Encoding Techniques

In the bus invert approach, we have to calculate the hamming distance between two consecutive codes and transmit an additional polarity signal, which changes the interfaces between the emitter and the receiver. An improvement consists of making the guess that if the most significant bit (MSB) is "1," the inverted code should be transmitted. The MSB = "1" can be used as the polarity information. This technique is efficient when the vector to transmit is a 2's complement arithmetic data bus, with MSB being the sign bit.

Another common technique takes advantage of the fact that, very often, the value transmitted on a bus (an address bus, for instance) is simply the previous value with an increment. Therefore, the lines can remain the same (i.e., no power consumption) as long as the codes are consecutive, which is mentioned to the receiver by an additional control signal.

Finally, we can also use the knowledge of the set of symbols or probability of sequences to encode them in order to reduce the switching activity on the bus. For instance, if the sequence "0101" ⇒ "1010" occurs 90% of the time, we can save power by recoding "0101" into "0000" and "1010" into "0001."

11.8 Conclusion

The RTL coding step is not too early in the design flow to address power consumption optimization. For each source of consumption and each type of digital block, appropriate solutions can be implemented. Although the theory behind some of these techniques can be complex, they are often easy to implement. RTL designers should be aware of these techniques and use their knowledge of the system not only to optimize the speed performance, but also to reduce the unnecessary switching activity.

11.9 Acknowledgments

The authors thank Dr. Zinai Karima and Dr. Thomas Ea for their helpful comments and suggestions.

References

[1] G. Gerosa et al., A 2.2-W 80-Mhz superscalar RISC microprocessor, *IEEE J. Solid-State Circuits,* vol. 29, no. 12, pp. 1440–1454, Dec. 1994.
[2] C. Piguet et al., Low-power design of 8-bit embedded CoolRISC microcontroller cores, *IEEE J. Solid-State Circuits,* vol. 32, no. 7, July 1997.
[3] N. Raghavan, V. Akella, and S. Bakshi, Automatic insertion of gated clocks at register transfer level, *Proc. 12th Int.l Conf. on VLSI Design,* January 1999.
[4] *Power Compiler Design Manual,* Synopsys Ltd.
[5] PowerTheater, SequenceDesign Ltd.
[6] F. Theeuwen and E. Seelen, Power reduction through clock gating by symbolic manipulation, *Proc. Symp. Logic and Architecture Design,* Dec. 1996, pp. 131–136.
[7] M. Ohnishi, A. Yamada, H. Noda, and T. Kambe, A method of redundant clocking detection and power reduction at RT-level design, *Proc. 1997 Int. Symp. Low-Power Electronics and Design,* Monterey, CA, Aug. 1997, pp. 184–191.
[8] C. Arm, J.-M. Masgonty, and C. Piguet, Double-latch clocking scheme for low-power IP cores, *PATMOS 2000,* Goettingen, Germany, September 13–15, 2000.
[9] T. Schneider, *VHDL: Méthodologie de Design et Techniques Avancées,* Dunod, Paris, France, 2001.
[10] M. Alidina, J. Monteiro, S. Devadas, A. Gosh, and M. Papaefthymiou, Precomputation-based sequential logic optimization for low power, *Proc. 1994 Int. Comput.-Aided Design,* San Jose, CA, Nov. 1994, pp. 74–81.
[11] V. Tiwari, S. Malik, and P. Ashar, Guarded evaluation: pushing power management to logic synthesis/design, *Proc. Low-Power Design Symp.,* Dana Point, CA, Apr. 1995, pp. 221–226.
[12] H. Kapadia, L. Benini, and G. De Micheli, Reducing switching activity on datapath buses with control-signal gating, *IEEE J. Solid-State Circuits,* vol. 34, pp. 405–414, Mar. 1999.
[13] G.K. Yeap, *Practical Low-Power Digital VLSI Design,* Kluwer Academic Publishers, Dordrecht, 1998.
[14] G. De Micheli, *Synthesis and Optimization of Digital Circuits,* McGraw-Hill, New York, 1994.
[15] M. Hikari, H. Kojima, et al., Data-dependent logic swing internal bus architecture for ultralow-power LSIs, *IEEE J. Solid-State Circuits,* vol. 30, no. 4, pp. 397–402, Apr. 1995.
[16] H. Yamauchi, H. Akamatsu, and T. Fujita, An asymtomatically zero-power charge-recycling bus architecture for battery-operated ultra-high data rate ULSIs, *IEEE J. of Solid-State Circuits,* vol. 30, no. 4, pp. 423–431, Apr. 1995.
[17] L. Benini, Designing advanced NoCs architectures, *Int. Seminar on Application-Specific Multi-Processor SoC,* Chamonix, France, July 7–11, 2003.
[18] M. Stan and W. Burleson, Bus-invert coding for low-power IO, *IEEE Trans. on VLSI Syst.,* vol. 3, no. 1, pp. 49–58, Mar. 1995.
[19] C. Tsui and M. Pedram, Low-power state assignment targeting two and multi-level logic implementation, *ACM/IEEE Int. Conf. on CAD,* pp. 82–87, Nov. 1994.
[20] L. Benini and G. DeMicheli, Transformation and synthesis of FSMs for low-power and gated-clock implementation, *ACM/SIGDA ISLP '95,* Apr. 1995.
[21] L. Benini, G. DeMicheli, and F. Vermulen, Finite-state machine partitioning for low power, *IEEE ISCAS '98,* pp. 5–8, May 1998.
[22] R. Shelar and M.P. Desai, Orthogonal partitioning and gated-clock architecture for low-power realization of FSMs, *IEEE ASIC/SOC '2000,* pp. 266–270, Sept. 2000.

12

Clocking Multi-GHz Systems

Vojin G. Oklobdzija
University of California—Davis

12.1 Introduction

The clock speed has been rising rapidly, doubling every 3 years as plotted in Figure 12.1. Currently, the highest microprocessor clock frequency is slightly above 3 GHz, while that number is changing rapidly upward. It is expected that we will reach 10 GHz in the next 5 years, and by the year 2010, the processors will be running at frequencies beyond 10 GHz. At that clock rate, several challenges may force us to reexamine standard approaches to clocking.

As the clock speed increases, the number of logic levels in the critical path diminishes. In today's high-speed processors, instructions are executed in one cycle, which is driven by a single-phase clock. In addition, the pipeline depth is increasing to 15 or 20 to accommodate the speed increase. Today, 10 levels of logic in the critical path are common; however, the amount of logic between the two stages is decreasing further. Thus, any overhead associated with the clock system and clocking mechanism that is directly and adversely affecting the machine performance is critically important.

At today's frequencies, the ability to absorb clock skew and to use a faster clocked storage element (CSE) results in direct and significant performance improvements. These improvements are very difficult to obtain through architectural techniques or micro-architecture levels. As the clock frequency reaches 5 to 10 GHz, traditional clocking techniques will be stretched to their limits. New ideas and new ways of designing digital systems are required.

12.1.1 Clock Distribution

The two most important timing parameters affecting the clock signal are: clock skew and clock jitter.

Clock skew is a spatial variation of the clock signal as distributed through the system. It is caused by the various RC characteristics of the clock paths to the various points in the system, as well as different loading of the clock signal at different points on the chip. In addition, we can distinguish between global clock skew and local clock skew. These are both equally important in high-performance system design.

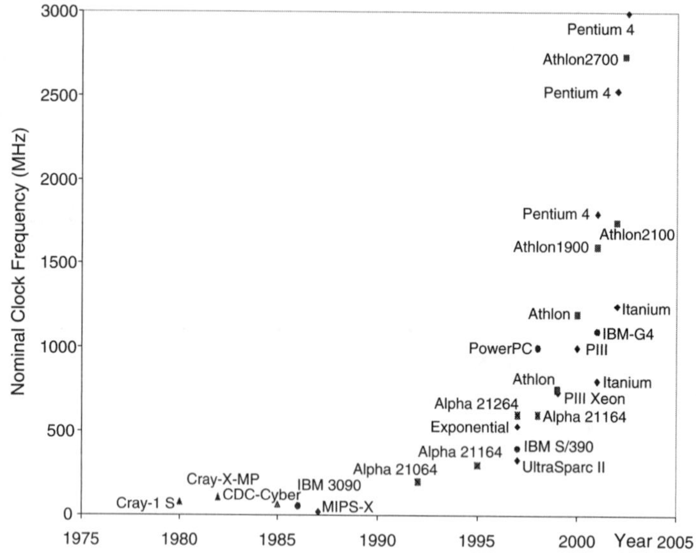

FIGURE 12.1 Clock frequency over the years for various representative machines and processors.

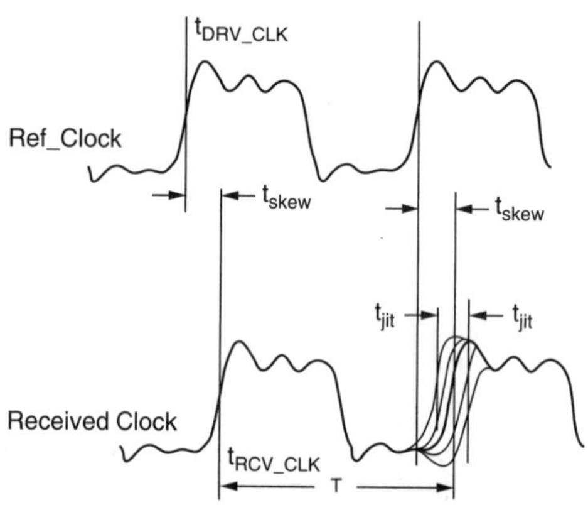

FIGURE 12.2 Clock parameters: period, width, clock skew, and clock jitter.

Clock jitter is a temporal variation of the clock signal with regard to the reference transition (i.e., reference edge) of the clock signal as illustrated in Figure 12.2.

Clock jitter represents edge-to-edge variation of the clock signal in time. As such, clock jitter can also be classified as long-term jitter and edge-to-edge clock jitter, which defines the clock signal variation between two consecutive clock edges. In the course of high-speed logic design, we are more concerned about edge-to-edge clock jitter because this phenomenon affects the time available for the logic operation.

Typically, the clock signal has to be distributed to several hundreds of thousands of the CSEs. Therefore, the clock signal has the largest fan-out of any node in the design, which requires several levels of amplification. Consequently, the clock system alone can use up to 40 to 50% of the power of the entire very large-scale integration (VLSI) chip [1,9]. We also must assure that every CSE receives the clock signal precisely at the same moment in time.

12.2 Clocking Considerations in Sequential Systems

A traditional view of the finite state machine (FSM) is represented by the Huffman model, which consists of combinational logic (CL) and CSEs. In this model, the next state, which is determined by the present

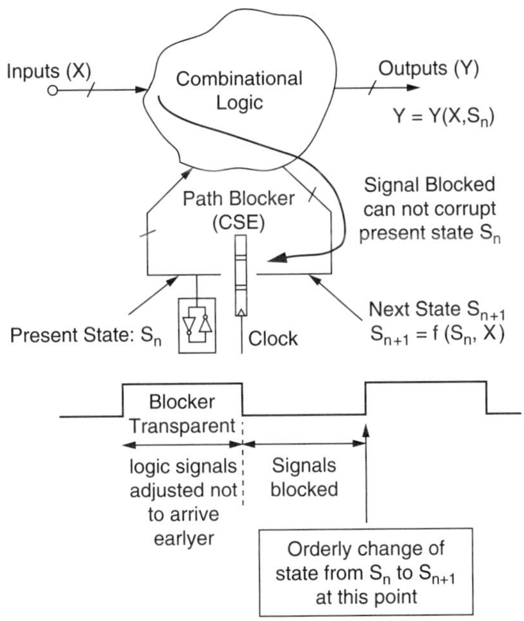

FIGURE 12.3 Different view of an FSM (Huffman model).

state and the input (as in the case of Mealy machine), is stored into the CSE by the triggering mechanism of the clock (i.e., edge or level). Following this model, we are used to thinking that the purpose of the CSE is to "hold" or "memorize" the state. This view is further supported by the level sensitive scan design (LSSD) methodology, which uses the storage elements to "scan-out" the state of the machine during the test and debug mode.

We want to present a slightly different view. The purpose of the CSE is to prevent the corruption of the next state as illustrated in Figure 12.3.

This model is broader and can represent wave pipelining [2], for example. In the case of wave pipelining, the signal is blocked from corrupting the present state S_n by a sheer delay of the wire. It simply cannot arrive in time, therefore, no blocking is necessary; however, this model also reveals problems of wave pipelining technique. Ideally, all the signals should arrive at the same point in time, which is not possible. Therefore, the fast-path problem becomes more difficult to control and stringent requirements are necessary. Thus, the system will run the risk of corrupting the state after several cycles.

The case of skew-tolerant domino logic [5,6], illustrated in Figure 12.4, conforms to the model presented in Figure 12.3.

Blocking of the signal is accomplished by the precharge phase of the clock. For example, while clock Φ_2 is "low" (precharge), data from stage 1 cannot be passed onto the stage 2. Only after the precharge phase has elapsed and clock Φ_2 has returned to "high" value can data from the stage 1 be passed to Stage 2. This transfer has to be completed while the clock Φ_l is "high." Obviously, the speed of this logic is determined by precise matching of the clocks. This is accomplished by having the clock signal travel along the data-path, while delaying the clock for the amount of time needed in the logic stage generates the local clocks. In some way, this is similar to the clocking used in the early mainframe computers [3].

12.2.1 Clocked Storage Elements

The function of a CSE (flip-flop or latch) is to block the signal path, thus preventing it from corrupting the present state. In addition, it may be used to capture the state information and preserve it as long as it is needed by the digital system. It is not possible to define a storage element without defining its relationship to the *clock*.

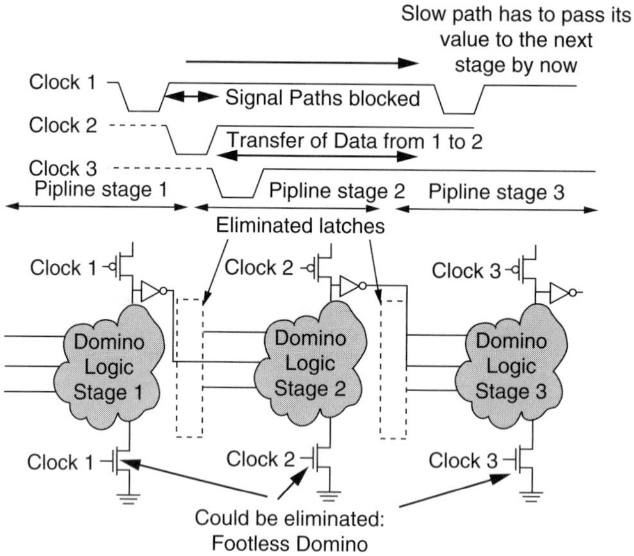

FIGURE 12.4 Skew-tolerant domino logic: no explicit latches.

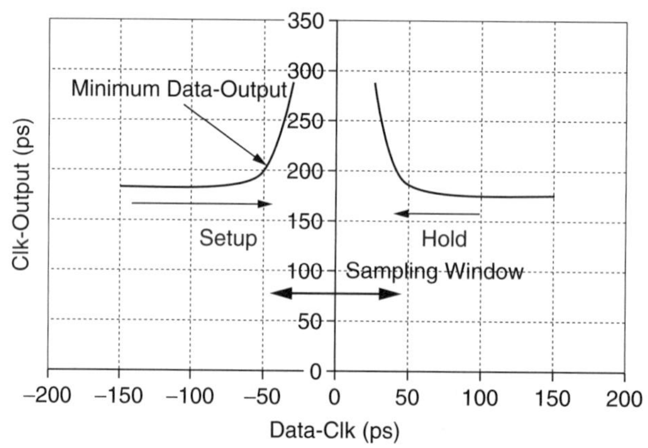

FIGURE 12.5 Setup and hold time behavior as a function of clock-to-output delay [8].

12.2.1.1 Timing Parameters

Data and clock inputs of a CSE must satisfy basic timing restrictions to ensure correct operation [4]. Fundamental timing constraints between data and clock inputs are quantified with setup and hold times, as illustrated in Figure 12.5 [8]. Setup and hold times define time intervals during which input has to be stable to ensure correct flip-flop operation. The sum of setup and hold times define the sampling window of the CSE. The sampling window is the period in which the CSE is sampling, and data is not allowed to change.

12.2.1.2 Setup and Hold Time Properties

Failure of the CSE due to the setup and hold time violations is not an abrupt process. This failing behavior is depicted in Figure 12.6.

Two opposing requirements exist with respect to the change of data signal as the locking event is approaching [8]:

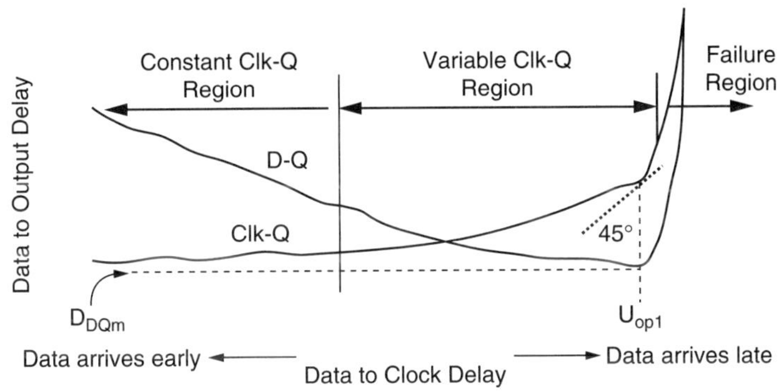

FIGURE 12.6 Setup and hold time behavior as a function of Data-to-Output delay.

1. The change should be kept farther from the failing region (Figure 12.6) for the purpose of design reliability.
2. The change should be as close as possible to increase the time available for the logic operation.

In industry, setup and hold times are specified as points in time when the Clk-Q (t_{CQ}) delay raises for an arbitrary number (commonly 5 to 20%). This reason is not valid. If we pay attention to D-Q (t_{DQ}) delay (instead of Clk-Q), we see a different picture. Despite the increase in Clk-Q delay, there are still benefits of getting closer to the locking event because D-Q delay (representing the time taken from the cycle) is reduced [16].

12.2.2 Time Borrowing and Absorption of Clock Uncertainties

Even if data arrives past the clock edge, the delay contribution of the storage element is still smaller than the amount of delay passed onto the next cycle. This allows for more time for useful logic operation. This is known as time borrowing or cycle stealing [7]. To understand the full effects of delayed data arrival, we have to consider a pipelined design where the data captured in the first clock cycle is used as input in the next clock cycle (see Figure 12.7).

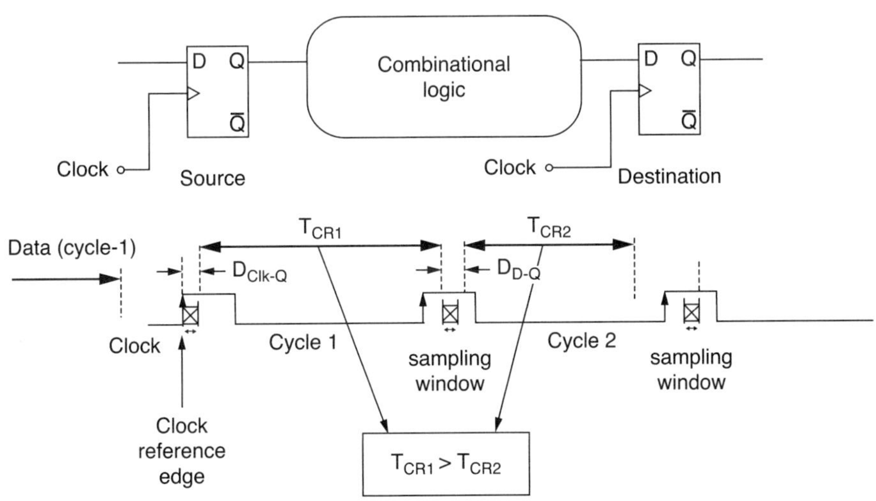

FIGURE 12.7 "Time borrowing" in a pipelined design.

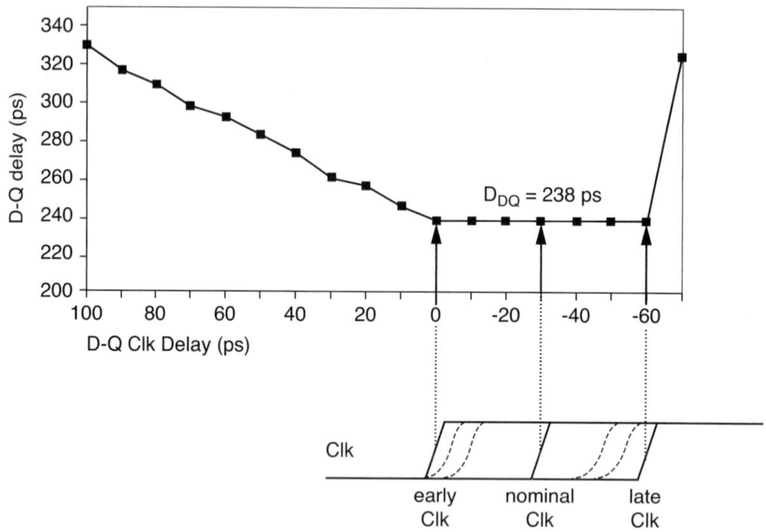

FIGURE 12.8 Clock skew absorption property: data-to-output delay vs. clock arrival time.

The sampling window is defined as the period in which CSE is sampling, and data is not allowed to change. The amount of time for which the T_{CR1} was stretched did not come for free. It was simply taken away (i.e., stolen or borrowed) leaving less time in the next cycle (Cycle 2) for T_{CR2}. As a result, of late data arrival in Cycle 1, less time is available in Cycle 2. Thus, a boundary between pipeline stages is somewhat flexible. This feature not only helps accommodate a certain amount of imbalance between the critical paths in various pipeline stages, but it helps in absorbing the clock uncertainties: skew and jitter.

Thus, time borrowing is one of the most important characteristics of today's high-speed digital systems. Absorption of the clock jitter is depicted in Figure 12.8 [7], and the effect on data arrival in the following cycle is illustrated in Figure 12.9. We observe how moderate amounts of clock uncertainties can be effectively absorbed, while the absorption property diminishes as the clock uncertainties become excessive.

The benefits of the "flat" data-to-output characteristic are presented in Figure 12.8 and Figure 12.9. We create the flat characteristic by expanding the transparency window of the CSE. Widening of the transparency window is equivalent to increasing the separation between the two reference events in time:

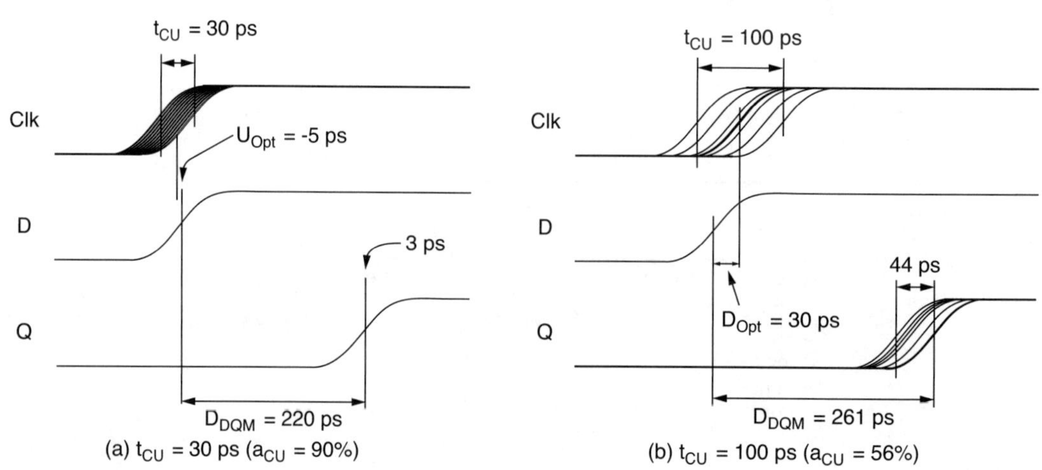

FIGURE 12.9 Effects of clock uncertainties to data arrival in the next cycle [20].

one that opens and other one that closes the CSE. In effect, the storage element behaves as a transparent latch for the short amount of time after the active clock edge [10]. Widening the transparency window can be achieved by intentionally creating wider capturing pulses of flip-flop and pulsed latch [11], or overlapping master and slave clocks in the master–slave latch. A consequence of increasing the transparency window is that the failure region of the data-to-output characteristic is moved away from the nominal clock edge. This results in the negative setup time but at the expense of increasing the hold time of the storage element. Large hold time makes fast path requirement harder to meet. Thus, the design for clock uncertainty absorption is often traded for a longer hold time. In many cases, however, these two requirements are not contradictory because different types of storage elements are used in fast and slow paths. The maximal clock skew that a system can tolerate is determined by CSEs. If the clock-to-output delay of a CSE is shorter than the hold time required and no logic exists in between two storage elements, a race condition can occur. A minimum delay restriction on the clock-to-output delay is given by:

$$t_{CLK-Q} \geq t_{hold} + t_{skew} \tag{12.1}$$

If this relation is satisfied, the system is immune to hold time violations.

The clock uncertainty absorption property shows how the propagation delay of a CSE is changing if the arrival of the reference clock is uncertain. Applying the clock uncertainty to a CSE is equivalent to holding reference clock arrival fixed and allowing data arrival to change.

More generally, uncertainty absorption should be treated as degradation of data-to-output delay for uncertain data-to-clock delay. As such, it can be used to describe the timing of the CSE if used in time borrowing, in exactly the same way if used for clock uncertainty absorption. Therefore, a "soft clock edge" designates a storage element where the output follows both early and late arrivals of the input, allowing slower stages to borrow time from the faster subsequent stages.

The time-borrowing capability and the clock uncertainty absorption are not mutually exclusive. They can be traded off for each other. Figure 12.10 illustrates a case where a wide transparency window, denoted

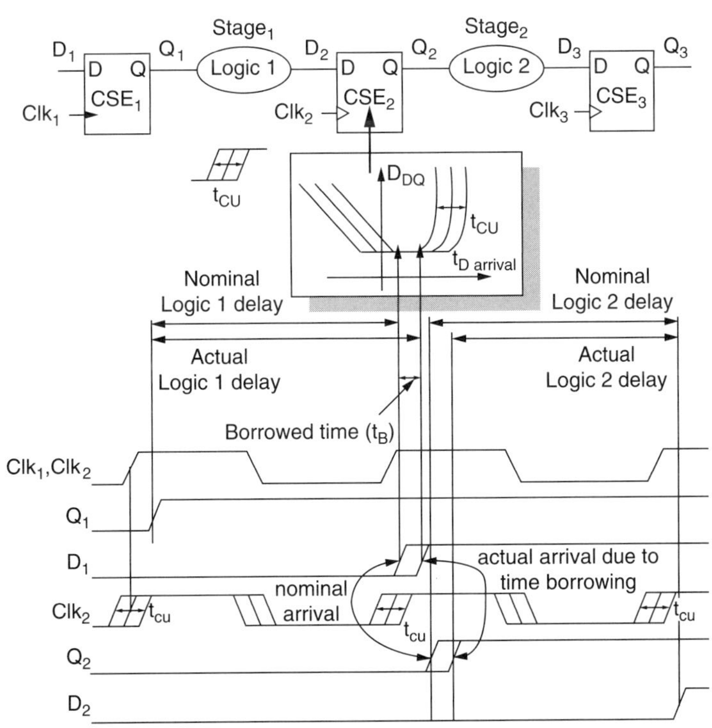

FIGURE 12.10 Time borrowing with uncertainty-absorbing CSEs [16].

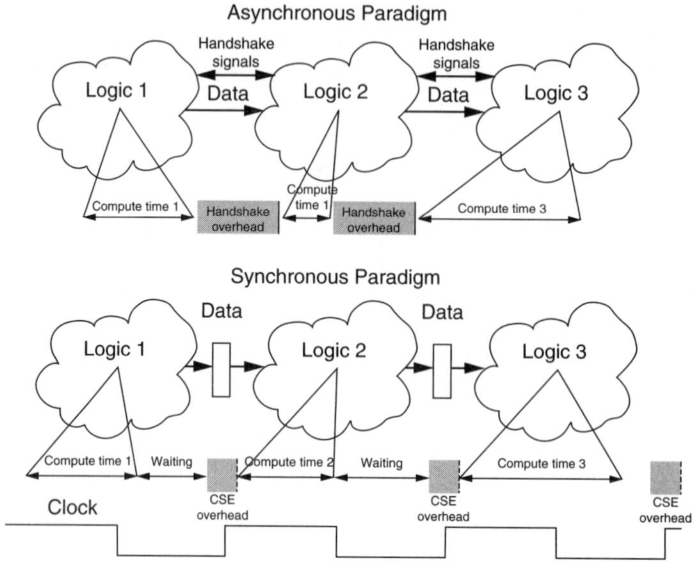

FIGURE 12.11 Data transfer in an asynchronous system vs. a synchronous system.

as a flat data-to-output characteristic, is used to absorb both the clock uncertainties, t_{CU}, and to borrow time, t_B, from the surrounding stages. Combinational logic of stage 1 takes more time than nominally assigned, and it borrows a portion of the cycle time from stage 2. In general, the storage element may not be completely transparent (i.e., data-to-output characteristics are not completely flat). The combination of clock uncertainty t_{CU} and time borrowing t_B causes an increase in the data-to-output delay of the flip-flop ΔD_{DQ}.

It should be noted that the practical values of the total borrowed time are approximately the width of the transparency window of the storage element and are, in any event, shorter than the hold time.

12.3 Asynchronous Systems

As the clock frequency increases, synchronous systems are facing serious problems such as the lack of ability to precisely control the clock, nonscaling clock uncertainties, wire delays, and the simple fact that the signal may need one or more clock cycles to reach its destination. Thus, asynchronous system design has been revisited.

The overhead imposed on the synchronous system by the clock uncertainties and CSE properties is simply traded for the overhead imposed by the handshake signaling in the asynchronous system (see Figure 12.11). Thus, the question really is: which one of the two can be designed so that it imposes lesser penalties on the data transfer as the speed of the logic keep rapidly increasing? As of today, it makes logical sense to use synchronous design in local domains, which can be clocked synchronously without considerable difficulties. Data transfer lasting several clock cycles could be accomplished using asynchronous communication. This opinion is supported by the fact that at 10 GHz or more, it would take several clock cycles to cross from one chip edge to another, as well as the fact that an entire processor in a 1-billion-transistor chip would occupy only a small portion of the chip.

12.4 Globally Asynchronous Locally Synchronous Systems

Following industry projections, VLSI chips will contain 1 billion transistors before the year 2010; however, the number of transistors used to build the logic of a single processor has not been increasing at the same rate. On the contrary, that number has remained relatively constant. Table 12.1 lists some

TABLE 12.1 Logic Transistors in Representative RISC Processors

Feature	Digital 21164	MIPS 10000	PowerPC 620	HP 8000	Sun UltraSparc
Frequency	500 MHz	200 MHz	200 MHz	180 MHz	250 MHz
Pipeline stages	7	5–7	5	7–9	6–9
Issue rate	4	4	4	4	4
Out-of-order execution	6 loads	32	16	56	none
Register renaming (int/FP)	none/8	32/32	8/8	56	none
Total transistors	9.3 M	5.9 M	6.9 M/	3.9 M*	3.8 M
Logic transistors	1.8 M	2.3 M	2.2 M	3.9 M	2.0 M

Source: Microprocessor Report 1998 issues.

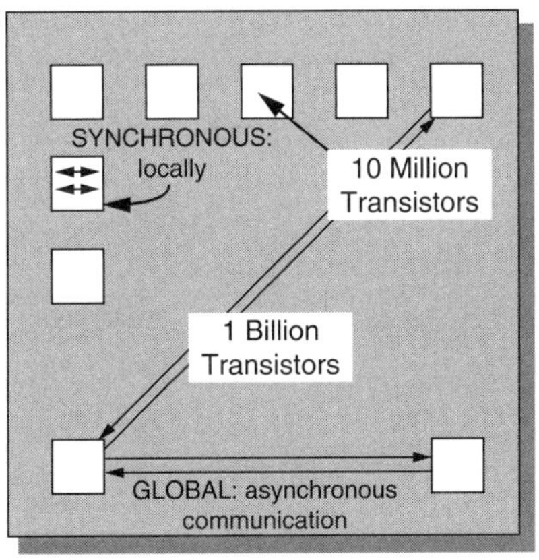

FIGURE 12.12 Projection of a 1-billion-transistor VLSI chip.

of the transistor numbers for a sample of typical super-scalar RISC architecture processors around the year 2000.

Figure 12.12 provides an illustration of a 1-billion-transistor chip. If we project the speed of the chip at that time and compare it with the projection for the speed of the interconnect, it becomes obvious that synchronous design on the entire chip may not be possible. Several clock cycles would be necessary for the signal to cross from one side of the chip to the other. From the design standpoint, it is also obvious that the future 1-billion-transistor chip will contain multiple cores in either multiprocessor or system on chip (SoC) arrangement. Therefore, globally asynchronous and locally synchronous clocking is considered as a promising technique for the SoC design.

In such a system, a number of independently synchronized modules communicate between each other through an asynchronous handshake mechanism. It is projected that interconnect effects will be manageable within a local synchronous module; therefore, a synchronous design would continue to be a viable option for the processor core.

Globally asynchronous and locally synchronous systems contain several independent synchronous blocks that operate with their own local clocks and communicate asynchronously with each other. The main feature of these systems is the absence of a global timing reference and the use of several distinct local clocks, or clock domains, possibly running at different frequencies.

This methodology is also viable when various intellectual property (IP) blocks are integrated in a single chip in an SoC environment because proven IP blocks can be reused without any modifications while relying on asynchronous interface between blocks. Such design preserves the benefit of synchronous design while avoiding problems caused by global wiring, especially a global clock signal.

Most conventional microprocessor designs are synchronous in their construction; that is, they have a global clock signal that provides a common timing reference for the operation of all the circuitry on the chip. On the other hand, fully asynchronous designs built using self-timed circuits do not have any global timing reference.

The globally asynchronous, locally synchronous design is not a novel approach, given that such a concept has long been used in mainframe computer systems, and this represents its logical migration into the VLSI chip.

12.5 Conclusion

Clocking for high-performance and low-power systems represents a challenge given the rapid increase in clock frequency, which has already reached multiple GHz rates. We expect that current clocking techniques will hold up to 10 GHz. Afterward, the pipeline boundaries will start to blur and synchronous design will be possible only in limited domains on the chip. A mix of synchronous and asynchronous design may emerge even in digital logic. This may represent the next design challenge in complex chips.

To Probe Further

For complete analysis of clocking, please see Oklobdzija [16]. Overviews of low-power circuit design techniques are available in Kuroda and Sakurai [9] and Oklobdzija [15]. Good references for asynchronous clocking including articles by Hauck and Sutherland [12,13]. Globally asynchronous, locally synchronous systems are described in the article by Hemani et al. [14].

References

[1] P.E. Gronowski, et al., High-performance microprocessor design, *IEEE J. Solid-State Circuits,* Vol. 33, No. 5, May 1998.

[2] W.P. Burleson, M.Ciesielski, F. Klass, and W. Liu, Wave-pipelining: a tutorial and research survey, *IEEE Trans. of Very Large-Scale Integration (VLSI) Systems,* Vol. 6, No. 3, September 1998.

[3] L.W. Cotten, Circuit implementation of high-speed pipeline systems, *AFIPS Proc., Fall Joint Comput. Conf.,* pp. 489–504, 1965.

[4] S.H. Unger and C.J. Tan, Clocking schemes for high-speed digital systems, *IEEE Trans. on Computers,* Vol. C-35, No. 10, October 1986.

[5] D. Harris and M.A. Horowitz, Skew-tolerant domino circuits, *IEEE J. Solid-State Circuits,* Vol. 32, No. 11, November 1997.

[6] D. Harris, et al., Opportunistic Time-Borrowing Domino Logic. U.S. Patent No. 5,517,136. Issued May 14, 1996.

[7] H. Partovi et al., Flow-through latch and edge-triggered flip-flop hybrid elements, *IEEE Int. Solid-State Circuits Conf. (ISSCC), Dig. of Tech. Papers,* San Francisco, CA, February 8–10, 1996.

[8] V. Stojanovic and V.G. Oklobdzija, Comaparative analysis of master-slave latches and flip-flops for high-performance and low-power VLSI systems, *IEEE J. Solid-State Circuits,* Vol. 34, No. 4, April 1999.

[9] T. Kuroda and T. Sakurai, Overview of low-power VLSI circuit techniques, *IEICE Trans. on Electron.,* E78-C, No. 4, April 1995, pp. 334–344. Invited paper, special issue on low-voltage low-power integrated circuits.

[10] F. Klass et al., A new family of semidynamic and dynamic flip-flops with embedded logic for high-performance processors, *IEEE J. Solid-State Circuits,* Vol. 34, No. 5, pp. 712–716, May 1999.

[11] T. James, S. Narendra, Z. Chen, S. Borkar, M. Sachdev, and V. De, Comparative delay and energy of single edge-triggered and dual edge-triggered pulsed flip-flops for high-performance microprocessors, *Proc. 2001 Int. Symp. on Low-Power Electron. and Design,* Huntington Beach, CA, August 6–7, 2001.

[12] S. Hauck, Asynchronous design methodologies: an overview, *Proc. IEEE,* Vol. 83, No. 1, January 1995.

[13] I.E. Sutherland, Micropipelines, *Commun. ACM,* Vol. 32, No. 6, June 1989.

[14] A. Hemani, T. Meincke, S. Kumar, A. Postula, T. Olsson, P. Nilsson, J. Oberg, P. Ellervee, and D. Lundqvist, Lowering power consumption in clock by using globally asynchronous locally synchronous design style, *Proc. 36th Design Automation Conf.,* June 21–25, 1999.

[15] V.G. Oklobdzija, Ed., *High-Performance System Design: Circuits and Logic,* John Wiley & Sons/IEEE Press series on microelectronics systems, 1999.

[16] V.G. Oklobdzija, V. Stojanovic, D. Markovic, and N. Nedovic, *Digital System Clocking: High-Performance and Low-Power Aspects,* John Wiley & Sons, New York, January 2003.

13

Circuit Techniques for Leakage Reduction

Kaushik Roy
Amit Agarwal
Chris H. Kim
Purdue University

13.1 Introduction

Semiconductor devices are aggressively scaled each technology generation to achieve high integration density while the supply voltage is scaled to achieve lower switching energy per device. To achieve high performance, however, commensurate scaling of the transistor threshold voltage (V_{th}) is needed. Scaling of transistor threshold voltage is associated with exponential increase in subthreshold leakage current [1]. Aggressive scaling of the devices in the nanometer regime not only increases the subthreshold leakage, but also has other negative impacts such as increased drain-induced barrier lowering (DIBL), V_{th} roll-off, reduced on-current to off-current ratio, and increased source-drain resistance [2]. DIBL increases the dependency of V_{th} on channel length. A small variation in channel length might result in large V_{th} variation, which makes device characteristics unpredictable. To avoid these short-channel effects (SCE), oxide thickness scaling and higher nonuniform doping needs to be incorporated [3] as the devices are scaled in nanometer regime. The International Technology Roadmap for Semiconductors (ITRS) predicts gate oxide thickness of 1.2 to 1.6 nm for sub-100nm CMOS [4]. The low oxide thickness gives rise to high electric field, resulting in considerable direct tunneling current [5]. This current destroys the classical infinite input impedance assumption of metal-oxide semiconductor (MOS) transistors and thus affects circuit performance severely. Higher doping results in high electric field across the p-n junction (source-substrate or drain-substrate), which causes significant band-to-band tunneling (BTBT) of electrons from the valence band of the p-region to the conduction band of the n-region. Peak halo doping (P+) is restricted such that the BTBT component is maintained reasonably small compared with the other leakage components.

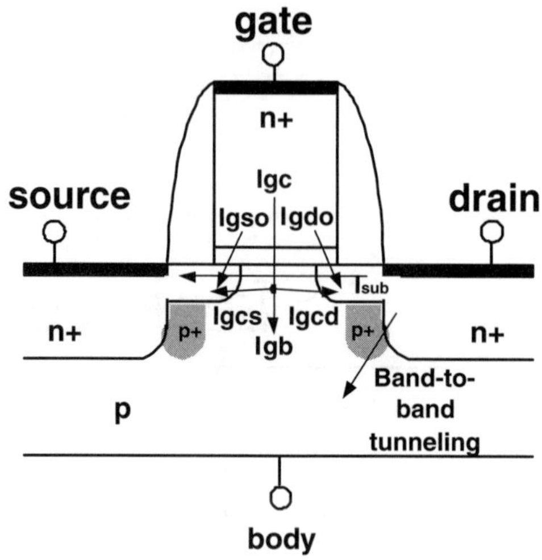

FIGURE 13.1 Leakage components in MOSFET.

This chapter proposes different integrated circuit techniques to reduce overall leakage in both logic and cache memories. A spectrum of circuit techniques including dual V_{th}, variable V_{th}, dynamically varying the V_{th} during runtime, sleep transistor, natural stacking, and multiple/dynamic supply circuits are reviewed. Based on these techniques, different leakage tolerant schemes for logic and memories are summarized.

13.2 Leakage Components

A metal-oxide semiconductor fluid-effect transistor (MOSFET) of the nanometer regime has three dominant components of leakage:

1. Subthreshold leakage, which is the leakage current from drain to source (I_{sub} in Figure 13.1).
2. Direct tunneling gate leakage, which is due to the tunneling of electron (or hole) from the bulk silicon through the gate oxide potential barrier into the gate.
3. The source/substrate and drain/substrate reverse biased p-n junction BTBT leakage; this leakage component is expected to be large for sub-50-nm devices [6].

Other components of leakage current described in Roy et al. [7], such as gate-induced drain leakage (GIDL) and impact ionization leakage, are not expected to be large for regular nanoscale CMOS devices.

13.2.1 Subthreshold Leakage

Subthreshold or weak inversion conduction current between source and drain in a MOS transistor occurs when gate voltage is below V_{th} [8]. Weak inversion typically dominates modern device off-state leakage due to the low V_{th} that is used. The weak inversion current can be expressed based on the Equation (13.1) [8].

$$I_{subth} = Ae^{\frac{q}{nkT}(V_{GS}-V_{TH0}-\gamma V_{SB}+\eta V_{DS})}(1-e^{-\frac{qV_{DS}}{kT}}) \tag{13.1}$$

where

$$A = \mu_0 C'_{ox} \frac{W}{L_{eff}} (\frac{kT}{q})^2 e^{1.8}$$

V_G, V_D, V_S, and V_B are the gate voltage, drain voltage, source voltage, and body voltage of the transistor, respectively. Body effect is represented by the term $\gamma' V_{SB}$, where γ' is the linearized body effect coefficient; η is the DIBL coefficient, representing the effect of V_{DS} on threshold voltage; C_{ox} is the gate oxide capacitance; μ_0 is the zero bias mobility; and n is the subthreshold swing coefficient of the transistor. This equation shows the exponential dependency of subthreshold leakage on V_{TH0}, V_{GS}, V_{DS} (due to DIBL), and V_{SB}. Each of the leakage reduction techniques described in the latter sections utilizes these parameters in a MOSFET to achieve a low leakage state.

13.2.2 Gate Leakage

Gate direct tunneling current is due to the tunneling of electron (or hole) from the bulk silicon through the gate oxide potential barrier into the gate. The direct tunneling is modeled as

$$J_{DT} = A\left(V_{ox}/T_{ox}\right)^2 exp\left(\frac{-B\left(1-\left(1-V_{ox}/\phi_{ox}\right)^{3/2}\right)}{V_{ox}/T_{ox}}\right) \quad (13.2)$$

where J_{DT} is the direct tunneling current density, V_{ox} is the potential drop across the thin oxide, ϕ_{ox} is the barrier height of tunneling electron, and t_{ox} is the oxide thickness [9]. The tunneling current increases exponentially with decrease in oxide thickness. It also depends on the device structure and the bias condition [10]. Figure 13.1 describes the various components of gate tunneling in a scaled NMOS device [11]:

- Edge-direct tunneling (EDT) components between gate and source-drain extension (SDE) region (I_{gso} and I_{gdo})
- Gate-to-channel current (I_{gc}), part of which goes to source (I_{gcs}) and rest goes to drain (I_{gcd})
- Gate-to-substrate leakage current (I_{gb})

Tunneling current increases with the increase in voltage drop across oxide (V_{ox}). The voltage across oxide in different regions (i.e., channel, gate-source overlap, and gate-drain overlap region) depends on biasing of the nodes representing the region.

13.2.3 Source/Substrate and Drain/Substrate P-N Junction Leakage

Drain and source-to-well junctions are typically reverse biased, causing p-n junction leakage current. A reverse biased p-n junction leakage has two main components: One is minority carrier diffusion/drift near the edge of the depletion region, and the other is due to electron-hole pair generation in the depletion region of the reverse biased junction.

In the presence of a high electric field (> 10^6 V/cm), electrons will tunnel across a reverse biased p-n junction. A significant current can arise as electrons tunnel from the valence band of the p-region to the conduction band of the n-region. Tunneling occurs when the total voltage drop across the junction is greater than the semiconductor band-gap. Because silicon is an indirect band gap semiconductor the BTBT current in silicon involves the emission or absorption of phonons.

In an NMOS device, when the drain or source is biased at a potential higher than that of the substrate, BTBT current flows through the drain-substrate or source-substrate junction. If both *n*- and *p*-regions are heavily doped, which is the case for scaled MOSFETs using heavily doped shallow junctions and halo doping for better SCE, BTBT significantly increases and becomes a major contributor to the total off-state current. Substantial increase in BTBT current is observed at high reverse biases. Reducing substrate doping near the substrate-drain/source junction is an effective way to reduce the BTBT current; however, this increases the SCE leading to considerable increase in the subthreshold current. Although circuit techniques specifically targeted at reducing BTBT have not been reported, forward substrate biasing can

TABLE 13.1 Circuit Techniques to Control Leakage in Logic

Design Time Techniques	Run Time Techniques	
	Standby Leakage Reduction	Active Leakage Reduction
Dual-V_{th} [12,13,14,15,16,17]	Natural Stacking [20,21,22,23]	DVS [37]
	Sleep Transistor [24,25,26,27,28,29,30,31]	
Multiple Supply Voltage [19]	VTCMOS [32,33,34,36]	DVTS [38,39]

be used to reduce BTBT in a MOSFET (because electric field reduces with reduction in the reverse bias across the junction).

13.3 Circuit Techniques to Reduce Leakage in Logic

Because circuits are mostly designed for the highest performance — for instance, to satisfy overall system cycle time requirements — they are composed of large gates and highly parallel architectures with logic duplication. As such, the leakage power consumption is substantial for such circuits; however, not every application requires a fast circuit to operate at the highest performance level all the time. Modules in which computation is bursty in nature (e.g., functional units in a microprocessor or sections of a cache) are often idle. It is of interest to conceive of methods that can reduce the leakage power consumed by these circuits. Different circuit techniques have been proposed to reduce leakage energy utilizing this slack without impacting performance. These techniques can be categorized based on when and how they utilize the available timing slack (Table 13.1) (e.g., dual V_{th} statically assigns high V_{th} to some transistors in the noncritical paths at the *design time* to reduce leakage current). The techniques, which utilize the slack in *runtime,* can be divided into two groups depending on whether they reduce standby leakage or active leakage. Standby leakage reduction techniques put the entire system in a low leakage mode when computation is not required. Active leakage reduction techniques slow down the system by dynamically changing the V_{DD} or V_{th} to reduce leakage when maximum performance is not needed. In the active mode, the operating temperature increases due to the switching activities of transistors. This has an exponential effect on subthreshold leakage (Equation (13.1)), making it the dominant leakage component during active mode and aggravating the leakage problem.

13.4 Design Time Techniques

Design time techniques exploit the delay slack in noncritical paths to reduce leakage. These techniques are static; once they are fixed, they cannot be changed dynamically while the circuit is operating.

13.4.1 Dual Threshold CMOS

In logic, a high V_{th} can be assigned to some transistors in the noncritical paths to reduce subthreshold leakage current, while the performance is not sacrificed by using low V_{th} transistors in the critical path(s) [12]. No additional circuitry is required, and both high performance and low leakage can be achieved simultaneously. Figure 13.2(a) illustrates the basic idea of a dual V_{th} circuit. The path distribution of dual V_{th} and single V_{th} standard CMOS for a 32-bit adder is illustrated in Figure 13.2(b). Dual V_{th} CMOS has the same critical delay as the single low V_{th} CMOS circuit, but the transistors in the noncritical paths can be assigned high V_{th} to reduce leakage power. Dual threshold CMOS is effective in reducing leakage power during both standby and active modes. Many design techniques have been proposed, which consider upsizing of high V_{th} transistor [13–15] in dual V_{th} design to improve performance. Upsizing of high V_{th} transistor affects switching power and die area that can be traded off against using a low V_{th} transistor that increases leakage power.

Domino logic can be susceptible to leakage — especially wide OR domino gates. Low threshold evaluation logic reduces noise immunity. Thus, for scaled technologies, domino may require larger keeper

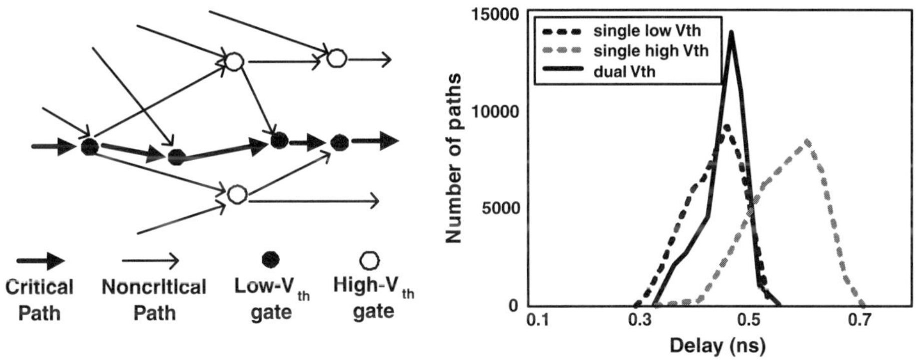

FIGURE 13.2 (a) A dual V_{th} CMOS circuit and (b) path distribution of dual V_{th} and single V_{th} CMOS.

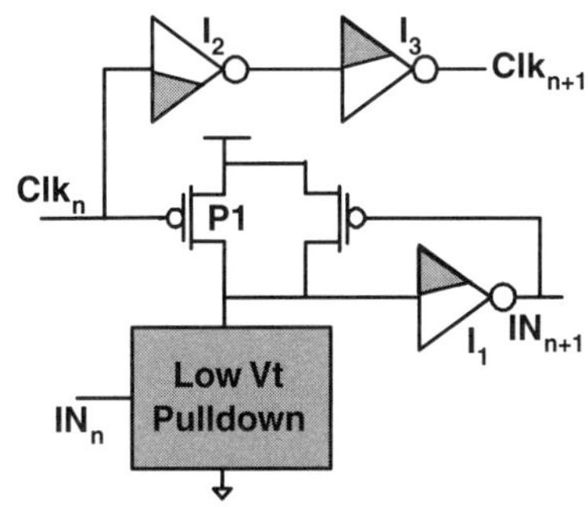

FIGURE 13.3 Dual V_{th} Domino gate [16] with low V_{th} devices shaded.

transistors that, in turn, can affect speed. Figure 13.3 depicts a typical dual V_{th} domino logic [16] for low leakage noise immune operations. Because of the fixed transition directions in domino logic, one can easily assign low V_{th} to all transistors that switch during the evaluate mode and high V_{th} to all transistors that switch during precharge modes. When a dual V_{th} domino logic stage is placed in standby mode, the domino clock needs to be high (evaluate) to shut off the high V_{th} devices (e.g., P1, I2 PMOS, and I3 NMOS). Furthermore, to ensure that the internal node remains at solid logic ZERO, which turns off the high V_{th} keeper and I1 NMOS, the initial inputs into the domino gate must be set high.

Instead of changing the channel doping profiles, a higher t_{ox} can be used to obtain a high V_{th} device for dual threshold CMOS circuits. In order to suppress the SCE, the high t_{ox} device needs to have a longer channel length as compared with the low t_{ox} device. Multiple t_{ox} CMOS (MoxCMOS) can optimize the power consumption due to subthreshold leakage, gate oxide tunneling leakage as well as switching power. An algorithm for selecting and assigning optimal transistor oxide thickness is derived in Sirisantana et al. [17]. The simulation results on IEEE International Symposium on Circuits and Systems (ISCAS) benchmark circuits for 70-nm technology show that the total power consumption of MoxCMOS circuits can be reduced by an average of 34% with over 70% reduction in gate oxide tunneling leakage compared with standard CMOS circuits.

13.4.2 Multiple Supply Voltage

Supply voltage scaling was originally developed for switching power reduction. It is an effective method for reducing switching power because of the quadratic dependency of switching power on supply voltage. Supply voltage scaling also helps reduce leakage power because the subthreshold leakage due to GIDL and DIBL decreases as well as the gate leakage component when the supply voltage is scaled down. In a 1.2-V, 0.13-μm technology, it is demonstrated that the supply voltage scaling has impacts in the orders of V^3 and V^4 on subthreshold leakage and gate leakage, respectively.

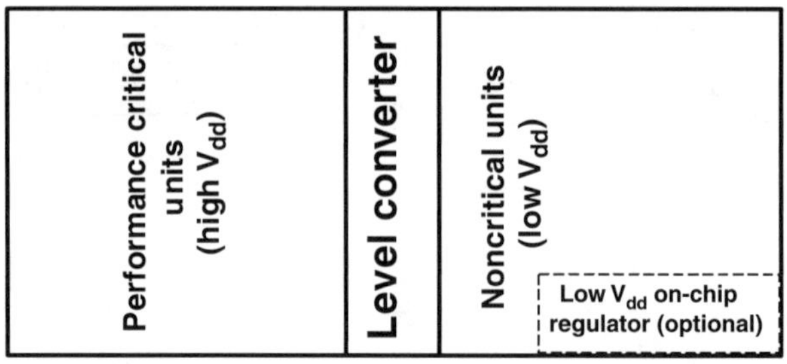

FIGURE 13.4 Two-level multiple supply voltage scheme [18].

To achieve low-power benefits without compromising performance, two methods of lowering supply voltage can be employed: static and dynamic (see Section 13.6) supply scaling. Static supply scaling is a multiple supply voltage approach [18] as depicted in Figure 13.4. Critical and noncritical paths or units of the design are clustered and powered by higher and lower supply voltages, respectively [19]. Because the speed requirements of the noncritical units are lower than the critical ones, supply voltage of noncritical unit clusters can be lowered without degrading system performance. Whenever an output from a low V_{DD} cluster has to drive an input to a high V_{DD} cluster, a level conversion is needed at the interface. The secondary voltages may be generated off-chip or regulated on-die from the core supply.

13.5 Runtime Standby Leakage Reduction Techniques

A common architectural technique to keep the power of fast, hot circuits within bounds has been to freeze the circuits — place them in a standby state — any time when they are not needed. Standby leakage reduction techniques exploit this idea to place certain sections of the circuitry in standby mode (low leakage mode) when they are not required.

13.5.1 Leakage Control Using Transistor Stacks (Self-Reverse Bias)

Leakage currents in NMOS or PMOS transistors depend exponentially on the voltage at the four terminals of transistor (Equation (13.1)). Figure 13.5 illustrates the variation of I_{DS} with respect to V_{GS} (V_G is tied to "0"). Increasing V_S of NMOS transistor reduces subthreshold leakage current exponentially due to the following three effects:

- Gate-o-source voltage becomes negative, thus the subthreshold current reduces exponentially.
- Negative body to source potential causes more body effect resulting in increased threshold voltage and thus reducing the subthreshold leakage.
- Drain-to-source potential decreases, resulting in less DIBL and thus lower subthreshold leakage.

This effect is also called self-reverse biasing of transistor. The self-reverse bias effect can be achieved by turning off a stack of transistors [20]. Turning off more than one transistor in a stack raises the internal

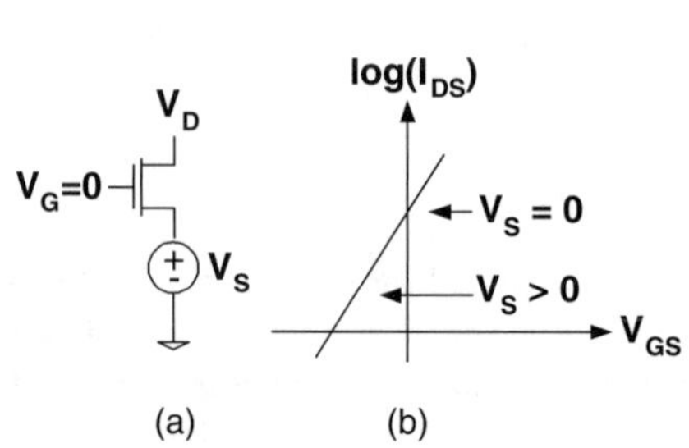

FIGURE 13.5 Leakage control using self-reverse bias.

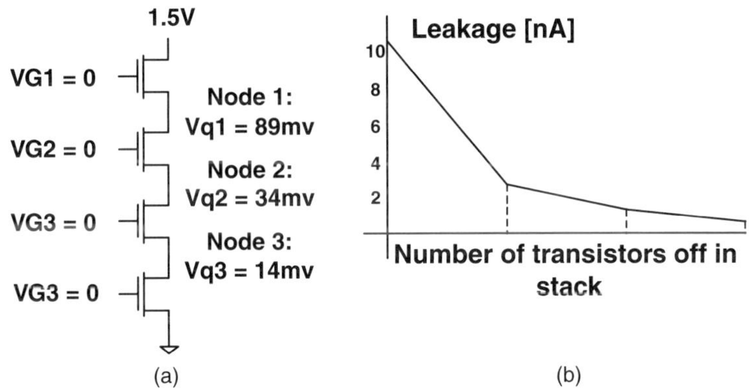

FIGURE 13.6 (a) Effect of transistor stacking on source voltage and (b) leakage current vs. number of transistors off in stack.

voltage (source voltage) of the stack, which acts as reverse biasing the source. Figure 13.6(a) depicts a simple pull-down network of a four input NAND gate. This pull-down network forms a stack of four transistors. If some of the transistors are turned off for a long time, the circuit reaches a steady state where leakage through each transistor is equal and the voltage across each transistor settles to a steady state value. In a case where only one NMOS device is off, the voltage at the source node of off transistor would be virtually zero because all other on transistors will act as short circuit. Thus, there is no self-reverse biasing effect, and the leakage across the off transistor is large. If more than one transistor is off, the source voltages of the off transistor, except the one connected to ground by on transistors, will be greater than zero, and the leakage will be determined mainly by the most negatively self-reverse biased transistor (because subthreshold leakage is an exponential function of gate-source voltage). The voltages at the internal nodes depend on the input applied to the stack. Figure 13.6(a) shows the internal voltages when all four transistors are turned off. These internal voltages make the off transistors self-reverse biased. The reverse bias makes the leakage across the off transistor very small. Figure 13.6(b) depicts the subthreshold leakage current vs. number of off transistors in a stack. A large difference in leakage current exists between one off transistor and two off transistors. Turning off three transistors does improve subthreshold leakage, however, there is a diminishing return.

It is evident from the preceding discussion that the leakage through logic gates depends on the applied input vector. Functional blocks such as NAND, NOR, or other complex gates readily have a stack of transistors. Maximizing the number of off transistors in a natural stack by applying proper input vectors can reduce the standby leakage of a functional block. A model and heuristic is proposed in [21] to estimate leakage and to select the proper input vectors to minimize the leakage in logic blocks. Table 13.2 presents the quiescent current flowing into different functional blocks for the best and worst case input vectors. All the results are based on HSPICE simulation using 0.18-μm technology with V_{DD} = 1.5 V. Results show

TABLE 13.2 Input Vector Control

Circuit	Input Vector	Iddq (nA)	Comments
4 input NAND	ABCD=000	0.60	Best
	ABCD=111	24.1	Worst
3 input NOR	ABC=111	0.13	Best
	ABC=000	29.5	Worst
Full adder	A,B,Ci=111	7.8	Best
	A,B,Ci=001	62.3	Worst
4 bit ripple adder	A=B=0000,Ci=0	91.3	Best
	A=B=1111,Ci=1	94.0	Best
	A=B=0101,Ci=1	282.9	Worst

that application of proper input vector can be efficient in reducing the total subthreshold leakage in the standby mode of operation [22].

Recent studies [23] demonstrate that as gate leakage is becoming a significant component of leakage, the input vector control technique using a stack of transistors needs to be reinvestigated to effectively reduce the total leakage. It has been reported that with gate leakage, the traditional way of using stacking fails to reduce leakage and in the worst case might increase the overall leakage. The gate leakage depends on the voltage drop across different regions of transistor (Section 13.2). Having "00" as the input in a two transistor stack has a high voltage drop across the gate-drain overlap region of the first transistor increasing the gate leakage, which may dominate the total leakage at room temperature. (Gate leakage is insensitive to temperature whereas subthreshold leakage is a strong function of temperature and increases with temperature [7].) Forcing inputs to "10" reduces this gate leakage component at the cost of subthreshold leakage. In scaled technology where gate leakage dominates the total leakage, using "10" might produce more savings in leakage as compare to "00." A three-transistor stack (NMOS) with input "100" can improve total leakage compare to "000" inputs for similar stack where subthreshold is the major component of leakage. The source/substrate and drain/substrate junction BTBT leakage is a weak function of input voltage, and thus it can be neglected from the analysis [6].

13.5.2 Sleep Transistor

This technique inserts an extra series-connected transistor (sleep transistor) in the pull-down/pull-up path of a gate and turns it "off" in the standby mode of operation [24]. During regular mode of operation, the extra transistor is turned on. This provides substantial savings in leakage current during standby mode of operation. As depicted in Figure 13.6(b), stacking of two off devices can significantly reduce leakage as compared with a single off device. Due to the extra stack transistor (sleep transistor), however, the drive current of forced-stack gate will be lower resulting in increased delay. Thus, this technique can only be used for paths that are noncritical. If the V_{th} of the sleep transistor is high, extra leakage saving is possible. This circuit topology is known as MTCMOS (multi-threshold CMOS) (Figure 13.7) [25].

In fact, only one type (i.e., either PMOS or NMOS) of high V_{th} transistor is sufficient for leakage reduction. The NMOS insertion scheme is preferable, because the NMOS on-resistance is smaller at the same width, and thus it can be sized smaller than a corresponding PMOS [26]. MTCMOS can be easily implemented on already existing circuits. A 1-V DSP (digital signal processing) chip for mobile phone applications has been recently developed using the MTCMOS scheme [27]; however, MTCMOS can only reduce leakage power in standby mode and the large inserted sleep transistors can increase the area and

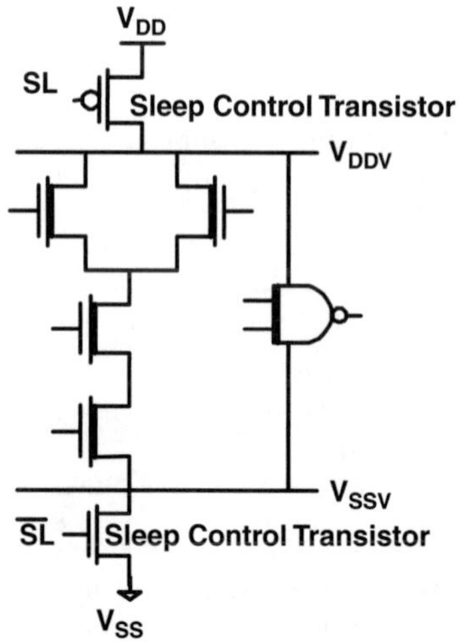

FIGURE 13.7 Schematic of MTCMOS circuit [25] with low V_{th} devices shaded.

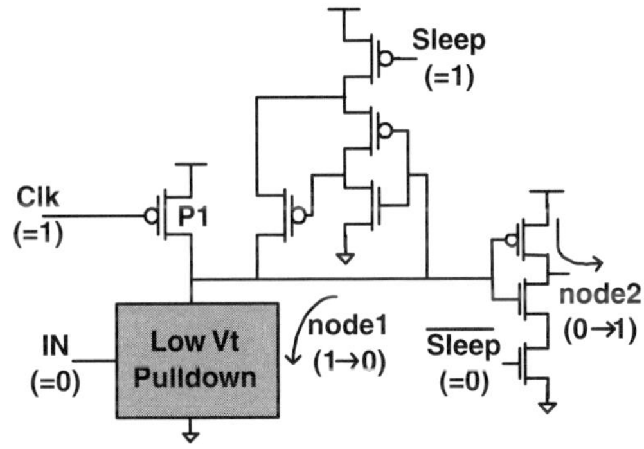

FIGURE 13.8 Domino gate with sleep transistor [31].

delay. Moreover, if data retention is required in standby mode, an extra high V_{th} memory circuit is needed to maintain the data [28]. Instead of using high V_{th} sleep transistors, a super cut-off CMOS (SCCMOS) circuit uses low V_{th} transistors with an inserted gate bias generator [29]. For the PMOS (NMOS) insertion, the gate is applied to 0 V (V_{DD}) in the active mode, and the virtual V_{DD} (V_{SS}) line is connected to the supply V_{DD} (V_{SS}). In standby mode, the gate is applied to V_{DD} + 0.4 V (V_{SS} − 0.4 V) by using the internal gate bias generator to fully cut off the leakage current. Compared with MTCMOS where it becomes difficult to turn on the high V_{th} sleep transistor at very low supply voltages, SCCMOS circuits can operate at very low supply voltages. Recent designs have been proposed using low-V_{th} devices for the sleep transistor to minimize the performance and area impacts [30].

In Figure 13.8, two small sleep transistors are added to conventional CMOS domino gate to save leakage [31]. In standby mode, clock is left high and sleep signal is asserted. If the data input were high, node 1 would have been discharged. If the data input was low, node 1 would be high, but leakage through NMOS dynamic pull-down stack would slowly discharge the node to ground. The NMOS sleep transistor is added to prevent any short circuit current in the static output logic while the dynamic node discharges to ground. Node 2 would rise as static pull-up turns on which would cause the NMOS transistors in the pull-down stacks of the following domino gates to turn on, accelerating the discharge of their internal dynamic nodes. Because sleep transistors are not in the critical path (evaluation path), minimal performance loss is incurred.

13.5.3 Variable Threshold CMOS (VTCMOS)

Variable threshold CMOS (VTCMOS) is a body-biasing design technique [32]. Figure 13.9(a) depicts the VTCMOS scheme.To achieve different threshold voltages, a self-substrate bias circuit is used to control the body bias. In the active mode, a zero body bias (ZBB) is applied. While in standby mode, a deep reverse body bias (RBB) is applied to increase the threshold voltage and cut off the leakage current. This scheme has been implemented in a two-dimensional discrete cosine transform core processor [32]. Furthermore, in active mode, a slightly forward substrate bias can be used to increase the circuit speed while reducing the SCE [33]. Providing the body bias voltage requires routing a body bias grid and this adds to the overall chip area. Keshavarzi et al. reported that RBB lowers integrated circuit (IC) leakage by three orders of magnitude in a 0.35-μm technology [34]. More recent data, however, demonstrates that the effectiveness of RBB to lower I_{OFF} decreases as technology scales due to the exponential increase in band-to-band tunneling leakage at the source/substrate and drain/substrate p-n junctions due to halo doping in scaled devices [34]. For scaled technologies, forward body biasing (FBB) can be used together with RBB to achieve better current drive with less SCE [35].

Raising the NMOS source voltage while tying the NMOS body to ground can produce the same effect as RBB. Forward body biasing can also be realized by applying a negative source voltage with respect to the body, which is tied to ground. Figure 13.9(b) illustrates the circuit diagram of this technique [36]. The main advantage is that it eliminates the need for a deep N-well or triple-well process because substrate

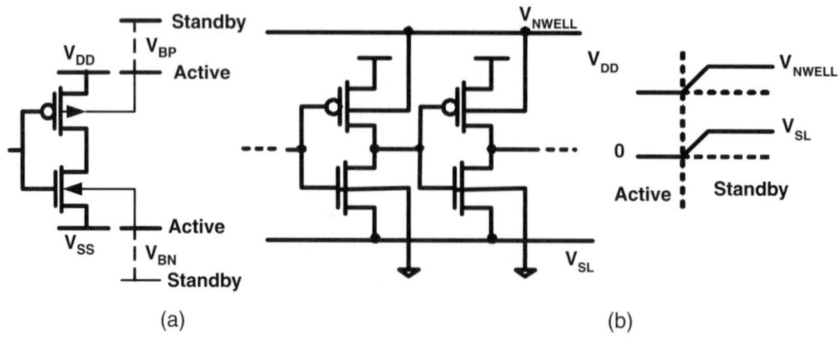

FIGURE 13.9 (a) Variable threshold CMOS [32], and (b) realizing body biasing by changing the source voltage with respect to body voltage, which is grounded [36].

of the target system and the control circuitry can be shared. The source voltage of the PMOS should also be raised if the charge in the storage node is to be kept constant while the NMOS source voltage is raised. A charge pump circuit is required if the body of the PMOS is to be raised higher than the source of the PMOS for RBB. In cases where V_{DS} can be further reduced, additional leakage improvement is possible. The smaller V_{DS} raises the transistor V_{th} (DIBL mechanism) and substantially reduces the subthreshold leakage component. GIDL component and gate leakage are also reduced due to the smaller gate-to-drain/source voltages.

13.6 Runtime Active Leakage Reduction Techniques

Not every application requires a fast circuit to operate at the highest performance level all the time. Active leakage techniques exploit this idea to intermittently slow down the fast circuitry and reduce the leakage power consumption as well as the dynamic power consumption when maximum performance is not required.

13.6.1 Dynamic V_{dd} Scaling (DVS)

Dynamic supply scaling overrides the cost of using two supply voltages (static supply scaling), by adapting the single supply voltage to the performance demand. The highest supply voltage delivers the highest performance at the fastest designed frequency of operation. When performance demand is low, supply voltage and clock frequency are lowered, just delivering the required performance with substantial power reduction. Implementing DVS in a general-purpose microprocessor system includes three key components:

1. An operating system that can intelligently vary the processor speed
2. A regulation loop that can generate the minimum voltage required for the desired speed
3. A microprocessor that can operate over a wide voltage range

Figure 13.10 depicts a DVS system architecture [37]. Control of the processor speed must be under software control, as the hardware alone may not distinguish whether the currently executing instruction is part of a computation-intensive task or a nonspeed-critical task. Supply voltage is controlled by hard-wired frequency-voltage feedback loop, using a ring oscillator as a critical path replica. All chips operate at the same clock frequency and same supply voltage, which are generated from the ring oscillator and the regulator.

13.6.2 Dynamic V_{th} Scaling (DVTS)

Similar to the dynamic VDD scaling (DVS) scheme, a dynamic V_{th} scaling (DVTS) scheme can be used to reduce the active leakage power in sub-100-nm generations where leakage power accounts for a large fraction of the total power consumption even during runtime. When the current workload is less than

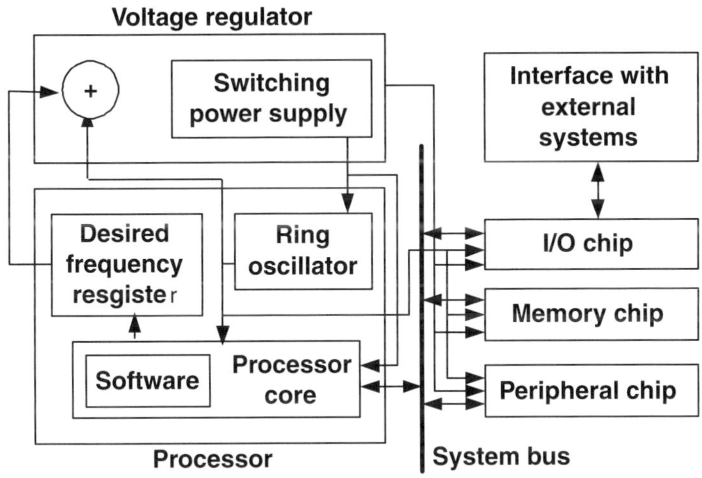

FIGURE 13.10 Dynamic voltage scaling architecture [37].

the maximum, the operating system commands a lower clock frequency to the hardware. Based on the given reference clock frequency, the DVTS hardware raises the transistor V_{th} via RBB to reduce the runtime leakage power dissipation. In cases when there is no workload at all, the V_{th} can be increased to its upper limit using body biasing, to significantly reduce the standby leakage power. "Just enough" throughput is delivered for the current workload by tracking the optimal V_{th} while leakage power is considerably reduced by intermittently slowing down the circuit.

Figure 13.11 plots the power consumption of DVTS and DVS systems for a speculative 70-nm process technology (only subthreshold leakage is considered) where leakage power accounts for 52% of total power dissipation (T = 125°C) [38]. By reducing the clock frequency without changing the V_{DD} or V_{th}, total power decreases in proportion to the operating frequency. This is because dynamic power is a linear function of clock frequency. The leakage power does not change with clock frequency, which makes 52% of total power wasted even when the clock frequency is zero. By dynamically scaling the V_{DD} together with the clock frequency, the speed requirement can be met while consuming significantly less power. Because the leakage power is dominant, DVTS appears to be comparable to DVS in saving total power for this technology. Figure 13.11 demonstrates that when the desired clock rate is 30% of the maximum operation frequency, 92% total energy savings can be achieved using DVTS. The following discussions address the merits and issues related to DVTS system designs.

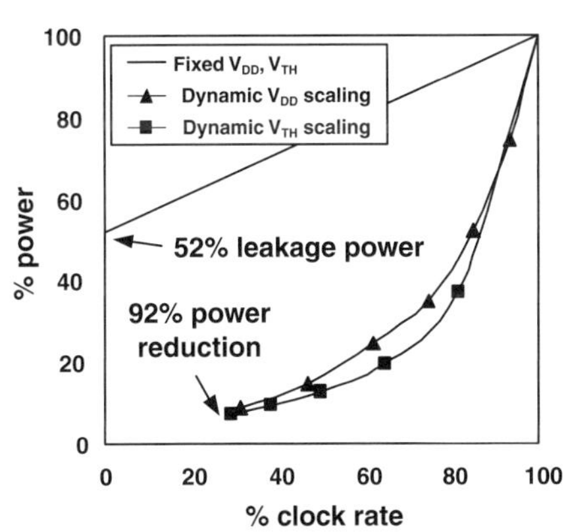

FIGURE 13.11 Total power vs. clock frequency for DVTS scheme and DVS scheme (BPTM, 70 nm, V_{DD} = 0.9 V, V_{th} = 0.15 V [38]).

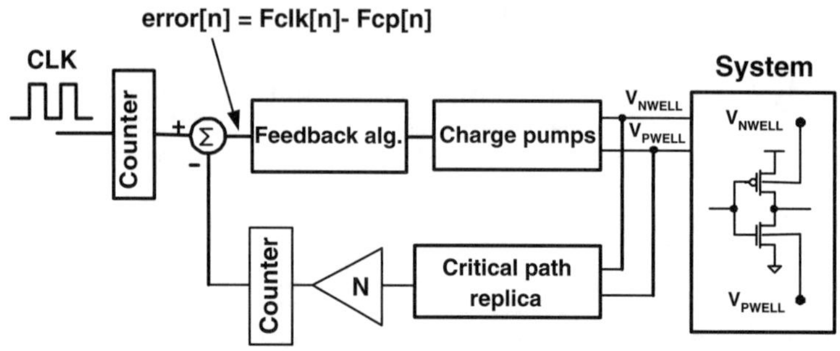

FIGURE 13.12 Dynamic V_{th} scaling system proposed in Kim and Roy [39].

- Simple hardware. Charge pumps are a simple solution for boosting voltages where current demand is low. No external inductors are needed and power consumption is very low compared with buck converters, which are used for DVS systems. Charge pumps are used in DVTS systems to generate the body bias voltages, which are outside the supply rail.
- Transition energy overhead. Results VTCMOS demonstrate that the energy overhead for a 120-K-transistor test chip in 0.3μm triple well technology consumes 10 nJ per V_{th} transition [32]. In case the V_{th} transition occurs frequently, transition energy overhead for DVTS systems becomes nonnegligible.
- Substrate noise. Charge pumps generate an unregulated body bias voltage due to the absence of external inductors. Any fluctuation in body bias will induce noise in logic.
- Process complexity. PMOS and NMOS body biases of the DVTS control circuit must be isolated from the target system in order to function as a reference. Thus, deep N-well or triple well technology is needed for the DVTS systems. The overall cost penalty by using these advanced processes is less than 5% [32].

Several different DVTS system implementations have been proposed in literature [39,40]. Figure 13.12 shows a DVTS hardware that uses continuous body bias control to track the optimal V_{th} for a given workload. A clock speed scheduler, which is embedded in the operating system, determines the (reference) clock frequency at runtime. The DVTS controller adjusts the PMOS and NMOS body bias so that the oscillator frequency of the critical path replica tracks the given reference clock frequency. The error signal, which is the difference between the reference clock frequency and the oscillator frequency, is fed into the feedback controller. The continuous feedback loop can also compensate for process, supply voltage, and temperature variations. A simpler method called "V_{th} hopping scheme," which dynamically switches between low V_{th} and high V_{th} depending on the performance demand, is proposed in [40]. The schematic diagram of the V_{th} hopping scheme is depicted in Figure 13.13. Compared with the continuous body bias control in Figure 13.12, the discrete control has two levels of V_{th}. If control signal VTHlow_Enable is asserted, the transistors in the target system are forward body biased and the V_{th} is low. When performance can be traded off for lower power consumption, VTHhigh_Enable is asserted and a high V_{th} is applied. The operating frequency of the target system is set to f_{CLK} when V_{th} is low and to $f_{CLK/2}$ when the V_{th} is high. An algorithm that adaptively changes the V_{th} depending on the workload is also verified and applied to an MPEG4 video encoding system. In future technology generations, the effectiveness of RBB is expected to be low due to the worsening SCE and increasing band-to-band tunneling leakage at the source/substrate and drain/substrate junctions. FBB can be applied together with RBB to achieve a better performance-leakage trade-off for DVTS systems.

13.7 Circuit Techniques to Reduce Leakage in Cache Memories

Figure 13.14(a) illustrates the seven available terminals in a conventional 6T SRAM cell; V_{SL}, V_{PWELL}, V_{NWELL}, V_{DL}, V_{WL}, V_{BL}, and V_{BLB}. Various SRAM cell architectures have been proposed in the past where

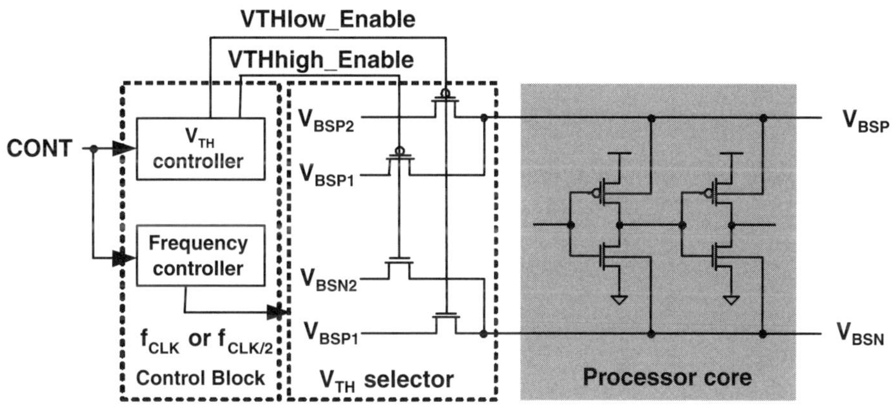

FIGURE 13.13 V_{th} hopping scheme proposed in Nose et al. [40].

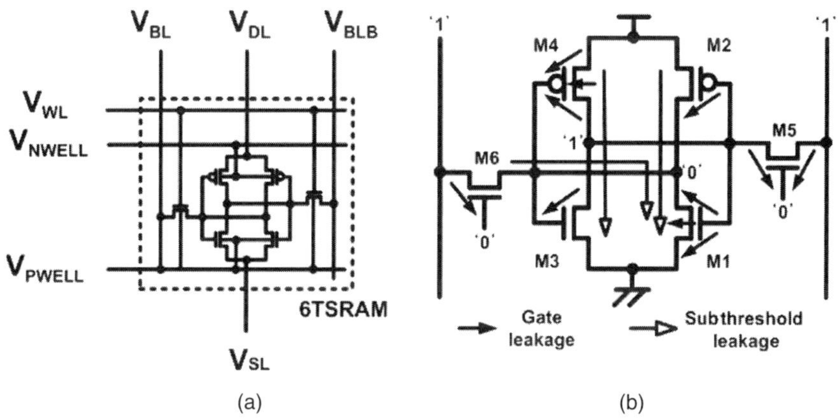

FIGURE 13.14 (a) Seven terminals of the 6-T SRAM cell, and (b) dominant leakage components in a 6-T SRAM.

one or more of the seven terminal voltages are controlled during standby mode for reducing the leakage components shown in Figure 13.14(b). Each technique exploits the fact that the active portion of a cache is very small, which gives the opportunity to put the large idle portion in a low-leakage sleep mode.

Effectiveness and overhead of each technique are evaluated based on the following discussions. First, the impact of the technique on various leakage components should be considered. Although subthreshold leakage still continues to dominate the I_{OFF} at high temperatures, ultra-thin oxides and high doping concentrations have led to a rapid increase in direct tunneling gate leakage and BTBT leakage at the source and drain junctions in the nanometer regime. Each leakage reduction technique needs reevaluation in scaled technologies where subthreshold conduction is not the only leakage mechanism. Second, the impact of the leakage reduction technique on SRAM read/write delay should be considered. Third, the transition latency/energy overhead should be taken into account, because of the limited time and energy budget for the mode transition. Last, the leakage reduction technique should not have a noticeable impact on SRAM cell stability or soft error rate (SER). Based on these discussions, the different low-leakage SRAM cells are summarized in Table 13.3.

A source biasing scheme raises the source line voltage (V_{SL}) in sleep mode [41–45], which reduces subthreshold leakage due to the three effects described in Section 13.5. The gate leakage in the cell is also reduced due to the relaxed signal rail, V_{DD}–V_{SL} (Section 13.2) [45]. An extra NMOS has to be series connected in the pull-down path to cut off the source line from ground during sleep mode, and this, in turn, imposes an extra access delay. The reduced signal charge in sleep mode also causes the soft error rate (SER) to rise, which requires additional error correction coding circuits [44].

TABLE 13.3 Low-Leakage SRAM Cell Techniques

	Source Biasing (V_{SL})	Body-Biasing (V_{PWELL}, V_{NWELL})	Dynamic V_{DD} (V_{DL})
Scheme	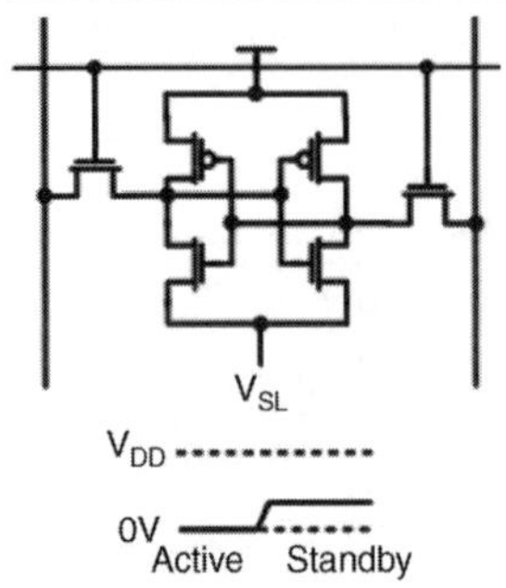	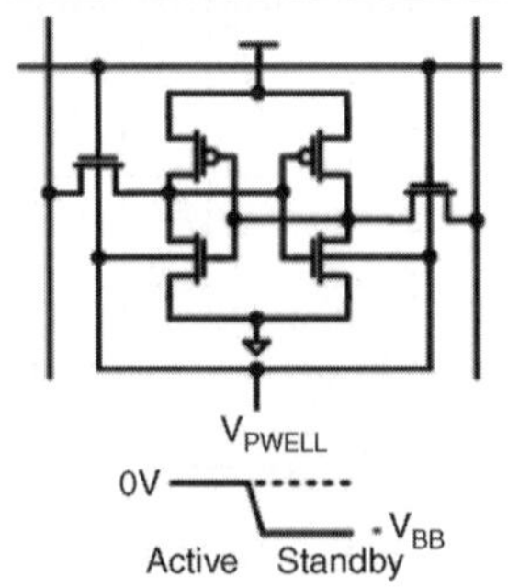	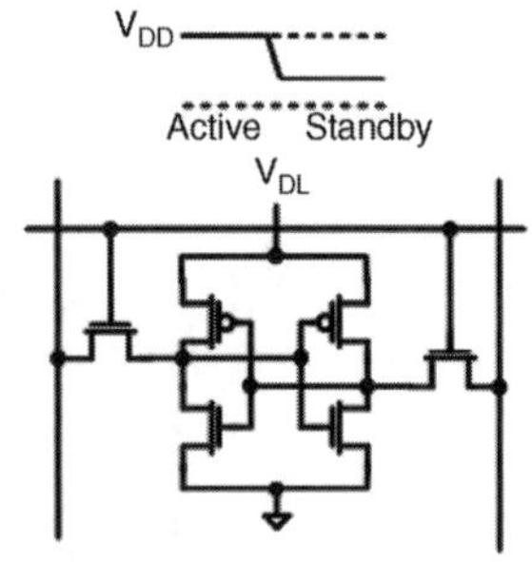
References	[41], [42], [43], [44], [45]	[43], [46], [47]	[43], [48]
Leakage reduction	subthreshold, gate: ↓↓	subthreshold: ↓↓ *BTBT: ↑	subthreshold, gate: ↓ bitline leakage: ↑
Performance	*Delay increase	No delay increase	No delay increase
Overhead	Medium transion overhead	Large transition overhead	Large transition overhead
Stability	Impact on SER	No impact on SER	*Worst SER
	Floating Bitlines (V_{BL}, V_{BLB})	Negative Word Line (V_{WL})	
Scheme	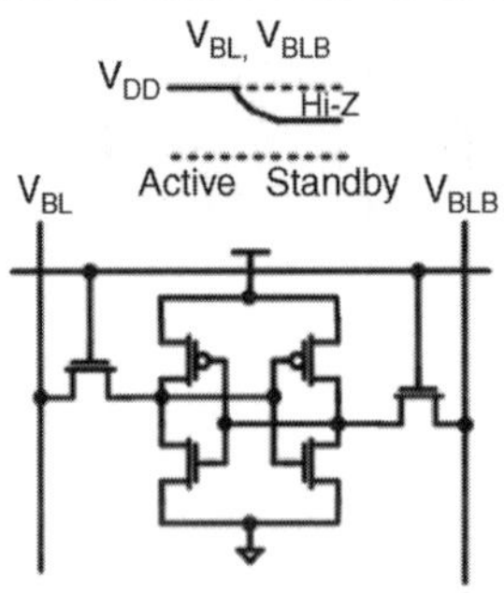	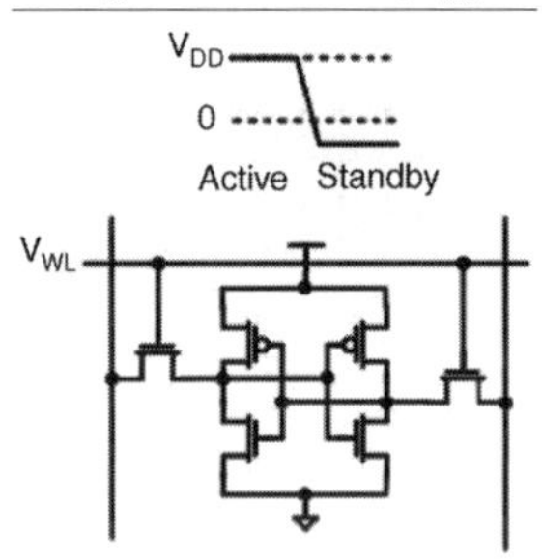	
References	[49]	[50]	
Leakage reduction	subthreshold, gate: ↓	subthreshold: ↓, *gate: ↑	
Performance	No delay increase	No delay increase	
Overhead	*Precharge latency overhead	*Low charge pump efficiency	
Stability	No impact on SER	No impact on SER, high voltage stress	

Reverse body-biasing (RBB) the NMOS (or PMOS) can reduce subthreshold leakage via body effect, while not affecting the access time by switching to zero body-biasing (ZBB) in the active mode [43,46,47]. A large latency/energy overhead is imposed for the body-bias transition due to the large V_{BB} swing and substrate capacitance. This scheme becomes less attractive in scaled technologies because the body coefficient decreases with smaller dimensions, and the source/drain junction BTBT leakage becomes enhanced by RBB.

Supply voltage is lowered in a dynamic V_{DD} SRAM (DVSRAM) [43,48], which, in turn, reduces the subthreshold, gate, and BTBT leakage. This scheme requires a smaller signal rail (V_{DL}–V_{GND}) compared with the SBSRAM for equivalent leakage savings. Although there is no impact on delay in the active mode, the large V_{DD} swing between sleep and active mode imposes a larger latency/energy transition overhead than the SBSRAM. Moreover, the greatest drawback of the DVSRAM is that it increases the bitline leakage in the sleep mode because the voltage level in the stored node also drops as the V_{DD} is lowered. Therefore, this scheme is not suitable for dual V_{th} designs where the speed-critical access transistors may already be using low V_{th} devices with high leakage levels.

A technique that biases the bitlines to an intermediate level has been proposed to reduce the access transistor leakage via the DIBL effect [49]. Because only the access transistors benefit from the leakage reduction, the overall leakage savings is moderate. Unlike the three previously mentioned techniques, this scheme has to be applied to the entire subarray because the bitline is shared across different cache lines. The main limitation comes from the fact that there is a precharge latency whenever a new subarray is accessed. This would mean that an architectural modification is required in order to resolve the multiple hit times in case the precharge instant is not known ahead of time.

The negative word line scheme [50] pulls down the V_{WL} to a negative voltage during standby in order to avoid the subthreshold leakage through the access transistors. However, it has issues such as increase in gate leakage and higher voltage stress in the access transistors. Although this technique has no impact on performance or SER, a power loss occurs due to the generation of a negative bias using charge pumps. This becomes more serious as the supply voltage is scaled.

References

[1] Borkar, S., Design challenges of technology scaling, *IEEE Microelectron.*, 19(4), 23, 1999.

[2] Brews, J., *High-Speed Semiconductor Devices,* Sze, S.M., Ed., John Wiley & Sons, New York, 1990, chap. 3.

[3] Roy, K. and Prasad, S.C., *Low-Power CMOS VLSI Circuit Design*, Wiley Interscience Publications, New York, 2000, chap. 5, 224.

[4] International Technology Roadmap for Semiconductors (ITRS), 2001 ed., Semiconductor Industry Association, http://public.itrs.net/Files/2001ITRS/Home.htm.

[5] Taur, Y. and Ning, T.H., *Fundamentals of Modern VLSI Devices,* Cambridge University Press, New York, 1998, chap. 2, 94.

[6] Mukhopadhyay, S. and Roy, K., Accurate modeling of transistor stacks to effectively reduce total standby leakage in nano-scale CMOS circuits, *Symp. of VLSI Circuits,* June 12–14, 2003.

[7] Roy, K., Mukhopadhyay, S., and Mahmoodi-Meimand, H., Leakage current mechanisms and leakage reduction techniques in deep-submicron CMOS circuits, *Proc. IEEE,* 2003.

[8] Taur, Y. and Ning, T.H., *Fundamentals of Modern VLSI Devices,* Cambridge University Press, New York, 1998, chap. 3, 120.

[9] Schuegraf, K. and Hu, C., Hole injection SiO_2 breakdown model for very low voltage lifetime extrapolation, *IEEE Trans. on Electron. Devices,* 41, 761, 1994.

[10] Choi, C., Nam, K., Yu, Z., and Dutton, R.W., Impact of gate direct tunneling current on circuit performance: a simulation study, *IEEE Trans. on Electron. Devices,* 48, 2823, 2001.

[11] Cao, K. et al., BSIM4 gate leakage model including source drain partition, *IEDM Tech. Dig.,* 815, 2000.

[12] Wei, L., Chen, Z., Johnson, M., Roy, K., Ye, Y., and De, V., Design and optimization of dual threshold circuits for low voltage low power applications, *IEEE Trans. on VLSI Syst.,* 16, 1999.

[13] Karnik, T. et al., Total power optimization by simultaneous dual-V_t allocation and device sizing in high-performance microprocessors, In *ACM/IEEE Design Automation Conf.*, 486, June 10–14, 2002.

[14] Pant, P., Roy, K., and Chatterjee, A., Dual-threshold voltage assignment with transistor sizing for low-power CMOS circuits, *IEEE Trans. on Very Large-Scale Integration (TVLSI) Syst.,* 9, 390, 2001.

[15] Sirichotiyakul, S. et al., Stand-by power minimization through simultaneous threshold voltage selection and circuit sizing, *Proc. 36th ACM/IEEE Conf. on Design Automation,* 436, June 21–25, 1999.

[16] Kao, J.T. and Chandrakasan, A.P., Dual-threshold voltage techniques for low-power digital circuits, *IEEE J. of Solid-State Circuits,* 35, 1009, 2000.

[17] Sirisantana, N., Wei, L., and Roy, K., High-performance low-power CMOS circuits using multiple channel length and multiple oxide thickness, *Proc. 2000 Int. Conf. on Computer Design,* 227, Sept. 17–20, 2000.

[18] Krishnamurthy, R.K., Alvandpour, A., De, V., and Borkar, S., High-performance and low-power challenges for sub-70-nm microprocessor circuits, *Proc. IEEE Custom Integrated Circuits Conf.*, 125, May 12–15, 2002.

[19] Takahashi, M. et al., A 60-mW MPEG4 video CODEC using clustered voltage scaling with variable supply-voltage scheme, *IEEE J. of Solid-State Circuits*, 33, 1772, 1998.

[20] Ye, Y., Borkar, S., and De, V., A new technique for standby leakage reduction in high performance circuits, *IEEE Symp. on VLSI Circuits*, 40, June 11–13, 1998.

[21] Chen, Z., Wei, L., Johnson, M., and Roy, K., Estimation of standby leakage power in CMOS circuits considering accurate modeling of transistor stacks, *IEEE Int. Conf. on Comput.-Aided Design*, Aug. 10–12, 1998.

[22] Chen, Z., Wei, L., Keshavarzi, A., and Roy, K., IDDQ testing for deep submicron ICs: challenges and solutions, *IEEE Design and Test of Comput.*, 24, 2002.

[23] Mukhopadhyay, S., Neau, C., Cakici, T., Agarwal, A., Kim, C.H., and Roy, K., Gate leakage reduction for scaled devices using transistor stacking, *IEEE Trans. on Very Large-Scale Integration Syst.*, 2003.

[24] Johnson, M.C., Somasekhar, D., and Roy, K., Leakage control with efficient use of transistor stacks in single threshold CMOS, *Proc. ACM/IEEE Design Automation Conf.*, 442, June 21–25, 1999.

[25] Mutoh, S. et al., 1-V Power supply high-speed digital circuit technology with multi-threshold voltage CMOS, *IEEE J. Solid-State Circuits*, 30, 847, 1995.

[26] Kao, J., Chandrakasan, A., and Antoniadis, D., Transistor sizing issues and tool for multi-threshold CMOS technology, *Proc. of ACM/IEEE Design Automation Conf.*, 495, June 9–13, 1997.

[27] Mutoh, S. et al., A 1-V multi-threshold voltage CMOS DSP with an efficient power management for mobile phone application, *Dig. Tech. Papers EEE Int. Solid-State Circuits Conf.*, 168, Feb. 8–10, 1996.

[28] Shigematsu, S. et al., A 1-V high-speed MTCMOS circuit scheme for power-down applications, *IEEE J. Solid-State Circuits*, 32, 861, 1997.

[29] Kawaguchi, H., Nose, K., and Sakurai, T., A CMOS scheme for 0.5-V supply voltage with pico-ampere standby current, *Dig. Tech. Papers IEEE Int. Solid-State Circuits Conf.*, 192, Feb. 5–7, 1998.

[30] Tschanz, J. et al., Dynamic-sleep transistor and body bias for active leakage power control of microprocessors, *IEEE Int. Solid-State Circuit Conf.*, 2003.

[31] Heo, S. and Asanovic, K., Leakage-biased domino circuits for dynamic fine-grain leakage reduction, *Symp. on VLSI Circuits*, 316, June 13–15, 2002.

[32] Kuroda, T. et al., A 0.9-V, 150-MHz, 10-mW, 4-mm^2, 2-D discrete cosine transform core processor with variable-threshold-voltage scheme, *Dig. of Tech. Papers IEEE Int. Solid-State Circuits Conf.*, 166, Feb. 8–10, 1996.

[33] Oowaki, Y. et al., A sub-0.1um circuit design with substrate-over-biasing, *Dig. of Tech. Papers of IEEE Int. Solid-State Circuits Conf.*, 88, Feb. 5–7, 1998.

[34] Keshavarzi, A., Hawkins, C.F., Roy, K., and De, V., Effectiveness of reverse body bias for low power CMOS circuits, *Proc. 8th NASA Symp. on VLSI Design*, 231, Oct. 1999.

[35] Keshavarzi, A. et al., Forward body bias for microprocessors in 130nm technology generation and beyond, *Symp. on VLSI Circuits*, 312, June 13–15, 2002.

[36] Mizuno, H. et al. An 18-μA standby current 1.8-V, 200-MHz microprocessor with self-substrate-biased data-retention mode, *IEEE J. Solid-State Circuits*, 34, 1999.

[37] Lee, S. and Sakurai, T., Run-time voltage hopping for low-power real-time systems, *Proc. IEEE/ACM Design Automation Conf.*, 806, June 5–9, 2000.

[38] UC Berkeley device group, Predictive technology model, http://www-device.eecs.berkeley.edu/~ptm /.

[39] Kim, C.H. and Roy, K., Dynamic V_{th} scaling scheme for active leakage power reduction, *Design, Automation and Test in Europe*, 163, March 5–8, 2002.

[40] Nose, K. et al., V_{th}-hopping scheme for 82% power saving in low-voltage processors, *Proc. IEEE Custom Integrated Circuits Conf.*, 93, May 6–9, 2001.

[41] Agarwal, A., Li, H., and Roy, K., A single-V_t low-leakage gated-ground cache for deep submicron, *IEEE J. of Solid-State Circuits,* 2003.

[42] Yamauchi, H. et al., A 0.8V/100MHz/sub-5mW-operated mega-bit SRAM cell architecture with charge-recycle offset-source driving (OSD) scheme, *Symp. on VLSI Circuits,* 126, June 13–15, 1996.

[43] Bhavnagarwala, A.J., Kapoor, A., and Meindl, J.D., Dynamic threshold CMOS SRAMs for fast, portable applications, *ASIC/SOC Conf.,* 359, 2000.

[44] Osada, K. et al., 16.7fA/cell tunnel-leakage-suppressed 16Mb SRAM for handling cosmic-ray-induced multi-errors, *Int. Solid-State Circuits Conf.,* 302, 2003.

[45] Agarwal, A. and Roy, K., Noise-tolerant cache design to reduce gate and subthreshold leakage in nanometer regime, In *Int. Symp. Low-Power Electronics and Design (ISLPED2003),* Aug. 25–27, 2003.

[46] Kawaguchi, H., Itaka, Y., and Sakurai, T., Dynamic leakage cut-off scheme for low-voltage SRAMs, *Symp. on VLSI Circuits,* 140, June 11–13, 1998.

[47] Kim, C.H. and Roy, K., Dynamic V_t SRAM: a leakage-tolerant cache memory for low-voltage microprocessors, *Int. Symp. on Low-Power Elecron. and Design,* 251, Aug. 12–14, 2002.

[48] Flautner, K. et al., Drowsy caches: simple techniques for reducing leakage power, *Int. Symp. on Comput. Architecture,* 148, May 25–29, 2002.

[49] Heo, S. et al., Dynamic fine-grain leakage reduction using leakage-biased bitlines, *Int. Symp. on Comput. Architecture,* 137, May 25–29, 2002.

[50] Itoh, K., Fridi, A.R., Bellaouar, A., and Elmasry, M.I., A deep sub-V, single power-supply SRAM cell with multi-V_t, boosted storage node and dynamic load, *Symp. on VLSI Circuits, Dig. of Tech. Papers,* 132, June 13–15, 1996.

14 Low-Power and Low-Voltage Communication for SoCs

Christer Svensson
Linköping University

14.1 Introduction

Integrated circuits (ICs) mainly consist of transistors and interconnects. Normally, we are more interested in the transistors and how they are combined to form logic gates, flip-flops, memories, and other functional units. Interconnects are easily overlooked because they are just nodes in a circuit diagram; but interconnects are responsible for all communication between logical gates, functional units, and subsystems and are, therefore, of crucial importance. In reality, interconnects are one or several wires with various lengths, which connect transistors and blocks over various distances. Their behavior strongly depends on their lengths. When discussing power consumption, interconnect tends to dominate the power consumption, due to their large total capacitance [1–3]. In Liu and Svensson [2], 30 to 40% of the power consumed by a chip (input/output, I/O, excluded) is estimated to be related to interconnect and an additional 40% to the clock distribution (of which some half is related to the wires). In Chandra et al. [3], about 70% of the power of a high-performance chip (microprocessor) is estimated to be related to interconnect and clock in the 180-nm technology node. It is, therefore, well motivated to consider interconnect power separately.

Very short wires, connecting transistors inside small blocks, as for example simple logic gates, are most easily treated as parasitic capacitance added to other parasitic capacitances of similar size, as gate or drain capacitances [4]. For interconnect between blocks, we normally see a broad distribution of wire lengths

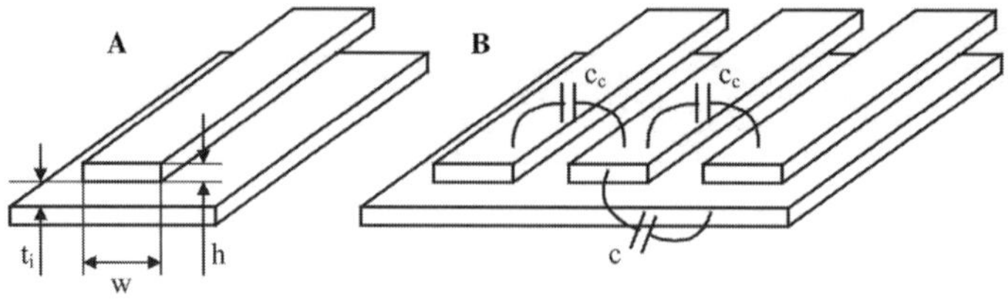

FIGURE 14.1 Single and multiple microstrip wire.

[4]. Here, interconnect capacitance easily dominates over other capacitances, making interconnect power consumption substantial.

When looking outside the chip, we normally have quite large capacitive or resistive loads, making the I/O power substantial. As much as 60% of the total chip power consumption can be related to I/O [2]. Because the outside load is largely controlled by the properties of interconnect between chips (on the circuit board), we will include a treatment of I/O power consumption. Outside the chip, interconnect may become electrically long, with the result that they behave more as transmission lines than as capacitive lines [4], making their behavior quite different, also in a power perspective. For large enough clock frequencies, on-chip wires may also approach transmission line behavior in the future.

The continuous scaling of IC processes also introduces changes in the metal stack [5]. Most important changes are an increased aspect ratio (metal height/width) of the wires and an increased number of metal layers, with larger dimensions of the upper layers. The increased aspect ratio leads to an increased wire-to-wire capacitance, compared to wire-to-ground capacitance, which will increase power consumption. The increased number of metal layers offers new opportunities, as the upper layers have much lower loss than the lower layers.

We start with a brief discussion of the properties of wires and interconnects, then describe power modelling of interconnects, and, finally, we discuss various methods to reduce interconnect power consumption.

14.2 Basics of Wires

14.2.1 General

The simplest model of a wire is the microstrip, that is a metal strip on top of a ground plane (Figure 14.1(a)). Such a wire can be modelled through its capacitance to ground per unit length, c, its resistance per unit length, r and its inductance per unit length, l. In the simplest case, very short wires, capacitance is sufficient as a model. For longer wires, we need to add resistance; for very long wires, inductance will also become important. For the microstrip, we may approximate the capacitance per unit length as [6]:

$$c = c_{pp} + c_{fringe} = \frac{\varepsilon_i (w - h/2)}{t_i} + \frac{2\pi\varepsilon_i}{\log(2 + 4t_i/h)} \tag{14.1}$$

where c_{pp} is the "plate capacitor" capacitance per unit length, c_{fringe} is the fringing capacitance per unit length, w is the wire width, t_i is the insulator thickness, h is the wire thickness, and ε_i is the insulator dielectric constant.

More generally, the wire is surrounded by many other wires (Figure 14.1(b)). We then have capacitances not only to ground, but also to the neighboring wires, the coupling capacitances per unit length, c_c. A reasonable model here is to use the principle of superposition of signals. We thus assume that all wires except the actual one are grounded, and describe it through its total capacitance to its surroundings. The effect of any signal on a neighboring wire is then treated by calculating the crosstalk from each neighboring wire to the actual one, and add this voltage to the voltage on the actual wire. Most important is thus the total capacitance per unit length, c_{tot}, which may replace c in Equation 14.1. In some cases, crosstalk effects may increase power consumption above the superposition model described above (see Section 14.3.1). The situation becomes even more complex if we have no ground plane or other metal in parallel with the actual wire. In such cases, the return current may take a complex path leading to a large inductance, which is quite hard to predict. Such situations are not discussed here.

For longer wires, we also need to consider the series resistance per unit length, r, given by

$$r = \frac{\rho}{A} \tag{14.2}$$

where ρ is the metal resistivity and A = hw the wire cross section. For longer wires at high speed, we may need to include also the inductance per unit length, l. In such a case, the wire behaves as a transmission line (assuming a ground plane or regular return path) [4,7]. A transmission line has the following properties: it is considered to carry two waves, one in each direction. Each wave moves with velocity v_d, given by:

$$v_d = \frac{c_0}{\sqrt{\varepsilon_r}} \tag{14.3}$$

where c_0 is the velocity of light in vacuum, and ε_r is the relative dielectric constant of the dielectric. Furthermore, for each wave its characteristic impedance, Z_0, relates voltage and current. Z_0 is real at high frequencies. An ideal transmission line has no resistance and transports any wave without attenuation or distortion. A transmission line with resistance is said to be lossy and will attenuate each wave as $e^{-\beta L}$, where β is the attenuation factor:

$$\beta = \frac{r}{2Z_0} \tag{14.4}$$

For βL > 1, we may consider the wire an RC-wire, as described previously. For βL < 1, we may consider it a lossy transmission line. A lossy transmission line also exhibits skin effect, for which the resistance becomes frequency dependent, making the wire distort high-speed signals [7].

Transmission lines with low loss should normally be terminated, that is, connected to the impedance Z_0, at least at one of its ends. The reason for this is to avoid reflections. If a wave travels in x-direction along the wire and meets Z_0, it will be completely absorbed; however, if it instead meets, for instance, $Z = \infty$, then a new wave in –x direction is generated (to fulfill the voltage to current relation), which travels in the opposite direction on the wire. This is also termed "full reflection." When the reflected wave reaches the terminated transmitter, it will however be absorbed with no further action. From this, we can understand that if the wire is incorrectly terminated in both ends, we will have waves bouncing back and forth several times, thus leading to uncontrolled "ringing." If the wire is terminated in one end, we control the signal much better. To ensure full control, we often prefer to terminate the wire in both ends.

On-chip wires can normally be treated as RC-lines. Very long global lines in thick upper level metal may sometimes be considered lossy transmission lines [7]. Off-chip lines (on printed circuits boards for example) normally behave as lossy transmission lines.

14.2.2 Interconnect Delays

Any wire-carrying signal is driven by some circuit. The simplest driver is the inverter. We will therefore use a simple inverter model when discussing driven wires in this section. The simple inverter is described by its Thevenin equivalent (Figure 14.2). For a short wire with length L, we may describe the wire with its total capacitance, $C_w = c_{tot}L$, and we may then estimate the delay to:

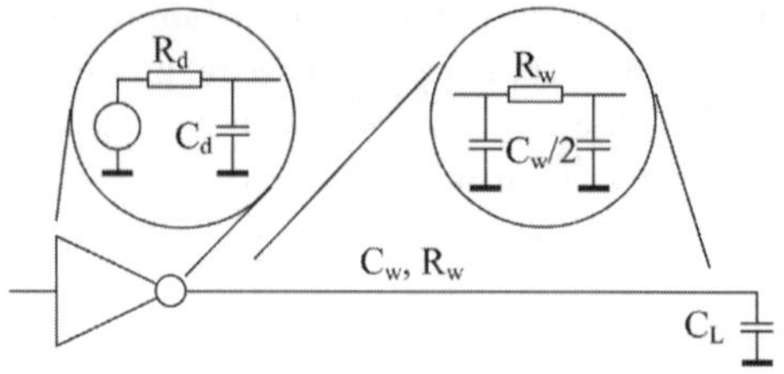

FIGURE 14.2 Inverter driver and wire with lumped element models in the inserts.

$$t_d = R_d(C_d + C_w + C_L)\log(2) \tag{14.5}$$

where R_d and C_d are the driver output resistance and capacitance, respectively, and C_L is the load capacitance.

For a longer wire, including wire resistance $R_w = rL$, we may approximate the delay to [4]:

$$t_d = \left(R_d(C_d + C_w + C_L) + R_w\left(\frac{C_w}{2} + C_L\right)\right)\log(2) \tag{14.6}$$

This expression includes a term, R_wC_w, which is proportional to wire length squared, L^2 (as each of R_w and C_w is proportional to wire length). This leads to very long delays for long wires.

In the case of crosstalk from neighboring wires an increased delay may occur [8]. Consider the case of a positive transition of amplitude ΔV on the actual wire interacting with a negative transition of the same amplitude on its neighbor. The coupling capacitance between the two wires, $C_c = c_cL$, thus change its voltage from ΔV to $-\Delta V$, thus $2\Delta V$, making the transient current through the capacitor double comparing to the noncrosstalk case. The capacitance thus appears as a capacitance to ground of value $2C_c$, increasing wire delay. For two neighbors (in a bus) we may experience a worst-case capacitance of $4C_c$, to be considered in worst-case delay calculations. This phenomenon is similar to the Miller effect in inverters and affects power consumption (see Section 14.3.1).

For a transmission line, finally, the interconnect delay is given by the velocity of light in the actual dielectric:

$$t_d = \frac{L}{v_d} \tag{14.7}$$

14.2.3 Wires with Repeaters

As noted earlier, wire delay grows fast with wire length. One way to mitigate this is to introduce repeaters along the wire [8]. Let us divide the wire into m sections of length $L_s = L/m$ each and attach a repeater (an inverter) at each section (Figure 14.3). Minimum delay occurs for a certain section length, given by:

$$L_{sopt} = \sqrt{\frac{t_{p1}}{0.38rc}} \tag{14.8}$$

where t_{p1} is the delay of an inverter driving another similar inverter.

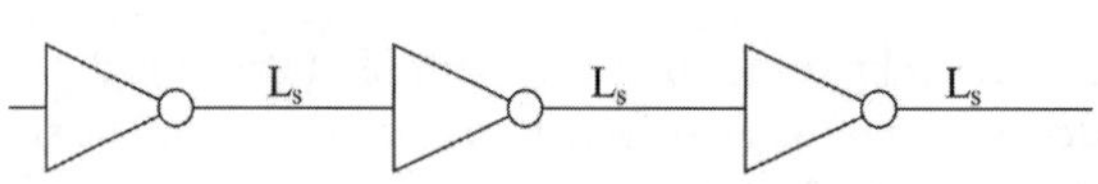

FIGURE 14.3 Wire with repeaters.

The minimum delay occurs at an inverter size (compared to the minimum size) given by:

$$s_{opt} = \sqrt{\frac{R_d c}{r C_d}} \tag{14.9}$$

and will have the value:

$$t_{d\min} = \left(1.02 + \frac{1.38}{\sqrt{1+\gamma}}\right)\sqrt{rct_{p1}}\,L \tag{14.10}$$

Here, R_d and C_d are the minimum inverter output resistance and input capacitance respectively, and γ is the ratio between the inverter input capacitance and its output capacitance. The total delay of the wire with repeaters is thus proportional to L instead of L^2, as described previously. The total capacitance of a section of the wire, C_s is the sum of the wire capacitance, $c_{tot}L_{sopt}$ and the inverter capacitances, sC_d and γsC_d. Calculating C_s/L_{sopt} yields:

$$\frac{C_s}{L_{sopt}} = c + 0.74\sqrt{(1+1/\gamma)c} \tag{14.11}$$

The total switching capacitance of the wire including the optimized repeaters is thus larger than the wire itself (c), leading to corresponding higher power consumption. With a typical value of $\gamma = 1$, the capacitance overhead is about 100%.

14.3 Power Consumption Related to Interconnect

14.3.1 Basics

As described earlier, short interconnects are just described by their total capacitance, C_w. For longer interconnects that are still electrically short, we describe them as RC-lines or by an RC-lumped model. In both cases, power consumption occurs as in logic, which is the interconnect capacitance is charged by the supply voltage and then discharged, making the power consumption related to the interconnect given by:

$$P_w = \tfrac{1}{2}\alpha f_c C_w \Delta V^2 \tag{14.12}$$

where α is the signal activity (the probability that the signal will change per clock cycle), f_c is the clock frequency, and ΔV the signal voltage swing. Here, we assumed that the driver is an ideal inverter, that is a switch connected either to ΔV or to ground. Equation (14.12) is also valid for a switch with series resistance (e.g., an inverter where the transistors have series resistance) driving an open wire. We included only C_w in the expression. For the full interconnect including driver and load, we should include the capacitances of these as well.

The fact that power consumption depends on signal activity (described by α in Equation (14.12)) leads to severe difficulties in power prediction and, therefore, in power optimization. Any prediction and optimization must consider signal activity, which depends on actual data statistics and therefore on actual architectures and the applications run on these. One consequence is that total interconnect length is not sufficient for power estimation. Instead, we need individual wire lengths and individual signal activities [9].

For wires with crosstalk, the worst-case power consumption may be larger than predicted by Equation (14.12), due to the Miller effect discussed previously. If there is a transition with opposite polarity on a neighboring wire, the effective value of the coupling capacitance, C_c, may double, thus increasing the power consumption through a larger C_w in Equation (14.12). This effect was analyzed in [10], where also methods to reduce the effect was discussed. The effect depends on the correlation between neighboring signals and is quite sensitive to the exact timing relation between the two transitions (see Sasaki et al. [11]).

For transmission lines, we have two main cases: a wire that is terminated to Z_0 at the far end and a wire that is open at the far end. The terminated case is very simple, as the input impedance to a terminated wire simply is Z_0. The wire thus behaves as a resistor with resistance Z_0. Assuming it is driven by a driver with output impedance Z_0, we get a power consumption of

$$P_w = \frac{V_{dd}^2}{4Z_0} \tag{14.13}$$

where V_{dd} is the supply voltage, and we assumed equal probability for ones and zeros. Note that the voltage swing is smaller than the supply voltage in this case, $\Delta V = V_{dd}/2$. The open case was treated in Svensson [7]. We may understand the behavior in terms of forward and reflected waves. If the driver output changes from low to high at $t = 0$, it creates a forward wave of amplitude $V_{dd}/2$ in the wire (assuming a driver output impedance of Z_0). This wave is reflected at the far end of the wire, creating a backward wave of amplitude $V_{dd}/2$. After time $2T_d$, where T_d is the wire delay, the voltage at the wire input becomes V_{dd}. If the input driver still is driving high, the current becomes zero and the total charge driven into the wire is $2T_dV_{dd}/Z_0$. This charge can be shown to be equal to the wire capacitance charged to V_{dd}. If the driver instead has changed to driving zero after time $2T_d$ (which may occur if the electrical wire length T_d is larger than half the data symbol length, T_s), the backward wave is terminated to ground and will partly discharge the wire. We can imagine that for a long open wire, the whole data sequence sent will return after time $2T_d$, and depending on which state the driver is in (output connected to V_{dd} or ground), the return current will either cancel the V_{dd} current or go to ground. The average current consumption therefore depends on the correlation between sent data at times t and $t+2T_d$. In Svensson [7], this average was calculated for random data. Thus, we have:

$$P_w = \frac{1}{4} f_c C_w V_{dd}^2 \qquad 2T_d < T_s \tag{14.14}$$

$$P_w = \frac{V_{dd}^2}{8Z_0} \qquad 2T_d > T_s \tag{14.15}$$

We may then conclude the transmission line case as follows (assuming a random sequence of binary data and a driver with output impedance Z_0). For transmission lines terminated by Z_0 in the far end, the power consumption is simply the same as that of a resistor of value $2Z_0$ (driver resistance in series with load resistance). For an open transmission line, we have two cases: for the electrically short line, the power consumption is the same as of a capacitor of value C_w. For an electrically long line, it is half of the terminated line case with the same supply voltage (note, however, that voltage swing in the terminated case is $V_{dd}/2$, and, in the open case, is V_{dd}) (see Figure 14.4).

14.3.2 Power Consumption Related to Drivers and Repeaters

Because wire capacitance often is quite large, the driver must be upsized to facilitate short delay and fast rise-time. For driving large loads, we normally use a tapered inverter chain in order to minimize delay. This means that we have an upsized inverter to drive the wire and then a multistage predriver to drive the upsized inverter. The power consumption related to the driver itself (i.e., on top of the power

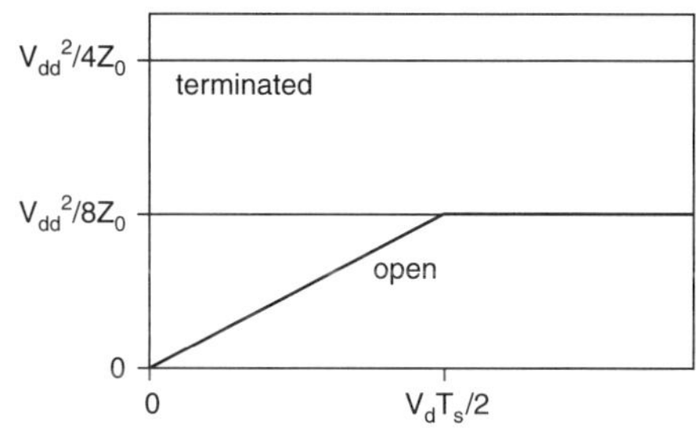

FIGURE 14.4 Power consumption of transmission lines vs. length for binary symbols with length T_s. Two cases are depicted: a line terminated by Z_0 and an open line.

related to its load) also becomes significant and proportional to the load capacitance. The dynamic portion of the driver power consumption can be expressed through the total switched capacitance of the driver, C_d [12]:

$$C_d = \frac{(1+1/\gamma)}{(f-1)}(C_w + C_L) \tag{14.16}$$

where *f* is the tapering factor (size ratio between following inverters in the chain; the value for minimum delay is about 3.5). This leads to a power overhead of about 80% of the power consumed by wire and load ($C_w + C_L$). Drivers are also vulnerable to additional power consumption due to short-circuit power, because of their large loads. In the case of drivers aimed for driving external loads (I/O), there may exist additional constraints, asking for more complex circuits than a simple inverter [13]. There may be needs for accurate output impedance, for tri-state outputs, for wired-or capability, for differential signals, for mitigating short-circuit current and for rise-time control. Such additional constraints normally leads to a larger power consumption compared with the simple inverter model discussed previously, either directly, through for example voltage loss in the circuit, or indirectly, through increased transistor sizes leading to larger capacitance and larger predrivers. These issues are further discussed in Section 14.4.5.

As mentioned earlier, repeaters are often used on chip to optimize wire delays. Repeaters are also utilized to mitigate crosstalk in long wires. The dominating power consumption in well-designed repeaters is dynamic power, thus following Equation (14.12) (assuming electrically short wires) with C_w replaced by the sum of C_w and the total switching capacitance of the repeaters:

$$C_{tot} = \frac{C_s}{L_s} L \tag{14.17}$$

As mentioned earlier, C_{tot} is about 100% larger than C_w for optimized delay. However, the delay minimum is very shallow, so in order to save power it is preferable to have smaller repeaters at longer distances than optimum, which saves considerable power at a small delay penalty [14].

14.3.3 Power Related to Precharged Buses

Sometimes wires are precharged instead of driven statically [8]. This may speed up a wire with large capacitance. It also mitigates the detrimental effect of coupling capacitance (i.e., the Miller effect) on both delay and power consumption. The reason for this is that signals on neighboring wires are monotonous, that is they cannot change in opposite direction. However, precharged nodes always have a large activity, about 0.5, independent of the signal activity, α, thus increasing the power consumption compared to a static bus [12].

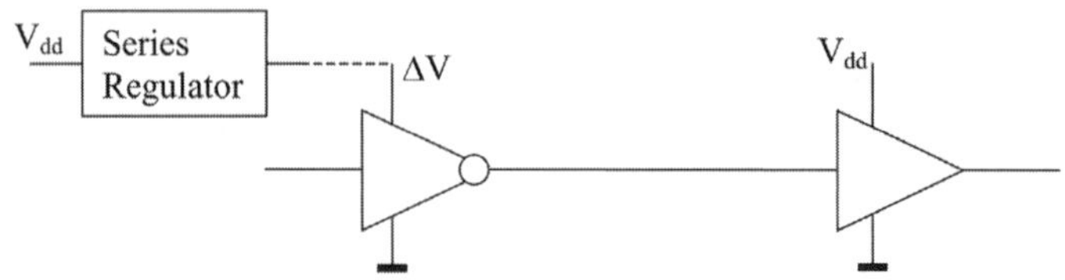

FIGURE 14.5 Transmitter, wire, and receiver.

14.4 Strategies for Power Savings in Interconnect

14.4.1 Introduction

As mentioned in the introduction to this chapter, power consumption related to interconnect is substantial. It may therefore be very profitable to find methods to reduce interconnect-related power consumption. An obvious method is of course to limit long distance communications. This is a very important route to reduced power, but an architectural issue and therefore outside the scope of this chapter. It may be important to minimize wire capacitance through changes in the fabrication process (Equation (14.12) and Equation (14.14)). Such work is ongoing, by searching for dielectrics with lower dielectric constants. At the same time, however, scaling tends to increase capacitance through changes in wire aspect ratios leading to larger coupling capacitances [5].

A very efficient power savings method is to reduce the signal voltage swing (Equation (14.12)). This method normally leads to some delay penalty, which may be hard to accept in high performance systems. Particularly in wires with repeaters, reduced swing is tricky and leads to delay penalties. A possible route is to avoid repeaters by utilizing the lower loss in an upper level, thicker, metal layer [15,16]. If repeaters must be used, their delay may be traded for lower power consumption.

Another powerful method for power saving is to reduce the product of data activity and capacitance. One way to accomplish this is to consider data activity during floorplanning and routing, thus facilitating power optimization based on wire activity/length product. Another way is to reduce data activity on buses, and minimize the amount of crosstalk-related power consumption. This can be accomplished through coding. Several coding methods have been evaluated for power saving in data buses. Utilizing precharged buses could be seen as a special case of data coding.

All methods mentioned can be utilized for on-chip as well as off-chip wires. Off-chip interconnect is, however, more vulnerable to noise, often motivating special solutions. Finally, some more exotic methods are charge redistribution, charge recovery, and adiabatic methods. Let us discuss each of these possibilities below.

14.4.2 Reduced Voltage Swing

Reduced voltage swing is often an effective way to save power in logic. This is true also for wires as can be seen in Equation (14.12) through Equation (14.15). If we consider interconnect between two logic blocks, we could reduce the voltage at the transmitter, thus saving power of the wire. To drive the logic at the receiving side, however, we normally need to restore the reduced voltage to the normal logic levels again [8]. We thus need an amplifying receiver that causes a delay penalty and needs power. For high data rates, we may find an optimum voltage swing, for which the total power (wire and receiver) is minimum [15,17]. For lower data rates, this optimum voltage becomes very small and will instead be limited by noise.

Let us study a complete link with driver, interconnect, and receiver (Figure 14.5). The simplest possible driver is an inverter, driven by a reduced supply voltage ΔV. Let us for simplicity assume that one voltage level is equal to ground, which means that the interconnect voltage is 0 or ΔV. The current consumption of the driver when driving an electrically short wire is then:

$$I_D = \frac{1}{2}\alpha f_c C_{tot} \Delta V \tag{14.18}$$

where C_{tot} is the total load of the driver (i.e., driver output capacitance, wire capacitance, and receiver input capacitance). The power consumption now depends on from where we take the current. If the current is taken from a separate power supply of voltage ΔV, then the power consumption becomes:

$$P_w = I_D \Delta V = \frac{1}{2} \alpha f_c C_w \Delta V^2 \tag{14.19}$$

If, on the other hand, the current is taken from the ordinary supply at voltage V_{dd}, assuming that ΔV is generated from V_{dd} through a lossless series regulator, we get:

$$P_w = I_D V_{dd} = \frac{1}{2} \alpha f_c C_w V_{dd} \Delta V \tag{14.20}$$

The latter case also describes circuits where the voltage swing reduction is obtained through a series diode or MOS diode (for example reducing the swing from V_{dd} to V_{dd}–V_T by an MOS diode in series with V_{dd}).

We thus have a transmitter (for simplicity assumed to be an inverter) and a receiver amplifier (Figure 14.5), which should amplify the reduced swing signal to full logic swing (assumed to be V_{dd}). The amplifier speed and gain will be limited by the gain-bandwidth product (or f_T) of the actual process. This leads to a substantial delay in the receiver if the gain is large (swing is low). Furthermore, the amplifier must be designed to fit the actual gain needed. It turns out that the number of amplifier stages and the transistor sizes of the amplifier depend on the gain needed, resulting in a power consumption that increases with gain and speed requirements [17]. A lower swing therefore saves power in interconnect but leads to a larger power consumption of the receiver amplifier. Therefore, we may see an optimum voltage swing corresponding to minimum total power consumption [17]. As an example, a 4-mm long open wire, run at 5 Gb/s, using an amplifier in a 0.18-μm process and using V_{dd} = 1.8 V as supply voltage show a minimum power consumption of 0.7 mW at a voltage swing of 200 mV [15]. Several implementation schemes have been proposed for low-swing interconnect [8,12,15].

If the interconnect use repeaters, a reduction in voltage swing is less obvious. We then need a repeater that accept the reduced swing input and generate the same reduced swing at the output. As repeaters normally are used to mitigate delay, the new repeater should not increase delay. The simplest solution to this problem is to use an inverter with a reduced supply voltage of ΔV as repeater. It will then automatically adjust to as well the reduced input voltage as to the reduced output voltage. We will then still follow Equation (14.8) through Equation (14.11), but we must note that the reduced supply voltage leads to increased values of t_{p1} and R_d because of a smaller effective gate voltage and, therefore, a reduced current driving ability of the transistors. Reducing voltage swing thus leads to an increase in L_{sopt} (which is good as we get less repeaters) and an increased minimum delay (Equation (14.10)). Some simulations in a 0.13-μm process indicate that a swing of 75% of the supply voltage leads to a power saving of 23% (assuming that V_{dd} is used as transmitter supply) at a delay penalty of 15% [18]. Larger power savings leads to considerably larger delay penalty, which is often not accepted. An alternative solution to the simple inverter could consist of a standard receiver amplifier followed by a standard driver. Such a solution will have considerable delay, however, because the amplifying receiver always has a large delay [15].

14.4.3 Reduced Interconnect Activity

For all electrically short interconnects, dynamic power consumption dominates, making the power consumption proportional to the logical activity on the interconnect. It is therefore important to reduce the length of interconnects with large activity. This can be accomplished by having the wire activity control the place and route process. The effect of such an optimization procedure was investigated in Prabhakaran et al. [19], using a few benchmark circuit examples. It was then shown that a reoptimization toward a minimum power goal (replacing a minimum latency goal) resulted in power savings of 25 to 70% with a latency penalty of 10 to 50% and no area penalty.

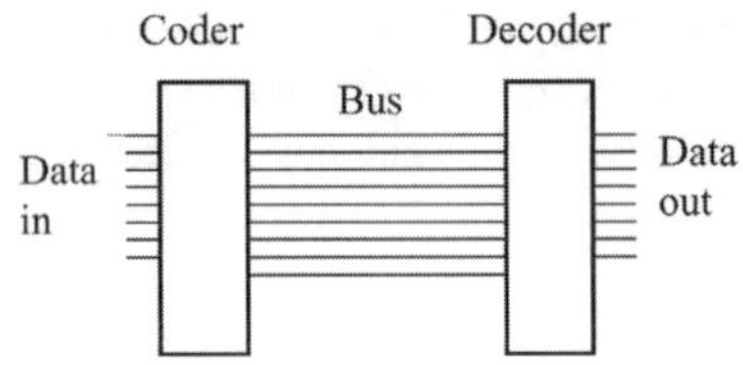

FIGURE 14.6 Using coding on a bus.

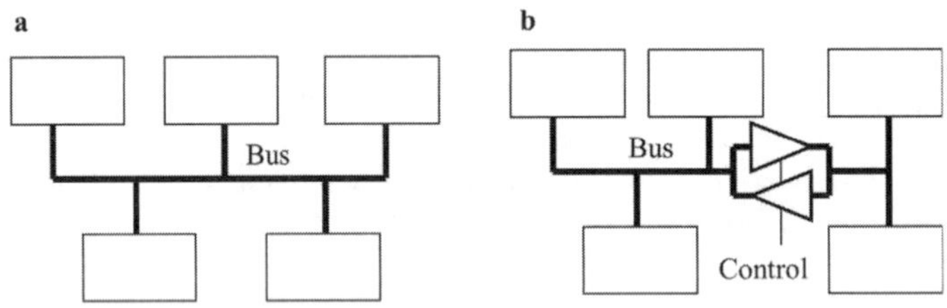

FIGURE 14.7 Going from a common bus (a) to a split bus (b).

Another method for reducing the power consumption is to reduce the logical activity on long interconnects (e.g., long on-chip or off-chip buses) [20]. Such a reduction of activity can be accomplished through coding (Figure 14.6). A simple example is the bus-invert-coding applied to an n-bit bus. The principle is to add one wire to the bus (the invert line, I) to give room for coding. On the rest of the bus we transmit the data, D, by the codeword C. For each new piece of data, D_{next}, we measure the Hamming distance to the previous codeword $C_{previous}$. If the Hamming distance is less than n/2, C_{next} is equal to D_{next} and I = 0, else C_{next} is equal to the inverted value of D_{next} and I = 1. The decoder on the receiver side then recovers D again. In this way, we minimize the number of transitions between two consecutive codes on the bus, thus minimizing the power consumption. Stan and Burleson [20] showed that this is an optimal solution for random data if we add just one extra wire, and finds that the power saving is 25% at best. The best savings occur for small bus widths and is reduced to about 15% at n = 16. In all cases, the bus-invert-coding reduces the maximum supply current with 50%, which is also a very important accomplishment.

In addition, introducing the effect of mutual capacitances (i.e., the Miller effect) makes coding even more efficient for power savings. Sotiriadis and Chandrakasan [10] demonstrated a power savings of 15 to 40% for buses with c_c/c varying from zero to infinity. Another way to mitigate the extra power consumption caused by the Miller effect is to code the bus in such a way that signals on neighboring wires are monotonous, that is we use a precharged bus. The drawback with such an approach is that precharging always increases the total activity (as also some of the lines which do not change data, will charge and discharge). Still, a combination of precharging and coding can give performance benefits without power penalty [21].

Considerably more power savings can be accomplished by context-dependent coding. If we have *a priori* knowledge of the bus traffic, we may use this knowledge to adapt the coding to the traffic. One example is memory bus coding for processor systems, where we know that addresses often are sequential. In such cases, coding at an architectural level may save as much as 64 to 85% [22]. In general, various coding and other architectural optimizations are quite effective for power savings [23].

Buses often use much power. One reason for this is that the whole bus is excited for each transaction, even if the transaction concerns blocks that are close together (Figure 14.7). One way to save power is to divide the bus into several segments, so all short-range transactions only excites part of the bus. Simple simulations indicate power savings of 16 to 50%, depending on the characteristics of the data transfer among the modules and the configuration of the split bus [24].

14.4.4 Power Savings in Drivers and Repeaters

The power overhead related to the wire driver may be quite large (80% or more) in high performance systems. This power overhead can be reduced with a delay penalty by increasing the tapering factor, f.

Increasing f from 3.5 to 9, for example, reduces the power overhead from 80 to 25% at a delay penalty of 20% [12].

As mentioned earlier, interconnect that is delay-optimized by repeaters has a power overhead of about 100%. As the number of repeaters is expected to increase with scaling, this problem will increase in the coming process generations. The delay-minimum is quite shallow, however, so there is a good opportunity to reduce power consumption with a relatively small delay penalty [14]. By optimizing power consumption for a given delay penalty, it has been demonstrated that repeater power dissipation can be decreased by 50%, with a very limited delay penalty of 5% [14]. Allowing a larger delay penalty facilitates larger power savings. A more drastic solution is, of course, to avoid repeaters by using an upper metal layer for long interconnects [15,16].

14.4.5 Off-Chip Interconnect

Off-chip interconnect is traditionally run at moderate speeds (i.e., a few hundred Mb/s) at the same voltage swing used by the ICs. They are, therefore, electrically short. The capacitance of these interconnects consists of the wire capacitance itself (C_w) and the capacitance of the loads. Board wire capacitance is of the order of 100 pF/m [13] and a chip input capacitance of the order of 1 pF. Normally, several loads are allowed, so the output driver is specified for quite a large capacitive load (e.g., 50 to 100 pF). The large specified load combined with a large voltage swing makes the off-chip interconnect have very large maximum power consumption, as mentioned in the introduction, 14.1. In addition, the peak current needed is particularly problematic in the case of I/O because the I/O current surge cannot be absorbed by decoupling capacitors either on-chip or off-chip, as the current pass the chip edge. Because of the large current spikes on the driver supply voltage, this supply is often separated from the supply for the rest of the circuit to protect the circuit from the I/O noise.

The wire power consumption discussed previously refers to a simple driver in the form of an inverter. This case also represents the minimum power case (for a given voltage swing). Quite often, I/O circuits are more complicated than the simple inverter, thus giving rise to larger power consumption. Let us discuss some of these circuits and the reasons for their use.

Often, we want several I/O drivers to drive the same wire (like a common external bus). This is accomplished either with a tri-state driver or with an open drain driver. In the tri-state case, we could use gating transistors in series with the driving transistors (Figure 14.8(a)) or we could use one p-MOS and one n-MOS transistor driver with some additional logic (Figure 14.8(b)). The second solution is preferable from the power consumption point of view, as we can keep the output transistors of minimum size. (In the first case, both output transistors must have double width to keep the output resistance.) Similarly, avoiding short-circuit power in the driver by "break-before-make-action" can be solved by logic in front of the driver transistors [13]. An open-drain output (Figure 14.8(c)) is simple but consumes static power (that is current is continuously consumed when data on the wire is low). Thus, assuming equal probabilities for ones and zeros on the wire:

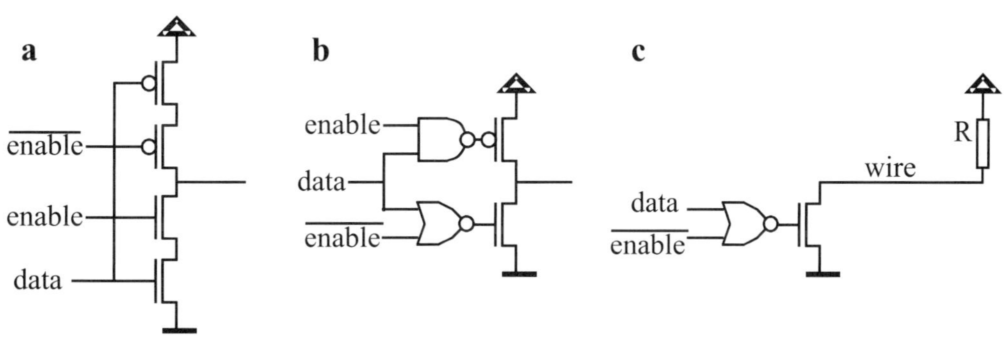

FIGURE 14.8 Tristate drivers (a and b) and open drain driver (c).

$$P = \frac{V_{dd}^2}{2R} \tag{14.21}$$

On the other hand, it saves transistor size, particularly because p-transistors are normally two to three times wider than n-transistors.

Another class of drivers is the current-mode drivers. Such a driver looks exactly as the open-drain driver, but the transistor operates as current generator and the voltage swing is reduced, $\Delta V = I_0 R < V_{dd}$, where I_0 is the current driven by the transistor. The size of the current from the transistor can be controlled by a current-mirror [13]. The power consumption thus becomes:

$$P = \frac{V_{dd} I_0}{2} = \frac{V_{dd}\Delta V}{2R} \tag{14.22}$$

For high data rates, reflection may become a problem. This problem is far more pronounced for off-chip communication than for on-chip wires of two reasons. First, off-chip wires are much longer than on-chip wires. Second, off-chip wires have much less loss and, therefore, behave more like ideal transmission lines. To mitigate reflections, we need to terminate the wire in at least one end. Using an inverter-driver and an open wire allows such a termination without power penalty (except if the transistors need to be wider to get the right on-resistance) by just adjusting the driver output impedance to Z_0. The nonlinear behavior of the transistor on-resistances may be unacceptable, however, calling for a linear resistor in series with each transistor [25]. Then the transistors must be much wider to have smaller resistance, thus increasing their capacitance, which leads to larger power consumption. Still, using a self-terminating driver like this is the most economical solution in terms of power savings. Using a current-mode driver also allows termination, as we can make the resistor in the far end equal to Z_0. The power-consumption is still given by Equation (14.22). In this case, the current-generator driver is considered to have an output impedance that is much larger than Z_0 (wire open on transmitter side). Sometimes, we want to have the wire terminated in both ends. For example, this can be obtained by a current driver with a load resistor of Z_0 in both ends of the wire. The total resistance is then $Z_0/2$, doubling the power consumption compared with Equation (14.22):

$$P = \frac{V_{dd}\Delta V}{Z_0} \tag{14.23}$$

Because of the large power consumption of off-chip interconnect, a reduced voltage swing on these is very profitable. The risk when using low voltage swing is that the interconnect becomes vulnerable to noise. Therefore, most low swing off-chip interconnect use differential signaling [13], which is much less sensitive to noise. A modern standard using differential signaling is LVDS [26]. Because LVDS is aimed for high speeds, it utilizes terminated wires.

A simple way to achieve low swing is to use current-mode drivers as described previously. The drawback is a higher power consumption than necessary. The most power-efficient way is to use a voltage-mode driver with a separate supply voltage as discussed earlier in this paper (Equation (14.13) with $V_{dd} = \Delta V$). Let us exemplify the saving opportunity by using low voltage swing from an example from Svensson [17]. Assuming a terminated wire (terminated by 50 Ω), a 0.13-μm process with a 1.3-V ordinary supply and with separate transmitter supply we arrive to a total power consumption of 0.25 mW at 10Gb/s and at an optimum swing of 120mV. This could be compared with an experimental example of a differential LVDS link in a 0.25-μm process with 2.5-V supply. Here, a total power consumption of 1 mW at 1 Gb/s and 100 mV swing was observed, corresponding to 0.5 mW per wire [27]. A similar experiment, with the voltage swing related to ground, demonstrates 0.8 mW per wire at 200 mV swing [28]. A full swing

(2.5 V) single interconnect, while still terminated to 50 Ω, would have a power consumption of 16 mW (Equation (14.13)).

14.4.6 Charge Recovery Techniques

When considering electrically short wires, which can be modeled as capacitors, the power dissipation is not really occurring in the capacitor, but in the driver. Therefore, could we avoid this dissipation? The answer is yes. If we charge the capacitor slowly with a voltage ramp, no energy is lost during charging. It is then possible to recover this charge back into the power source by discharging the capacitor to a down-ramping supply. This is the adiabatic or energy-recovery principle [29]. Such principles, together with various principles for charge reuse by redistribution, can be utilized for power savings in interconnects [30–32] (see also Chapter 15).

14.5 A Comment about Optical Interconnect

It has been suggested that optical interconnect consumes less power than electrical interconnect [33]. This was based on the observation that also a small photocurrent from a photodiode can give rise to a large voltage, comparable to the logical swing, if the impedance level is large enough. On the other hand, a full logical swing on a wire causes large dynamic power consumption; however, this conclusion did not consider the opportunity for reduced voltage swing on the electrical interconnect and is therefore wrong. A more accurate comparison between optical and electrical interconnect should be done at the same signal-to-noise ratio and consider the receiver power consumption in both cases. This was done in Berglind et al. [34] with the conclusion that electrical interconnect use less or the same power compared with an optical interconnect. A similar analysis performed in Yoneyama et al. [35] considers very long interconnect and estimates the breakeven length, below which electrical interconnect is superior. They find that electrical interconnect is superior for distances less than about 5 m at a data rate of 20 Gb/s. An extrapolation to 100 Gb/s indicates that electrical interconnect is superior up to about 3 m in length. This latter observation demonstrates the most important difference between electrical and optical interconnect, that is optical interconnect is superior for long distances because of a very low attenuation. Then, there may also be other benefits with optical interconnect, such as less electromagnetic emission and less sensitivity to electrical noise, as well as drawbacks as technical complexity and alignment problems.

14.6 Conclusion

The power consumption related to on-chip and off-chip communication is substantial. It is, therefore, very profitable to seek power reductions of interconnect. We have described the most important properties of wires and their design, including the impact of drivers, receivers, and repeaters. We have further described how to model power consumption of wires, including long transmission lines. Power saving can be accomplished by various methods, from the architectural level to the physical level. Reducing the need for communication between distant blocks based on the choice of system architecture is probably the most efficient method. Other architectural methods include utilizing coding on buses to save transitions. Such methods may save about 50% power. On the physical level, floor planning and layout based on communication activity may save up to 70% of power. Reduction of the voltage swing on the interconnects is very effective, particularly for the off-chip case. Savings can be anything from a few percent to maybe 20×. Power savings using voltage swing reduction always leads to some delay penalty, which often can be kept quite small. In the case of off-chip interconnect it may also be very profitable to choose a low-power circuit technique for the wire driver. Some further savings can be achieved through pure technological changes, such as low dielectric constant insulators; however, this savings is expected to barely compensate for capacitance increase caused by scaling. Finally, several more advanced techniques have been proposed, such as charge recovery (adiabatic) techniques or optical interconnects. Among these techniques, the charge recovery technique has potential, whereas optical interconnect is not expected to lead to power savings.

References

[1] H.B. Bakoglu, *Circuits, Interconnections, and Packaging for VLSI,* Addison-Wesley, Reading, MA, 1990.

[2] D. Liu and C. Svensson, Power consumption estimation in CMOS VLSI chips, *IEEE J. Solid-State Circuits,* vol. 29, pp. 663–670, June 1994.

[3] G. Chandra, P. Kapur, and K.C. Saraswat, Scaling trends for the on-chip power dissipation, *Proc. IEEE 2002 Int. Interconnect Technol. Conf.,* pp. 154–156, 2002.

[4] J.M. Rabaey, A. Chandrakasan, and B. Nicolic, *Digital Integrated Circuits, 2nd ed.,* Prentice Hall, Upper Saddle River, New Jersey, 2003, chapter 4.

[5] K. Soumyanath, S. Borkar, C. Zhou, and B.A. Bloechel, Accurate on-chip interconnect evaluation: a time-domain technique, *IEEE J. Solid-State Circuits,* vol. 34, pp. 623–631, May 1999.

[6] Corrected version of Equation (4.2) in Rabaey et al. [4].

[7] C. Svensson, Electrical interconnects revitalized, *IEEE Trans. VLSI Syst.,* vol. 10, p. 777, Dec. 2003.

[8] J.M. Rabaey, A. Chandrakasan, and B. Nicolic, *Digital Integrated Circuits, 2nd ed.,* Prentice Hall, Upper Saddle River, New Jersey, 2003, chapter 9.

[9] A. Alvandpour, P. Larsson-Edefors, and C. Svensson, GLMC: interconnect length estimation by growth-limited multifold clustering, *Proc. Int. Symp. on Circuits and Syst.,* vol. 5, pp. 465–468, 2000.

[10] P.P. Sotiriadis and A. Chandrakasan, Low-power bus coding techniques considering inter-wire capacitances, *IEEE Custom Integrated Circuit Conf.,* pp. 507–510, 2000.

[11] Y. Sasaki, M. Sato, M. Kuramoto, F. Kikuchi, T. Kawashima, H. Masuda, and K. Yano, Crosstalk delay analysis of a 0.13-μm node test chip and precise gate-level simulation technology, *IEEE J. Solid-State Circuits,* vol. 38, pp. 702–708, 2003.

[12] C. Svensson and D. Liu, Low-power circuit techniques, in J. M. Rabaey and M. Pedram, Eds., *Low-Power Design Methodologies,* Kluwer, Dordrecht, 1996.

[13] W.J. Dally and J.W. Poulton, *Digital Systems Engineering,* Cambridge University Press, Cambridge, 1998.

[14] P. Kapur, G. Chandra, and K.C. Saraswat, Power estimation in global interconnects and its reduction using a novel repeater optimization methodology, *Proc. 39th Design Automation Conf.,* pp. 461–466, 2002.

[15] P. Caputa and C. Svensson, Low-power, low latency global interconnect, *15th Annu. IEEE Int. ASIC/SOC Conf.,* pp. 394–398, 2002.

[16] C. Svensson and P. Caputa, High-bandwidth, low-latency global interconnect, *Proc. SPIE, VLSI Circuits and Syst.,* vol. 5117, pp. 126–134, May 2003.

[17] C. Svensson, Optimum voltage swing on on-chip and off-chip interconnects, *IEEE J. Solid-State Circuits,* vol. 36, p. 1108, July 2001.

[18] C. Svensson, unpublished data.

[19] P. Prabhakaran, P. Banerjee, J. Crenshaw, and M. Sarrafzadeh, Simultaneous scheduling, binding and floorplanning for interconnect power optimization, *Proc. 12th Int. Conf. on VLSI Design,* pp. 423–427, Jan. 1999.

[20] M.R. Stan and W.P. Burleson, Bus-invert coding for low-power I/O, *IEEE Trans. VLSI Syst.,* vol. 3, pp. 49–58, March 1995.

[21] M. Anders, N. Rai, R.K. Krishnamurthy, and S. Borkar, A transition-encoded dynamic bus technique for high-performance interconnects, *IEEE J. Solid-State Circuits,* vol. 38, p. 709, May 2003.

[22] Y. Aghaghiri, F. Fallah, and M. Pedram, BEAM: bus encoding based on instruction-set-aware memories, *Proc. Asian and South Pacific Design Automation Conf.,* pp. 3–8, 2002.

[23] E. Macii, M. Pedram, and F. Somenzi, High-level power modeling, estimation, and optimization, *IEEE Trans. Comput.-Aided Design,* vol. 17, p. 1061, Nov. 1998.

[24] C.-T. Hsieh and M. Pedram, Architectural power optimization by bus splitting, *IEEE Trans. Comput.-Aided Design,* vol. 21, pp. 408–414, 2002.

[25] M. Haycock and R. Mooney, 3.2 GHz 6.4 Gb/s per wire signaling in 0.18-μm CMOS, *48th Int. Solid-State Circuits Conf., Dig. Tech. Papers,* pp. 62–63, Feb. 2001.

[26] ANSI/TIA/EIA-644 LVDS Standard. See also J. Goldie, The many flavors of LVDS, http://www.national.com/nationaledge/feb02/flavors.html

[27] S. Hirsch and H.-J. Pfleiderer, CMOS receiver circuits for high-speed data transmission according to LVDS standard, *Proc. SPIE, VLSI Circuits and Syst.,* vol. 5117, pp. 238–244, 2003.

[28] M. Hedberg and T. Haulin, I/O family with 200 mV to 500 mV supply voltage, *44th Int. Solid-State Circuits Conf., Dig. Tech. Papers,* pp. 340–341, Feb. 1997.

[29] W.C. Athas, L.J. Svensson, J.G. Koller, N. Tzartzanis, and E.Y.-C Chou, Low-power digital systems based on adiabatic-switching principles, *IEEE Trans. VLSI Syst.,* pp. 398–407, Dec. 1994.

[30] M. Hiraki, H. Kojima, H. Misawa, T. Akasawa, and Y. Hatano, Data-dependent logic swing internal bus architecture for ultralow-power LSIs, *IEEE J. Solid-State Circuits,* vol. 30, pp. 397–401, April 1995.

[31] H. Yamauchi, H. Akamatsu, and T. Fujita, An asymptotically zero power charge recycling bus architecture for battery-operated ultrahigh data rate ULSIs, *IEEE J. Solid-State Circuits,* vol. 30, pp. 423–431, April 1995.

[32] L.J. Svensson, W.C. Athas, and R.S.-C. Wen, A sub-CV2 pad driver with 10-ns transition time, *IEEE Int. Symp. on Low-Power Electron. and Design,* pp. 105–108, 1996.

[33] D.A.B. Miller, Optics for low-energy communications inside digital processors: quantum detectors, sources and modulators as efficient impedance converters, *Opt. Lett.,* vol. 14, p. 146, Oct. 1996.

[34] E. Berglind, L. Thylén, B. Jaskorzynska, and C. Svensson, A comparison of dissipated power and signal-to-noise ratios in electrical and optical interconnects, *J. Lightwave Technol.,* vol. 17, p. 68, Jan. 1999.

[35] M. Yoneyama, K. Takahata, T. Otsuji, and Y. Akazawa, Analysis and application of a novel model for estimating power dissipation of optical interconnections as a function of transmission bit error rate, *J. Lightwave Technol.,* vol. 14, pp. 13–22, Jan. 1996.

15

Adiabatic and Clock-Powered Circuits

Lars Svensson
Chalmers University

15.1 Introduction

The integrated digital circuitry ubiquitous in the electronic equipment that surrounds us is mainly implemented with complementary metal-oxide semi-conductor (CMOS) technologies. The commonly used logic styles operate in binary voltage mode, where each logic gate drives its output to one of the end points of the available voltage range. Simple circuit styles exist where the supply voltage rails define the voltage range. These full-swing logic styles, including the well-known static-CMOS and domino styles, dominate CMOS logic; special-purpose circuits occupy important niches such as memories. In summary, full-swing voltage-mode CMOS logic styles have been extremely successful, both technically and in terms of market share. In this chapter, they will sometimes be referred to as "conventional" logic.

A lower limit on the dynamic power dissipation of a capacitively loaded conventional logic gate is easy to calculate. Each transition will at least cause a dissipation that depends only on the load capacitance and the voltage swing. Any attempt to seriously reduce this dissipation must reduce the number of transitions, the load capacitance, the voltage swing, or some combination of these.

This chapter describes *adiabatic charging,* a family of techniques to design logic and other switching circuits which circumvent this lower limit of dynamic power dissipation. The principle of adiabatic charging is wide-reaching — it is grounded in very generic models of the switching elements — but implementations have so far been completely dominated by CMOS. Future implementations may benefit from the availability of other manufacturing technologies.

15.2 The Adiabatic-Charging Principle

Consider the conventional, capacitively loaded CMOS inverter depicted in Figure 15.1(a). For the purpose of this example, the idealized resistive-switch network depicted in Figure 15.1(b) can represent the inverter. (Because this discussion concerns lower limits for the dissipation, short-circuit currents and other second-order effects are ignored.) When the input, V_{in}, is pulled low, the pMOS device turns on, and the linear load capacitance (C) is charged from 0 to V. The charging process causes energy dissipation in the pMOS device, because the charge experiences a potential drop on its way from the supply node

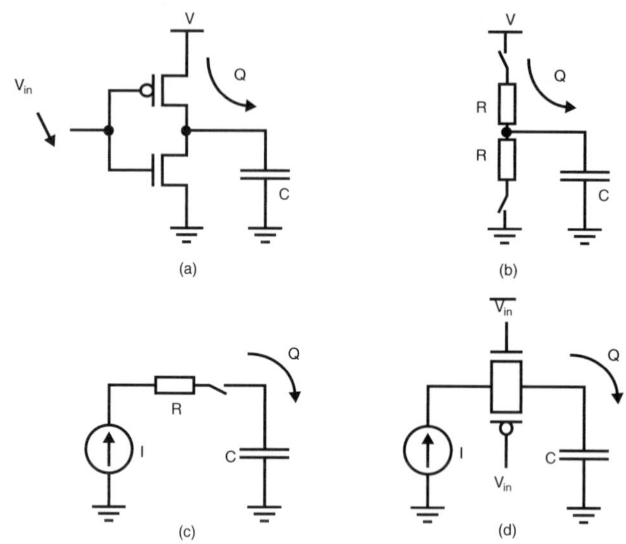

FIGURE 15.1 (a) A static CMOS inverter charging a capacitance C to the supply voltage V. (b) A simplified resistive-switch representation of the same circuit. (c) Resistive-switch representation of constant-current charging of C. (d) CMOS implementation of the same circuit.

to the load. At the outset, no charge is yet stored in C, so the potential drop is V; at the end of the transition, the potential drop is zero. The average potential drop traversed by the charge, V_{avg}, is therefore V/2. The amount of charge stored, Q, is equal to CV. The energy dissipation is therefore $E_{conv} = V_{avg}$ Q = (V/2) (C V) = $CV^2/2$: a familiar result. The node energy, E_{cap}, which is stored on C after charging, is also $CV^2/2$. This energy will dissipate during the subsequent discharging of C.

It is clear from the expression for E_{conv} that dissipation is reduced if V or C are made smaller. At first sight, it seems that nothing can be done if both C and V are fixed. Closer inspection reveals, however, that the dissipation would be reduced if V_{avg} could somehow be reduced below V/2. Actually, the energy dissipation could be reduced to an arbitrary degree, if V_{avg} were reduced commensurately.

Consider now the idealized resistive-switch network in Figure 15.1(c). The constant supply voltage level, V, has been replaced by a constant current source, I; the voltage drop across the switch is IR throughout the charging, so the energy dissipation is E_{curr} = IR CV. The charging current, I, is equal to CV/T, where T is the charging time used to charge C from 0 to V. When this expression is substituted for I, the result is the expression $E_{curr} = (RC/T)\ CV^2$.

We may now make several observations:

- E_{curr} can be lower than E_{conv}, if T is long enough. Actually, E_{curr} may be made arbitrarily small by further extending the charging time. In the limit, charge is moved onto C with no dissipation (i.e., with no heat exchange with the environment). This observation is the motivation for the term "adiabatic charging" [1].
- It can easily be demonstrated that the constant-current charging is the most energy-efficient way to charge a linear capacitance though a resistance in a given time, be it short or long. (Note, however, that the constant voltage drop across R must be maintained throughout the charging. For short-enough T, it would be much larger than V.)
- A lower path resistance R brings a lower dissipation. This result is in contrast to the conventional case: the expression for E_{conv} does not contain R.
- If the current direction is reversed, C is discharged through the same path through which it was charged. The node energy, E_{cap}, which was moved onto C during charging, is then removed, again with a dissipation that depends inversely on the charging time. The node energy minus the

dissipated part is recovered by the current source, and may be reused in the next charging. This mechanism is the motivation for the term "energy-recovery CMOS" [2].

A simple implementation of the resistive-switch network of Figure 15.1(c), in terms of metal-oxide semiconductor (MOS) devices, is depicted in Figure 15.1(d). The switch and the resistance R are implemented as a transmission gate. This choice allows the control signals for the switch to swing from 0 to V, just like the output. A single device of either kind would need a higher-swing control signal to allow the output to be charged all the way from 0 to V.

In conventional static-CMOS circuits, such as the inverter in Figure 15.1(a), a supply-voltage reduction brings lower dynamic dissipation. The dissipation of the circuit of Figure 15.1(d), however, has a minimum value at a certain voltage swing [1]. Further reduction of the swing will bring a dissipation increase caused by the reduced gate-to-channel voltages of the transmission-gate devices; E_{curr} will rise with the on-resistance of the transmission gate.

The constant-current generator presents implementation problems. It is not clear how to build individual controllable constant-current generators for each capacitive load in a large circuit without wasting more power than was gained by introducing them. Thus, all adiabatic-charging circuits presented to date use some approximation of the constant-current source. A time-dependent voltage source that generates periodic positive- and negative-going linear voltage ramps will create current waveforms similar to those generated by the current source, if the (RC/T) factor in the expression for E_{curr} is small enough. In addition to the operating power, the ramp signals naturally provide timing information to the circuits, and, therefore, are often referred to as power-clocks or simply as clock signals. This is the motivation for the term "clock-powered circuits" [22].

15.3 Implementation Issues

The adiabatic-charging principle outlined in the previous section describes how dynamic energy dissipation may be reduced below the $CV^2/2$ per switching event required for conventional switching circuits. Many engineering problems must be solved to utilize the principle in the design of logic circuits. As already indicated, approximately constant currents must be generated and distributed to those circuit nodes that are to be charged. It is not trivial to accomplish this current generation; most published approaches have centralized the current generation in an adiabatic power supply (APS) and solved the logic-design problem separately. The timing of charge and discharge phases of the logic circuits ties together the APS design with the logic and pipeline design. All these subproblems must be solved in any realization of the adiabatic-switching principle; an efficient adiabatic logic style that lacks an efficient APS will not provide a low-power solution.

Two main approaches dominate the adiabatic-logic styles presented to date. The more ambitious approach aims to recover the node energies of *all* circuit nodes, including nodes inside logic gates. The other, more pragmatic approach applies the adiabatic principle to nodes with large capacitance and, therefore, large node energies. The APS designs for these approaches will be somewhat different, as described next.

15.3.1 Adiabatic Logic

Section 15.2 described how adiabatic techniques could be used to reduce the energy dissipation caused by charging a capacitive load. This section presents the incremental construction of a logic style, which allows fully adiabatic operation [1]. The construction starts from the buffer circuit in Figure 15.1(d), but adds a clamp device to ensure that the output is securely grounded when it is not to be charged. The resulting buffer circuit is depicted in Figure 15.2(a).

The transmission gate used to connect the load to the APS requires a dual-rail control signal, as do all logic gates based on transmission gates. It is not admissible to introduce a conventional inverter to derive one of these control signals from the other. At each transition, the inverter would dissipate energy, which would not scale with the transition time, as required for fully adiabatic operation. In a logic style

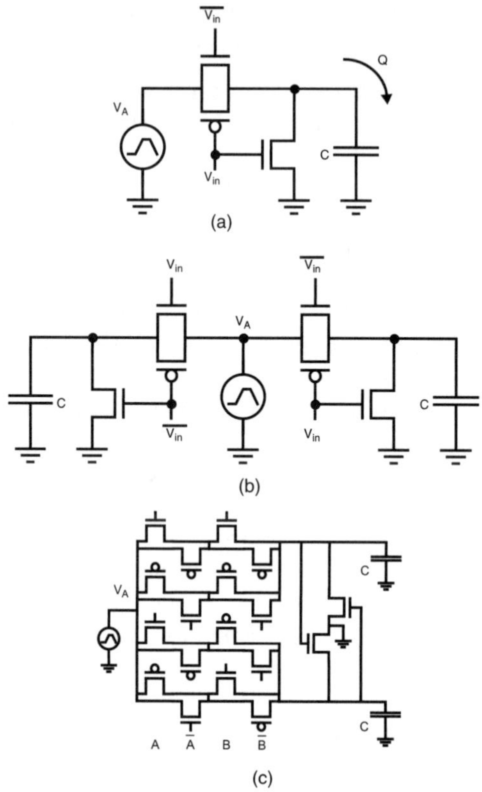

FIGURE 15.2 (a) The adiabatic buffer of Figure 15.1(d), with an additional clamp device. The current generator has been replaced with a time-dependent voltage source, V_A. (b) A dual-rail version of the same circuit. (c) A dual-rail EXOR circuit.

based on transmission gates, therefore, all logic gates must generate dual-rail signals. The dual-rail version of the buffer is depicted in Figure 15.2(b).

The circuit of Figure 15.2(b) is trivially extendable to implement a logic function other than duplication of the input signal to the output. Figure 15.2(c) presents an EXOR gate with two inputs. The function controlling the clamping of an output to ground must be the inverse of the function implemented by the transmission-gate network driving that output. Cross-coupled clamp devices, as depicted in the figure, can replace a full pull-down network controlled by the inputs to the gate. The combinational gate of Figure 15.2(c) can drive its output fully adiabatically, but because of the transmission gates, its device count is twice that of a corresponding static-CMOS counterpart.

Signal timing presents a difficult problem. For fully adiabatic operation to be possible, the inputs must be held static throughout the charging and discharging of the load. It would be possible to charge the load capacitances, latch the result, and discharge the load; but conventional latches invariably cause nonadiabatic dissipation [3] and are therefore not permissible. Instead, a discharge path for each load capacitance, separate from the charge path, must be used to relax the static-input requirement (Figure 15.3). The inputs must now be held static throughout the charging of the load, whereas the subsequent discharge is controlled by another set of signals. It is tempting to try to derive these control signals directly from the gate outputs; but such signals will, by necessity, not be stable throughout the discharge. Any schemes based on the same idea are bound to fail for thermodynamic reasons [4]. The solution is more complex: each stage in a fully adiabatic pipeline must generate signals that control the charging path of the following stage, as well as signals that control the discharging path of the previous stage in the pipeline. Such a pipeline is illustrated in Figure 15.4; the figure is still simplified, in that all signals, including the voltage ramp signals, must be dual-rail-encoded [1]. For the approach to be workable, all combinational

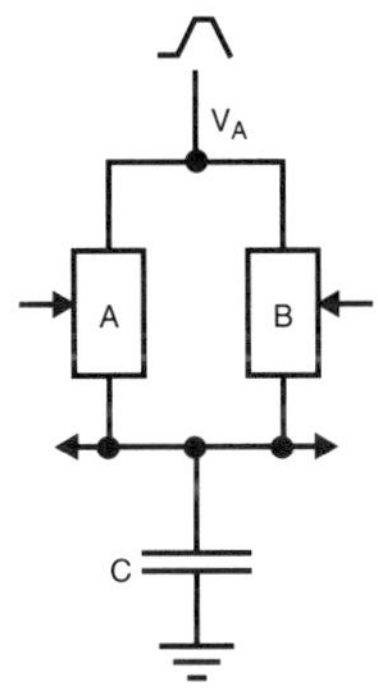

FIGURE 15.3 The capacitance C is charged though one path, A, and later discharged through another path, B. The inputs to the path A must be stable throughout the charging, whereas the inputs to path B must be stable throughout the discharging.

functions used in the pipeline must be invertible: there must exist an inverse function, which can compute the inputs to the original function from its results. The resulting pipeline is reversible and runs backward if the clocks are reversed in time. This reversibility property is characteristic of all fully adiabatic pipelines.

The complexity of two mutually inverse function blocks are usually similar, so the separate discharge path doubles the amount of hardware. Additionally, the restriction to use only invertible functions adds overhead which may be quite large. In a benchmark experiment, a fully reversible, bit-level-pipelined three-bit adder required 20 times as many devices as a conventional one, and 32 times the silicon area [5]. Other styles of reversible logic have less overhead [6,7], but none are as compact as the conventional, nonadiabatic logic styles. Nevertheless, it is possible to design fully reversible processors [8], which would dissipate very little power when operated slowly enough.

The hardware overhead of reversible logic has encouraged many researchers to seek ways to apply adiabatic techniques also for nonreversible logic. Such solutions will not be fully adiabatic, so their dissipation will not scale to the very lowest levels; but at higher performance levels, they may well offer lower dissipation than a corresponding reversible implementation. For a fair evaluation, any such partially adiabatic logic style should be benchmarked against logic styles according to best conventional practice, and the conventional control case should be optimized for power (typically by supply-voltage selection) at the same performance level as the adiabatic style.

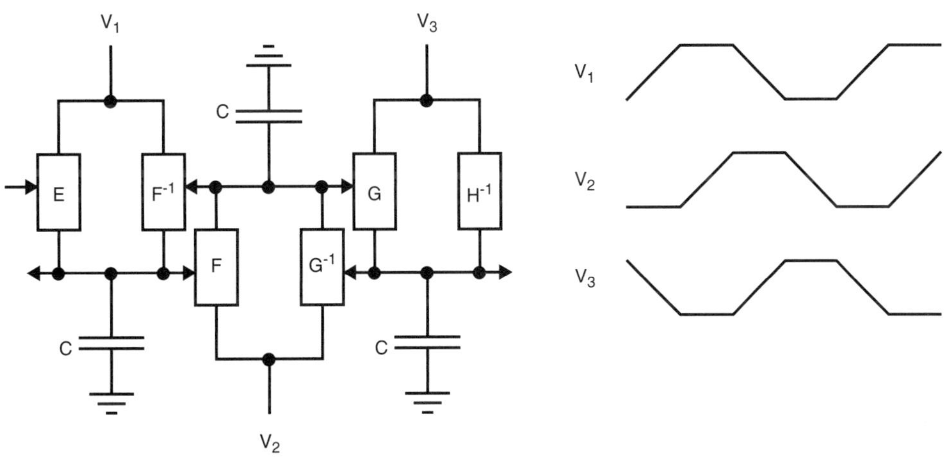

FIGURE 15.4 Node capacitances in this reversible pipeline are charged and discharged through two different paths. Each discharge path implements the inverse logic function to the charge path of the next pipeline stage. Thus, the discharge path has full information of the state of the variables to erase.

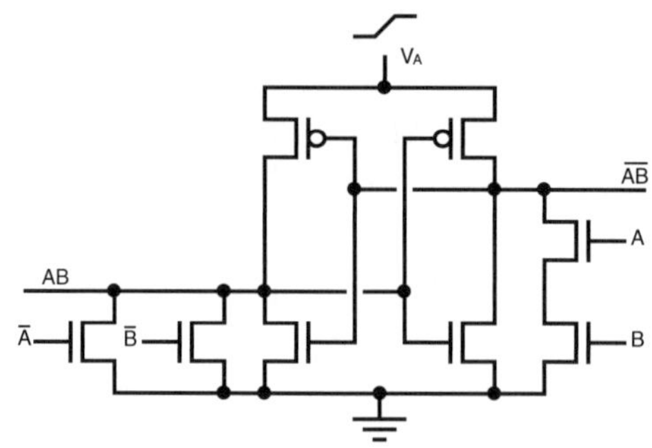

FIGURE 15.5 A partially adiabatic logic gate, based on two cross-coupled inverters. One of the two output nodes is held low for each combination of input values. The other output node is charged from the APS through one of the pMOS devices when the voltage ramp is applied.

Most of the partially adiabatic logic styles presented to this date use cross-coupled devices connecting two nodes that form the true and inverted outputs of a gate. The gate settles in one of two stable states when a voltage ramp is applied. The behavior of the outputs depends on a small imbalance between the two nodes, which in turn has been caused by the input values. Figure 15.5 illustrates an early example of the type [3,9]. The figure presents an AND/NAND gate. When the clock signal rises, the right output will be held low if both inputs are high; the output not held low will follow the clock signal and be charged adiabatically once the corresponding pMOS device has turned on. The change from one stable state to the other (when the output changes from the value in the previous cycle) is associated with a nonadiabatic dissipation of approximately $C\,V_{th}^2$, where V_{th} is the device threshold voltage and C is the driven capacitance. The clock signal must be held high to provide static inputs for the next pipeline stage while its clock is ramped up. A pipeline of such gates can be operated with four identical clock signals ninety degrees out of phase [9].

Newer partially adiabatic logic styles seek to minimize the number of separate clock signals needed. Several examples have been demonstrated to work with a single sinusoidal clock. Further reduction in the number of clocks appears unlikely.

15.3.2 Adiabatic Buffering

The previous section outlined some of the issues to be addressed when designing adiabatic-charging CMOS logic circuits. Although many of the problems have been solved in principle, no consensus has been reached on how best to design circuits that recover most of the charge and energy of the internal nodes of logic gates. This section will describe a more limited approach, where mainly nodes that contribute a large amount of dynamic power (that is, nodes with a large capacitance and a high switching frequency) are driven with adiabatic-switching techniques. In contrast, most nodes inside blocks of combinational logic have a rather small capacitance; many of the complications of adiabatic switching can be circumvented if such combinational blocks are implemented with conventional circuitry.

The largest node capacitances in a conventional chip (aside from the supply and ground nodes) typically belong to the clock distribution network. A continuously running, global clock signal is particularly suitable to energy recovery: the clock node capacitance may be readily resonated with on- or off-chip inductances. Because the charge path is identical from cycle to cycle, there is no need for a resistive switch to be inserted to direct the charge flow. The result would be a sinusoidal clock signal.

Present-day low-power digital chips use clock gating as an essential tool for power minimization. By disabling the clock signal for an unused subblock, the designer ensures that any flip-flops served by the clock stay idle and therefore consume no dynamic power. Combinational logic blocks whose inputs emanate from these flip-flops are also kept in an idle state. The power, which would have been used to distribute the disabled clock, is also saved with clock gating.

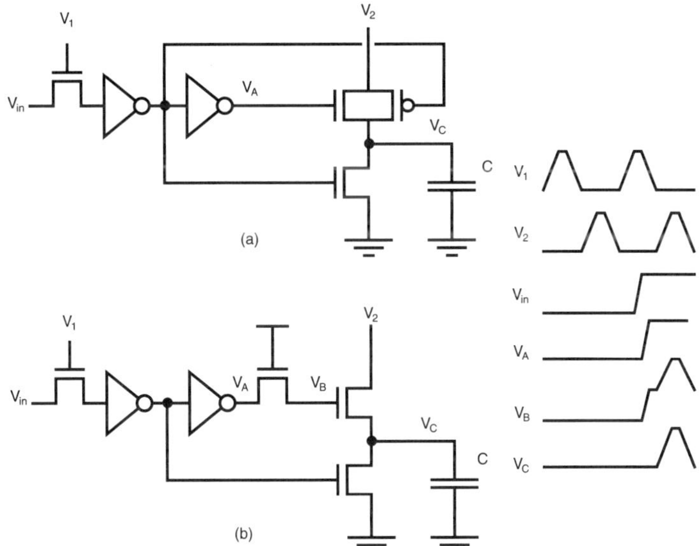

FIGURE 15.6 (a) A latch and energy-recovery driver circuit for two-phase, nonoverlapping clocks. (b) The Athas-Tzartzanis latch and energy-recovery driver, where a bootstrapped nMOS device replaces the transmission gate of (a).

Clock gating is still possible with a resonant clock driver. The subblock clock network must be connected to the global clock network through a low-resistance switch, such as the one given in Figure 15.1(d). The same principle may be further extended. A high-capacitance signal line will be charged and discharged if it is connected to the clock network for an appropriate time. If the RC time constant of the load capacitance and the connecting-switch resistance is small compared with the rise time of the clock, the driven signal will faithfully follow the clock signal. Figure 15.6(a) depicts a combined latch and energy-recovery driver circuit, suitable for this purpose. It is intended for two-phase, nonoverlapping clocks. The input value is latched on the falling edge of phase 1 and buffered through the inverter pair. When the latched value is low, the clamp device holds the output at ground potential, and the transmission gate devices are turned off. When the latched value is high, the transmission gate connects the load to V_2 in time for the phase-2 clock pulse.

The transmission-gate pMOS device must be quite wide to maintain a low switch resistance at the peak voltage. Consequently, its gate capacitance (which is driven without the benefit of energy recovery) will be uncomfortably large. A variation of the circuit, introduced by Tzartzanis and Athas [10] and further developed by Athas et al. [2], is depicted in Figure 15.6(b). When the latched value is low, it works as the circuit in Figure 15.6(a). When the latched value is high, the second inverter will charge the gate of the bootstrap device through the isolation device. The positive edge of phase 2 will be capacitively coupled from the channel of the bootstrap device to the boot node, which will rise, immediately turning off the isolation device. The boot node will now follow the phase-2 waveform through its positive and negative transitions, if the isolation device has been properly sized with respect to the bootstrap device (the boot-node voltage is subject to capacitive voltage splitting). In the process, the capacitive load will first be charged from the phase-2 clock line and will then deliver its charge back to the same clock line, all through the bootstrap device.

It is instructive to study the design of the Athas–Tzartzanis energy-recovery latch, or "ER latch." Its simplicity and efficiency stems from several considerations. The bootstrapping technique allows a single nMOS device to replace the transmission gate, which saves both power and layout area. The drawback is that the boot node rises to a voltage larger than that of the clock. With a full-swing clock, careful analysis is needed to ensure long-term survival of the substrate-diffusion junction at the boot node. Furthermore, the latch produces pulse rather than level: it will transfer a clock pulse to its output when the input signal is high. If a high value is latched repeatedly, several clock pulses will be transferred to

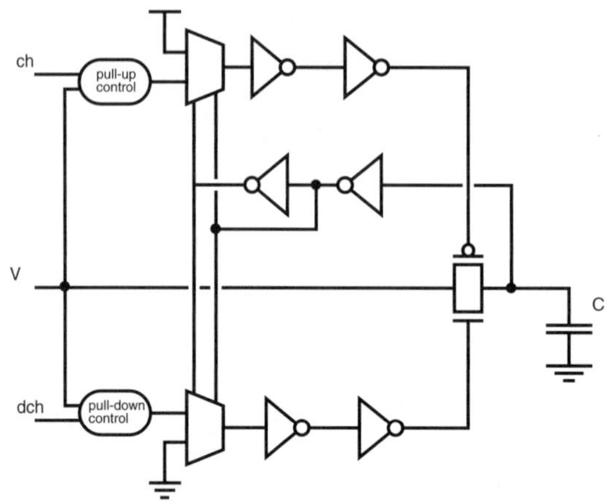

FIGURE 15.7 Energy-recovery driver according to Kim [11]. Signals marked ch and dch control whether the load, C, is to be charged or discharged (see Figure 15.6).

the output, with each pulse causing energy dissipation. Such a signaling scheme is not a good choice when the receiving circuits are implemented in a conventional static-CMOS style; dynamic, domino-style circuits are more suitable, and need no footer devices because their input signals are already gated with the clock signal.

Figure 15.7 is another circuit proposed for selectively connecting data lines to the clock line. This circuit, introduced by Kim et al. [11], retains the transmission gate of Figure 15.1(d) but adds a feedback path, which allows the driver to release the driven node after a one-way transition. The benefit is that a long run of a single value will cause dissipating transitions only at the beginning and the end of the run. (The output node seems to be left floating between these transitions, which could however easily be corrected with two clamp devices connecting the output to the supply and ground nodes, respectively.) The circuit is clearly more complex than the Athas–Tzartzanis ER latch. The transistor-count overhead and the extra parasitic capacitance at the output indicate that the Athas–Tzartzanis ER latch may be used with smaller loads, whereas the Kim ER driver would likely be more efficient at long runs of high values.

Both these drivers have been used in implementations of static memories [11,12], an application that offers rich opportunities for signal buffering due to the large-capacitance bit lines and word lines. The implementations differ in the details: Tzartzanis uses a current-mode sense amplifier, whereas Kim uses a voltage-mode design; Kim uses energy recovery for bit- and word-line drivers only, whereas Tzartzanis uses the principle also for row decoders and internal data buses. Despite all these differences, both designs reach similar speeds in simulation (Kim: 300 MHz for a 256 × 256 memory in 0.35 μm; Tzartzanis: 200 MHz for a 256 × 256 memory in 0.8 μm) and offer similar energy savings (Tzartzanis: 2.4×–4.2×; Kim: 2.66×) when compared with a conventional control case.

Neither design uses adiabatic charging to read values out of the memory cells. It would seem that a memory cell could connect a resonantly driven word line with a bit line, depending on the value stored in the memory cell, maybe with circuits similar to those in Figure 15.8. This approach would, however, require the connecting switch to be enabled already when the word line is driven, or nonadiabatic dissipation would result. The switches of the unselected words must however *not* be enabled because the pulse on the bit line could travel backwards through these switches and onto other unrelated word lines, as presented in the figure. The requirement to preenable the switches in only the selected word leads to the introduction of an auxiliary word line, which would go high before the power-delivery word line. The memory cell itself would be quite large, as the readout circuitry in Figure 15.8 would need to be augmented with enable logic.

This problem points to a general design difficulty with the energy-recovery circuit styles described here: they are ill-suited to circuit nodes with large fan-ins (such as memory bit lines). The reason is that the switch that connects a gate output to a power-clock is bidirectional by design. Charge is supposed

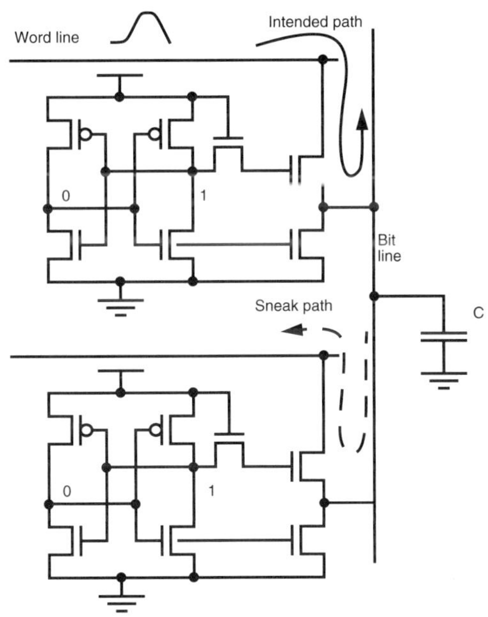

FIGURE 15.8 A first attempt at a memory-cell readout circuit that is able to charge the bit lines adiabatically. The upper cell is selected when its word line is adiabatically driven. A bootstrapped driver as in Figure 15.6 drives the bit line when the cell content is as indicated. The bootstrap device in the lower cell, however, is also enabled, allowing the pulse to enter the second word line.

to be able to flow both ways through the switch. Many such switches are connected to a node with a large fan-in. It will often be necessary to explicitly turn off all but one of these switches before driving the fan-in node. This requirement significantly complicates both circuits and timing. In the case of memory readout, another low-dissipation alternative is readily available because memory sense amplifiers can work with small differential swings. Thus, there appears to be little to gain in terms of dissipation by recovering bit-line node energies during read operations.

Clock lines and memory bit-lines (during write operations) are good candidates for adiabatic driving, because of their large capacitance. Outside the confines of a single chip, more opportunities appear. A trace capacitance on a printed circuit board (PCB) is likely to be larger than any on-chip capacitance (with the exception of a clock net in a large microprocessor). Some energy-recovery drivers targeted to this application are described at the end of Section 15.3; they do not work by tying the output to a clock line transition, but rather charge the load gradually in several steps.

Large capacitances and a nonnegotiable voltage swing can also be found in liquid-crystal display (LCD) drivers. The column drivers of an active-matrix LCD dominate the dissipation of the display (with the possible exception of the backlight). Because the speed requirements are modest, the column drivers would seem to be prime candidates for adiabatic charging, but the situation is complicated by the requirement for driving the column line to a controllable voltage level. Two solutions have been published, both targeting chip-sized rather than laptop-sized displays. Ammer et al. [13] distributes digital pixel values to each column driver and a common, adiabatically generated voltage ramp to all columns. Initially, all columns are connected to the voltage ramp; each column is then disconnected from the ramp when a counter determines that the proper voltage has been reached. Lal et al. [14] distributes an analog video signal, which is sampled at each column; the sampled value is used to control the final voltage on the column.

Throughout this section, it has been implicitly assumed that the driven load can be considered purely capacitive. Adiabatic charging was invented for such cases, but node energies may be recovered also when the load has a significant resistive component, as is likely when driving a long signal line off chip or on chip. In these cases, the formulae for the dissipation are more complex than those given in Section 15.2 because of the wire-resistance influence. The qualitative behavior of the energy dissipation can be summarized in the following points:

- The switch resistance, R_S, and the wire resistance, R_W, must both be much smaller than T/C_W for the dissipation to approach 0.
- If T is approximately R_WC_W, the wire properties limit the amount of energy recovery possible. It will be useful to reduce R_S to a certain degree, but further reductions are to no avail. In contrast, all switch-resistance reduction is useful when the load is purely capacitive.
- For a wire with a uniform resistance and capacitance per unit of length, a maximum length exists at which some given percentage of its energy can be recovered. This maximum length depends on the charging time T. Thus, global signals in large, fast chips must be split in parts and adiabatically rebuffered if most of their node energy is to be recovered.

15.3.3 Adiabatic Power Supplies

Adiabatic switching can reduce overall power dissipation only if some part of the switching-circuit node energies can be recovered and reused. As exemplified in the previous sections, the switching circuits (which may be logic circuits or simply drivers for large capacitances) must be designed with this requirement in mind. Additionally, the recovered node energies must somehow be stored in the APS while waiting to be reused. The two principal alternatives for energy-storing circuit elements are inductors and capacitors. Both have been used for APS energy storage; some examples are described next. Other suggestions include transmission lines [15] and piezo-electric resonators [16].

A conceptually simple method to repeatedly deliver energy to a capacitive load is to connect the capacitance to an inductance, thus forming an LC resonance circuit. Once the circuit has been excited, the energy will oscillate between the inductance and the capacitance, with a frequency proportional to the inverse of the square root of the LC product. The sine-shaped voltage waveform on the capacitance will be damped because the inevitable resistive losses will convert part of the circuit energy to heat in each cycle of oscillation. A sustained near-sinusoidal waveform may be produced if the energy is replenished regularly, as indicated in Figure 15.9. This simple arrangement, which requires a logic style compatible with a single-phase sinusoidal clock, was used by Maksimovic et al. [17]; a more evolved version was presented by Ziegler et al. [18].

The LC circuit of Figure 15.9 displays some properties that are inherent to all simple LC-resonance APSs. First, the frequency of operation is set by the total driven capacitance C and the resonance inductor L. It may be possible to operate the APS at a frequency slightly different from its self-resonance frequency, but at a reduced efficiency.

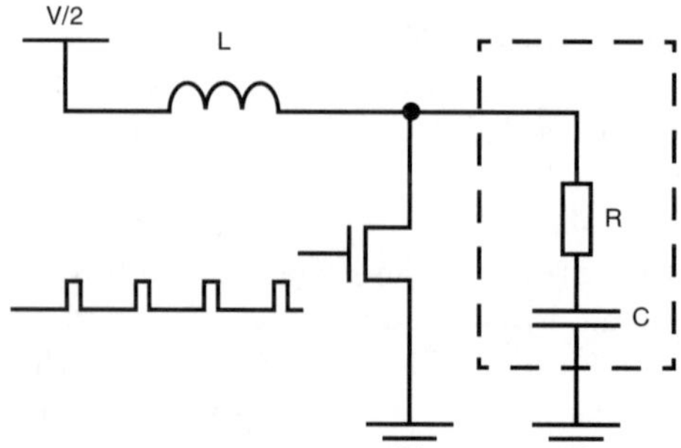

FIGURE 15.9 A simple adiabatic power supply (APS). The logic circuit, in the dashed box, is represented as an RC link, where the C corresponds to the node capacitances and the R corresponds to the on-resistances of the logic gates. The RC link is periodically shunted by the nMOS device. When the control signal frequency agrees with the LC resonance, the voltage across the RC link approximates a sine wave with a peak-to-peak value of V.

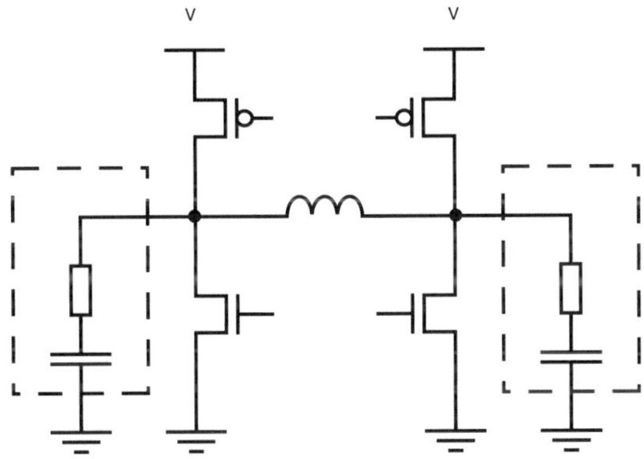

FIGURE 15.10 A dual-rail version of the circuit in Figure 15.9. Each pMOS device is used to tie one adiabatic power rail to V when the other is tied to ground by its nMOS device.

Second, the overall efficiency is determined by the Q value of the LC resonance circuit, which in turn is set by the average resistance in series with the driven capacitances. If L can be increased without introducing more parasitic resistance, efficiency will be increased, but at the cost of a lower frequency of operation: with the same voltage swing, the average charging currents follow the frequency, and because $P \sim I^2R$, power will fall as the square of the frequency.

Third, if the driven capacitance would change instantaneously, as would be typical for a design using single-rail logic or clock gating, both the self-resonance frequency and the amplitude of the waveform could be immediately affected. (To avoid the amplitude change, the auxiliary capacitance must be switched into or out of the circuit when the inductor current is zero; also, the capacitance must be charged to the same voltage as the in-circuit capacitances when it is switched in.) The changes in frequency and amplitude are related to the square root of the relative capacitance change, which suggests that a "ballast" capacitance could be added in parallel with the payload to keep the variations small. This approach works, but adds to the overall capacitance driven and therefore to the dissipation.

The almost-sinusoidal waveforms of the circuit in Figure 15.9 are easy to produce, but they restrict the choice of logic styles available to the designer. A two-phase version of the same circuit generates a sinusoidal waveform and its inverse, using only one inductor, as presented in Figure 15.10 [19]. (Large inductors with high Q-values cannot presently be integrated on a silicon chip, so with current packaging techniques, the number of inductors must be kept small to reduce cost.) With slightly modified control-signal timing, the same circuit can produce a good approximation of the "ideal" clock signal for many adiabatic logic styles: linear voltage ramps interspersed with periods of constant voltage. When the circuit is operated in this mode, the switches first connect the inductor across the supply rails, allowing a current to build up. When the switches are released, the current continues to flow and moves charge from one of the load capacitances to the other. When both load capacitances have reached the opposite supply voltage, the inductor is connected to the rails again, but now in the opposite direction. The inductor current will shrink, change direction, and build up again, and another cycle can start. Clearly, this circuit requires equal capacitances to deliver equal rise times at both ends of the inductor.

As described previously, logic circuits of very high efficiency require that the input voltages of a logic gate are held constant while charge is transported to and from its output. Logic styles suitable for reversible computing therefore require several interleaved clock signals. Several circuits such as that in Figure 15.10 can be used to provide four, six, eight, or more equally spaced clock signals. It is also possible to generate any number of such clock signals with only one inductor, if only two signals transition at the same time [7].

An important property of the circuit in Figure 15.10 may not be immediately obvious. The inductor is connected with switches to a constant voltage to allow current to build up. This current will cause ohmic losses in the switches. Wider switch devices have lower resistances and thus cause proportionally lower ohmic losses. It would seem that to minimize the losses, the devices should be chosen as wide as

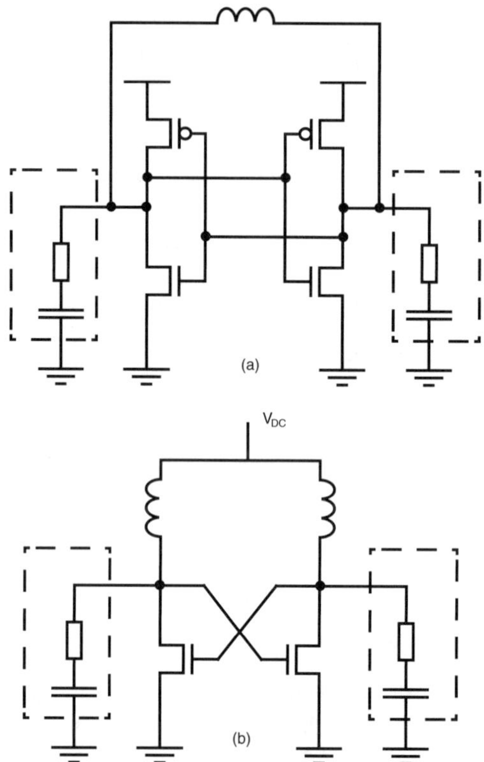

FIGURE 15.11 (a) The APS of Figure 15.10, but with resonant driving of the gates of the inductor switches. (b) Simplified version of (a), which avoids crowbar current through the inductor switches.

practically possible; but the gate capacitance is also proportional to device width, and the circuits driving these gate capacitances will dissipate proportionally larger amounts of energy per cycle. This relationship changes the system-level power dependence on transition time from $P \sim T^{-1}$ to $P \sim T^{-1/2}$ [1]. To avoid this degradation, the driving signals for the ohmic switches must actually be resonantly driven, as in Figure 15.11(a) [19]. A simplified version of this circuit, depicted in Figure 15.11(b), does not suffer from crowbar current; it generates waveforms where sinusoidal "blips," suitable for the Athas–Tzartzanis ER latch of Figure 15.6(b), alternate on the two output lines [20]. The peak voltage is approximately 3 times the DC voltage.

The ohmic switches can be made much wider when they are resonantly driven. A first-order analysis indicates that the overall dissipation is minimized when their gate capacitance is equal to the total payload capacitance. Thus, the switches act as "ballast" capacitances, minimizing the influence of fluctuations in the payload capacitance. The price paid is that the system is now free-running, without control signals, and its frequency cannot be easily controlled.

All APS designs described previously use inductors for energy storage. Inductors have several practical drawbacks: high-Q inductors cannot be integrated on silicon chips; timing errors can cause ringing or damaging voltage spikes; the inductor-based APS solutions suffer from jitter when subjected to variable capacitive loads. It is possible to avoid these drawbacks by using capacitors for energy storage [21]. Consider the circuit in Figure 15.12, where the tank capacitors C_{Ti} are charged to the evenly distributed voltages (i/N) V. When the load C_L is charged from 0 to V, it is connected to each of these tank capacitors in sequence and receives from each of them a charge C_LV/N before finally being connected to the supply voltage V. To discharge C_L, it is again connected to all the tanks, but in opposite sequence, and then finally connected to ground. Each tank capacitor receives during the discharge procedure the same amount of charge as it provided during the charge procedure, so the tank capacitor voltages are self-

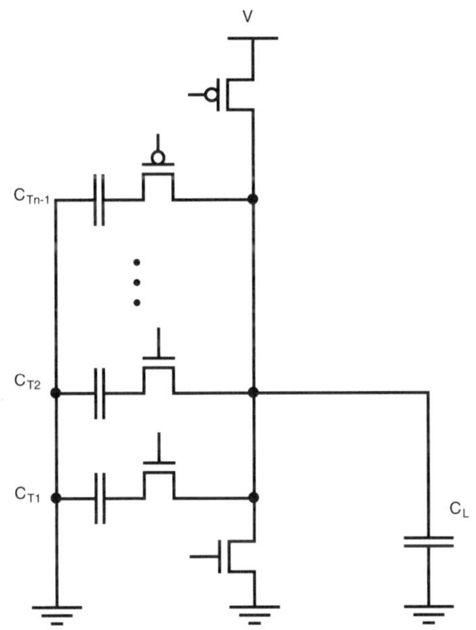

FIGURE 15.12 A stepwise driver for the load capacitance CL. The load is connected to each of the large "tank" capacitors, C_{Ti}, in sequence during charging, and in reverse sequence during discharging. The tank capacitor voltages converge to i (V/N).

sustaining. Additionally, the tank capacitances converge automatically towards the evenly distributed voltages when the circuit is exercised, so no auxiliary circuits are needed to maintain the tank voltages. The reduced dissipation is due to the smaller voltage drop traversed by the charge. In addition, it is clear that only a charge C_LV/N is drawn from the supply line for each cycle: a reduction from the conventional case by a factor of N. The energy drawn from the supply line, C_LV^2/N, is also reduced by the same factor.

The circuit in Figure 15.12, known as a stepwise driver, may be used as an APS to drive adiabatic logic circuits. It is, however, not as energy-efficient as an inductor-based circuit with the same transition time, because of all the switches that must be turned on and off for each step (which itself causes additional dissipation). This overhead limits the useful number of steps to below 10 in almost all cases.

The most natural place for a stepwise driver is not as an APS for a computing circuit, but rather as an off-chip pad driver: such circuits rarely use the maximum speed possible in the technology, because the resulting high current derivatives would cause large voltage spikes across the package inductances. A stepwise driver of three steps can offer 50% lower dissipation with little performance impact [21]. It may also be introduced without system-level redesign, as would often be required for the more ambitious inductive solutions.

15.4 Conclusion

Adiabatic and energy-recovery techniques offer new possibilities to trade dynamic power dissipation for delay in switching circuits. In some cases, the voltage swing is fixed for external reasons, such as in display drivers and micro-mechanical actuators, and in pad drivers that must produce industry-standard voltage levels. In these cases, adiabatic switching is the only known technique that allows this fundamental trade-off to be made at the circuit level.

Most published research in adiabatic-circuit techniques has focused on novel logic styles. The application examples for each style have been few, however, and many of them have been arithmetic blocks of various types (adders, multipliers, etc.). A wider range of application examples would make it easier to understand where each of these the adiabatic techniques outperform the conventional ones.

Some system-level experiments have been carried out where a moderately complex digital system has been implemented with energy-recovery techniques. The AC-1 and MD-1 microprocessors [2,22] used the less-ambitious approach to energy recovery: no attempt was made to recover signal energies inside logic gates. The results were encouraging, in that the processor-APS combinations displayed power levels under resonant drive, which were significantly lower than for conventional control cases. The improvements were, however, not large enough to cause an immediate paradigm shift in digital circuit design. Further system-level experiments, building on these experiences, should be able to improve significantly on these initial results.

The very lowest dynamic dissipation figures for a digital logic block can be reached only if all logic functions are invertible and connected to form a reversible apparatus. The circuit overhead for building fully reversible logic pipelines in present-day CMOS appears large enough to be prohibitive in most cases. Logic-style breakthroughs may still occur which would reduce the overhead; but it is equally important not to overlook the power-supply influence on the overall dissipation. It is fruitless to build a reversible-logic chip from which 99.999% of the node energies can be recovered if the power supply is only 99% efficient. In addition, device leakage through nominally off devices in the logic circuits themselves must be better analyzed for ultimate-low-power claims to be credible.

References

[1] W.C. Athas, L.J. Svensson, J.G. Koller, N. Tzartzanis, and E.Y.-C. Chou. Low-power digital systems based on adiabatic-switching principles. *IEEE Trans. on VLSI Systems,* Vol. 2, No. 4, pp. 398–407, Dec. 1994.

[2] W.C. Athas, N. Tzartzanis, L.J. Svensson, and L. Peterson. A low-power microprocessor based on resonant energy. *IEEE Journal of Solid-State Circuits,* Nov. 1997, pp. 1693–1701.

[3] J.G. Koller and W.C. Athas. Adiabatic switching, low-energy computing, and the physics of storing and erasing information. *Proc. of the Workshop on Physics and Computation, PhysCmp '92,* Oct. 1992, IEEE Press, 1993.

[4] R. Landauer. Irreversibility and heat generation in the computing process. *IBM J. Res. Dev.,* Vol. 5, pp. 183–191, 1961.

[5] W.C. Athas and L. Svensson. Reversible logic issues in adiabatic CMOS. *Proc. 1994 Workshop on Physics and Computation,* Nov. 1994.

[6] S. Younis and T.F. Knight. Asymptotically zero-energy split-level charge recovery logic. *Proc. Int. Workshop on Low-Power Design,* Napa, CA, 1994, pp. 177–182.

[7] J. Lim, D.-G. Kim, and S.-I. Chae. nMOS reversible energy recovery logic for ultra-low-energy applications. *IEEE J. Solid-State Circuits,* June 2000, pp. 865–875.

[8] C. Vieri. Reversible computer engineering and architecture. Ph.D. thesis, Massachusetts Institute of Technology, Cambridge, MA, June 1999.

[9] J.S. Denker. A review of adiabatic computing. *Symp. on Low-Power Electronics,* San Diego, CA, Oct. 10–11, 1994, pp. 94–97.

[10] N. Tzartzanis and W.C. Athas. Design and analysis of a low-power energy-recovery adder. *Proc. 5th Great Lakes Symp. on VLSI Design,* Buffalo, NY, Mar. 16–18, 1995, pp. 66–69.

[11] J. Kim, C.H. Ziesler, and M.C. Papaefthymiou. Energy recovering static memory. *Proc. 2002 IEEE ISLPED,* Monterey, CA, Aug. 12–14, 2002, pp. 92–97.

[12] N. Tzartzanis, W. Athas, and L. Svensson. A low-power SRAM with resonantly powered data, address, word, and bit lines. *Proc. ESSCIRC 2000,* Stockholm, Sweden, Sep. 19–21, 2000.

[13] J. Ammer, M. Bolotski, P. Alvelda, and T.F. Knight, Jr. A 160 × 120 pixel liquid-crystal-on-silicon microdisplay with an adiabatic DACM. *ISSCC 1999 Dig. of Tech. Papers,* pp. 212–213.

[14] R. Lal, W. Athas, and L. Svensson. A low-power adiabatic driver system for AMLCDs. *Symp. on VLSI Circuits Dig. of Tech. Papers,* Honolulu, June 15–17, 2000, pp. 198–201.

[15] S. Younis and T. Knight. Non-dissipative rail drivers for adiabatic circuits. *Proc. 16th Conf. on Advanced Research in VLSI, 1995,* Los Alamitos, CA, March 27–29, 1995, pp. 404–414.

[16] P. Solomon and D. Frank. The case for reversible computation. *Proc. Int. Workshop on Low-Power Design,* Napa, CA, April 24–27, 1994, pp. 93–98.

[17] D. Maksimovic et al. Clocked CMOS adiabatic logic with integrated single-phase power-clock supply. *IEEE Trans. VLSI Syst.,* Vol. 8, No. 4, Aug. 2000, pp. 460–463.

[18] C.H. Ziesler, S. Kim, and M.C. Papaefthymiou. A resonant clock generator for single-phase adiabatic systems. *Proc. Int. Symp. on Low-Power Electronics and Design,* Huntington Beach, CA, Aug. 6–7, 2001, pp. 159–164.

[19] A. Dickinson and J.S. Denker. Adiabatic dynamic logic. *IEEE J. Solid-State Circuits,* Vol. 30, No. 3, Mar. 1995, pp. 311–315.

[20] W.C. Athas, L.J. Svensson, and N. Tzartzanis. A resonant clock driver for two-phase, almost-non-overlapping clocks. *Proc. IEEE ISCAS '96,* Atlanta, GA, May 12–15, 1996.

[21] L.J. Svensson, W.C. Athas, and R.S.-C. Wen. A sub-CV^2 pad driver with 10-ns transition time. *Proc. Int. Symp. on Low-Power Electronics and Design,* Monterey, CA, Aug. 12–14, 1996.

[22] W. Athas, N. Tzartzanis, W. Mao, L. Peterson, R. Lal, K. Chong, J.-.S Moon, L. Svensson, and M. Bolotski. The design and implementation of a low-power clock-powered microprocessor. IE*EE J. Solid-State Circuits,* Nov. 2000, pp. 1561–1570.

16

Weak Inversion for Ultimate Low-Power Logic

Eric A. Vittoz
CSEM

16.1 Introduction

In digital circuits, power is needed to charge the load capacitance C of each logic node at the switching frequency f. This dynamic power consumption can be expressed as

$$P_{dyn} = fC\Delta VV_B \tag{16.1}$$

where V_B is the supply voltage and ΔV the logic voltage swing, smaller or equal to V_B. Thus, the dynamic power can be reduced by reducing ΔV, but this gate voltage swing is needed to ensure a sufficient current ratio I_{on}/I_{off} in the transistors producing the transitions. Indeed, the on-current I_{on} must be large enough to ensure transitions at the required speed, and I_{off} should be as small as possible to limit the static power consumption $P_{stat} = I_{off}V_B$ between transitions.

The swing ΔV needed to achieve a given value of I_{on}/I_{off} can be reduced by reducing the gate voltage overhead, until it becomes minimum when weak inversion is reached. Logic circuits based on transistors operated in weak inversion (also called subthreshold) therefore offer minimum possible operating voltage, and thereby minimum P_{dyn} for a given P_{stat}. This is only possible, however, if the threshold voltage of the transistors can be precisely adapted to this very low value of supply voltage V_B. The feasibility of CMOS inverters with supply voltages as low as 200 mV was already demonstrated more than 30 years ago [1], with the possibility of reducing it to 100 mV if fast surface state would be negligible (which has become true for more than 20 years). However, the minimum channel length was still on the order of 5 μm, limiting the maximum frequency to just a few hundred kHz. Therefore, the idea was buried for several decades dominated by the struggle for maximum speed. It has been revived recently [2] and applied to complete subsystems operated below 200 mV [3,4]. In the meantime, weak inversion was used extensively for very low-power analog circuits [5–7], and a special model was developed to better describe the behavior of a MOS transistor from weak to strong inversion [8,9]. This chapter relies on this experience of weak inversion and on this model to derive the analytical results needed to optimize such low-voltage digital circuits and to identify their ultimate limits.

16.2 MOS Model in Weak Inversion and Basic Assumptions

The drain-to-source current I_{DS} of n-channel MOS transistors operated in weak inversion can be expressed as [6,8,9]:

$$I_{DS} = I_S \exp\frac{V_{GS} - V_T}{nU_T}\left[1 - \exp\frac{-V_{DS}}{U_T}\right] \tag{16.2}$$

where $n > 1$ is the slope factor (practically always below 1.6), V_{GS} and V_{DS} the gate to source and drain to source voltages, V_T the gate to source threshold voltage and I_S the specific current given by

$$I_S = 2n\mu C_{ox} U_T^2 W/L \tag{16.3}$$

where μ is the carrier mobility, C_{ox} the gate oxide capacitance per unit area, $U_T = kT/q$ and W/L the width-to-length ratio of the channel. The threshold voltage V_T depends on the source-to-substrate voltage V_{SB} according to

$$V_T = V_{T0} + (n-1)V_{SB} \tag{16.4}$$

where V_{T0} is the threshold voltage for $V_{SB} = 0$. Because V_{SB} must be larger than about $-4U_T$ (source junction reverse biased or only slightly forward biased, to avoid parasitic bipolar effects), V_T can only be increased or just slightly decreased with respect to V_{T0}.

By introducing the saturation current for $V_{GS} = 0$:

$$I_0 = I_S \exp\frac{-V_T}{nU_T} = I_S \exp\frac{-(V_{T0} + (n-1)V_{SB})}{nU_T} \tag{16.5}$$

which is also controllable by V_{SB}, Equation (16.2) is reduced to

$$I_{DS} = I_0 \exp\frac{V_{GS}}{nU_T}\left[1 - \exp\frac{-V_{DS}}{U_T}\right]. \tag{16.6}$$

The same equations are valid for p-channel transistors if the sign of current and voltages is inverted. Thus, I_{DS}, V_{GS}, V_{DS} become I_{SD}, V_{SG}, V_{SD}, and V_T is the threshold value of V_{SG}.

For the following analysis, two basic assumptions will be made about the process:

1. The native threshold V_{T0} for p- and n-channel transistors does not exceed $4U_T$, even in the worst case of process and temperature variation. Practical ways of adjusting the value of V_T according to Equation (16.4) will be discussed in Section 16.8.
2. True twin wells (often called triple-well) are available, to allow separate adjustment of p- and n-channel V_T.

16.3 Static CMOS Inverter

Consider the simple CMOS inverter of Figure 16.1 with input voltage V_i and output voltage V_o. We will assume that the local p- and n- substrates are properly biased with respect to the $V+$ and $V-$ rails of the power supply to control V_T and I_0, according to Equation (16.4) and Equation (16.5). To simplify the analysis, we will further assume that the two types of transistors are symmetrical, with same values of n and I_0.

By normalizing currents and voltages according to

$$x_k = V_k / U_T \text{ and } y_k = I_k / I_0 \tag{16.7}$$

the normalized values of current I_p and I_n flowing through the two transistors can be expressed from Equation (16.6) as

$$y_n = e^{x_i/n}(1 - e^{-x_o}) \tag{16.8}$$

$$y_p = e^{(x_B - x_i)/n}(1 - e^{x_o - x_B}) \tag{16.9}$$

The inverter is only loaded by a capacitance C, thus the static transfer function is obtained by equating these two currents, which yields:

$$x_i = \frac{x_B}{2} + \frac{n}{2}\ln\left[(1 - e^{x_o - x_B})/(1 - e^{x_o})\right] \tag{16.10}$$

which can be inverted to give

$$x_o = x_B + \ln\frac{1 - G + \sqrt{(G-1)^2 + 4Ge^{-x_B}}}{2} \quad \text{where } G = e^{(2x_i - x_B)/n} \tag{16.11}$$

This transfer function is plotted in Figure 16.2(a) for $n = 1.6$ and several values of x_B.

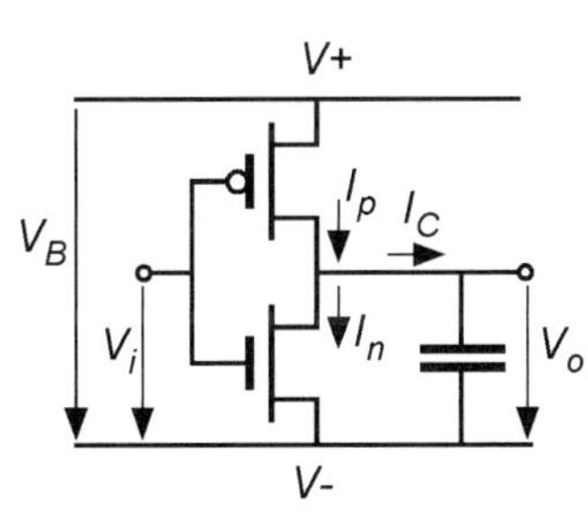

FIGURE 16.1 CMOS inverter.

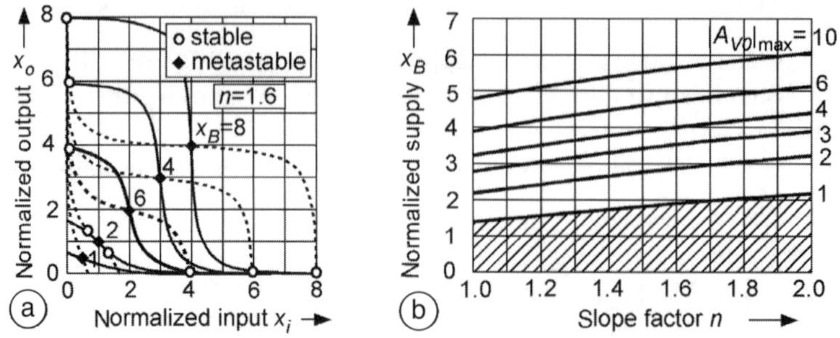

FIGURE 16.2 Stables states: (a) transfer function; (b) voltage required for gain.

The plots are repeated with horizontal and vertical axes permuted, to represent two inverters connected as a flip-flop. If x_B is not too small, there are two stable solutions corresponding to the logic states. Now the latter only exist if the maximum value of the inverter gain $|A_{V0}|$ exceeds 1. Differentiating Equation (16.10) or Equation (16.11) yields

$$-A_{V0} = \frac{2(1-e^{x_o-x_B}-e^{-x_o}-e^{-x_B})}{n(2e^{-x_B}-e^{x_o-x_B}-e^{-x_o})} \tag{16.12}$$

the maximum of which occurs for $x_o = x_i$:

$$|A_{V0}|_{max} = (e^{x_B/2}-1)/n \text{ or } x_B = 2\ln(n|A_{V0}|_{max}+1) \tag{16.13}$$

The latter is plotted in Figure 16.2(b) as a function of n. The lowest curve with $|A_{V0}|_{max} = 1$ depicts the absolute minimum voltage required for bistability. For a realistic value of $n = 1.6$, this absolute minimum is 1.91 ($V_B \cong 50$ mV at ambient temperature).

The normalized high value x_H and low value x_L of the stable points can be calculated by iteration in a series of inverters described by Equation (16.11), driving them by any value outside the metastable state. The result is represented in Figure 16.3(a) for $n = 1.6$.

The static current I_{stat} flowing through both transistors at the stable states can easily be calculated by introducing the values of x_H and x_L as input and output voltages in Equation (16.8) or Equation (16.9). The result is plotted in Figure 16.3(b). It shows that this static current is only slightly larger than I_0.

For $n = 1.6$ or less, $x_B = 4$ ($V_B = 100$mV at ambient temperature) is sufficient to ensure a swing almost equal to V_B and a static current virtually equal to I_0.

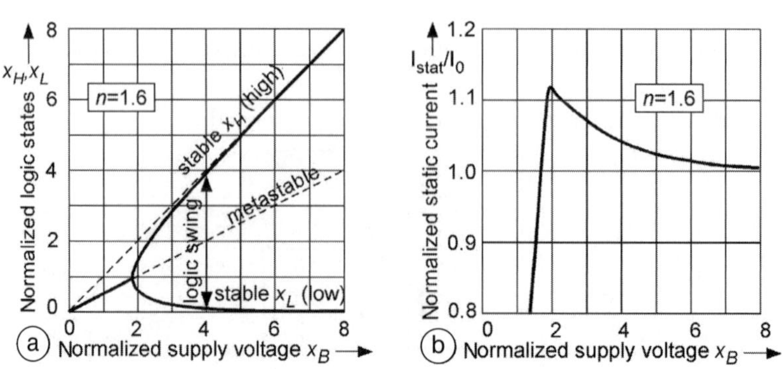

FIGURE 16.3 Logic states: (a) evolution with supply voltage; (b) static current.

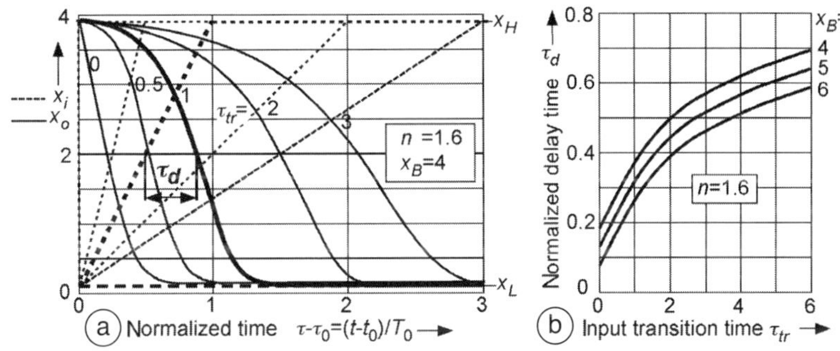

FIGURE 16.4 Gate transition: (a) input and output transitions; (b) delay time.

16.4 Dynamic Behavior of the CMOS Inverter

16.4.1 State Transition

Referring again to Figure 16.1, the normalized current flowing into capacitor C during a transition is obtained from Equation 16.8 and Equation 16.9:

$$y_C = I_C / I_0 = y_p - y_n = e^{(x_B - x_i)/n}(1 - e^{x_o - x_B}) - e^{x_i/n}(1 - e^{-x_o}) \tag{16.14}$$

The evolution of the output voltage of the inverter is thus obtained by integrating I_C into C:

$$x_o = x_{o0} + \frac{I_0}{CU_T}\int_{t_0}^{t} y_C(x_i, x_o)dt = x_{o0} + \int_{\tau_0}^{\tau} y_C(x_i, x_o)d\tau \tag{16.15}$$

$$\text{where } \tau = t / T_0 \text{ and } T_0 = CU_T / I_0 \tag{16.16}$$

are the normalized time and the characteristic time, respectively, and x_{o0} is the initial value of x_o at $t = t_0$. This evolution depends on that of the input voltage $x_i(t)$.

It is plotted in Figure 16.4(a) for various constant slopes of $x_i(t)$ characterized by their normalized transition time $\tau_{tr} = T_{tr} / T_0$ between the two logic states $x_i = x_L$ and $x_i = x_H$. The delay time $T_d = \tau_d T_0$ defined on this figure is minimum for a step input and increases with input transition time T_{tr}. This variation is plotted in Figure 16.4(b) for various values of supply voltage x_B.

16.4.2 Currents and Charges

During transitions, neither of the two transistors is as blocked as in the static states, therefore, some additional current y_{sc} flows directly through them. For a rising input, this short-circuit current corresponds to an increase of y_p, the normalized current flowing through the p-channel device, as can be seen in Figure 16.5(a).

Integrating this surplus current during the whole transition gives the short-circuit charge per transition Q_{sctr}, or its normalized value:

$$q_{sctr} = \frac{Q_{sctr}}{CU_T} = \int_{before}^{after} (y_p - y_{stat})d\tau \tag{16.17}$$

which is also presented in the figure, amplified by a factor of 10. This charge is delivered by the power supply, in addition to the main charge Q_{Ctr} flowing from the capacitor, given by:

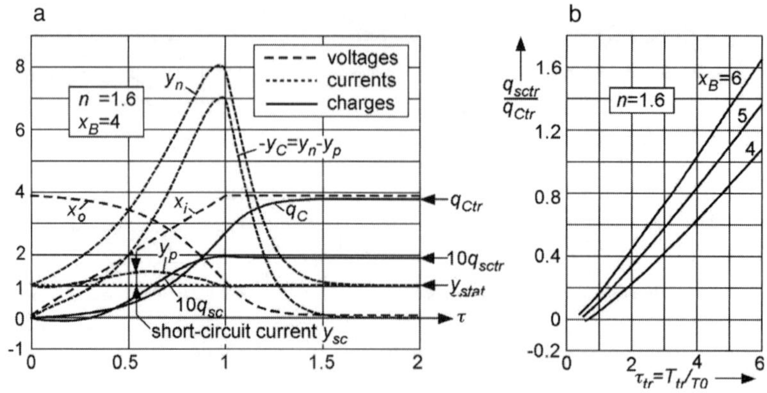

FIGURE 16.5 Transition of the inverter: (a) currents and charges; (b) proportion of short-circuit charge.

$$q_{Ctr} = \frac{Q_{Ctr}}{CU_T} = \int_{before}^{after} (y_n - y_p) d\tau \tag{16.18}$$

The total charge delivered by the power supply for each cycle of up and down transitions (and added to that due to static current I_{stat} is

$$Q_{tr} = Q_{Ctr} + 2Q_{sctr} \tag{16.19}$$

The calculated charge ratio Q_{sctr}/Q_{Ctr} as a function of the normalized input transition time $T_{tr} = \tau_{tr} T_0$ is plotted in Figure 16.5(b) for various values of normalized supply voltage x_B. It increases approximately proportionally to τ_{tr}, but remains below 20% for $\tau_{tr} < 1$ ($T_{tr} < T_0$).

16.5 Behavior of the Inverter for Standard Transitions

16.5.1 Definition and Delay Time

In large digital circuits, the inputs of most gates are outputs of other gates. If all gates have same fan-in and approximately same capacitive load, then they all tend to the same transition behavior. Such standardized transitions can be emulated in a long chain of inverters identical to the single inverter analyzed in Section 16.4. The result is illustrated in Figure 16.6(a) for a cascade of eight stages.

With the first stage driven by a rising step, the transitions become perfectly standardized after just a few stages. The standard delay time for a pair of gates $2\tau_d = 2T_d / T_0$ can then be defined as presented in the figure: such a definition would still be valid if the inverters are not be symmetrical. The variation of

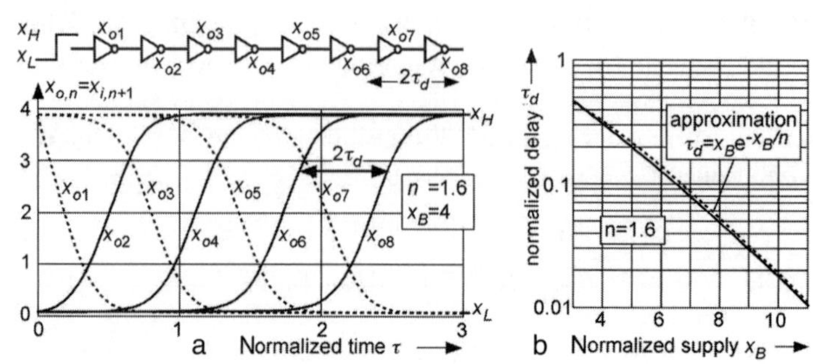

FIGURE 16.6 Cascade of inverters: (a) transitions of successive stages; (b) standard delay time.

this delay with supply voltage, calculated between x_{o6} and x_{o8} is represented in Figure 16.6(b). An approximative analytical expression of T_d can be obtained as follows: assuming that, at each transition, the whole on-current I_{on} of a fully saturated transistor driven by the full supply voltage V_B is flowing through capacitor C and changes the voltage across it by V_B, then:

$$T_d = \frac{CV_B}{I_{on}} = \frac{CV_B}{I_0 \exp\frac{V_B}{nU_T}} \text{ or } \tau_d = x_B e^{-x_B/n} \tag{16.20}$$

where the saturated on-current I_{on} is obtained from Equation (16.6) for $V_{DS} >> U_T$.

This expression is also plotted on Figure 16.6(b), demonstrating that is is a relatively good approximation. Because the exponential increases much faster than its argument, the delay time decreases approximately exponentially with increasing supply voltage, this for a constant characteristic time $T_0 = CU_T/I_0$. As can be seen, T_d is always smaller or much smaller than T_0 for such standard transitions.

16.5.2 Currents and Charges

The various voltages, currents, and charges of a standard transition calculated at the eighth stage of the cascade of inverters are plotted in Figure 16.7(a) for a particular small value of normalized supply voltage x_B.

As a consequence of the short standard transition time, the short-circuit current y_{sc} (now shown amplified by a factor of 100) is always a very small fraction of current y_C in the capacitor. It results in a short-circuit charge q_{sctr} much smaller than the charge q_{Ctr} flowing in or out of the capacitor, as can be expected from Figure 16.5(b) for values of $\tau_d << 1$. This proportion, represented in Figure 16.7(b) for various values of normalized supply voltage x_B, is always smaller than 1% for $x_B > 4$. The dynamic power consumption in homogeneous circuits with standard transitions is therefore very close to that due to the charge supplied to the capacitor:

$$P_{dyn} = f(Q_{Ctr} + Q_{sctr})V_B \cong fQ_{Ctr}V_B = fC(V_H - V_L)V_B \cong fCV_B^2 \tag{16.21}$$

where f is the frequency of transitions.

The short-circuit charge is always a very small fraction of the charge flowing in the capacitor. As a result, the short-circuit current can be practically neglected, the only current flowing (permanently) through the two transistors being I_{stat}.

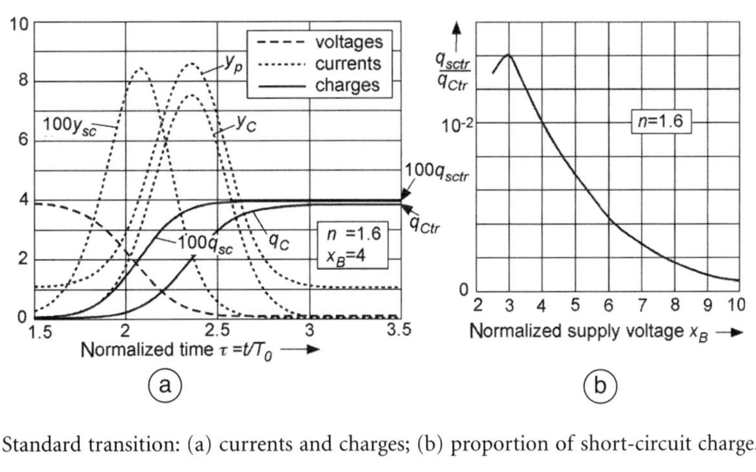

FIGURE 16.7 Standard transition: (a) currents and charges; (b) proportion of short-circuit charge.

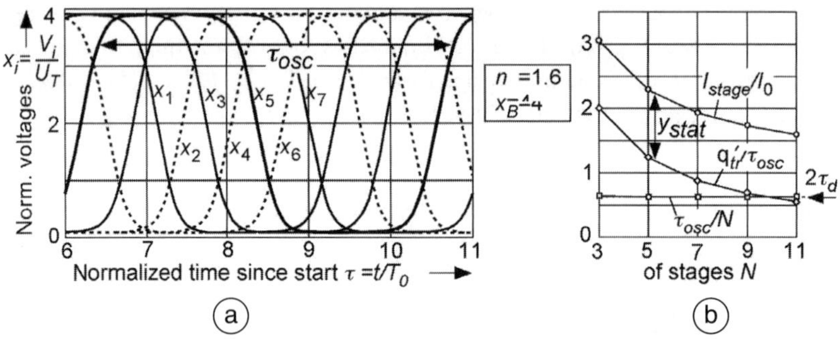

FIGURE 16.8 Ring oscillator: (a) voltages of 7-stage ring; (b) current per stage and period of oscillation.

16.5.3 Ring Oscillator

The dynamic behavior of inverters can also be examined by connecting an odd number N of them in closed loop to form a ring oscillator. The normalized output voltages of the different stages are plotted in Figure 16.8(a) for seven stages.

Because each stage must switch up and down once per period of oscillation T_{osc}, the period of oscillation for N stages should be given by

$$T_{osc} = 2NT_d \text{ or } \tau_{osc} = 2N\tau_d \tag{16.22}$$

As plotted in Figure 16.8(b), T_{osc} is slightly larger for the shortest ring ($N = 3$) because the swing does not reach the full static swing $x_H - x_L$, but Equation (16.22) becomes valid for longer rings. Figure 16.8(b) also shows that the average current I_{stage} consumed by each stage is given by

$$I_{stage} = Q_{tr} / T_{osc} + I_{stat} = (q_{tr} / \tau_{osc} + y_{stat})I_0 \tag{16.23}$$

Again, this equation becomes exact when the ring is long enough to achieve full static swing. When N is increased, corresponding to a decrease of the frequency of oscillation, the activity of each stage decreases, resulting in an increased proportion of static current I_{stat}.

16.5.4 Power-Delay Product

As was shown in Figure 16.3(b), the static current I_{stat} consumed in each stable state is negligibly higher than I_0, the saturation current for $V_{GS} = 0$. Therefore, the static power consumption per inverter can be simply approximated by

$$P_{stat} = V_B I_{stat} = V_B I_0 \tag{16.24}$$

Furthermore, it has been shown in Figure 16.7(b) that, for standard transitions obtained in an homogeneous system, the short-circuit charge Q_{sctr} is only a negligible fraction of the charge Q_{Ctr} flowing in the capacitor. The dynamic power consumption can thus reasonably be approximated as in Equation (16.21), and the total power P by:

$$P = P_{stat} + P_{dyn} = V_B I_0 + fCV_B^2 \tag{16.25}$$

Now, the frequency of operation f cannot be larger than $1/(2T_d)$. For $f = 1/(2T_d)$, the inverter (or in general the gates) are always in transition. A duty factor α can thus be defined as:

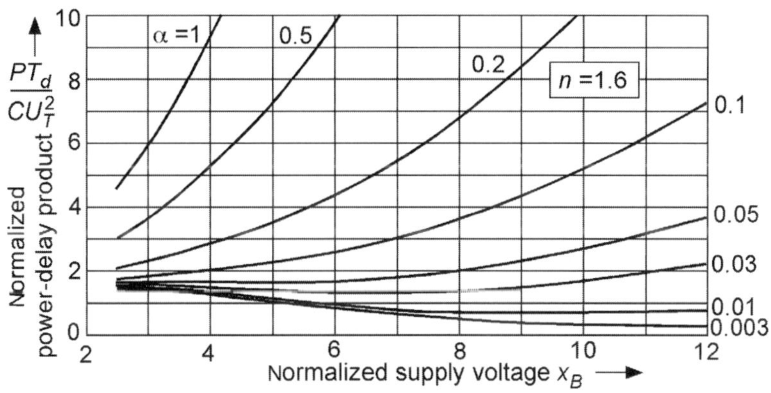

FIGURE 16.9 Power-delay product.

$$\alpha = 2fT_d \leqq q1 \tag{16.26}$$

Introducing Equation (16.26) in Equation (16.25) with the definitions from Equation (16.7) and Equation (16.8) yields the power-delay product:

$$PT_d = CU_T^2 \left(x_B \tau_d(x_B) + \frac{\alpha x_B^2}{2} \right) \tag{16.27}$$

where the first term results from the static power and the second from the dynamic power. This product can be calculated as a function of the normalized supply voltage x_B by using the previous calculations of $\tau_d(x_B)$ represented in Figure 16.6(b). Results are plotted in Figure 16.9 for various values of duty factor α.

Using the approximation Equation (16.20) for $\tau_d(x_B)$, the power-delay product can be expressed from Equation (16.27) as

$$PT_d = CU_T^2 x_B^2 \left(e^{-x_B/n} + \frac{\alpha}{2} \right) \tag{16.28}$$

giving results very close to those of Figure 16.9. Using this expression with the definition of α by Equation (16.26) of α to replace the gate delay T_d by the (average) frequency of operation *f*, the power to frequency ratio can be expressed as:

$$\frac{P/f}{C(nU_T)^2} = (x_B/n)^2 \left(\frac{2}{\alpha} e^{-x_B/n} + 1 \right) \tag{16.29}$$

where the first term is again the static power and the second term the dynamic power. This result is represented in Figure 16.10 as a function of the normalized supply voltage x_B divided by *n*, for various values of duty factor α.

Except for duty factors approaching unity, these curves have a minimum for an optimum value of x_B. This optimum value is obtained by differentiating Equation (16.29) with respect to x_B and equating the result to zero. This yields:

$$\alpha = (x_{Bopt}/n - 2) e^{-x_{Bopt}/n} \tag{16.30}$$

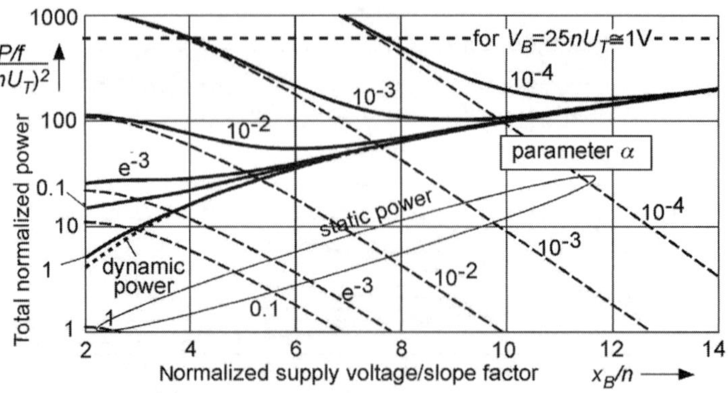

FIGURE 16.10 Power/frequency ratio.

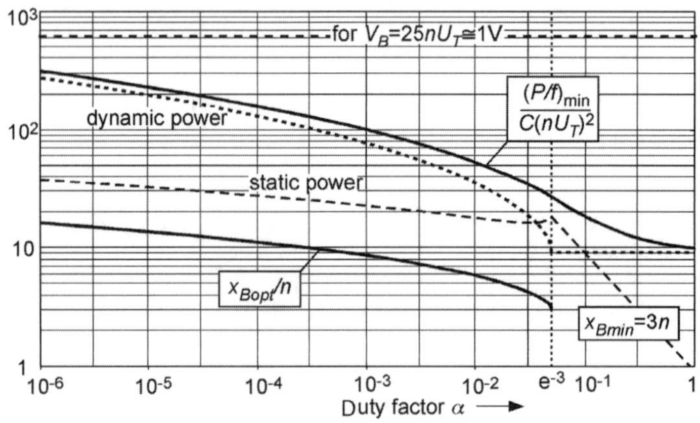

FIGURE 16.11 Optimum value of supply voltage and corresponding minimum power/frequency ratio.

which only has a solution for $\alpha \geqq e^{-3}$. Figure 16.10 shows that this optimum is not very critical. It can be obtained by numerically inverting Equation (16.30). The result can then be introduced into Equation (16.29) to provide the minimum possible value of P/f for a given duty factor α plotted in Figure 16.11.

Because no optimum value of x_B is available, for $\alpha > e^{-3}$, its minimum value must be fixed to ensure a logic swing close to the supply voltage, according to Figure 16.3(a). A value of $x_B = 3n$ has been chosen for this figure. The dynamic power for a supply voltage of 25 nU_T (approximately 1 V) is also indicated on the figure. As can be seen, a 50-fold reduction would be possible for $\alpha > 0.3$; the possible gain compared with 1-V operation is smaller for more realistic lower values of the duty factor, but it remains larger than a factor 10 for $\alpha > 3/1000$. For $\alpha << 1$, the minimum power occurs for $P_{stat} << P_{dyn}$. Indeed, according to Equation (16.20), increasing I_0 does not allow to reduce V_B significantly if delay time T_d must be maintained constant.

16.5.5 Minimum Delay Time in Weak Inversion

The degree of inversion of a transistor can be characterized by the inversion coefficient IC [6] defined as

$$IC = I_{DSsat} / I_S \qquad (16.31)$$

where I_S is the specific current of the transistor given by Equation (16.3). Introducing this definition in the approximation of the delay time given by Equation (16.20) results in

$$T_d = \frac{CV_B}{I_S IC_{on}} = BV_B / IC_{on} \tag{16.32}$$

where $B = C/I_S$ has a minimum value given by the process. The only possibility to reduce T_d is thus to increase the inversion coefficient in the on-state of the transistor; but weak inversion and its features according to the model introduced in Section 16.2 is limited to $IC < 1$. Therefore, the minimum delay time and the maximum average frequency achievable in weak inversion are simply given by

$$T_{dmin} = BV_B \text{ and } f_{max} = \frac{\alpha}{2BV_B} \text{ (in weak inversion)} \tag{16.33}$$

16.6 Effect of Entering Moderate and Strong Inversion

16.6.1 Transistor Model

With the definitions of currents and voltages introduced in Section 16.2, the following simple expression can be used to describe the drain current in and above weak inversion [6,8,9]:

$$I_{DS} = I_S \left(\ln^2 \left[1 + \exp\frac{V_{GS} - V_T}{2nU_T} \right] - \ln^2 \left[1 + \exp\frac{V_{GS} - V_T}{2nU_T} \exp\frac{-V_{DS}}{2U_T} \right] \right) \tag{16.34}$$

which reduces to Equation (16.2) for weak inversion when the exponential terms are much smaller than unity. In saturation, the second term is negligible, and this equation becomes

$$IC = \frac{I_{DSsat}}{I_S} = \ln^2 \left[1 + \exp\frac{V_{GS} - V_T}{2nU_T} \right] \tag{16.35}$$

This equation can be inverted to provide

$$V_{GS} = V_T + 2nU_T \ln(e^{\sqrt{IC}} - 1) \tag{16.36}$$

16.6.2 Required Voltage Swing

The gate voltage swing required to obtain a ratio K between the on-current I_{on} and the off-current I_{off} of a transistor can be obtained by application of Equation (16.36):

$$\Delta V_{GS} = 2nU_T \left[\ln(e^{\sqrt{IC_{on}}} - 1) - \ln(e^{\sqrt{IC_{on}/K}} - 1) \right] \tag{16.37}$$

This equation is plotted in Figure 16.12(a) for various values of K. It can be seen that the required swing is minimum and constant in weak inversion (as can be expected from the exponential behavior), but increases drastically when the maximum inversion coefficient IC_{on} is increased beyond unity.

Solving Equation (16.37) with respect to current ratio K yields

$$K = \frac{IC_{on}}{\ln^2 \left[1 + \exp\frac{-\Delta V_{GS}}{2nU_T} (\exp\sqrt{IC_{on}} - 1) \right]} \tag{16.38}$$

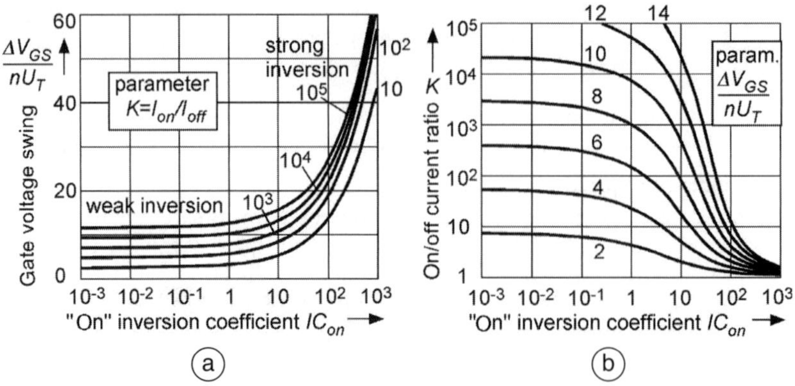

FIGURE 16.12 Gate voltage swing required for an on/off current ratio K.

which is plotted in Figure 16.12(b). This representation clearly demonstrates that the on/off current ratio produced by a given value of gate voltage swing is drastically reduced when the inversion coefficient is increased beyond unity. Thus, if the transistors are pushed into strong inversion to further reduce the delay time T_d according to Equation (16.32), the swing and thus supply voltage V_B must be increased to maintain the static power. The dynamic power is thus increased and a part of the expected reduction of T_d is lost.

16.6.3 Degeneration of Logic States

Assuming that the two types of transistors have the same specific current I_S and the same slope factor n, the current flowing through each of them in a CMOS inverter operated above weak inversion can be expressed by means of Equation (16.34):

$$I_n / I_S = \ln^2\left[1 + \exp\frac{x_i - x_T}{2n}\right] - \ln^2\left[1 + \exp\frac{x_i - x_T - nx_o}{2n}\right] \tag{16.39}$$

$$I_p / I_S = \ln^2\left[1 + \exp\frac{x_B - x_i - x_T}{2n}\right] - \ln^2\left[1 + \exp\frac{(1-n)x_B - x_i - x_T + nx_o}{2n}\right] \tag{16.40}$$

where $x_T = V_T/U_T$ is the normalized threshold voltage. The maximum value IC_{on} of the inversion coefficient is reached when the transistors are saturated with the gate voltage equal to the supply voltage. Thus, from Equation (16.35):

$$IC_{on} = \ln^2\left[1 + \exp\frac{x_B - x_T}{2n}\right] \text{ or } x_T = x_B - 2\ln(e^{\sqrt{IC_{on}}} - 1) \tag{16.41}$$

which can be introduced in Equation (16.39) and Equation (16.40) to eliminate x_T. These two equations can then be used to calculate a long chain of inverters to obtain the values of the low- and high-logic states x_L and x_H and that of the static current I_{stat}. The results are presented in Figure 16.13 for two values of normalized supply voltage.

As can be seen, for $x_B = 4$ (part a of the figure), the logic states degenerate rapidly when the inversion factor IC_{on} is increased beyond unity; they vanish for $IC_{on} > 5$. Moreover, the static current I_{stat} increases drastically, which will increase the static power dissipation. If x_B is doubled to 8 (part b of the figure), more speed can be obtained by increasing IC_{on} beyond unity, but the dynamic power will be increased by a factor 4. This demonstrates again that the minimum power-delay product increases significantly as soon as the devices are operated beyond weak inversion to obtain more speed than the approximate limit given by Equation (16.33).

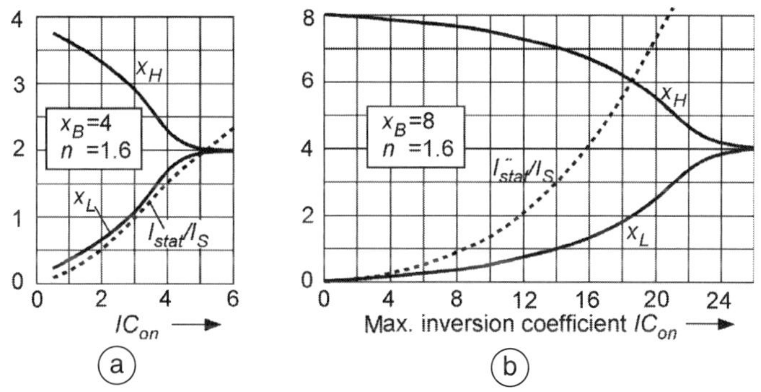

FIGURE 16.13 Degeneration of stable states and increase of static current above week inversion.

16.7 Extension to Logic Gates and Numerical Examples

To build useful logic circuits, the simple inverter of Figure 16.1 must be replaced by a logic gate, with at least two inputs. Let us assume that all gates of the system have the same number N of inputs, which means that the fan-in and fan-out of each gate are both equal to N. Let us further assume that N has the reasonable value 2 or 3, and that positive logic is used, in which each NAND gate is built by connecting N n-channel transistors in series and N p-channel transistors in parallel, as depicted in Figure 16.14.

If all transistors have the same (minimum) dimensions, then such a gate is approximately equivalent to the symmetrical inverter assumed in the previous analysis because the mobility of electrons is two to three times larger than that of holes, and the positive current I_C charging load capacitor C is normally that of only one p-channel transistor (except if several input transit simultaneously). According to Equation (16.3), the equivalent specific current I_S is then given by

$$I_S = \frac{2nW\mu_n C_{ox} U_T^2}{NL} \tag{16.42}$$

The equivalent capacitance C of the loaded gate can be expressed as

$$C = 2N(C_G + C_D) + C_{int} \tag{16.43}$$

where C_G and C_D are the gate capacitance and drain capacitance of a single transistor, and C_{int} is the interconnection capacitance.

Values of I_S and C given by Equation (16.42) and Equation (16.43) are evaluated in Table 16.1 for two typical technologies with $N = 3$. All transistors are of minimum size, and it is assumed that $C_{int} = C/2$.

Numerical values of the most important results derived previously are also given in this table. As can be seen, the minimum energy per transition (P/f for $\alpha = 1$ and $V_B = 4U_T$) is only about 0.23 fJ for

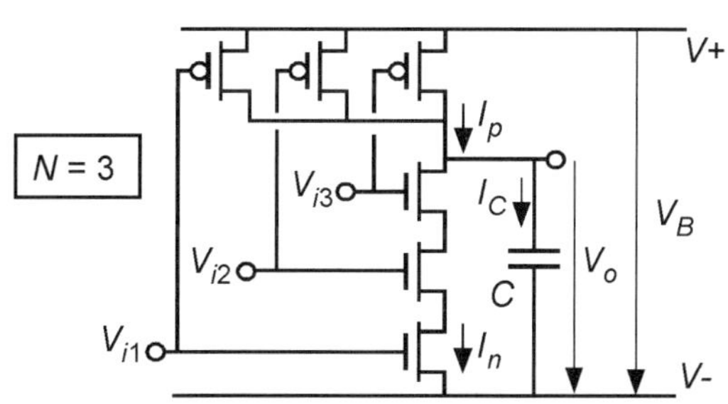

FIGURE 16.14 3-input CMOS NAND gate.

TABLE 16.1 Numerical Values for Two Different Processes ($N = 3$)

Parameter	Process A	Process B	Unit	Reference
L_{min}	500	180	nm	
U_T	25	25	mV	
n	1.5	1.3		
I_S	200	400	nA	equ. (16.42)
C	20	4	fF	equ. (16.43)
$C(nU_T)^2$	2.8.10^{-17}	4.2.10^{-18}	J	
B	1.10^{-7}	1.10^{-8}	V/s	equ. (16.32)
P/f for $\alpha = 1$ and $V_B = 4U_T$	2.28.10^{-16}	4.37.10^{-17}	J	equ. (16.29)
$(P/f)min$ for $\alpha = 10^{-2}$ and $V_B = V_{Bopt} = 6nU_T$	1.46.10^{-15}	2.20.10^{-16}	J	Fig. 16.11
P_{dyn}/f at $V_B = 1$V	2.00.10^{-14}	4.00.10^{-15}	J	
T_{dmin} in weak inversion for $V_B = 4U_T$	10	1	ns	equ. (16.33)
f_{max} for $\alpha = 1$ and $V_B = 4U_T$	50	500	MHz	equ. (16.33)
f_{max} for $\alpha = 10^{-2}$ and $V_B = V_{Bopt}$	0.22	2.56	MHz	equ. (16.33)
P_{min} at f_{max} above	32.5	56.3	nW	Fig. 16.11

conservative process A, and is further reduced by a factor 5 with the more advanced process B. If the duty factor α is reduced to 1%, the minimum equivalent energy per transition is increased by a factor 5 to 6, due to the increased importance of static power and to the necessary increase of supply voltage. Still, these values are 14 to 18 times lower than the dynamic power of the same gate operated at 1 V.

The minimum delay time in weak inversion is respectively 10 and 1 ns for the two processes, corresponding to a maximum possible clock frequency of 50 MHz and 500 MHz. If the duty cycle α is only 1%, the maximum average frequency is reduced to 220 kHz, respectively 2.56 MHz; it is decreased more than proportionally to α because V_B must be increased to limit the static power.

As discussed in Section 16.6, increasing the average frequency beyond this limit for weak inversion rapidly increases the necessary equivalent energy per transition.

16.8 Practical Considerations and Limitations

The ideal situation discussed in the previous sections is based on some basic assumptions. It is now necessary to examine if and how these assumptions can be made valid in practice.

16.8.1 Low-Voltage Power Source

Very low power is possible in weak inversion because the supply voltage V_B can be reduced to a very small value. If the duty factor is large ($\alpha > e^{-3}$), this value should be at least about $4U_T$ to ensure a full logic swing, according to Figure 16.3. For smaller α, the power consumption for a given average frequency f of transition is minimum for an optimum value of V_B that depends on α, as plotted in Figure 16.11. This optimum ranges from 4 to $15U_T$ for realistic values of α.

To take advantage of the scheme, the supply voltage must thus be adapted to this low value by means of a high-efficiency voltage converter. Although V_B should ideally be proportional to U_T and therefore proportional to the absolute temperature (PTAT), a fixed value could be used because the optimum is not very critical as demonstrated by Figure 16.10. Different supply voltages V_B should be used for blocks with different levels of duty factor α.

16.8.2 Low-Threshold and Threshold Adjustment

Once $P/f = 2PT_d/\alpha$ has been minimized by using the optimum value of V_B, the total power consumption P can be minimized by adjusting T_d to the maximum value compatible with f. Examination of the approximation (Equation (16.20)) of T_d shows that because C is imposed by the process, the only remaining possibility is to adjust I_0, which itself depends on threshold voltage V_T according to Equation

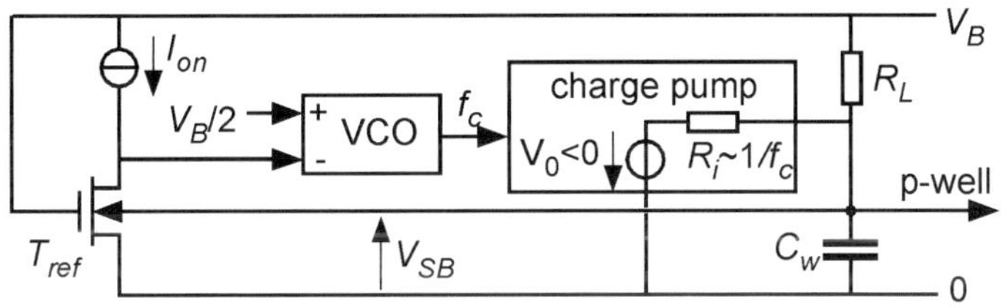

FIGURE 16.15 Principle of threshold adjustment for n-channel transistors [10].

(16.5). The latter can be adjusted by the source-to-bulk voltage V_{SB} according to Equation (16.4). More precisely, what must be adjusted by V_{SB} is the on-current I_{on}, in order to eliminate the exponential dependency on V_B exhibited by Equation (16.20). A possible implementation of such an adjustment is depicted in Figure 16.15 [10].

The on-current of reference transistor T_{ref} is compared with the required value of I_{on}. This produces a voltage that controls the frequency f_c generated by a VCO. The output of the VCO drives a capacitive charge pump, which produces a negative voltage V_0. The internal resistance R_i of this capacitive charge pump is inversely proportional to control frequency f_c. The output of the charge pump is connected to the p-well of T_{ref} and of all the transistors to be controlled, and is loaded by a resistor of fixed value R_L. As a result, the value of V_{SB} depends on frequency f_c and the loop is closed. It stabilizes when the on-current of T_{ref} (and that of all transistors to be controlled) is equal to the required value I_{on}. The large well-to-substrate capacitance C_w stabilizes the loop.

The effective value of threshold V_T can only be slightly reduced by $V_{SB} < 0$ (until a parasitic bipolar is activated by the forward-biased source-to-bulk junction). Thus, most of the possible adjustment is an increase of V_T by $V_{SB} > 0$. Because the required value of V_T is never more that about $10U_T$ (a few hundred mV), the nominal native threshold $V_{T0} = V_T(V_{SB} = 0)$ should be close to zero.

As depicted by Figure 16.10 and Figure 16.11, the minimum value of P/f is proportional to n^2; however, a value of n sufficiently above unity is needed to control V_T according to Equation (16.4).

The scheme must be symmetrically duplicated to control all p-channel devices in their separate n-well (true twin well is needed). The entire complementary scheme must be repeated for each different value of supply voltage V_B, if any.

16.8.3 Symmetry and Matching

The entire preceding analysis has assumed a perfect p/n symmetry of the gates. This situation can be approached by using NAND gates in a positive logic (see Figure 16.14) to approximately compensate for the difference of hole and electron mobility. The residual asymmetry causes a difference between rise and fall time, which may result in a reduced value of maximum frequency. The short-circuit charge Q_{sc} might also be increased, while remaining essentially negligible for homogeneous gates (see Figure 16.7(b)). Further adjustment of symmetry can be obtained by adapting the width of p- or n-channel devices, at the cost of an increased minimum power.

The mismatch of currents in weak inversion is dominated by the effect of threshold mismatch [6], with

$$\sigma(\Delta I / I) = \sigma(\Delta V_T) / (nU_T) \tag{16.44}$$

This spread of currents may reach several tens of percent from gate to gate, resulting in a proportional reduction of maximum frequency. Because mismatch is approximately proportional to $1/(WL)^{1/2}$, it may be reduced by increasing gate width W, but this would drastically increase capacitance C and thereby the power consumption. Increasing length L might have a lesser effect on power (if the drain capacitance dominates), but would drastically reduce the speed.

16.8.4 Process Scaling and Short-Channel Effects

Scaling of a process amounts to reducing all its geometrical dimensions by a factor S. In order to maintain constant fields, this geometrical scaling should be associated with a down-scaling of voltages and an up-scaling of doping concentrations by the same factor S. Now, because in most cases the absolute temperature T cannot be scaled down, $U_T = kT/q$ remains constant. As a result, approaching and even entering weak inversion should come naturally with scaled-down processes.

Such a constant-field scaling should bring no qualitatively new effect, until the apparition of quantum effects. Because all capacitors would be scaled-down by S, the power-delay product PT_d should be reduced by the same factor according to Equation (16.28). In weak inversion, the channel is equipotential and the current is carried by diffusion. As long as this is true, the mobility remains essentially constant and specific current I_S given by Equation (16.3) scales up with S. Therefore, the minimum delay time T_{dmin} given by Equation (16.33) should be reduced by S^2.

In practice, submicron scaling has been more aggressive in striving to reach the maximum possible speed by reducing voltages slower than dimensions. The fields have therefore increased dramatically, resulting in the need for special means to avoid high-field effects. As a result, some basic characteristics such as the slope n in weak inversion have been degraded in some processes. Whether this trend will be pursued with further scaling is not clear nowadays.

The gate current due to tunneling becomes a very important problem with the very thin oxide of deep submicron processes; however, this problem should be strongly alleviated at supply voltages of just a few hundred mV.

The evolution of threshold mismatch with scaled-down processes depends on the dominant cause of mismatch. If it is due to random fluctuations of the fixed interface charge, and if these fluctuations can be maintained constant in absolute value, then the increase of mismatch due to the reduction of gate area should be compensated by the increase of C_{ox}.

16.8.5 System Architecture and Applications

As explained in Section 16.5, all gates of the same digital block should be homogeneous in speed to keep the short-circuit charge per transition Q_{sctr} negligible. This is not different for strong inversion logic operated at a supply voltage larger than the sum of p and n thresholds [11,12], but new architectural approaches are needed to fully exploit the potential of weak inversion logic.

Traditional approaches in CMOS digital design exploit the fact that a CMOS gate consumes power only during transitions. Thus, the total power consumption can be minimized by minimizing the overall switching activity [13], independently of the total number of gates because idling gates consume negligible power. On the contrary, idling gates should be avoided in weak inversion logic because the minimum power increases when duty factor α is decreased as shown by Figure 16.11.

Ideally, the delay time T_d of each gate given by Equation (16.20) should be adjusted so that the gate is always in transition, corresponding to $\alpha = 1$. This is, of course, not possible in practice, but Figure 16.10 and Figure 16.11 demonstrate that a considerable improvement with respect to 1-V operation is still possible for α as low as 0.01. Architectures allowing the largest possible value of α should be developed. For example, small logic depth, pipelined and possibly asynchronous architectures should be favored. Large groups of gates should have similar values of T_d and α, so that the number of separate threshold adjustment loops of Figure 16.15 can be limited.

Dynamic adjustment of speed (adjustment of threshold voltage in time) might be considered, including sleep modes with a drastic reduction of I_0, but this is expected to complicate the implementation of the adjustment loop, especially if fast wake-up is required. Such a dynamic adjustment might be the only way to deal with the very low duty factor of read/write memory blocks.

Weak inversion logic represents the ultimate limit of the present trend to lower the supply voltage in order to reduce the power consumption and/or to limit the electric fields in scaled-down processes. Indeed, with the low threshold voltages characteristic of deep submicron processes, the residual channel current I_0 of blocked transistors increases according to Equation (16.5). Therefore, a reduction of the the average idling time of gates (increase of α) is becoming a priority to avoid a situation where power consumption is dominated by its static component.

Because the maximum frequency is considerably lower than for strong inversion, weak inversion logic is best applicable to systems that do not require high local speed. It seems ideally suited to implement massively parallel digital architectures, for example, in applications related to image processing [7].

16.9 Conclusion

Digital circuits may in principle be operated at a supply voltage as low as $4U_T$ (100 mV at ambient temperature), while maintaining well-defined stable states (Figure 16.3) and a sufficient on/off current ratio (Figure 16.12), thanks to the exponential behavior in weak inversion. However, this requires a special CMOS process with values of native threshold voltages close to zero and separate wells (truly twin wells) for both types of transistors, in order to electrically adjust these threshold voltages against process spreading and temperature variations.

The dynamic power consumption is reduced with the square of the supply voltage, thus by as much as 100 compared with a 1-V operation. Static power consumption cannot be neglected anymore, as is the case for any CMOS digital circuits using low-threshold devices. The relative importance of this static power depends on the duty factor α, defined as the average percentage of time the gates are in transition. Still, the overall power reduction (with respect to 1 V) remains a factor larger than 30 for $\alpha = 0.1$ and larger than 10 for $\alpha = 0.01$ (see Figure 16.10 and Figure 16.11).

Despite the limited drain current density available in weak inversion, the maximum frequency of operation may reach several hundred MHz for a deep submicron process ($0.18\mu m$). Moreover, the very low voltage of operation is expected to prevent or to alleviate most high-field effects that plague short-channel devices operated at higher voltages.

The approach is substantially different from traditional CMOS digital circuits. Therefore, new architecture should be developed to best exploit its potential for low power. It appears best suited for very low-power digital systems running at moderate clock frequency, such as parallel image processing.

Even if it is not used directly, weak inversion logic represents the lower limit of supply voltage and power-delay product that are achievable with a given process at a given temperature. It can therefore be used as a reference to assess the merit of any digital circuit.

It is interesting to notice that nature has also found this voltage limit in its processing capability because the action potentials in neurons is also in the range 50 to 100 mV (2 to $4U_T$). The local speed of neurons is much lower, however, because of the lower mobility of ions. Still, it is compensated by a huge degree of parallelism that results in fast and low-power capabilities for processing very complex global tasks.

References

[1] R.M. Swanson and J.D. Meindl, Ion-implanted complementary MOS transistors in low-voltage circuits, *IEEE J. Solid-State Circuits,* vol. SC-7, April 1972, pp. 146–153.

[2] H. Soeleman and K. Roy, Ultra-low power digital subthreshold logic circuits, *Proc. Int. Symp. on Low-Power Electronics and Design,* pp. 94–96, 1999.

[3] J. Burr and J. Shott, A 200-mV self-testing encoder/decoder using Stanford ultra-low-power CMOS, *ISSCC Dig. of Tech. Papers,* 1994, pp. 84, 85, 316.

[4] M. Miyazaki et al., A 175-mV multiply-accumulator unit using an adaptive supply and body bias (ASB) architecture, *ISSCC Dig. of Tech. Papers,* 2002, pp. 58, 59, 444.

[5] E. Vittoz and J. Fellrath, CMOS analog integrated circuits based on weak inversion operation, *IEEE J. Solid-State Circuits,* vol. SC-12, June 1977, pp. 224–231.

[6] E. Vittoz, Micropower techniques, *Design of VLSI Circuits for Telecommunications and Signal Process.,* chapter 3, J. Franca and Y. Tsividis, Eds., Prentice Hall, Englewood Cliffs, NJ, 1994.

[7] E. Vittoz, Low-power design: ways to approach the limits, *ISSCC Dig. of Tech. Papers,* 1994, pp. 14–18.

[8] E. Vittoz, Very low power circuit design: fundamentals and limits, *Proc. Int. Symp. on Circuits and Syst.,* 1993, vol. 2, pp. 1439–1442.

[9] C. Enz, F. Krummenacher, and E. Vittoz, An analytical MOS transistor model valid in all regions of operation and dedicated to low-voltage and low-current applications, *Analog Integrated Circuits and Signal Process.,* vol. 8, July 1995, pp. 83–114.

[10] V. von Kaenel et al., Automatic adjustment of threshold and supply voltage for minimum power consumption in CMOS digital circuits, *Proc. IEEE Symp. on Low-Power Electron.,* San Diego, CA, 1994, pp. 78–79.

[11] J.R. Burns, Switching response of complementary-symmetry MOS transistor logic circuits, *RCA Review,* Dec. 1964, pp. 627–661.

[12] H.J.M. Veendrick, Short-circuit dissipation of static CMOS circuitry and its impact on the design of buffer circuits, *IEEE J. Solid-State Circuits,* vol. SC-19, August 1984, pp. 468–473.

[13] A.P. Chandrakasan and R.W. Brodersen, Minimizing power consumption in digital CMOS circuits, *IEEE Proc.,* vol. 83, April 1995, pp. 498–523.

17

Robustness of Digital Circuits at Lower Voltages

Harry Veendrick
Philips Research Labs

17.1 Introduction

Over the last two decades, there has been a change in the drive for the continuous scaling of devices and circuits. Figure 17.1 plots the scaling dependence of different parameters.

The presented diagram is divided into two regions: constant-voltage scaling and constant-field scaling. The constant-voltage scaling timeframe reflects the period in which (C)MOS devices were still supplied by a constant voltage of 5 V. After a certain point in time (in the diagram: about 1997 for volume production) the supply voltage was reduced at the same pace as the transistor's channel length, thereby keeping a constant field across the transistor channel: constant-field scaling. During the first period, in each new technology node, the average fabrication costs increased with a factor of $s^{-0.5}$ (about 1.2 times; s ≈ 0.7), while the intrinsic speed improvement as obtained from the technology scaling was about a factor of s^{-2} (≈ 2). In the same period, the power efficiency (= $1/\tau D$-product: how much speed do I get per watt?) improved only by a factor of s^{-1} (≈ 1.4). Therefore, the rise in fabrication costs could easily be compensated by the speed and density increase in that period.

After that period, the voltage scales at the same pace as the technology feature sizes and causes a reduction in the speed increase to only a factor of s^{-1}. The fabrication costs now also increase with the same factor, however, meaning that the speed increase has become less of a drive for further scaling. Nevertheless, voltage scaling has one big advantage: it improves the power-efficiency increase to a factor of s^{-3} (≈ 2.8) per technology node. This means that if a function consumes a certain amount of power in a given technology node, it will consume a factor of s^{-3} less power in the next technology node. Or,

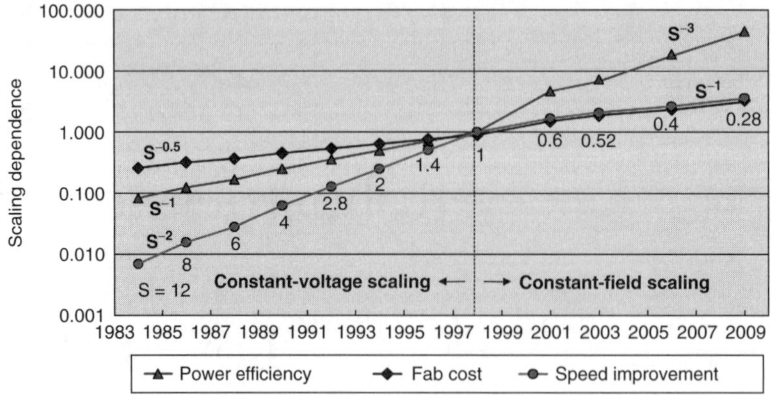

FIGURE 17.1 From speed drive to power-efficiency drive.

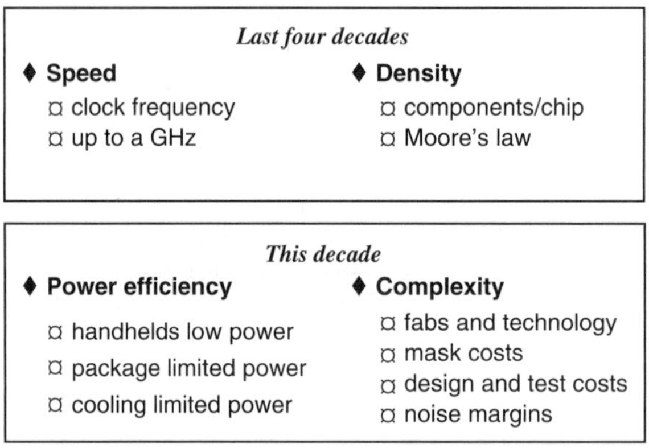

FIGURE 17.2 Summary of changing scaling trends.

to put it another way, in the next technology node, we can have about 2.8 times more functionality (2 times the number of transistors running at 1.4 times the speed) at the same power consumption. Therefore, power efficiency has become the major performance drive in this decade. Figure 17.2 is a rough summary of these scaling trends over the past and near future.

Whereas the speed and density improvements drove the semiconductor scaling process according to Moore's law during the last couple of decades, today, the performance focus has changed toward power efficiency. Yet, the complexity has become another focus to manage (read: limit) the excessive increase of the costs in all semiconductor disciplines. This holds for the production costs (i.e., fabs and the lithography), the mask costs, and, finally yet importantly, the design and test costs. The increased design costs are not only due to the higher complexity of the circuits, it is also a result of increased noise and reduced noise margins. This noise margin reduction is an issue that we, as integrated circuit (IC) designers, should be worried about mostly. Today, many so-called deep-submicron (DSM) effects manifest themselves more as the minimum feature sizes and the voltages are reduced. These DSM effects cause an increase of physical design aspects that have to be taken into account, mostly during the back-end of an IC design. Figure 17.3 depicts a heterogeneous system on chip (SoC) including both system-design and physical-design aspects. As an example, we take the global bus/switch-matrix. Whereas the protocol and bandwidth are typical system design aspects of a bus, signal propagation and cross talk are more physics related. In designing or choosing the I/O pads, the bandwidth and the interface standard are system-design aspects, while di/dt noise and EMC behavior have a closer relation to the physical layer. An overall conclusion from this SoC representation is that, as we scale further, more of these physical aspects pop up and start threatening the robustness of operation of the IC.

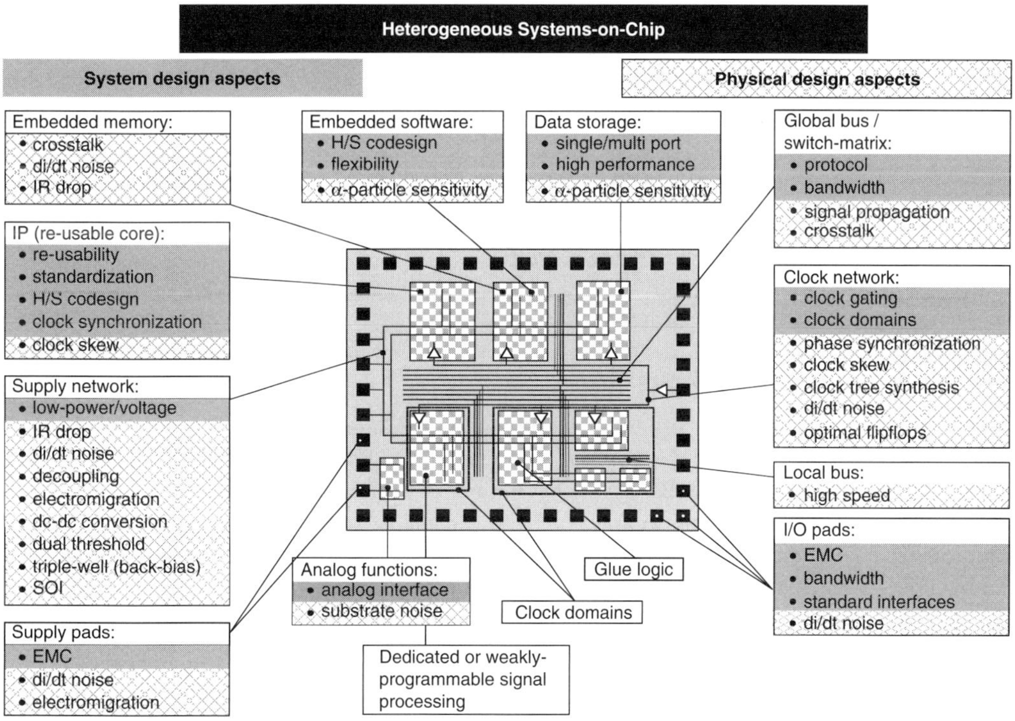

FIGURE 17.3 System and physical design aspects in an SoC [1].

This chapter deals with the robustness of digital circuits in relation to the continuous scaling process.

It covers most topics related to signal integrity (e.g., cross talk, signal propagation, voltage drop, supply and substrate noise, soft-errors, and electromagnetic compatible [EMC]) as well as such reliability issues as electromigration, electro-static discharge (ESD), latch-up, hot-carrier effect, and negative temperature bias instability (NBTI). The reducing signal integrity is a result of two conflicting effects: the increase of noise and the reduction of the noise margins (V_{dd} and V_t). The next section, therefore, begins with a discussion on noise issues and ways to maintain signal integrity at a sufficiently high level. A continuous reduction of the noise margins also has a severe impact on the quality of the IC test. The increasing discrepancy between chip operation during test and in the application will result in more customer returns and design spins. The next paragraph will therefore also include some remarks on the effect of scaling on test coverage and complexity. The third paragraph in this section is devoted to trends in reliability and ways to maintain it. Paragraph four presents some conclusive remarks and deals with different scaling scenarios. It presents scaling tables for constant-voltage scaling, constant-field scaling, and constant-size scaling and focuses on the challenges that DSM effects imply on the robustness of operation of future ICs.

17.2 Signal Integrity

Signal integrity indicates how well a signal maintains its original shape when propagating through a combination of circuits and interconnections. On-chip effects from different origin may influence this shape. Signals can be influenced by switching of nearby neighbors (cross talk), by voltage changes on the supply lines (voltage drop and supply noise), by local potential changes in the substrate (substrate noise), or when the signal node is hit by radioactive or cosmic particles (soft-error). In addition, the speed at which a signal propagates through bus lines is heavily affected by the switching behavior of neighboring bus lines.

The next subsections will focus on each of these signal-integrity topics individually and present ways to limit the noise level or the influence of the potential noise sources that threaten the signal integrity.

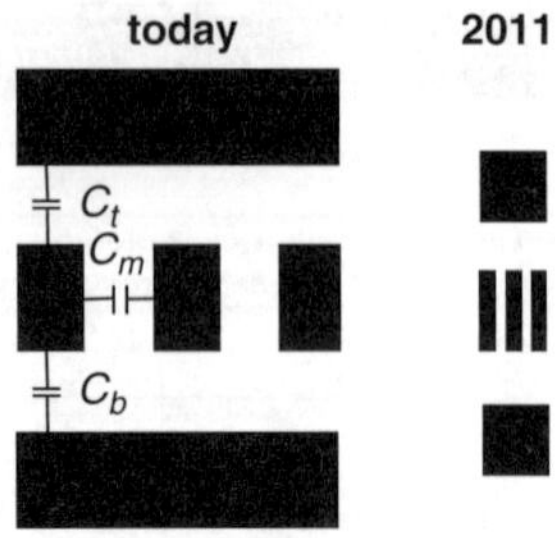

FIGURE 17.4 Expected scaling of metal track width and spacing.

17.2.1 Cross Talk and Signal Propagation

Due to the scaling of the transistors, their density has almost doubled every new technology node for more than four decades already. This requires the metal lines (width and spacing) to be scaled in the same order to be able to connect this increasing number of devices per unit of area. Per unit of area, however, the total length of the interconnections in one metal layer only increases with a factor of 1.4. This means that additional metal layers are needed to allow a high-density connection of all logic gates. The metal layers are also used to supply the current from the top metal layer all the way down to the individual devices. As will be discussed in the subsection on electro-migration, the current density also increases with a factor of 1.4 every new technology node, meaning that the thickness of the metal layers cannot be scaled at the same pace as the width and spacing. Consequently, the mutual capacitance between neighboring signal lines is dramatically increasing.

Figure 17.4 presents two cross sections of three parallel metal lines: one in a 120-nm CMOS technology and the other one in a 35-nm process (year 2011). It clearly shows that the bottom (C_b) and top capacitances (C_t) reduce while the mutual capacitances (C_m) increase. This increase in mutual capacitance has dramatic effects on the performance and robustness of integrated circuits. The first one is the growing interference between two neighboring interconnect lines, which is usually referred to as cross talk. The second one is the growing signal propagation delay because of the increasing *RC* times across the interconnect. Third, the increased interconnect capacitances also affect the overall IC's power consumption. We will discuss each one of these effects in more detail now. Figure 17.5 depicts the trend in the cross talk over several technology nodes.

The used model refers to two minimum-spaced interconnect wires in the same metal layer. A signal swing ΔV_{M1} on metal track M1, causes a noise pulse ΔV_{M2} on a floating metal track M2, as defined by:

$$\Delta V_{M2} = \Delta V_{M1} * Cm/(Cm + Cground) \tag{17.1}$$

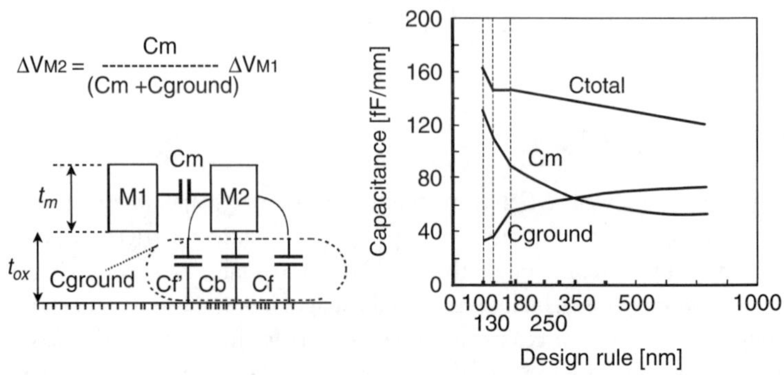

FIGURE 17.5 Interconnect capacitances across various technology nodes [1].

TABLE 17.1 Capacitance Values for Second Metal Layer in Different Technology Nodes

	180 nm CMOS	130 nm CMOS	90 nm CMOS
C_m	89	110	132
C_{ground}	58	36	32
C_{total}	147	146	164
$\Delta V_{M2} =$	$0.6 * \Delta V_{M1}$	$0.75 * \Delta V_{M1}$	$0.8 * \Delta V_{M1}$

Table 17.1 lists the capacitance values for different technology nodes. The bottom row in this table presents the amount that one signal propagates into the other one through cross talk. For the 90-nm node this means that 80% of the switching signal propagates into its floating neighbors. Because of this, all floating lines (e.g., precharged bit lines in a memory and tri-state buses are very susceptible to cross talk noise).

Even nonfloating (driven) lines in digital cores are becoming increasingly susceptible to cross talk causing spurious voltage spikes in the interconnect wires. Traditional design flows only deal with cross talk analysis in the back-end part, to repair the violations with manual effort, after the chip layout is completed. Because timing and cross talk are closely related, they should be executed concurrently with the place-and-route tools. The introduction of multi-V_{dd} and multi-V_t pose a challenge for the physical synthesis and verification tools because both design parameters affect timing and signal integrity.

In memory design, scaling poses other challenges to maintain design robustness. The layout of a static random-access memory (SRAM), for example, includes many parallel bit lines and word lines at minimum spacing in different metal layers. It is clear that these will represent many parasitic capacitances; however, there might be even more mutual capacitance between the various contacts (pillars) than between the metal tracks. Memories in DSM technologies therefore require very accurate three-dimensional (3-D) extraction tools in order to prevent that the silicon will, unexpectedly, run much slower than derived from circuit simulations.

Next to the cross talk between metal wires, the signal propagation across metal wires is also heavily affected by scaling. In a 32-bit bus, for example, most internal bus lines (victims) are embedded between two neighbors (aggressors). The switching behavior of both aggressors with respect to the victim causes a large dynamic range in signal propagation across the victim line. In case both aggressors switch opposite from the victim, the signal propagation across the victim lasts about six times longer than in case the aggressors and victim all switch in the same direction, for 20-mm long bus lines in a 180-nm CMOS technology. Figure 17.6 plots the increasing propagation delay (in nano-seconds) with the technology node for a 20-mm long bus line, embedded between two quiet (nonswitching) aggressors.

Although the introduction of copper with the 120-nm node provides some relief in the increase of the propagation delay, it will only help for about one technology node. This means that in the 120-nm node, with an aluminum backend, the interconnect propagation delay would reach the same order of magnitude as the 90-nm node with a copper backend. The diagram also indicates that the propagation delay will

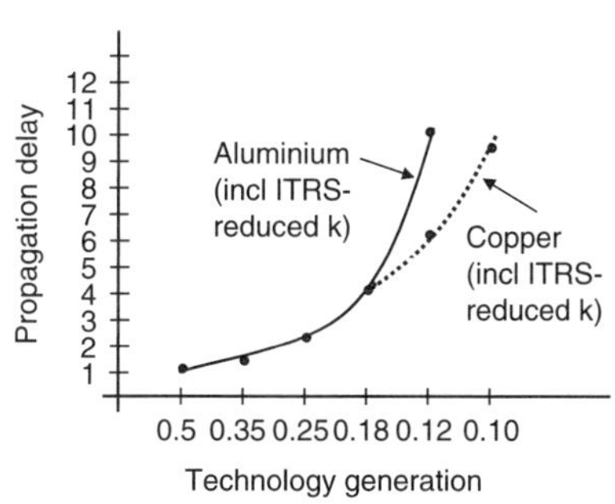

FIGURE 17.6 Propagation delay versus technology node in case aggressors are quiet.

further increase. This requires different design architectures, in which the high-speed signals are kept local. Such architectures must allow latency in the global communication or communicate these global signals asynchronously (i.e., islands of synchronicity; globally asynchronous, locally sychronous (GALS)).

In the preceding discussions, self- and mutual inductances were not taken into account; however, with the advances in speed and clock frequencies, the influence of these inductances becomes increasingly pronounced. The resistances of most of today's signal lines still exceed the values of inductance by more than one order of magnitude. For one reason this is because the resistance increases every technology node. The second reason is that the inductance contribution is linearly proportional to the frequency. As a rule of thumb, we can state that at 10 GHz, the inductance contribution to the total impedance of a metal wire reaches about the same value as the resistance contribution. This means that we need to change from an *RC* interconnect model to an *RLC* model for designs that exceed 1 GHz (at this frequency the inductance value is about 10% of the resistance value and can thus no longer be neglected). Generally, two effects determine the difference in accuracy between an *RC* and an *RLC* model: the damping factor and the ratio between the input signal rise time and the signal propagation speed across the line. Therefore, even in designs that do not yet reach 1GHZ, the wider metal lines (with lower resistance) (e.g., in clock distribution networks) and upper metal layers can exhibit significant inductive effects [2]. Because also the rise times of signals on interconnect lines are reducing with the advance of the technologies, RLC models need to be included in our computer-aided design (CAD) tools soon, in order to avoid inaccurate performance predictions or underestimate signal integrity effects, which may also lead to a reliability problem.

Finally, a number of methods exist to reduce cross talk and/or improve signal propagation. We will summarize them here, without discussing them in more detail:

- Use fat wires to reduce track resistance.
- Increase spacing to reduce mutual capacitance.
- Use shielding between the individual bus lines.
- Use staggered repeaters to compensate for noise.
- Use tools that can detect, replace, and reroute critical nodes.
- Use current sensing or differential signaling for improved speed and noise compensation.

17.2.2 Supply and Ground Bounce

Every new technology node allows us to almost double the number of transistors. Next to this, the bus widths have also gradually grown over the last couple of decades: from 4-bit in the late 1970s to 64-bit, or even 128-bit, today. The interface to a 256-Mbit DDR SDRAM, for instance, requires communicating 32 data bits — about 23 address bits plus a few additional control bits — totally adding up to some 60 parallel bits. In addition, due to the increased speed requirements, more flip-flops/pipelines are used within the logic blocks. All these individual trends contribute to a dramatic increase of simultaneously switching activity in an IC causing huge currents (i) and current peaks (δi). These currents cause voltage drop across the resistance (R) of on-chip supply network, while the current peaks cause relatively large voltage drops across the self-inductances (L) in the supply path. As is discussed in the previous subsection, most of the self-inductance is still in the bond wires and the package leads, instead of in the on-chip metal supply lines.

Another trend that keeps pace with technology advances is the reduction in switching times (δt) of the logic gates and driver circuits. The combination of these two trends leads to a dramatic increase of $\delta i/\delta t$, which term is mainly responsible for the supply and ground bounce generated on chip.

In total, we can summarize the voltage drop by:

$$\Delta V = i.R + L.\delta i/\delta t \tag{17.2}$$

The impact of this voltage drop on the behavior of the chip is twofold. First, the average supply voltage throughout the complete clock period determines the speed of a circuit. Let V_{dd} be the nominal supply voltage of a chip. Most commonly, this means that the chip is specified to operate within a 5 to 10% margin in this supply voltage. In case of a 0.18-μm CMOS design, this means that it should operate between 1.65 V and 1.95 V. So, in the application, the IC should operate correctly, even at 1.65 V. Because the logic synthesis is done using the gate delays specified at this lower voltage, an additional voltage drop ΔV within the chip could be disastrous for proper functionality. In other words, the designer should limit the total average voltage drop within stringent limits to assure the circuit operates according to the required frequency spec. It is commonly accepted that this voltage drop is limited to just a small percentage of the supply voltage (less than 5%). Second, ΔV introduces noise into the supply lines of the IC. The current is supplied through the V_{dd} supply lines and leaves the circuit through the V_{ss} ground lines. When the impedances of the supply and ground lines are identical, which is most commonly the case, the introduced bounce on the respective lines show complementary behavior and are identical in level. The total inductance (L) consists of on-chip contributions of the supply and ground networks and off-chip contributions of the bond wires, package leads, and board wires. Usually, the damping effect of high resistive narrow signal wires reduces the effect of on-chip inductive coupling. To reduce the contribution of the first term in the Equation (17.2), however, the supply and ground networks require wide metal tracks in the upper metal layers with very low sheet resistance. Particularly for designs operating at GHz frequencies, inductance in IC interconnects is therefore becoming increasingly significant.

The supply noise can be reduced in several ways. When using n supply pads for the supply connection, which are more or less homogeneously distributed across the IC periphery, the self-inductance will reduce to L/n. Both the use of a low-resistive supply network and multiple supply pads, however, contribute to a reduction of the overall impedance of the supply network. Because the bond wires, package leads, and board wiring, all act as antennae, the resulting increase of the current peaks ($\delta i/\delta t$) lead to a dramatic rise of interference with neighboring ICs on the board and may cause EMC problems in the system. Therefore, it is also required to keep the peak currents local within the different cores on the IC. In other words, it is necessary to lower the global $\delta i/\delta t$ contribution in the Equation (17.2) as well. The use of staggered driver turn-on, to limit the amount of simultaneous switching activity, as well as encouraging the use of "slow" clock transients will directly contribute to a lower $\delta i/\delta t$. Another measure to limit the global $\delta i/\delta t$ is the use of decoupling capacitors within each of the different cores. Figure 17.7 depicts two implementations of decoupling capacitor cells [1]. Figure 17.7(a) is a complementary set of transistors connected as an nMOS and pMOS capacitor, directly between V_{dd} and V_{ss}. Figure 17.7(b) is a "tie-off" cell used as decoupling capacitor. In several applications, a tie-off cell supplies dummy V'_{dd} and V'_{ss} potentials to circuits, which, for reasons of ESD, are not allowed to have an input directly connected to the V_{dd} and V_{ss} rails. The channel resistances R_n and R_p (Figure 17.7(c)) of the nMOSt and pMOSt, respectively, serve as additional ESD protection for the transistor gates connected to the V'_{ss} and V'_{dd}. This advantage can also be exploited when we use this cell only as a capacitor cell between V_{dd} and V_{ss}, however, without using the dummy V'_{dd} and V'_{ss} terminals. When a supply dip occurs, the charge stored on the gate capacitance C_n (C_p) of the nMOSt (pMOSt) must be supplied to the V_{dd} (V_{ss}) in a relatively

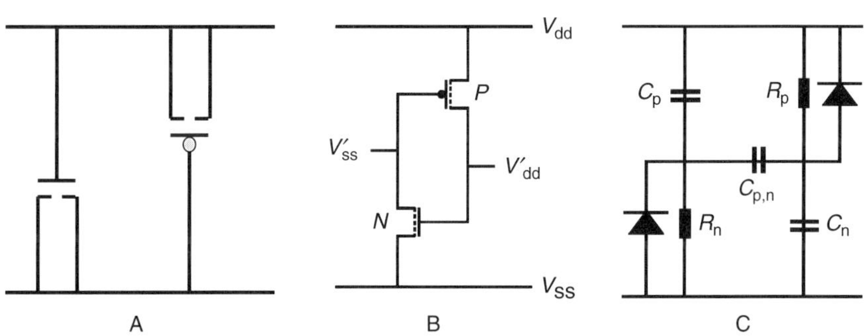

FIGURE 17.7 (a) normal decoupling capacitor, (b) tie-off cell decoupling capacitor, and (c) equivalent circuit.

short time, which puts some constraints to the value of R_n (R_p). Therefore, decoupling capacitor cell *b* presents a better ESD behavior compared with cell *a*.

These decoupling capacitors are charged during steady state (e.g., at the end of the clock period when the total switching activity has almost or completely come to an end). The additional charge, stored in these capacitors is then redistributed to the supply network during moments of intense switching, particularly at the clock transient that initiates the next signal propagation through the logic paths. These decoupling capacitor cells are designed as standard cells and are usually available in different sizes. The amount of decoupling capacitance that needs to be added in each core depends on the number of flip-flops in it and on the switching activity of its logic. The switching activity α is defined as the average number of gates that switch during a clock cycle. When a logic core has an activity factor of $\alpha = 1/3$, it means that the average gate switches one out of every three clock periods. Different algorithms require different logic implementations, which show different switching activities. It is known that average telecom and audio algorithms show less switching activity ($0.05 < \alpha < 0.3$) than an average video algorithm ($0.15 < \alpha < 0.4$), for example. The operating frequency of a logic block is a major component in determining the total amount of decoupling capacitance that needs to be added in a logic core. The higher the frequency, the less voltage drop can be allowed and the more decoupling capacitance needs to be added. As an example, the total additional decoupling capacitance in an average logic block, performing a video algorithm, running at 350 MHz in a 0.12-μm CMOS core in a digital chip, may occupy about 10 to 15% of its total area. When the standard-cell block utilization is less than 85%, this amount of decoupling capacitance fits within the empty locations inside a standard-cell core. In mixed analog/digital ICs, however, this amount could grow dramatically because the noise in these ICs is more restricted by the sensitivity of the analog circuits. Due to further scaling, δi will increase, while the δt will just do the opposite, requiring an increasing amount of decoupling capacitance every new technology node.

17.2.3 Substrate Bounce

Substrate noise is closely related to the ground bounce. On a mixed analog/digital IC, usually the digital circuits are responsible for most of this bounce, while the analog circuits are most sensitive to it (Figure 17.8). The substrate bounce has several contributors. The transistor substrate current injection is responsible for only a few mV. Junction and interconnect capacitances account for several tens of mV. The highest noise levels (several hundred mV), however, are introduced through the current peaks in the supply network, also causing the previously discussed supply noise.

In most CMOS circuits, it is common practice to connect the substrate to the V_{ss} rail, meaning that the ground bounce that is generated in the V_{ss} rail is directly coupled into the substrate. This is even a bigger problem, when the chip is realized on epitaxial wafers (see Section 17.3.4) with a low-ohmic substrate because it propagates the noise through the substrate to the analog part almost instantaneously and with hardly any loss of amplitude. Because the noise margins reduce with reducing supply voltages, the use of high-ohmic substrates is becoming increasingly important. Triple-well technology allows improved isolation of analog circuits from digital cores. The use of a silicon-on-insulator (SOI) technology

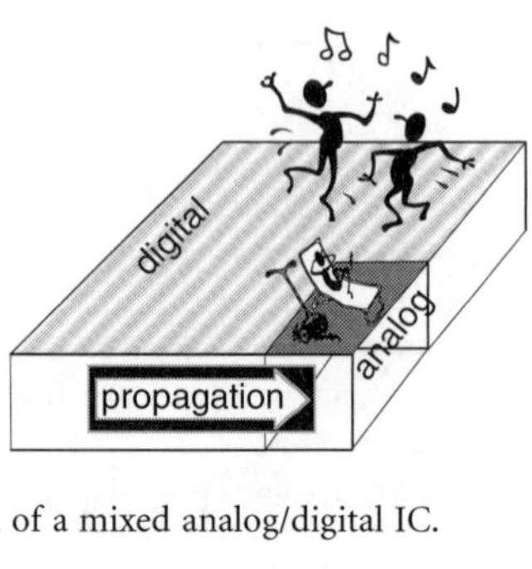

FIGURE 17.8 Symbolic representation of a mixed analog/digital IC.

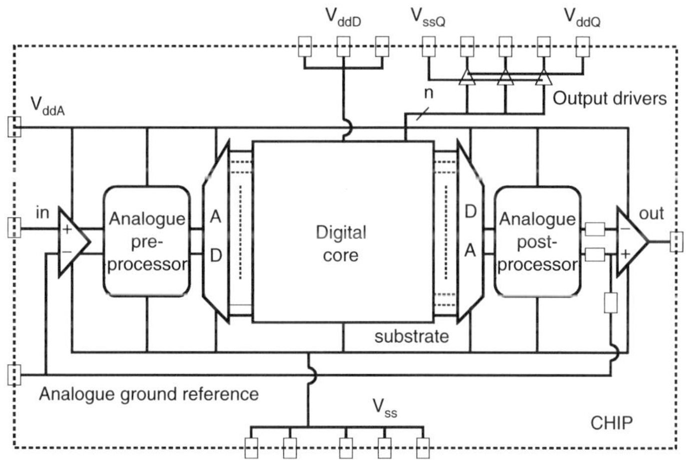

FIGURE 17.9 Proposed supply connections in a mixed analog/digital IC.

allows even a complete separation of the analog and digital circuits. Several other measures exist to reduce the level of substrate bounce. First, the measures that help reduce the supply and ground bounce, as discussed in the previous sub-section, are also beneficial for substrate bounce reduction. Second, a physical separation of the core and I/O supply nets from the analog supply net, according to Figure 17.9, prevents the relatively large noise introduced in these nets to propagate directly into the analog net [3].

The figure also illustrates that most digital and analog circuits share the same ground (V_{ss}) because it also serves as a reference for the communicated signals. Usually, the impedance of the internal and external V_{dd} and V_{ss} networks are almost symmetrical, meaning that they have equal widths and the same number of bonding pads. An increase in the impedance of the V_{dd} network with respect to the impedance of the V_{ss} network would increase the bounce in the V_{dd} supply network, while reducing it in the V_{ss} ground network. Because the analog and digital V_{dd} where separated anyway, this additional digital supply bounce is not coupled into the analog V_{dd}. Because the analog and digital circuits share the same ground, the lower V_{ss} ground bounce also reduces the substrate bounce. Therefore, to increase the margins and robustness of mixed analog/digital ICs, it may be advantageous to dedicate more supply pads to V_{ss} and less to the V_{dd}. Finally, particularly in the case of high-ohmic substrates, circuits with the highest switching activities, and driving strengths (e.g., I/O pads, clock drivers, and drivers with a high fan-out) must be located as far away from the analog circuits as possible.

17.2.4 EMC

The problem of supply and ground bounce caused by large current changes is not restricted to on-chip circuits. High current peaks may also introduce large electromagnetic disturbances on a printed-circuit board (PCB) because of the electromotive force and threatens the off-chip signal integrity. Because bonding pads, package, and board wiring act as antennae, they can "send" or "receive" an electromagnetic pulse (EMP), which can dramatically affect the operation of neighboring electronic circuits and systems [4].

When realizing EMC circuits and systems, the potential occurrence of EMPs must be prevented. The use of only one or a few pins for supply and ground connections of complex high-performance ICs is one source of EMC problems. Even the location of these pins is very important with respect to the total value of the self-inductance. The use of three neighboring pins for V_{dd}, for instance, results in an electromagnetic noise pulse that is twice as large as when these supply pins were equally divided over the package. The best solution is to distribute the power and ground pins equally over the package in a sequence such as V_{dd}, V_{ss}, V_{dd}, and V_{ss}. Bidirectional currents compensate each other's electromagnetic fields in the same way as twisted pairs do in cables. Another source of EMC problems is formed by the outputs. They can be many (about 60 for the address and data bits in a 256-Mbit DDR SDRAM interface), contain relatively large drivers with high current capabilities, and often operate at higher voltages than the cores. Actually, each output requires a low-inductance current return path, such that the best position for an output is right between one pair of V_{dd} and V_{ss} pads. This results in the smallest electromagnetic disturbances at PCB level and reduces the supply noise at chip level. Because this is not very realistic in many designs, however, more outputs will be placed between one pair of (V_{dd} and V_{SS}) supply pads. The limitation of this number is the designer's responsibility (simulation). In addition, the $\delta i/\delta t$, generated by these outputs, must be limited to what is really needed to fulfill the timing requirements. Finally, all measures that reduce on-chip supply and ground bounce, also improve the electromagnetic compatibility of the chip and result in a more robust and reliable operation.

17.2.5 Soft Errors

Because of the continuous shrinking of devices on an IC, the involved charges on the circuit nodes have been scaled down dramatically. Particles, independent of their origin, do have an increasing impact on the behavior of these shrinking devices. Several categories of particles can be distinguished, which all generate free electron-hole pairs in the semiconductor bulk material [5]:

- Alpha particles, originating from radioactive impurities (mainly uranium and thorium) in materials; these materials can be anything near the chip: solder, package, or even some of the materials used in the production process of an IC (metals or dielectrics). These so-called *α-particles* can create many electron-hole pairs along their track.
- High-energy cosmic particles, particularly neutrons, can even fracture a silicon nucleus. The resulting fragments cause the liberation of large numbers of electron-hole pairs.
- Low-energy cosmic neutrons, interacting with boron-10 (^{10}B) nuclei; when a ^{10}B nucleus breaks apart, an α-particle and a lithium nucleus are emitted, which are both capable of generating soft errors.

In all cases, the generated electrons and holes can be captured by capacitors (in dynamic logic and DRAMs) and may flip states of both dynamic and static storage circuits (e.g., memories, latches, and flip-flops). The resulting incorrect state is called a soft error because the next clock period its data may have been restored, meaning that the flipped state has not caused permanent damage to any of the circuit nodes. In addition, in static CMOS logic (or SRAM cells) the total charge of a node is an important criterion for the possibility of flipping its state after being hit by an ionizing particle. In a first-order approximation, the total critical charge (Q_{crit}) needed to flip a circuit node to its complementary state is defined by:

$$Q_{crit} = C.V \tag{17.3}$$

where V equals the supply voltage and C the total capacitance of the node. Due to the continuous scaling of sizes and voltages, both C and V reduce with about a factor of 0.7 (which is the average scale factor between two successive technology nodes). Furthermore, design complexity and memory size increase, whereas, conversely, the charge collection efficiency reduces. As a net result, the soft-error sensitivity of integrated circuits dramatically increases with scaling. Particularly the large capacity, densely packed (embedded) memories show an increasing failure in time (FIT). The previous considerations particularly hold for dynamic circuits and DRAMs. In static storage cells (e.g., SRAM cell, latch, or flip-flop), the critical charge is not only dependent on the capacitance of the nodes in these cells, but also on the drive strengths of the transistors that try to maintain the logic state. In this case, the critical charge varies with the width of the transient current pulse induced by a particle hit.

Several measures can prevent or limit the occurrence of soft errors:

- Careful selection of purified materials (i.e., package, solder, and chip manufacture) with low α-emission rates
- Usage of a shielding layer, most commonly polyimide; this layer must be sufficiently thick (~ 20 μm) in order to achieve about three orders of magnitude reduction of the soft-error rate (SER) caused by α-particles. This does not help against cosmic particles.

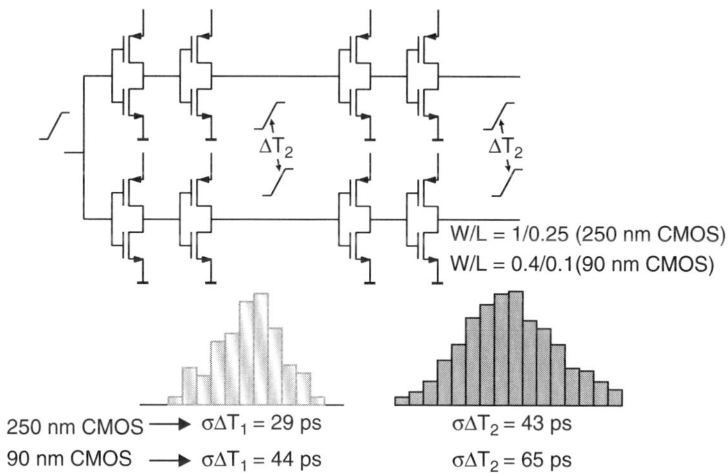

FIGURE 17.10 Spread in signal arrival times due to transistor mismatch.

- SER hardening of the circuits by changing memory cells, latches, and flip-flops
- Usage of process options or alternative technologies, such as SOI, to reduce the volume in which charges are generated along a particle track
- Inclusion of error-detection/correction circuits or making the designs fault tolerant

Currently, much effort is being put into the evaluation and prevention of soft-errors, particularly in systems containing large amounts of densely packed memories.

17.2.6 Transistor Matching

Matching of transistors means the extent to which two identical transistors (i.e., identical in type, size, and layout topology) show equal device parameters, such as β and V_t. Particularly in analog circuits (a memory is also an analog circuit) where transistor pairs are required to have a very high level of matching [6], the spread in V_t due to the doping statistics in the channel of the MOS transistors results in inaccurate or even anomalous circuit behavior. For minimum transistor sizes (area), this effect increases every new IC process generation, such that both the scaling of the physical size and the operating voltage of analog CMOS circuits lag one or two generations behind the digital CMOS circuits. In addition, for digital CMOS circuits (logic), matching of transistors is becoming an important issue, resulting in different propagation delays of identical logic circuits. Figure 17.10 depicts two identical inverter chains (e.g., in a clock tree), but due to the V_t spread, they show different arrival times of the signals at their output nodes. For circuits in a 90-nm CMOS technology, this time difference is in the order of several gate delays. Particularly for high-speed circuits, for which timing is a critical issue, transistor matching and its modeling is of extreme importance to maintain design robustness at a sufficiently high level.

17.2.7 Statistical Timing Analysis

In the preceding subsection, the influence of device parameter spread with respect to transistor matching is discussed; however, process-induced parameter spread in both the device and interconnect structures are also increasingly challenging chip-level timing behavior and analysis. Transistors vary in relation to oxides, doping, V_t, width and length. Interconnects vary in relation to track width, spacing, and thickness and dielectric thickness. So far, this spread was included in simulators in the so-called worst-case, nominal, and best-case parameter sets to provide sufficient design margins. For example, in worst-case timing analysis it is assumed that the worst-case path delay equals the sum of the worst-case delays of all individual logic gates from which it is built. This produces pessimistic results, incorrect critical paths and over-design. Static timing analysis is a means to optimize and estimate timing across the chip. Current static timing analysis tools use the previously mentioned deterministic values for gate and wire delays, which are appropriate for inter-die parameter variations, but does not account for in-die variations. Particularly these in-die variations show significant impact on the overall timing behavior. Delay faults caused by noise sources (e.g., cross talk and supply noise) are also unpredictable with respect to the induced delay. Statistical timing analysis is therefore needed in order to cope with these local variations, which cause random gate and wire delays.

An objective of statistical timing analysis is to find the probability density function of the signal arrival times at internal nodes and primary outputs. Traditionally statistical timing analysis has suffered from extreme run times. Related research is therefore focused to reduce run times [7,8]. Statistical timing analysis is just taking off. For the 90-nm technology node and below, statistical timing analysis is considered necessary, particularly for the complex and higher performance categories of ICs.

17.2.8 Signal Integrity Summary and Trends

From the previous subsections, it can be seen that all noise components increase because of scaling and integrating more devices onto the same die area. At the same time that noise levels in digital CMOS ICs increase with scaling, the noise margins reduce due to reducing supply voltages (Figure 17.11). Because they deal with large current peaks, high-performance ICs, such as the PowerPC (IBM, Motorola), the Pentium (Intel), and the α-chip (DEC/Compaq/HP), have faced signal-integrity problems already in the early 1990s. The average application-specific integrated circuit (ASIC), however, consumes more than a factor of ten less power (and current) and therefore faces these problems a couple of technology generations later in time.

When a certain noise level has reached a limit, a design or technology measure is required to reduce the noise level.

Examples of technology measures are:

- The use of copper instead of aluminum allows reduction of the metal height, thereby reducing the cross talk (see Section 17.2).
- The use of low-*k* dielectrics in the back-end of the technology has the same effect.

Examples of design measures are:

- The increase of space between long signal lines (buses) also reduces the cross talk.
- The use of on-chip decoupling capacitors reduces supply, ground, and substrate bounce.

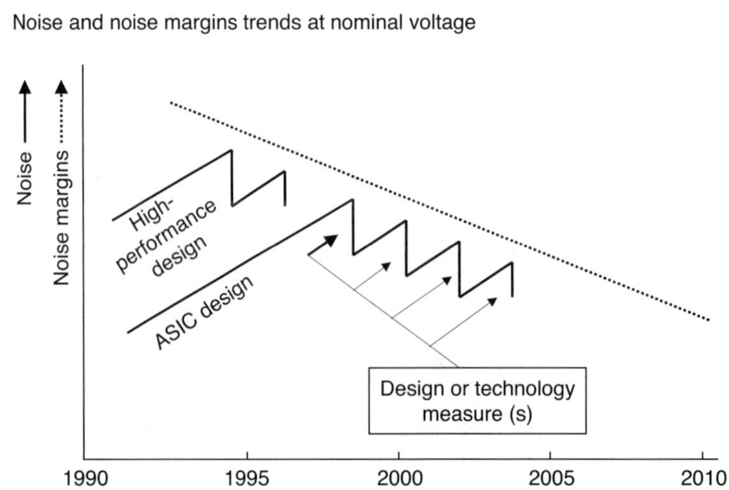

FIGURE 17.11 Noise and noise margin trends over the past and current decade.

Whatever technology or design measure is taken, it only fulfills the requirements in that technology node. The next technology node offers twice the number of transistors, which can intrinsically switch about a factor of 1.4 times faster. This results in a huge increase in the noise levels. In addition, the noise margin has reduced. Therefore, in every new technology node, it becomes more difficult to limit the noise within shrinking boundaries. In other words, the line (in Figure 17.11) that represents the increasing noise must be bent in the direction of the line that represents the reducing noise margins. This can only be obtained by applying more and more design and/or technology measures. In example: in today's ASIC designs, the decoupling capacitors occupy between 5 to 15% of the total area within a standard-cell block. It is expected that this number will have increased to almost 50% by the end of this decade, which means that, by that time, 50% of all transistor equivalents on a chip is needed to support the other 50% in their functional and storage operations. This is yet another factor that adds up to the already rocketing semiconductor development costs.

Another increasingly important topic is the relation between signal integrity and test. Because noise has the tendency to increase, while noise margins reduce (again Figure 17.11), there is not much room left for a reliable operation of an IC. Different operating vectors introduce different local and global switching activities. In many complex ICs, the operation and switching activity during testing are different from the operation and switching activity in the application. As a result, the noise, generated during a test, is different from the noise generated in the application. Because of the reducing noise margins, this increasing discrepancy between "test noise" and "application noise" cause devices that were found correct during testing to operate incorrectly in the application. This is because, in many cases, scan tests are performed to verify the IC's functional operation. These tests are mostly performed locally and in many cases at different frequencies causing a lower overall switching activity and less noise than in the application. On the other hand, depending on the design, different scan chain tests may run in parallel, synchronous and at the same frequency, causing much more simultaneous switching and noise than in the application. These ICs may be found to operate incorrectly during testing while showing correct functional behavior in the application.

Because the voltages continue to decrease, this trend is expected to continue, at least until the end of this decade. Provisions should therefore be taken in the designs, such that, during test, inactive IP cores should run dummy operations in order to emulate application activity. This poses additional challenges to the design, increases its complexity, and adds up to the total development costs.

17.3 Reliability

The continuous scaling of both the devices and interconnect has severe consequences for a reliable operation of an IC. Reliability topics, such as electro-migration, hot-carrier effects, NBTI, latch-up, and ESD are all influenced by a combination of physical and electrical parameters: materials, sizes, dope, temperature, electrical field, and current density. Improving reliability therefore means choosing the right materials, the right sizes and doping levels, as well as preventing excessive electrical fields, temperatures, and currents. This section discusses the effects of scaling on each of the aforementioned reliability issues.

17.3.1 Electromigration

The increase in current density associated with scaling may have detrimental impact not only on circuit performance, but also on the IC's reliability. High currents, flowing through the metal lines, may cause metal atoms to be transported through the interconnection layers due to the exchange of sufficient momentum between electrons and the metal atoms. For this effect, which causes a material to physically migrate, many electrons are required to collide with its atoms. Because of this physical migration of material from a certain location to another location, we get open circuits or voids (Figure 17.12(a)) on locations where the material is removed, and hillocks (Figure 17.12(b)) on locations where material is added. This "electromigration" effect damages the layer and results in the eventual failure of the circuit. Electromigration may therefore dramatically shorten the lifetime of an IC. Preventing excessive current

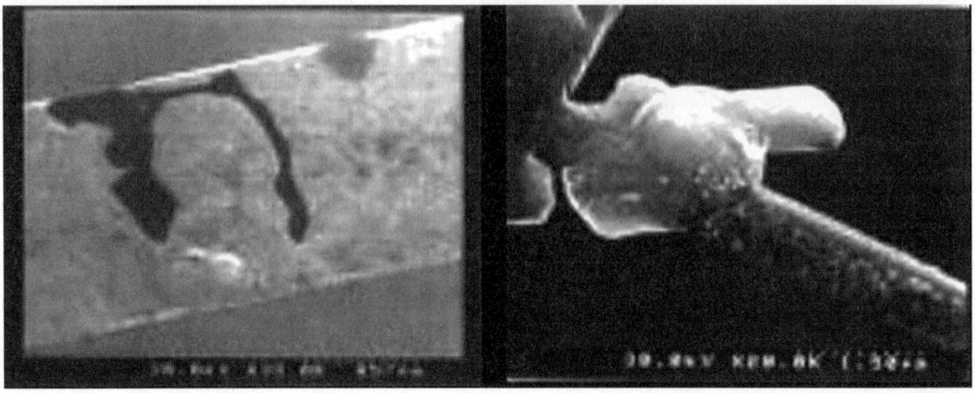

FIGURE 17.12 Electromigration damage in metal interconnect lines: voids (left); hillocks (right). (Courtesy of R. Frankovic, X. Pang, and G.H. Bernstein, University of Notre Dame, Indiana.)

densities eliminates the impact of electromigration. Electromigration design rules are therefore part of every design kit. These rules specify the minimum required metal track width for the respective metal (e.g., aluminum or copper) for a certain desired current flow at given temperatures. Electromigration effects increase with temperature because of the temperature dependence of the diffusion coefficient. This dependency causes a reduction of the maximum allowed current density (J_{max}) at higher temperatures in on-chip interconnect. The required metal width for electromigration roughly doubles for every 20°C to 25°C increase in temperature. Because most IC data sheets indicate a maximum ambient temperature of around 70°C or higher, the real worst-case junction temperature of the silicon itself may exceed 100°C in many applications. Therefore, it is common design practice to use the value for J_{max} at 125°C.

The minimum allowed width W_{em} of a metal wire with height H, to carry a current I, according to this electromigration requirement, is then equal to:

$$W_{em} = I/(J_{max} \cdot H) \tag{17.4}$$

Table 17.2 lists some parameter values, which are characteristic for metal layers in 0.18-μm and 0.12-μm CMOS technologies.

Because most of the currents on an IC flow through the supply lines, it is obvious that these are often implemented in the upper metal layer(s), which usually have a larger height (Table 17.2). Similarly, currents through contact holes and vias must be limited to eliminate electromigration-induced damage of the contact conductor. A typical maximum current density value for a 0.2 × 0.2-μm contact or via in a 0.12-um CMOS technology is around 5 mA/μm². The increase in the aspect ratios of the contacts and vias, in combination with a reduction of maximum currents through them, makes them an incremental part of the overall IC reliability.

TABLE 17.2 Metal Characteristics for 0.12-μm and 0.18-μm Bulk-CMOS Technologies

Technology and metal layer	R_{sheet}	H	J_{max} @ 125°C
0.18 μm CMOS second metal (aluminium)	72 mW/□	550 nm	2.3 mA/μm²
0.18 μm CMOS upper metal (aluminium)	35 mW/□	900 nm	2.3 mA/μm²
0.12 μm CMOS second metal (copper)	85 mW/□	350 nm	3.5 mA/μm²
0.12 μm CMOS upper metal (copper)	26 mW/□	900 nm	3.5 mA/μm²

The continuous scaling of feature sizes and voltages (constant-field scaling) by about a factor of 0.7, every new technology node, did not change the intrinsic power density of most standard-cell designs. Due to the reduction in supply voltage, however, the supply current per unit area of logic increases with about a factor of 1.4 every generation. This puts severe constraints to maintaining electromigration reliability across complex designs.

Due to the expected increase in currents through the metal layers, more Joule heating is expected in these layers. This, in combination with low-*k* dielectrics, which show a higher thermal resistance, made designers start worrying about this so-called "wire self-heating" mechanism; however, the width of a metal wire is not only specified by the appropriate electromigration requirements, but also by the maximum allowed voltage drop across the wire in order to limit speed loss of the connected circuit(s). Suppose an active logic block draws a supply current of 100 mA. When this block is located near the supply pads of the chip, the width of the supply lines is determined only by the electromigration requirement for this 100-mA current. When this block is near the center of the chip, for instance, at a 5-mm distance from the supply pads, the supply lines must be much wider in order to limit the voltage drop across it. Therefore, above a certain distance from the supply pads, the width of the metal (and thus its cooling area) grows with its length, keeping the voltage drop across the line constant. As a result, the resistance of the line (and thus its total I^2R Joule heating) will also be constant. In other words, the maximum wire self-heating occurs in wires with length equal to a cross-over length L_{co}, which is defined to be the length at which the metal-width required by electromigration is identical to the width required by the maximum allowed voltage drop. In Veendrick [9], it is shown that for 0.18-μm and 012-μm bulk-CMOS technologies, wire self-heating in supply lines causes only a limited temperature rise of the wires of just a few degrees. This temperature rise is by far negligible compared to the temperature rise due to the power consumption of the silicon part of the chip. From this result, it can be concluded that wire self-heating in supply lines should not be a real issue in current (and near future) properly designed CMOS VLSI chips.

17.3.2 Hot-Carrier Degradation

When carriers in the MOS transistor channel are given enough energy, these carriers collide with the substrate atoms and generate electron-hole pairs. These, in turn, are also accelerated and may collide with substrate atoms. This so-called impact ionization may cause large substrate currents, device breakdown and/or degradation of the silicon-to-gate oxide interface. Electrons actually collide with the gate oxide. When electrons achieve sufficient energy, they may cross this silicon-to-silicon-dioxide (Si/SiO_2) interface barrier (with a barrier energy of about 3.1 eV for electrons and 3.8 eV for holes) and will then be injected into the gate oxide. Injected carriers lead to the degradation of the Si/SiO_2 interface (i.e., electrically active interface defects are generated), to the generation of defects in the gate oxide film and to charge trapping in the oxide interface (both preexisting and newly generated). Oxide charge trapping and interface state generation induce a shift of the transistor threshold voltage and cause a degradation of the device drive current. This effect is called the hot-carrier effect (HCE) and leads to degraded device performance and reliability problems. Due to the lower mobility of holes with respect to electrons in the transistor channel, impact ionization in p-channel metal-oxide semiconductor field-effect transistors (MOSFETs) is much less. Therefore, the hot-carrier effect is much more severe in n-type MOSFETs.

Theoretically, in a 0.18-μm CMOS technology with a supply voltage of 1.8 V, an electron can only get an energy level of 1.8 eV during its flow through the channel from source to drain. This is less than the previously mentioned barrier energy to create hot electrons. Due to multiple collisions, however, some electrons may collect more energy than the required barrier energy and become "hot." From these considerations, it was generally accepted that when supply voltages are reduced, the chance to generate hot carriers in the transistor channel would reduce as well and the hot-carrier effect was expected to eventually disappear totally.

With the continuous scaling process, critical-dimension (CD) control becomes more difficult leading to transistors with different channel lengths showing different hot-carrier behavior. Shorter channel lengths easier introduce punch-through. Punch-through prevention requires different doping profiles around sources and drains, with increased doping levels. This has some negative effects on the hot-carrier behavior.

When voltages across the transistor are scaled at the same pace as the transistor feature sizes, the electrical fields remain almost constant, and the chance for impact ionization would hardly change. Particularly now, however, with 90-nm and smaller CMOS technologies, the effective channel length is scaling faster than the supply voltage, so that the increase in electrical field may lead to increased impact ionization. Because of this, hot-carrier effects may manifest themselves again more in sub 100-nm technologies than in the last couple of technology nodes, especially in the early development phase due to bad transistor drain engineering. Assuming the transistor is stressed under a worst-case condition (i.e., V_g such that the substrate current is maximal), the hot-carrier lifetime is described by a well-accepted empirical expression (Takeda) as:

$$\tau_{\text{drift}} = A.\ L_{eff}^{\ C}.\ e^{B/Vds} \tag{17.5}$$

where τ_{drift} represents the lifetime (usually at 10% degradation), L_{eff} the effective channel length, and A, B, and C are process-related coefficients. Practical values for B and C, in current technologies, are 60 and 10, respectively. It is clear that the hot-carrier lifetime reduces with decreasing channel length and increasing voltage. Therefore, when we scale the supply voltage with the same factor as the feature sizes, still this lifetime may increase, dependant on the constants A, B, and C.

An additional effect is that for future technologies the silicon dioxide will be replaced by high-k dielectrics. Most of them, however, have a significantly lower barrier [10] and the hot-carrier effects are not just slowly fading away due to reducing supply voltages below the barrier. Results from literature [11,12] stress the importance of continuous monitoring deep submicron technologies for hot carrier degradation, in order to maintain functional reliability at a sufficiently high level.

17.3.3 Negative Bias Temperature Instability (NBTI)

NBTI is a result of a negative bias applied to the gate of a p-channel MOS transistor with respect to the bulk. The mechanism is temperature activated. NBTI results in the degradation of many transistor parameters (drive current, trans-conductance, and threshold voltage), but the threshold voltage appears to be the most degrading one. NBTI was first reported in 1967, but the attention devoted to this mechanism has been escalating over the last couple of years, due to the introduction of gate-oxide nitridation [13] that enhances NBTI and the fact that other oxide wear-out mechanisms, such as HCE and oxide breakdown, were expected to become less severe as the gate oxide scales down. NBTI is strongly process dependent. It has been reported that a higher nitrogen concentration in the oxide [13], boron penetration [14], and plasma processing can enhance NBTI, while fluorine incorporation in the gate dielectric is beneficial against NBTI [15]. The physical nature of the wear-out mechanism induced by NBTI is not fully understood yet. The most accepted models imply positive charge build-up in the oxide-to-bulk and at the Si/SiO_2 interface (donor-like interface states) [16,17].

Whereas hot-carrier injection mostly affects n-channel MOSFETs and depends on the transistor channel length, NBTI mostly affects the pMOS transistor and is only slightly dependent on the transistor geometry, although it has also been reported that in shorter channel devices NBTI can be more severe [18]. Furthermore, the NBTI does not imply a current flow in the transistor channel and can occur at zero drain to source bias. This would mean that NBTI stress could even occur in the standby mode. Design configurations in which matched p-channel MOSFET pairs are subjected to unbalanced stress are reported as most sensitive to NBTI degradation because the threshold voltages of the transistor pair change differently with the stress [19]. Also matched p-channel MOSFET pairs operated symmetrically can lead to reliability fails due to NBTI when the transistors are subjected to different biases in power-down mode. Burn-in can also be a source of NBTI-induced circuit fails, due to the involved high temperature.

Even when an IC is produced in different fabs that run the same process, it may perform differently with respect to NBTI, because not all individual processing steps are identical. NBTI is therefore a technology issue, but critical design configurations, such as matched p-channel MOSFET pairs subjected to unbalanced stress, in either operation or power-down mode, should be avoided. For NBTI there is not a well-accepted model. Assuming a power-law dependence on the stress voltage (field), then the change in *Vt* is proportional to:

$$\Delta V_t = D.\, F_{ox}{}^{m} \tag{17.6}$$

where D is a process dependant parameter, F_{ox} represents the electrical field across the oxide and m a coefficient dependant on the dielectric material and the dielectric thickness [an approximate value is $m \approx 4$].

V_t shifts of 50 mV and more have been reported, so designers need to be convinced to build enough tolerance in their designs. The occurrence of NBTI can be lowered when a device is not subjected to voltage overshoot and/or high temperatures, either from its own heat dissipation or from its application environment. Therefore, reduced power consumption is also beneficial to reduce the chance for NBTI stress.

17.3.4 Latch-Up

The presence of nMOS and pMOS transistors in a CMOS process leads to the creation of parasitic thyristors, as shown in Figure 17.13. In this figure, R_1 and R_2 represent the substrate and n-well resistances, respectively.

Relatively high currents through the bipolar transistors will create relatively high voltages in the substrate and/or n-well. When a sufficiently high positive voltage is present somewhere in the substrate (e.g., at position A), it will switch on the parasitic NPN transistor, or a local voltage (e.g., at position B) within the n-well that is sufficiently lower than the V_{dd} will switch on the parasitic PNP transistor. When both transistors conduct, they are connected into a feed-forward loop, which means that they enhance each other's conduction state, which will finally be latched (maintained) in the thyristor. This state can only be recovered when the supply is completely switched off. This undesirable effect is called "latch-up" and leads to incorrect circuit behavior or even damage. Inductive effects or coupling capacitances may also cause the node connected to the drain to have overshoots and/or undershoots, thus forward biasing the drain substrate junction, which may initiate latch-up. This requires a controlled start up of ICs.

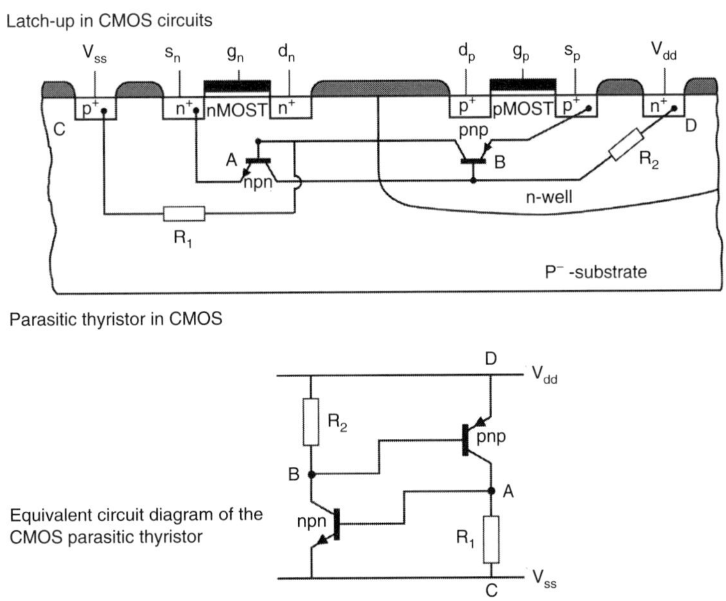

FIGURE 17.13 Parasitic thyristor in CMOS and its equivalent circuit diagram.

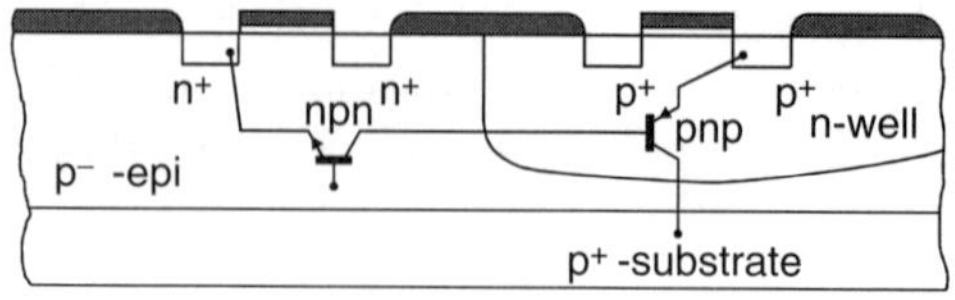

FIGURE 17.14 Cross section of a wafer with a thin p⁻-epi-layer on a thick p⁺-substrate.

Latch-up in CMOS circuits can be avoided by applying the following technological and/or design remedies:

- Minimize the substrate and/or n-well resistances. This can be done in two ways. One is the use of many substrate and n-well contacts in the design, which will reduce the values for *R1* and *R2*, respectively. The parasitic thyristor is then unlikely to turn on. Reducing both resistances by increasing the substrate and n-well doping is not an option because it also changes the threshold voltages and overall transistor behavior. A good alternative is the use of so-called epitaxial wafers (Figure 17.14).
- Epitaxy is a layer of single-crystalline silicon deposited/grown onto a single-crystalline silicon wafer. The crystalline structure of the substrate is reproduced in the growing material. This epitaxial layer, in which the devices are formed and whose thickness is usually between 1 to 5 μm, can be doped, as it is deposited, to the required doping type and concentration (usually with a resistivity of ≈ 10–20ΩCM) while continuing the substrate's crystalline structure. Therefore, we can create a thin p⁻-epitaxial layer on top of a p⁺-substrate. Because the current wafer thickness is between 200 and 700 μm, the p⁺-substrate is relatively thick and has a low resistivity (≈ 0.01–0.05 ΩCM). Such low-ohmic substrates show very low values for R_1. A large part of the PNP collector current will therefore flow through this substrate and only a small part will flow into the base of the NPN transistor. This transistor can no longer be turned on easily and is then largely excluded from the latch circuit. Epitaxial wafers with low-ohmic substrates have been massively used for CMOS products in 0.25-μm and older technologies. Due to decreasing supply voltages and increasing noise levels, the combination of analog and digital circuits onto one single chip has made its design a difficult and cumbersome task. Particularly the substrate noise sensitivity of analog circuits requires a good isolation from the digital noise "generators," which is why a high-ohmic substrate is preferred for mixed analog digital circuits.
- The use of guard rings is another way to make strong (low-ohmic) connections of local substrate and/or n-well areas to V_{ss} and V_{dd}, respectively. Moreover, the distance between n-type and p-type areas is also a matter of concern during the design phase and is particularly of interest in I/O circuits, which are usually supplied by higher voltages. Guard rings are more effective on high-ohmic substrates.
- Apply a back-bias voltage to the substrate. When the p⁻-substrate in Figure 17.13 is connected to a negative voltage instead of to V_{ss}, the base voltage V_A of the NPN transistor will be lowered. Therefore, this transistor can no longer be turned on easily. This technique is more a theoretical option and is not frequently used for latch-up prevention.
- Use SOI technology to completely isolate the nMOS from the pMOS transistors. In this technology, the NPN and PNP transistors are completely isolated from one another and so the connections to create latching thyristor circuits are missing.

The application of one or more of the preceding remedies has increased latch-up immunity to a very high level. The highest chance of occurrence for latch-up is during testing. Standard testing requirements include immunity to 100 mA or more, depending on what the IC can and should withstand from an application point of view. This means that with epi-wafer material, 100 mA can be supplied to the output of an output buffer (driver) even though no output transistor is conducting. This current, then, directly flows into the substrate, thereby raising the substrate voltage and possibly turning the thyristor on (Figure

17.13). In practice, some latch-up tests are done with 150–200 mA at a maximum ambient rated temperature for the device.

In future technologies, the latch-up phenomenon is likely to disappear inside electronic circuits, as the supply voltages will be reduced in every new technology node. At the chip I/Os, however, the requirements on latch-up remain relatively high because many applications still require a higher interface voltage (e.g., 1.8 V, 2.5 V, or 3.3 V). More on latch-up basics can be found in Troutman [20].

17.3.5 Electro-Static Discharge (ESD)

ICs are exposed to many possible sources of damage, both during and after the manufacturing process. The principle cause of damage is ESD, due to the transfer of charge between bodies at different electrical potentials. ESD pulse durations are very short and normally range from 1 to 200 ns, but they may introduce very large power spikes. The high impedance of MOS input circuits makes them particularly vulnerable to physical damage when they are exposed to these spikes. This may result from operations during the fabrication process or from handling (un)packaged dies and bonding. It may also occur during testing and maintenance or in the application. Although only a few devices or connections may be severely damaged, many more may suffer damage that is not immediately apparent. These latent failures will result in customer returns, which is one of the biggest worries of semiconductor vendors. Thus, an ESD is one of the most important factors that determine the reliability of an IC.

The damage caused by an ESD is irreversible. The human body is one of the main sources responsible for ESD. Just by walking on a carpet on a low-humidity day, for instance, a person, wearing shoes with highly insulating soles can build up a voltage in excess of 30,000 V. The resulting charge can then be transferred via an ESD to an electronic circuit during touching. It is also very important that precautions need to be taken to prevent ESD damage during IC fabrication. In addition, protective measures must be included in an IC's design to ensure that it can withstand acceptably large ESD pulses. On-chip MOS protection circuits are used to increase the immunity of an IC to ESD pulses. These circuits are designed to provide input and output circuits with low-impedance shunt paths, which prevent the occurrence of excessive voltages on the chip.

17.3.5.1 ESD Test Models and Procedures

ESD sources are emulated in several different ways. The human-body model is currently the most popular industry model and simulates the direct transfer of electrostatic charge from the human body to a test device. It is internationally accepted as a standard (JEDEC Standard No. 22-A114-B). Figure 17.15 is a human-body test setup. The basic requirement for this model, in combination with the parasitics (L) of the tester interface cables, is to generate ESD pulses with rise times between 10 to 15 ns.

The test is normally done on an ESD tester. This human-body model has not changed much over the last decade. A 100-pF capacitor is charged to the test voltage, and then discharged through a 1.5-KΩ resister across any combination of pins A and B (Table 17.3) of the device under test (DUT). The chip may consist of several supply (V_{dd}) and ground (V_{ss}) domains. Each domain may be supplied by more than one pin. The V_{ss} and V_{dd} in Table 17.3 refer to just one of the respective pins of a supply domain. In other words: each pin is then tested with respect to all grounded V_{ss} and V_{dd} domains and not to all grounded V_{ss} and V_{dd} pins, to save test time. Each signal pin is also tested with respect to all other grounded signal pins. The maximum test voltage ranges from 2 kV to 8 kV and depends on the application area of the chip. Because production environments are well controlled, a maximum voltage of 2 kV is usually

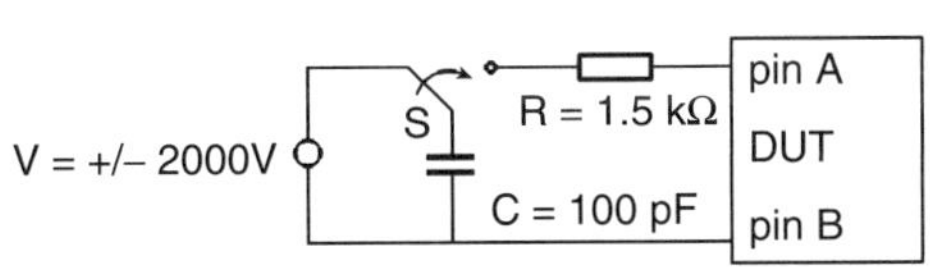

FIGURE 17.15 A typical equivalent circuit based on the human-body model.

TABLE 17.3 Different ESD Test States

State	DUT Pin A	DUT Pin B
1	input	V_{ss}
2	V_{ss}	input
3	input	V_{dd}
4	V_{dd}	input
5	output	V_{ss}
6	V_{ss}	output
7	output	V_{dd}
8	V_{dd}	output
9	input	output
10	output	input
11	V_{ss}	V_{dd}
12	V_{dd}	V_{ss}

required. However, because more and more IC pins can be touched in daily life (i.e., plug-ins such as USB ports, chip cards, SIM cards, memory sticks, and flash cards), the ESD-test requirements tend to increase. The 8-kV requirement is no longer the exception. The devices are classified when meeting a particular sensitivity criterion. A class-2 device, for instance, has passed the 2 kV, but fails after exposure to an ESD pulse of 4 kV (see the previously mentioned standard at http://www.jedec.org).

Generally, three to five positive and negative pulses are applied at 300ms intervals in all test states. Stressed pins are tested after application of each ESD pulse series. If no failure is observed for a sequence through the pins, then the ESD voltage level is increased by 100V and the sequence is repeated. The process continues until a failure occurs or the required maximum voltage is reached. The ESD is complete when a failure is observed or when all pins on the DUT have been stressed as described. Generally, the following (example) criteria may be used to determine failure:

- Incorrect functional operation or a violation of the device specifications
- A change of more than 5% in the forward voltage drop and breakdown voltage in the diode characteristic
- An increase of more than 10% in the I_{ddq} leakage current

Another standardized and popular ESD test model is the machine model, which emulates the rapid direct transfer of electrostatic charge, from a charged conductive object (tool or equipment) to a test device. Compared with the human-body model of Figure 17.15, the machine model specifies a discharge of a 200-pF capacitor through a 0.75-μH inductor. Due to the absence of the current limiting resistor, this model is considered more severe, and tests are run at lower voltages. The charged-device model is an alternative ESD test set up, which is most commonly used to emulate rapid electrostatic charge transfer during packaging and assembly. More details on the latter two models can be found in http://www.esd-lab.com/others.htm [21] or directly from the JEDEC Web site: http//:www.jedec.org.

17.3.5.2 On-Chip ESD Protection Circuits

Although much ESD and ESD-protection knowledge has been built over the last couple of decades, the design of on-chip ESD protection circuits is both scientific and experimental. This is because in every new semiconductor node, device architectures and feature sizes (e.g., width, spacing, and oxide thickness) have changed with respect to the previous node, which requires new protection solutions. Usually, several alternative protection circuits are explored in each new technology node and often semiconductor process development goes hand in hand with ESD protection development.

The purpose of a protection circuit is that it provides a low-ohmic shunt path in parallel with the MOS input and output transistors during the occurrence of an ESD pulse. MOS input protection circuits usually comprise a voltage spike filter and diode clamps. Because MOS inputs are connected to high-ohmic transistor gates, the protection of input circuits is more critical than that of output circuits. Output pads are connected to drain areas. Usually these drain areas are relatively large, because outputs usually have to drive large capacitance (10–50 pF) and the complementary drain junctions act as intrinsically available diode clamps. Of course, also the outputs must fulfill certain ESD design rules.

The behavior of MOS protection circuits depends very much on their size and layout and on various process parameters. Each manufacturing process has its own specific design rules for ESD protection circuits. Therefore, the design of such circuits should be done in cooperation with specialists in the field of protection devices.

Future technologies, particularly those for high-performance designs, require different substrates such as SOI or silicon germanium (SiGe). SOI technologies need a different approach for the development of ESD protection devices because their devices are built on an isolating substrate. The implementation of ESD protection diodes on SOI needs to change from the high-perimeter bulk CMOS diodes to an SOI lateral-gated diode structure. SiGe technology has become another important alternative for high-speed communications and wireless applications. Because the change in material and mobility will also influence ESD, developing an ESD strategy for SiGe circuits will be very challenging. More about ESD can be found in Ameraskera and Duvvury [22].

17.3.6 Charge Injection during the Fabrication Process

Many IC processing steps use plasma or sputter-etching techniques, in which charge particles are collected on conducting surface materials (e.g., polysilicon, metals). This, so-called antenna effect can create significant electrical fields across the thin gate oxides which can be stressed to such an extent that the transistor's reliability can no longer be guaranteed. It can also cause a threshold-voltage shift, which affects the matching behavior of transistor pairs in analog functions. It is industry practice to introduce additional "antenna design rules" to limit the ratio of antenna area to gate-oxide area. The back-end design tools can handle these design rules by limiting the maximum wire (antenna) length in the different metal layers. In addition, protection diodes are used in the library cells to shunt the transistor gates. Due to the trend in gate-oxide thickness scaling, the appearance of the antenna effect is expected to increase. The use of high-k gate dielectrics in building the transistor stack would therefore also be beneficial to reduce this antenna effect.

17.3.7 Reliability Summary and Trends

Most of the previously discussed reliability topics depend on size, doping profiles and levels, voltages, temperatures, and device materials. Scaling requires a change in many of these parameters and will therefore have dramatic effects on the reliability of CMOS devices and circuits. Moreover, in technologies with channel lengths below 45 nm, the transistors are expected to be built from a completely different stack of materials as compared with today's high-volume products. SOI and/or SiGe will probably replace the bulk-silicon substrate; due to the high leakage current, the SiO_2 gate oxide is expected to be replaced by a high-k dielectric and, because of gate depletion, a metal gate may replace the polysilicon gate. This has an additional impact on the reliability of the devices and vice versa. Maintaining reliability at a sufficiently high level will put severe demands on this new transistor stack and makes the choice for the right materials a very difficult and cumbersome one.

17.4 Conclusion

For many previous technology generations the supply voltage has been constant and equal to 5V. The scaling process over that period was called constant-voltage scaling. Over the last decade, the advances in CMOS technology were not just related to scaling of the devices and the minimum features sizes, but also of the supply voltages. This is called constant-field scaling. If a chip is fabricated in a certain technology, then it should be operated at the nominal supply voltage for which that technology and libraries are developed. Certain applications, particularly those driven by low-power requirements, need further reduced supply voltages. This can be seen as maintaining the size and reducing only the supply

TABLE 17.4 Different Scaling Scenarios

	Topic	Relation	Scaling Factor ($s \approx 0.7$)			
			$p \neq s \neq q$	p=1	p=s	s=q=1
Basic parameters	Voltages	V	p	1	s	p
		V_t	q	q	q	1
	Feature sizes	W, L, T_{is}, dist, t_{ox}(EOT)	s	s	s	1
	Devices per unit area	$\div 1/A$	$1/s^2$	$1/s^2$	$1/s^2$	1
	Transistor bias current	i	p	1	s	p
	Average current/unit area	I	p/s^2	$1/s^2$	1/s	p
	Capacitance	$C = \varepsilon_0 \varepsilon_r A/t_{is}$	s	s	s	1
	Metal resist. (top metals)	$R = \rho\ell/(t_m W)$ ($t_m \approx$const)	1	1	1	1
Performance	Gate delay τ, ($\div 1/f$)	CV/i	s	s	s	1
	Power dissipation/gate, D	CV^2f	p^2	1	s^2	p^2
	Power-delay product, τD	$\div CF^2$	$p^2 s$	s	s^3	p^2
	Power density, P	CV^2f/A	p^2/s^2	$1/s^2$	1	p^2
	Subthr. leakage current	Espon. with V_t and V	$s^{-1}12^{10(1-q)Vt+0.1(p-1)V}$	$s^{-1}12^{10(1-q)Vt}$	$s^{-1}12^{10(1-q)Vt+0.1(s-1)V}$	$12^{0.1(p-1)V}$
	Gate leakage current	Expon. with V and t_{ox}	$s^2 \cdot 10^{5(1-s)t_{ox}+2\log p}$	$s^2 \cdot 10^{5(1-s)t_{ox}}$	$s^2 \cdot 10^{5(1-s)t_{ox}+2\log s}$	$10^{2\log p}$
Reliability	Electromigr. (curr. dens.)	I = P/V	p/s^2	$1/s^2$	1/s	p
	Latch-up (for Vdd >> 1V)	$\div$V/dist	p/s	1/s	~1	p
	ESD susceptibility	$\div 1/t_{ox}$	1/s	1/s	1/s	1
	Hot-carrier lifetime	f(V,V/distance)	$s^C \cdot e^{B/V(1/p-1)}$	—	—	$e^{B/V(1/p-1)}$
	NBTI V_t-shift	$\delta Vt \div f(V,L,t_{ox})$	$(p/s)^m$	—	—	p^m
Signal integrity	Cross-talk/unit length	$\div$1/dist.	1/s	1/s	1/s	1
	Induct. noise/unit area	(di/dt)/A	p/s^3	$1/s^3$	$1/s^2$	p
	Voltage drop/unit length	IR/ℓ	p/s^3	$1/s^3$	$1/s^2$	p
	Soft-error rate ($\div$1/Q)	Q = CV	~1/ps	1/s	$1/s^2$	1/p
	Noise margin	V_{dd} and V_t	p and q	1 and q	s and q	p and 1
			With velocity saturation	↑ constant-voltage scaling	↑ constant-field scaling	↑ constant-size 'scaling'

voltage. This scaling scenario will be referred to as constant-size scaling. It is obvious that these different scaling scenarios have a different impact on the basic transistor parameters and on the performance and robustness of CMOS ICs. Table 17.4 shows how the transistor performance, reliability and signal integrity parameters depend on the scaling factor *s* for the sizes (an average value for $s \approx 0.7$ between successive technology generations) and *p* for the voltages and the impact of the different scaling scenarios, when we continue the scaling process as we did for more than four decades now. It means that no dramatic design and technology measures/changes have been taken into account.

The first scaling column ($p \neq s$) demonstrates how a parameter scales, when the voltages scale with a different factor than the sizes. In the constant-voltage scaling column ($p=1$), only the sizes scale, while the voltages are kept constant. In the constant-field-scaling column ($p=s$), both the sizes and the voltages scale with the same factor. Finally, in the constant-size scaling column ($s=1$), only the supply voltage scales, while keeping the sizes (= technology) constant.

In the table, the carrier mobility degradation due to velocity saturation is taken into account. It means that it is assumed that the transistor bias current (i) has a linear instead of the quadratic relation with the voltage. In understanding the table, a few more assumptions need to be explained. First, the thickness of the upper metal layers, which are commonly used for supply lines is assumed to stay almost constant and does not scale with s. This is because this thickness has hardly been scaled over many technology generations and is not expected to scale much further, because of electromigration requirements. However, metal layers 2 to 5 (or 8, depending on the technology node and metal layer options), which are used for routing, are assumed to almost scale with *s*. This is done, in combination with a slowly decreasing dielectric constant, to reduce the mutual wire capacitance. This not only reduces crosstalk, but it also helps to reduce the active power consumption per gate. It is also assumed, that the size of the cores (standard-cell blocks) remain almost constant as well. Therefore, the total capacitance per logic gate, which is defined by the fan-in capacitance of the connected logic gates, and the capacitance of the metal interconnections, is also assumed to scale with *s*.

The expression for the scaling of the subthreshold leakage current (per transistor) is based on the subthreshold slope, which is assumed to be 80mV/decade for bulk CMOS. This means that the subthreshold leakage current increases with about a factor of 12 for every 100mV reduction in the threshold voltage. The relation with the voltage scaling factor *p* originates from the drain-induced barrier lowering (DIBL) effect on the Vt, which is assumed to have a linear relation with the change in Vt (ΔVt = -γVds = -0.1V, where γ represents an emperically determined constant, which is assumed to stay close to 0.1).

For the gate leakage current scaling expression (also per transistor), it is assumed that it increases by a factor of 10 for every 0.2nm reduction of the gate-oxide thickness and it also increases by a factor of 10 for every doubling of the supply voltage.

For several parameters, the relation with the scaling factors is not completely clear.

Particularly the expressions for hot-carrier lifetime and NBTI depend heavily on the technology node, by the values used for B, C, and m.

Since these values only hold for one technology node, the expressions cannot be used to reflect the scaling trends and are therefore not included in the constant-voltage and constant-field scaling columns. In 45nm CMOS technology the device architecture is expected to completely change [e.g. (double) metal gate, high-k dielectric, strained silicon, etc.], which may dramatically affect NBTI as well as most of the other reliability parameters.

Latch-up depends on both voltage and size scaling. When only the supply voltage is reduced, the chance for latch-up is also reduced. If only the sizes shrink, however, the latch-up is expected to increase due to the smaller n^+ to p^+ spacings. Therefore, in the constant-field scaling column, a scaling factor of $\approx$1 is assumed for latch-up.

When going from the 90nm CMOS technology node to the 65nm and 45nm nodes, the voltages, particularly for low-leakage applications, no longer, or only hardly scale with the feature sizes. This means that, for these applications, we are slowly moving back from the constant-field scaling column towards the constant-voltage scaling column in Table 17.4. This has severe consequences, particularly for the active power consumption, the reliability and signal integrity topics. Parameters, such as the power and current density, as well as the inductive noise and the voltage drop all increase dramatically and will have severe design, package and application consequences. The drive for low leakage in standby operation then becomes a real burden to limit the power consumption during active operation.

Creative solutions, both in technology and design, are needed to keep the IC's robustness at a sufficiently high level in order to extend Moore's law for yet another decade; however, this will lead to a major increase of the complexity and total development and production costs of an IC. It is the author's opinion that, for many applications, the 32nm CMOS technology node, plus or minus one generation, is expected to be the last economically viable one.

17.5 Acknowledgment

The author thanks Dr. Andrea Scarpa for reviewing the hot-carrier and NBTI sections, Dr. Theo Smedes for the ESD and latch-up sections, Dr. Yuang Li for the section on electromigration, and, finally, Dr. Dick Klaassen and Dr. Ronald van Langevelde for discussions on the scaling table.

References

[1] H.J.M.Veendrick, *Deep-Submicron CMOS ICS: FROM Basics to ASICs, 2nd ed.*, Dec. 2000, Kluwer Academic Publishers, Dordrecht, 2000.

[2] Y.I. Ismail, On-Chip Inductance Cons and Pros, *IEEE Trans. on VLSI Syst.*, vol. 10, no. 6, Dec. 2002.

[3] B. Nauta and G. Hoogzaad, How to deal with substrate noise in analog CMOS circuits, *European Conf. on Circuit Theory and Design*, Budapest, September 1997.

[4] H.B. Bakoglu, *Circuits, Interconnections, and Packaging for VLSI,* Addison-Wesley, Reading, MA, 1990.

[5] E. Dupont et al., Embedded Robustness IPs for transient-error-free ICs, *IEEE Design Test of Comput.*, vol. 19, no. 3, pp. 56–70, May/June 2002.

[6] M. Vertregt, Embedded analog technology, *IEDM Short Course on System-on-a-Chip Technology,* Dec. 5, 1999.

[7] A. Agarwal et al., Statistical timing analysis using bounds, *DATE,* March 2003.

[8] J.-J. Liou et al., Fast statistical timing analysis by probabilistic event propagation *DAC 2001*, June 2001, Las Vegas, NV.

[9] H.J.M. Veendrick, Wire self-heating in supply lines on bulk-CMOS ICs, *ESSCIRC 2002 Dig. of Tech. Papers,* pp. 199–202, Sept. 2002.

[10] G.D. Wilk et al., High-k dielectrics: current status and materials properties considerations, *J. Applied Physics,* vol. 89, no. 10, pp. 5243–5275, May 15, 2001.

[11] A. Kottantharayil, Low-voltage hot-carrier issues in deep-sub-micron MOSFETs, Thesis, Universitat der Bundeswehr (München), 2001, http://137.193.200.177/ediss/kottantharayil-anil/inhalt.pdf, June 21, 2004.

[12] S. Mahaptra et al., Device scaling effects on hot-carrier induced interface and oxide-trapped charge distributions in MOSFETs, *IEEE Trans. on Electron. Devices,* vol. 47, no. 4, April 2000.

[13] K. Kushida-Abdelghafar et al., *Appl. Physics Lett.*, vol. 81, no. 23, 2002.

[14] Y. Hiruta et al., Interface state generation under long-term positive-bias temperature stress for a p^+ poly gate MOS structure, *IEEE TED 36,* p. 1732, 1989.

[15] T. B. Hook et al., The effect of fluorine on parametric and reliability in a 0.18-μm 3.5/6.8-nm dual gate oxide CMOS technology, *IEEE TED,* vol. 48, no. 7, p. 1346, 2001.

[16] C.E. Blat et al., Mechanism of negative-bias-temperature instability, *Journal of Applied Physics (JAP),* vol. 69, no. 3, 1991.

[17] Ogawa et al., Interface-trap generartion at ultrathin (4–6 nm) interfaces during negative-bias temperature aging, *JAP 77,* 3, 1995.

[18] A. Scarpa et al., Effect of the process flow on negative-bias-temperature-instability, *Proc. 8th Int. Symp. on Process- and Plasma-Induced Damage,* p. 142, 2003.

[19] P. Chaparala et al. NBTI in dual gate oxide PMOSFETs, *Proc. 8th Int. Symp. on Process- and Plasma-Induced Damage,* p. 138, 2003.

[20] R.R. Troutman, *Latchup in CMOS Technology,* Kluwer Academic Publishers, Dordrecht, 1986.

[21] http://www.esdlab.com/others.htm, June 21, 2004.

[22] A. Ameraskera and C. Duvvury, *ESD in Silicon Integrated Circuits,* John Wiley & Sons, New York, 2002.

Part III

CAD Tools for Low-Power

18

High-Level Power Estimation and Analysis

Wolfgang Nebel
Oldenburg University

Domenik Helms
OFFIS

18.1 Introduction

"Big things always start small." This wisdom also applies to microelectronic design; and it is at the beginning, when the complexity is still small and can well be understood under different aspects, that the important decisions are made, which will lead to success or failure. Once a design has been developed to a large structure of logic and wires, it is difficult to cure problems, which, in many cases, also started small and eventually became large, hard to solve, and without major design respins, these problems may cost months of design time, major engineering resources, and can be responsible for missed marketing opportunities.

This chapter covers the area of early system-level power analysis and algorithmic-level power estimation. The techniques presented here shall enable the reader to understand the underlying concepts as well as the chances and limitations of tools, which shall guide the designers in optimizing the global system architecture for low power and help them selecting and further optimizing the algorithms to be implemented at lower levels. The figure of merit in reducing the power consumption by making the right decisions during this early phase covers several orders of magnitude. Just to illustrate the potential: there exist dozens of known and well-understood sorting algorithms. They all perform exactly the same task: take a set of objects and put them in an order according to the chosen sorting criterion. Despite the exactly same functional behavior, however, they all perform differently with respect to the computation time, memory usage, and the power consumption. Similarly, different algorithms with equivalent functionality are known for Fourier transform, compression, and many other functions, which are copiously used in mobile multimedia applications. Selecting the most power efficient one can be a product-differentiating factor.

TABLE 18.1 Breakdown of Power Consumption

Technology	Switched Cap. Power	%	Short-Circuit Power	%	Leakage Power	%
150 nm	439 nW	71.1	173 nW	28.0	5.6 nW	0.9
130 nm	317 nW	71.8	118 nW	26.7	6.7 nW	1.5
100 nm	236 nW	73.5	75 nW	23.4	10 nW	3.1
90 nm	183 nW	70.1	67 nW	25.7	11 nW	4.2
70 nm	139 nW	56.3	55 nW	22.3	53 nW	21.4
45 nm	74 nW	36.8	30 nW	14.9	97 nW	48.3
32 nm	51 nW	23.3	28 nW	12.8	140 nW	63.9
22 nm	20 nW	13.9	14 nW	9.7	110 nW	76.4

18.1.1 Analysis vs. Estimation

Although the terms estimation and analysis are frequently used in the low-power community without careful distinction, we would like to clarify the terminology here. Analysis is based on an existing design at any level (i.e., the structure is given, typically in terms of a netlist of components). These modules are predesigned and for each one a power model exists. These power models can be evaluated based on the activation of the modules. Thus, power analysis is the task of evaluating the power consumption of an existing design at any level. It is used to verify that a design meets its power and reliability constraints (e.g., no electromigration occurs, no hot spots will burn the device, and no voltage drops will cause spurious timing violations). Power analysis finally helps to select the most cost efficient chip package.

In contrast, estimation builds on incomplete information about the structure of the design or part of the design under consideration. The design does not yet exist and can only be generated based on assumptions about the later physical implementation of the design, its modules, its interconnect structure, and physical layout. In summary, estimation requires design prediction followed by analysis; for instance, if the floorplan of a design is not yet available, interconnect power estimation first requires a floorplan prediction. Power estimation is applied to assess the impact of design decisions and compare different design alternatives on incomplete design data. It allows efficient exploration of the design space without the need for a detailed implementation of all different design options.

18.1.2 Sources of Power Consumption

This section briefly revisits the physical basics of power consumption, which is also the basis for high-level power analysis. Table 18.1 lists a breakdown of the estimated power consumption of a single transistor for high performance logic. The data are our own calculations based on the 2002 update of the International Technology Roadmap for Semiconductors [1]. Our assumptions include a 1% expected switching activity compared with the maximum transistor operation frequency, which is typical for processing components.

These data clearly demonstrate that today, for high performance applications, the switched capacitance power consumption is still dominating. Considering the fact that the short-circuit power is typically captured as part of the power models for dynamic power, and further, that the data in Table 18.1 do not include dynamic power related to interconnect capacitances, we can safely assume that the power consumption of computation intensive devices at 70 nm and larger technology nodes will be dominated by the dynamic power consumption. For mobile applications, language power is an important source of power consumption already at 90 nm.

18.1.2.1 Switched Capacity Power

Equation (18.1) allows for the calculation of the power consumption of a switched capacitor. At the transistor-level C_{load} includes the parasitic gate overlap and fringing capacitances as well as the Miller capacity. α models the switching probability of the transistor during a cycle of the clock toggling at frequency f. V_{dd} is the supply voltage. We will demonstrate later that applying this formula at higher levels of abstraction will require a revised interpretation of some of these parameters.

$$P_{swcap} = \frac{1}{2} C_{load} \cdot \alpha \cdot V_{dd}^2 \cdot f \tag{18.1}$$

Equation (18.1) holds for unnecessary transitions (glitches), while it needs refinement for modeling a sequence of n incomplete transitions within a period of T and with a voltage swing of ΔV_n (Equation (18.2)).

$$P_{incompleteswcap} = \frac{1}{2T} \cdot C_{load} \cdot V_{dd} \cdot \sum_{i=1}^{n} \Delta V_n \tag{18.2}$$

18.1.2.2 Short-Circuit Power

Short-circuit power is the second part of the dynamic power consumption. It occurs when during a short period both the pull-up and the pull-down networks of static CMOS-gates are conducting. Equation (18.3) gives a simple model of the short-circuit power with β modeling the transistors' conductivity per voltage factoring the linear region, T is the inputs' rise/fall time, and τ is the gate delay.

$$P_{shortcircuit} = \frac{\beta}{12} \left(V_{dd} - 2V_{th} \right)^3 \frac{\tau}{T} \tag{18.3}$$

Equation (18.3) is an overestimation by up to a factor of three. For an accurate analysis, transistor level models and transient analyses are needed [2]; however, $P_{shortcircuit}$ within modules can be captured as part of the dynamic power models of the modules.

18.1.2.3 Leakage

The leakage power consumption is mostly due to leakage currents flowing through the channel in weak inversion even when the gate-source voltage is below threshold and due to carriers tunneling through the gate oxide. The leakage power depends on the state of the circuit. Analysis, modeling, and optimization of the leakage power are currently subject of intensive research. It is covered in depth in Chapter 3 of this book.

18.2 Generic Design Flow for Low-Power Applications

This section introduces the common principles of power analysis at any level before we present the details of system-level power analysis and algorithmic-level power estimation in the following sections.

18.2.1 Generic Power Estimation and Analysis Flow

Generally, any analysis tool for the dynamic power consumption needs to evaluate Equation (18.1) and Equation (18.2). This can be done at different degrees of abstraction. For instance, for high-level power analysis, the entire dynamic power consumption of a module will be described by a single power model instead of by all individual capacitances inside the module. Similarly, the switching probability of all capacitances is lumped together into an activity model for the module. Figure 18.1 depicts a generic power estimation and analysis flow that can be applied at any level of abstraction.

The upper part of Figure 18.1 needs to be applied when a preimplementation power estimate is needed. In this case, the architecture of the design is unknown yet. The set of components to be allocated to implement the device is still to be determined. Consequently, neither their interconnection and communication structure nor their activation patterns are defined. Thus, an evaluation of Equation (18.1) and Equation (18.2) is not possible yet, even if higher-level capacitance and activation models were applied.

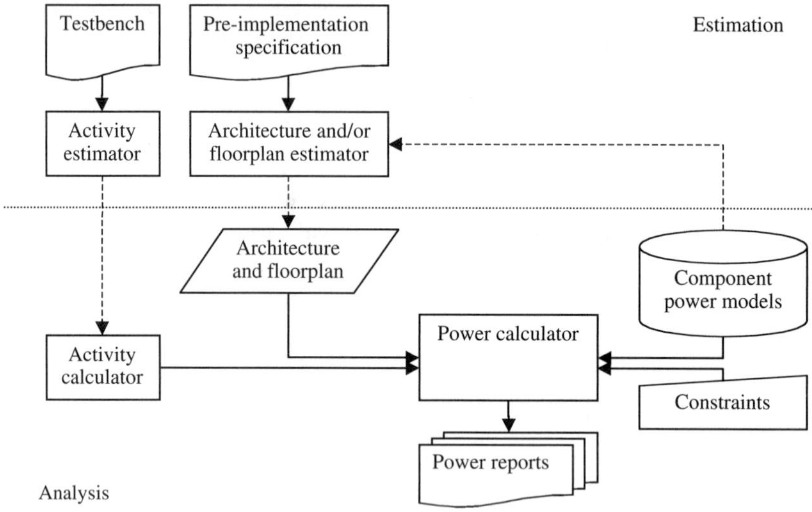

FIGURE 18.1 Generic power estimation and analysis tool flow.

It is the task of the architecture and the floorplan estimators to predict the component allocations, their physical floorplan, which is needed for the estimation of the interconnect and clock tree, and the scheduling of the operations, which again is needed to predict the activation of the components. These estimation techniques will be discussed in more detail in Section 18.4.

Once this is done, the kind of information is adequate for a power analysis of this predicted architecture. The relevance of the results of the following power analysis step, however, strongly depends on the quality of the predicted architecture.

Let us discuss two application scenarios for high-level power estimation. First, we are interested at the first possible instance to get an estimate of the to-be-expected final power consumption of a system (e.g., to validate the feasibility of a certain package). In this case, a reasonable absolute accuracy is needed. A reliable power estimate for this case requires that the predicted architecture is very similar to the final architecture once the system has been fully designed and optimized through all levels of abstraction. An architecture estimator for this application has to take into account the specific design styles, circuit technologies, design skills, and the tool flow applied to generate a sufficiently accurate architecture prediction.

In the second use case, we are interested in a fast comparison of different design options. We want to know which out of several paths through the design space to follow. Thus, we are looking for relative power figures for each of the options we have in mind. In this case, it is important, that the solution, which had been estimated to be the most power efficient one, really proves to be the least power consuming one. For this scenario, the predicted architectures for each of the options should be similar in their power efficiency, even if they are not exactly identical to the final implementation. The predicted power figure of the different solutions may differ to some extent from the final power figure after implementation as long as the order between them is maintained.

The lower part of Figure 18.1 depicts the generic power analysis flow. The power calculator collects the parameters of Equation (18.1) at the respective level of abstraction. The floorplan and the architecture of modules, each having a power model attached, determine the physical and structural architecture and represent the load capacitance C_{load} of Equation (18.1) — maybe at an abstract level. The activity calculator produces an activation profile for each of the components modeling the switching probability α of Equation (18.1). Finally, the supply voltage V_{dd} and the clock frequency f are part of the constraints provided by the designer.

As we can easily see, power analysis at any level requires three main input models: The architecture model and its component models as well as an activation model.

18.2.2 Low-Power Design Flow

This section exemplifies the generic power estimation and analysis flow of the previous section for each of the highest levels of abstraction. Figure 18.2 is a generic power conscious design flow, which, however,

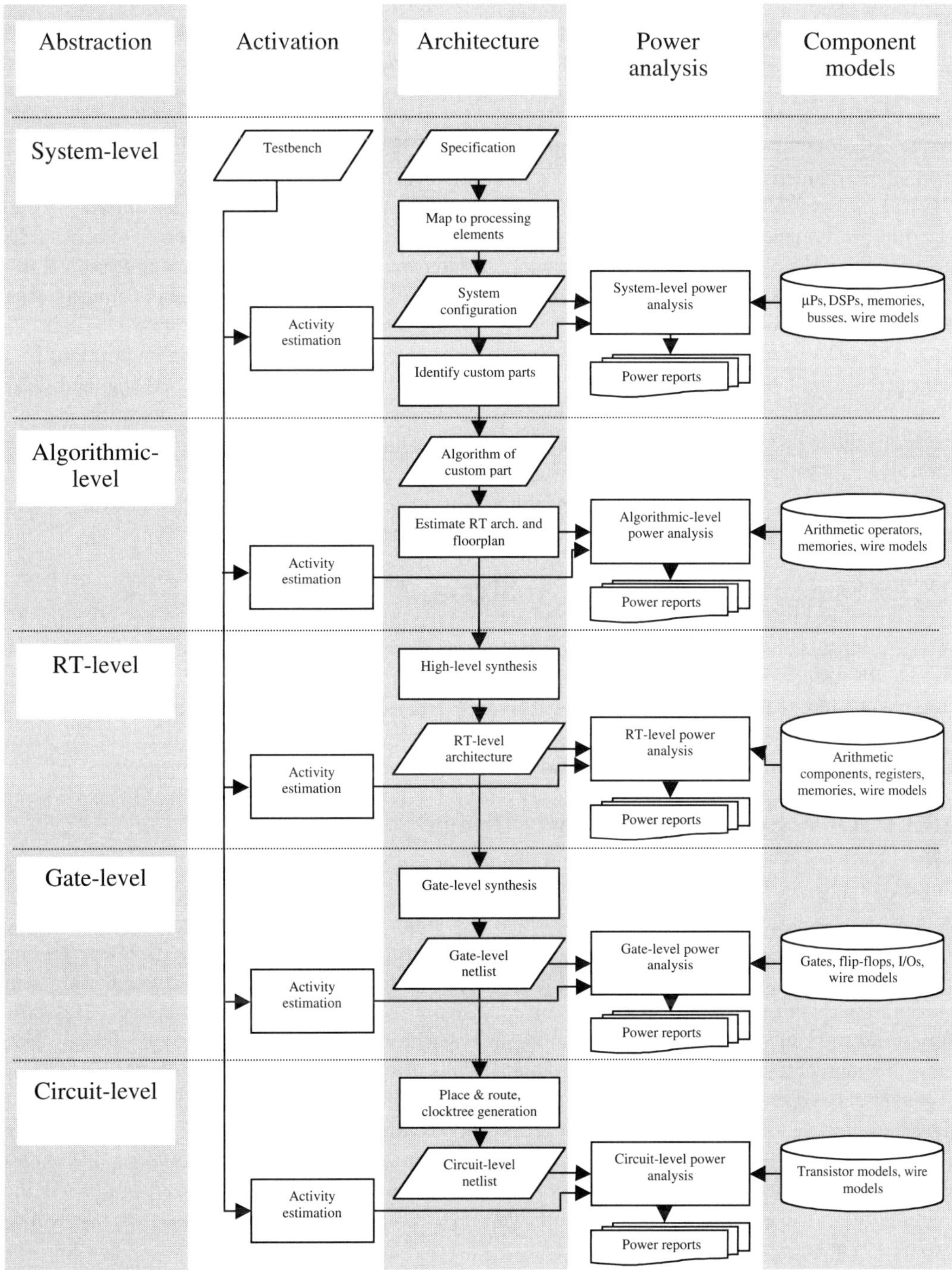

FIGURE 18.2 Generic low-power design flow.

is also applicable to platform based design and incremental designs. In the sequel, we will walk through this design flow in a top-down manner.

At the system-level, the design objective is to map a given, possibly informal, specification onto a target architecture. In many cases, this architecture will be constrained to a platform consisting of fixed architecture elements such as processors, micro-controllers, digital signal processors (DSPs), and a bus standard. The design objective is to find an optimized mapping, which meets functional and performance constraints at least cost and power consumption. A system-level power management function can be included at this level and needs to be considered during power analysis. The most frequently used tool at this level is a spreadsheet program.

The largest reduction in power consumption can most likely be gained by implementing the most computation intensive parts of the system by application specific logic. Due to the custom character of this part, no predefined module exists. Its functionality is best defined by an executable model (e.g., an algorithm written in a programming or a hardware description language). Because no architecture model exists yet, this has to be predicted together with an interconnect and clock tree model. Based on this estimated design combined with real power models for the allocated predefined components, a power analysis can be performed.

The lower levels of design follow by consecutively generating more detailed design descriptions by a sequence of synthesis steps, refining the test-bench by including bit width, data encoding, and delay information, and by providing the respective lower-level power models. The design objectives at these levels include further local optimizations of the same cost function already applied in a more abstract form at the higher levels of abstraction. A variety of commercial tools are available including, for instance, at the algorithmic-level: ORINOCO by ChipVision [3]; and at the RT-level: PowerChecker by BullDAST [4], PowerTheater by Sequence [5], and PrimePower by Synopsys [6].

18.3 System-Level Power Analysis

A system consists of a set of components, which jointly perform a common task. Definitions like this one describe the essence of system-level design: allocation of components, partitioning of the system's task onto these subsystems, and organization of the cooperation of the components. This section presents methods and tools that can be applied to exploit the largest possible gain in power reduction by partitioning the system in a power-optimal way as well as by introducing power management.

18.3.1 Objectives of System-Level Design

System-level design starts from a specification, some environmental constraints, and possibly a restriction of the design space. The specification can be given informally and, in this case, requires formalization. A well-established formalism is a task graph [7]. A task graph is a representation of a task depicting the subtasks (processes) and their data as well as control flow dependencies. It consists of vertices representing the subtasks and edges representing the data flow and control flow dependencies. A task graph is a system specification exhibiting parallelism and concurrency. Figure 18.3 is an example of a task graph. The start-vertex and the end-vertex are needed to model the synchronous beginning and termination of each execution loop of the task, the other vertices represent the processes P1 to P6 of the task. The solid edges represent data dependencies, which are important to exploit resource sharing. For instance, P6 is data-dependent on the results of P2. Thus, P6 cannot be executed in parallel to P2 (i.e., they can share resources). On the other hand, P3 and P2 are concurrent processes (i.e., it is up to the designer's choice whether he or she allows resource sharing between these processes or not). The dotted edges model the control flow. In the example of Figure 18.3, the edge between end and start specify that this task will be executed in a loop.

The environmental constraints typically include minimum performance requirements, maximum cost, and power constraints, as well as some form factors and I/O loads.

Finally, the third input can be a restriction of the design space (e.g., by requesting to use a given set of processors, DSPs, memories, available custom area, and bus structures). These elements, together with

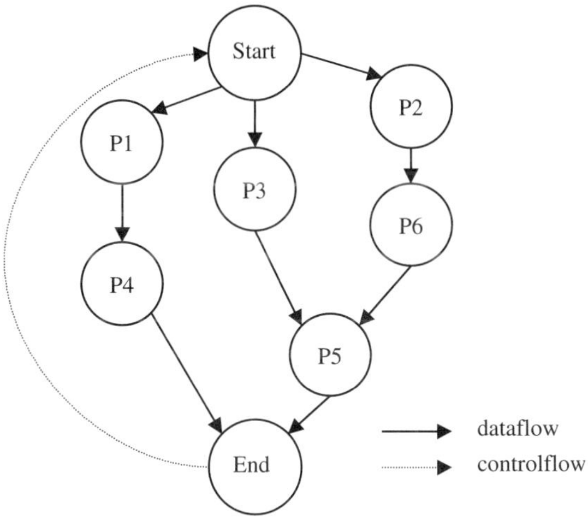

FIGURE 18.3 Task graph.

the respective software infrastructure, make a platform. Figure 18.4 is a generic system-level architecture, which could be a platform or an existing design. It is the objective of the system-level designer, the so-called system-architect, to allocate a set of components, map each process of the task graph onto exactly one component (frequently called the processing element), and define the necessary control and communication structure to implement the task graph. The optimization criterion is to achieve an architecture within the given design space, which meets all performance constraints and is an optimal trade-off between cost and power consumption. Note that at this level, the basic structures of a power management policy need to be defined.

A straightforward way for power and performance optimization is to identify so-called computational kernels [8]. These are the inner loops of computation intensive processes. Implementing them by application specific hardware will not only increase the performance and allow using a less expensive processor, but also reduce the power consumption because of the optimized datapath and hard-wired control. Although such a decision may be obvious, however, it requires a detailed understanding of the implications. For example, the communication between the processor and the application specific hardware needs to be considered. The memory architecture may become more complicated and require multi-port

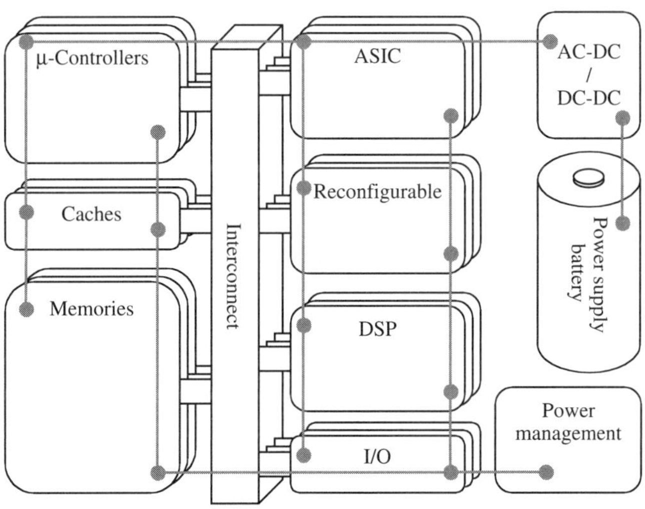

FIGURE 18.4 System-level architecture.

memories. Thus, an adequate tool support is needed to enable sound decisions. An excellent overview of the optimization techniques at this level can be found in Benini and de Micheli [9].

18.3.2 Analysis of an Implementation Model

A frequent design scenario is a platform based design flow. In this case, the designers have to use a given hardware (HW)-/software (SW)-platform to implement the specification. In many cases, an executable model of the system does not yet exist; however, experience with previous systems on the same platform is available. The power estimation can only be based on the general application know-how of the design team, existing power data of the architecture and its components, assumptions about application specific logic, and an intended mapping of the processes of the task onto the various processing elements of the platform, such as processors, DSPs, application specific logic, or memories.

The frequently used straightforward spreadsheet approach of power estimation is based on collecting power related information from data sheets, semiconductor vendors, the application, and experience in a spreadsheet, which implements a power model of the entire system. Often, the experience or data are not yet available for the intended target semiconductor technology of the new design, but only for previous technologies. In these cases, technology-scaling models need to be applied to estimate the power consumption of a module to be processed in a new technology from data of a recent one.

The sources of the power figures for the spreadsheet depend on the kind of the module under consideration. For processors and DSPs, power figures are available from the data sheets of the processor vendors. At this level, power data given by the vendors are typically not detailed to the instruction level or even data dependent, but a single figure in terms of power per megahertz for the processor working at a given supply voltage. Similarly, power data for embedded memories need to be collected from the memory provider or semiconductor vendor.

The interfaces of the design are typically well defined at this level. The system specification should exactly name the I/Os of the system and their required load. The data sheets of the semiconductor vendor offer detailed power figures for these cells.

So far, the power model is an analysis one. Even with rough power models of the components, there was no need to make assumptions about the system architecture. Estimation starts with the consideration of application specific logic. This part may only consume a small fraction of the entire power of the system, but it is the key to reduce the processor power and possibly downgrade the processor to a less expensive and less power-consuming version. At this stage, little is known about the application specific logic. At best, an algorithmic description of this part is available, which can be used for algorithmic-level power estimation and will be covered in Section 18.4.2. If such an algorithm does not yet exist, power estimation is based on the experience of the designer who can predict the number of gates and registers most likely needed to implement the required function within the given performance constraints. Application knowledge can help to estimate the expected activity of such a module. ASIC vendors can provide average power figures for logic in a given technology based on their experience and characterizations. This figure will come in terms of milliwatts per megahertz per kilogate and needs to be weighted with an activity ratio expected for the application. The number of registers can be a useful input to an estimate of the clock tree power.

Table 18.2 presents an example of such a spreadsheet. A refined approach to this principle has been developed as the Web-based tool PowerPlay [10]. Another concept at this level is the power state machine [11], which includes a dynamic model of the activation and deactivation of the various system components. It is well suited for evaluating and optimizing power management policies.

Given the rough granularity of the power models used to create a spreadsheet system-level power model, this can be useful to support a first check whether power constraints of the system will be met or not. It can be used to analyze the impact of moving the design to a new technology node or replacing a processor by another one; however, even these conclusions from the model have to be drawn carefully because many important parameters could not yet be captured. These include the communication power consumed by the data transfers between processors, memories, application specific logic, and I/Os. The

TABLE 18.2 Example Spreadsheet for System-Level Power Estimation

	Design				Reference Technology				Scaled Technology			
Example	No. Inst.	Complexity (k gates)	Registers	Activity	Frequency (MHz)	V_{dd} (V)	Power (µ W/ MHz/k gate)	Module Power (mW)	Frequency (MHz)	V_{dd}	Scaling Factor	Module Power [mW]
Processors												
Proc.1	1				200	1,8		160	250	1,2		90,0
DSP	1				200	1,8		110	250	1,2		33,0
Memories												
SRAM1	1				200	1,8		70,0	250	1,2		21,0
ROM1	1				200	1,8		55,0	250	1,2		16,5
ROM2	1				200	1,8		55,0	250	1,2		16,5
ASIC logic												
Mod. 1	1	25	500	0,3	200	1,8	4,0	1,5	250	1,2	0,72	0,6
Mod. 2	3	60	260	0,4	200	1,8	4,0	14,4	250	1,2	0,72	5,8
…												
Mod. N	2	18	200	0,1	200	1,8	4,0	0,7	250	1,2	0,72	0,3
I/O												
Inputs	18			0,2	200	1,8		0,2	250	1,2		0,1
Outputs	18			0,5	200	1,8		70,2	250	1,2		46,8
Total								537,0				230,5

clock network, which may consume a considerable part of the total power, is not yet designed. Issues like cross coupling and the second order effects of the scaling theory are out of the scope of such a model. Finally, neither the impact of the software structure nor of the data has been considered.

18.3.3 Analysis of an Execution Model

A more accurate power analysis at the system-level is possible once executable models of the system processes to be implemented exist. An executable model can have the form of a program written in a programming language, such as C, a hardware description language, such as VHDL or verilog, or a system-level language such as SystemC. Alternatively, a heterogeneous model, combining several languages and models of computation into a single framework, can model the system. In processor design, the system model also can be an executable performance model. By executing any of these models, an understanding of the dynamic behavior of the system can be achieved. This allows a more detailed power analysis under consideration of the real activity in the system. Still, power models of the various components of the system must be at hand, which can be combined with the system architecture and the component activation patterns to a power analysis as given in Figure 18.1 and Figure 18.2. The components of a system-level design include: software, memories, and other existing or yet to be designed modules.

Software-implementing algorithms and running on predefined and power-characterized cores are covered in Section 18.4, Algorithmic-Level Power Estimation and Analysis. If the actual processor is still being developed and optimized, a more detailed bus functional model of the processor execution is needed. Besides functional models, this includes activation models of the processor components (e.g., the issue queue, the branch prediction unit, the execution units, the cache, and the register file). During a simulation of this model, the various components are activated and this activity information is captured. It can be used to evaluate the power models of the components and provide a power analysis of the intended processor architecture [12]. In the case of multipurpose modules controlled by control signals, different power models are required associate to each of the operation modes. Their runtime percentages can be used to calculate the total module power consumption [13].

Memories may consume a considerable percentage of the total power consumption. Consequently, they offer a large power reduction potential. Research has proposed a number of methodologies for memory power optimization. The probably most holistic methodology, called DTSE, has been developed by Catthoor [14]. Arguing that memory power is the dominating part of the system power consumption in signal processing applications, Catthoor advocates that a memory power optimization should be performed before any other power optimizations. The key idea is to apply a sequence of optimizations on the specifications, which, partly automated, perform global loop and control-flow transformations, a data-reuse analysis, a storage cycle distribution, memory allocation and assignment, and, finally, an in-place optimization. The objective is to increase data locality, avoid memory access, and design an optimized memory hierarchy. The result is a system and memory description that can be synthesized.

To assess the power consumption regardless of the optimization methodology requires power models of the various memory types. Power models for memories are difficult to create. They shall be flexible and parameterized (at least with respect to their size), they shall be accurate, and they shall be generated efficiently. The task of characterization of a parameterized memory power model requires a simulation of different instances of the memory. Consequently, simulation models need to be available. Due to the flat structural hierarchy of memories, typically only transistor level models are available. They cause a prohibitive simulation time when executed for memories of practical size. Thus, abstractions have to be used in memory power modeling. These abstractions can be used to model parts of the memory cell array, particularly its capacitive load; however, this abstraction requires access to the internal data of the memory, which is sensitive proprietary information of the memory vendor. To overcome the confidentiality issue, the power model itself should not disclose any information about the internal structure of the memory. Thus, functional power models are adequate, which can be generated using regression techniques. An approach by Schmidt et al. [15] includes nonlinear terms in the regression, which are needed to accurately model, for instance, the address decoder, which is a logarithmic structure.

For other (nonsoftware) components, the activation has to be captured from the executable system model and mapped to a state dependent power macro-model. If application-specific logic (e.g., a block of standard or reconfigurable cells), is to be part of the architecture, this is not yet power characterized. The executable model in this case is a pure functional model (e.g., an algorithm), and the problem of power estimation is the same as discussed with the spreadsheet approach: it requires architecture estimation (see Section 18.4).

Due to long wires and heavily loaded system busses, interconnect power can exhibit a significant percentage of the system power. Analyzing the power consumption of interconnect requires input of physical layout and material properties. This can be partly available for a platform based on measurements or simulations. Off-chip interconnect capacitive loads, which can easily be several orders of magnitude larger than on-chip loads, can be derived from the system specification. The power analysis becomes more difficult for on-chip interconnect and in case of complex bus encoding schemes. The interconnect prediction problem for on-chip wires are discussed in more detail in Section 18.4.

Complex bus encoding schemes have been proven to allow a significant reduction of the switched capacitance of busses (e.g., Fornaciari et al. [16] report a 48% reduction in address busses with Gray Code. Because bus encoding is defined at this level, the impact on the power consumption needs to be taken into account including the overhead for the encoders and decoders.

Similarly, power management needs to be included in system-level power estimation. Its optimization is part of the system-level design. It requires models for the power management policies under consideration as well as for the shutdown and wake-up power penalty. The power management policies can be integrated into the execution model of the system or they can be modeled by a power state-machine [11]. Similarly, dynamic power management techniques, which are typically implemented in the software or the real time operating system (RTOS), require respective models of the policy [9].

18.4 Algorithmic-Level Power Estimation and Analysis

The design tasks at the algorithmic-level of abstraction include optimizing algorithms, which are to be implemented either by software, application specific hardware, or by a combination of both. The objectives include performance, cost, and power optimizations. Means of improvement include selection of the most suitable algorithm performing the requested function, optimizing this algorithm, and partitioning the algorithm into parts, which will finally be implemented in software, and others, which will be realized by application specific hardware.

Selecting the most power-efficient algorithm out of a repertoire of available and functionally equivalent ones requires an estimate of the expected power consumption of an implementation of the different algorithms. Of course, the comparison must be based on power-efficient realizations of these algorithms without the need to really implement them.

Once an algorithm has been chosen, it can be optimized for low power. First, the control flow can be optimized to reduce the number of control statements (e.g., by different kinds of loop unrolling strategies). Additionally, these transformations extend the scope of local statement reordering and pave the way to local memory access optimizations. An example of a sequence of such optimizations is presented in Sarker et al. [17]. The data of the algorithms is typically specified in terms of floating-point variables and arrays. For a hardware implementation, a more efficient data representation is possible (e.g., fixed-point data types of adequate precision for the intended application). Algorithmic-level power estimation is applied to evaluate the impact of the algorithmic transformations and design decisions mentioned in Stammermann et al. [18].

These optimizations, however, have to be made, while considering the target hardware. Moving the computational kernels of the algorithms to power optimized application specific hardware is the most promising path to the largest gain in power consumption. The reasons are simple: the application specific hardware has a hard-wired controller and no need for consecutive control steps to perform a single

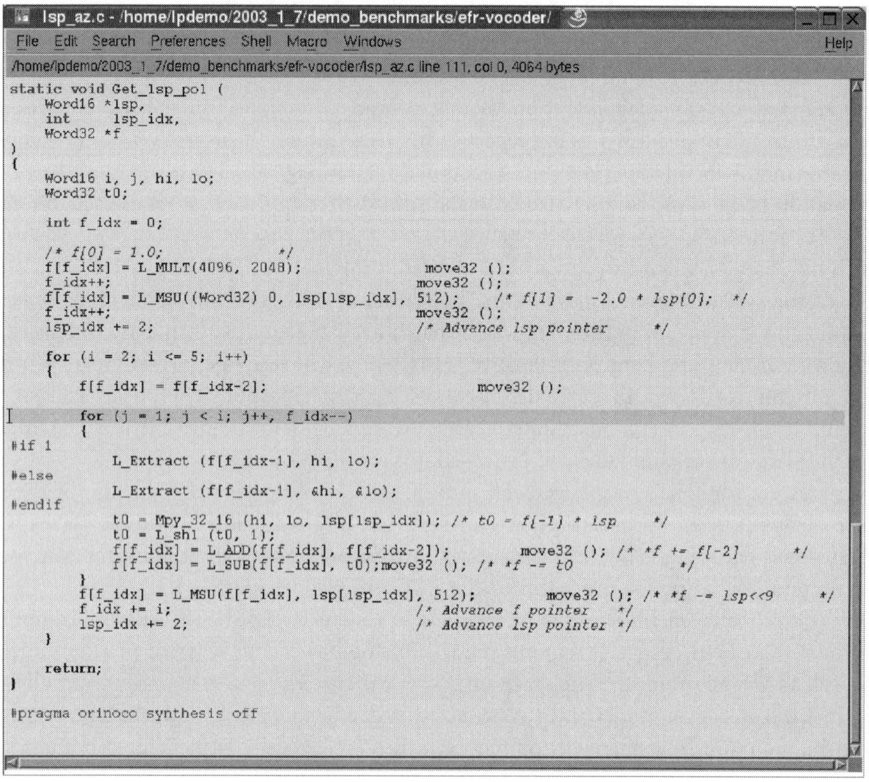

FIGURE 18.5 Algorithmic specification (C-source-code).

instruction. No memory access is needed to find out what to do next. The datapath just contains the minimum amount of hardware to perform the operation, and, finally yet importantly, concurrency can be exploited to a much larger degree than this is possible on a processor core. All this avoids wasting energy [19]. Thus, HW/SW partitioning is another important design step, which requires algorithmic-level power estimation to support a trade-off analysis between application specific hardware implementations of parts of the design vs. software implementations. Due to the different nature of software and hardware, dedicated tools are needed for software power analysis and algorithmic-level power estimation for hardware implementations. Both are covered later in this section.

Design input at the algorithmic-level is an algorithmic description, typically executable, or a functional model describing the I/O relation, and a set of constraints. It is important to note that the algorithm is not yet meant as an implementation, but just as a prototype, which needs optimization and implementation. Figure 18.5 is an example of parts of a vocoder design [20]. The function `Get_lsp_pol` is invoked as part of the entire design. It shall serve as an example of a process, which shall be implemented by application specific hardware. It consists of two nested for-loops with some arithmetic operations in the inner loop.

The algorithm can formally be represented by a control and data flow graph (CDFG) [21,22]. The vertices of the CDFG represent either the arithmetic or logic statements of the algorithm, or the control statements. The edges model the data and control flow dependencies. A CDFG implies a partial order on the execution of the statements as required by the data and control dependencies of the algorithm. Figure 18.6 presents the CDFG of the function pictured in Figure 18.5. Because the function contains a nested loop, a hierarchical CDFG is useful, which allows to partly unfolding the loops as demonstrated in Figure 18.6(B). Comparing Figure 18.6(A) and Figure 18.6(B) demonstrate that the CDFG does not yet imply a schedule but only a partial order.

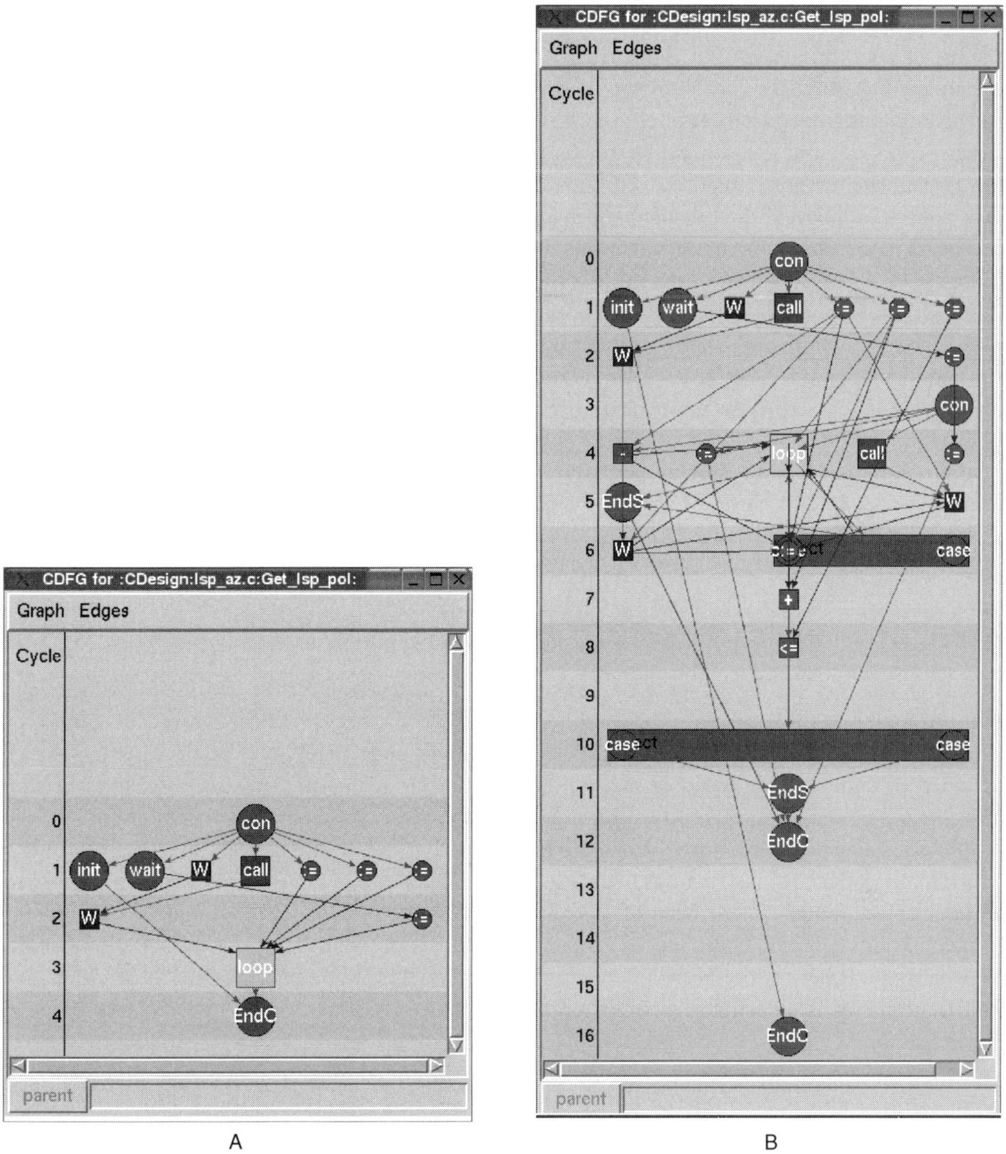

FIGURE 18.6 Control data flow graph (CDFG): (A) CDFG with folded outer loop, (B) CDFG with unfolded outer loop.

The output of the algorithmic-level design phase is an algorithm that can be compiled to the target architecture by a software compiler in case the target is software implementation, or an architectural synthesis tool in case of a custom hardware.

18.4.1 Software Power Analysis

Software power analysis is applied to processes to be implemented as software on embedded processors, which can be μ-controllers or DSPs. Software power analysis can be performed at three different levels of granularity:

1. The source-code-level
2. The instruction-level

3. Functional-bus-model-level

They are all based on power models of the target processor and an execution of the software to capture the dynamic behavior. The execution can be performed on the target processor, on another processor, or on a simulator. At any of these levels, the power analysis can be used to compare different programs, to select processors, and to optimize the software. The different levels offer a trade-off in accuracy versus effort to generate the power reports.

The source-code-level, which is the highest abstraction level, provides the fastest turn around times for software power estimation because it avoids the generation of the machine code for the target machine. Brandolese et al. [23] have demonstrated that the execution time of a program on a given processor can be used as a measure of its energy dissipation. Following this idea, the problem of source-code-level power analysis can be reduced to the estimation of the number of execution cycles. This number can be estimated by mapping the source code on instruction classes, which have been empirically characterized with respect to the instructions per cycle of each class. The total energy $E_{program}$, needed for the execution of a program, can thus be estimated using Equation (18.4) with $T_{execution}$ being the total execution time of the program and E_{proc} the energy per MHz clock frequency of the processor. The accuracy of the approach for the average power of single issue processors without considering memory power is within 20% of an instruction level power analysis.

$$E_{program} = T_{execution} \cdot E_{proc} \cdot f \tag{18.4}$$

A higher accuracy can be achieved by working on the instruction set for which code is generated. Power estimation at this level was pioneered by Tiwari et al. who measured the power consumption of individual instructions and the effects of inter-instruction dependencies [24]. The measurements can be performed by running long loops of the same instruction or sequence of instructions and physically measuring the power consumed by the processor. Through these measurements, a power figure for each pair of consecutively executed instructions can be obtained. A power analysis can thus be performed by capturing the sequence of instructions being executed and combining this information with the instruction-level power model. It has been observed, that an abstraction of the large number of different instructions and their addressing modes is possible by clustering the instructions into classes of similar power behavior. It has further been observed [23] that the relative power consumption of instructions of different classes is similar for different processors. This significantly simplifies the task of power characterization of a large set of processors.

Software power analysis at the levels mentioned so far is limited in accuracy because many aspects of the program execution cannot be considered. These aspects become an increasingly important with the deployment of more complex embedded μ-controllers and hierarchical memory architectures for systems on chip (e.g., pipelined RISC processors, multi-threaded CPUs, out-of-order execution, and embedded caches). Ideally, the processor power models should include the complex relationships between issue queue, execution unit, multiple threads, speculative execution, data dependencies, and cache hit- and miss-rates. Accurately analyzing the power consumption of such architectures requires a profiling of the software on an instruction set simulator with access to the system bus. For instance, such a model has been developed for the ARM processor [25] with an accuracy of 5%. Consideration of these architectural aspects during power analysis allows optimizing either the program for a given processor or the processor configuration for a given program. For example, Simunic et al. [25] could achieve a power reduction of more than 50% by replacing an L2 cache by a burst synchronous dynamic random access memory (SDRAM), while even improving the throughput of a signal processing system. The disadvantage of working at this low level is the long execution time of the simulation. Generating synthetic programs, demonstrating the same performance, and power consumption as the original program but with fewer executed instructions, can speed up the analysis by several orders of magnitude [26].

18.4.2 Algorithmic-Level Power Estimation for Hardware Implementations

As we have discussed, implementing an algorithm or part of it in application specific hardware can significantly reduce the power consumption and relieve the processor from computation intensive task. This may allow downgrading the processor to a cheaper and less power-consuming type. The problem of power estimation at this level, however, is different from the software power analysis problem. The main difference is that the target hardware is not designed yet. The building blocks of that hardware are not yet allocated; the control and data communication between these components is yet to be defined. Thus, before being able to predict the power consumption, it is necessary to estimate the target architecture (see Figure 18.1). The problem of algorithmic-level power estimation for hardware implementations is thus: given the CDFG of an algorithm (Figure 18.6), predict the power expected to be consumed by a power optimized custom hardware implementation of this algorithm. That is, predictions of the target architecture and the activation of the components of the architecture are needed as well as the prediction of the communication and storage. At the algorithmic-level, Equation (18.1) can be replaced by Equation (18.5) [27]:

$$P_{dynamic} = N_a \cdot C_{avg.} \cdot V^2 \cdot f_{comp} \tag{18.5}$$

N_a is the number of activations of the respective module per computation iteration (per sample), $C_{avg.}$ is the average switched capacitance of the module per activation, V the supply voltage of the component, and f_{comp} is the iteration (sampling) frequency of the algorithm. The number of modules and their activation strongly depend on the scheduling, allocation, and binding, which have not yet been performed at the algorithmic-level. To evaluate Equation (18.5), assumptions about the scheduling, the allocation, and binding, as well as the interconnect and storage architecture have to be made.

Additionally, power models for the components must be available. In the case of standard components, these models can be generated by simulation and power characterization based on lower-level power analysis tools [4] and appropriate power models [28,29]. Thus, algorithmic-level power analysis includes the following steps: architecture estimation (i.e., scheduling, allocation, binding of operations and memory accesses, and communication architecture estimation, including wire length prediction), activation estimation, and power model evaluation.

The main challenge of algorithmic-level power estimation for hardware implementations is the difficulty to predict the structural and physical properties of a yet to be designed power optimized circuit. Existing approaches to solving this problem rely on a power-optimizing architectural synthesis of the design before power analysis. The accuracy of the power analysis depends on how well the assumed architecture matches the final architecture. This final architecture is subject to many parameters (e.g., the design style specific architecture templates), which are the main differentiating factors in times of fabless semiconductor vendors, or the tool chain applied at the later phases of the design process (e.g., RT-level synthesis, floorplanning, routing, and clock tree generation). Thus, an architecture estimator should either consider the design flow and style applied to the real design, or generate an architecture of such high quality that it can be implemented without further changes.

18.4.2.1 Target Architecture

Architectural synthesis maps a CDFG onto an architecture template. Figure 18.7 depicts such a generic target architecture for the hardware implementation of a CDFG. It consists of three parts:

1. The datapath, which implements the dataflow of the CDFG
2. The controller, which organizes the dataflow and the control flow
3. The clock tree

It is the task of the architecture synthesis to schedule the operations under timing and resource constraints, as well as to allocate the required resources in terms of operation units. The operation units can be arithmetic or logic modules as well as memories. One result of the architecture synthesis is the

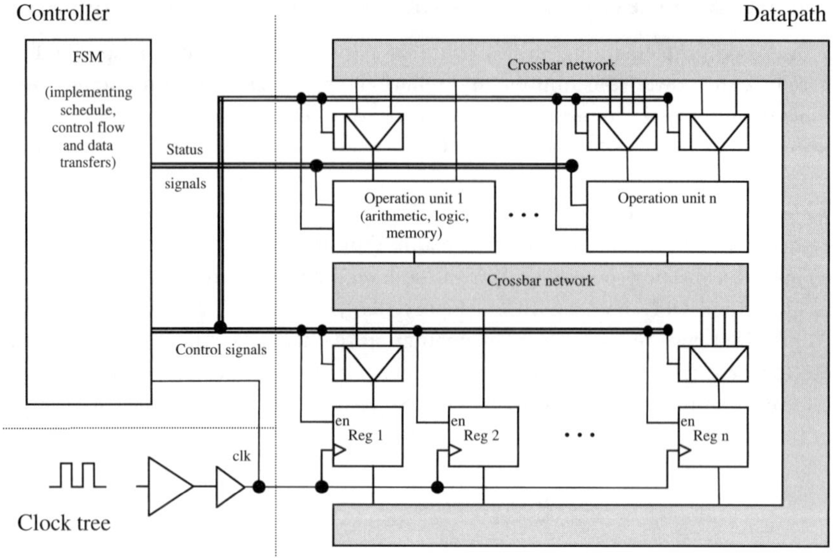

FIGURE 18.7 Generic target architecture.

set of operation units and registers allocated as well as the steering logic, which implements the data transfer connections between the operation units and the registers. The second output is the controller, which is a state machine generating the necessary control signals to steer the multiplexers, operation units, and enable signals of the registers. To do so, it needs to implement the control flow and the schedule based on the status signals of comparator operation units in the datapath. Early work on architectural synthesis for low power has analyzed the impact of binding and allocation during high-level synthesis on the power consumption and integrated power optimizations into high-level synthesis tools [22,30].

18.4.2.2 Scheduling

The schedule of a datapath defines at which control step each of the operations is performed. It has an impact on the power consumption. It defines the level of parallelism in the datapath and thus the number of required resources. The schedule determines the usage of pipelining and chaining. Although pipelining can be a means to reduce power by isolating the propagation of unnecessary signal transitions even within one operation unit, chaining causes the propagation of such glitches through several operation units in one clock cycle and thus increase the power consumption. Musoll and Cortadella [31] have proposed an approach to utilize operations of the CDFG with multiple fan-outs to reduce the power consumption by binding the successor nodes of the CDFG to the same resource in consecutive control steps if they are operationally compatible. This reduces the input activity of these operation units.

Figure 18.8 is the scheduled CDFG of the vocoder example introduced earlier. Comparing the scheduled CDFG with Figure 18.6(b) reveals that some operations have been moved to other control steps and that the additions and subtractions have been bound to specific instances of operation units.

18.4.2.3 Resource Allocation Binding and Sharing

The allocation of resources defines which and how many resources are to be used to implement the CDFG. The binding assigns exactly one operation unit to each of the operations of the CDFG. Several operations can be assigned to the same operation unit if they are scheduled into disjoint control steps and the operation belongs to a subset of the operations that can be implemented by the same unit. These operation units are predesigned and power-characterized modules, such as multipliers, memories, adders, ALUs, comparators, and subtractors.

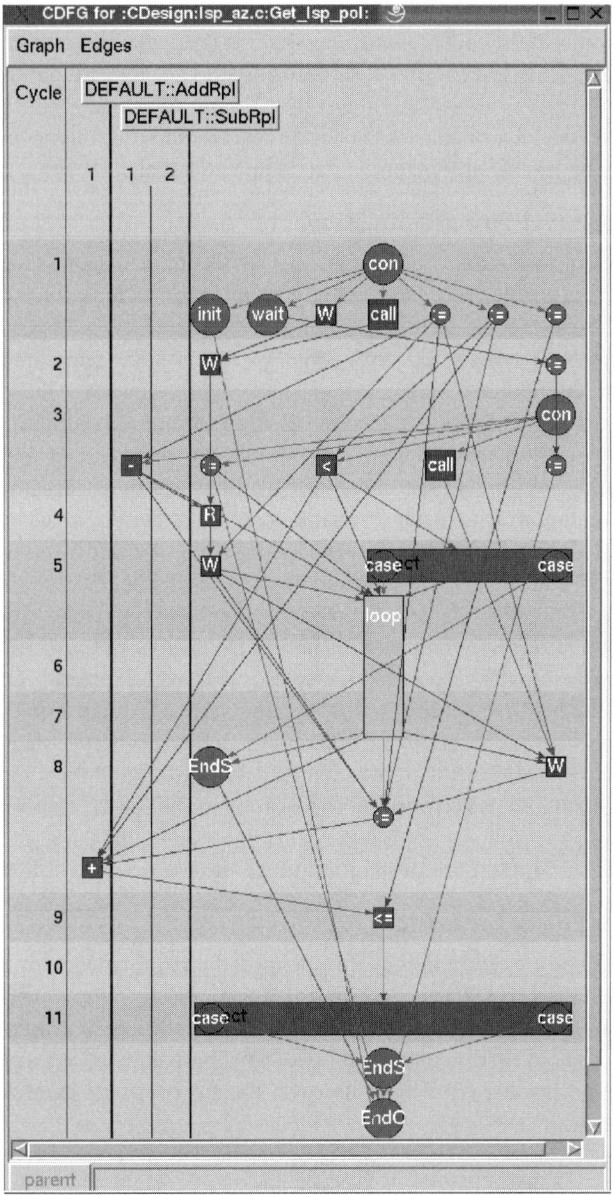

FIGURE 18.8 Scheduled CDFG.

The valid set of target units of the resource binding depends on the set of operations these units can perform. This opens further possibilities for power optimization because more than one type of operation unit can be chosen as target unit, influencing the resulting power consumption. For example, an addition can be bound to a carry-look-ahead adder, a carry-save adder or an arithmetic logic unit (ALU). Similarly, variables and arrays can be mapped to registers or memories. Typically, arrays are mapped to memories, while single variables are mapped to registers.

The resource allocation and binding affects the power consumption of the datapath due to several effects. The power consumption of each operation unit strongly depends on the switching activity of its inputs. In a routine processing real input data, the internal data applied to the operation units will usually not be independent, but highly correlated in a similar way over a wide range of input data. Applying consecutive input data of high correlation to an operation unit reduces its power consumption. An established measure for the input switching activity is the average hamming distance of a sequence of input patterns [28]. Analyzing the input streams of the operations allows assigning the operations to operation units in a power optimized way by exploiting these data correlations. Because this assignment is an NP-complete problem, different heuristics have been proposed. Khouri et al. [32] use an activity matrix to capture this data dependency and include control flow information and state transition probabilities into the power analysis, while Kruse et al. focus on the iterative nature of data dominated designs [33].

18.4.2.4 Behavioral-Level Power Estimation

In addition to the operations discussed in the previous subsection, an algorithmic specification and its CDFG may contain calls to nonstandard functions (e.g., combinational logic functions, which are defined by their I/O behavior). Because these are not part of the power-characterized library, they require a special treatment during algorithmic-level power estimation. Two main approaches are possible in principle: synthesis or complexity estimation.

Surely, the most accurate results could be achieved by a fully optimized synthesis of the function under observation. This large synthesis effort may be prohibitive for quick turnaround times desired when exploring the algorithmic-level design space. A quick synthesis can be a workaround if it is combined with a calibration procedure, which reliably estimates possible further improvements from the outcome of the quick synthesis. It is obvious that this approach lacks accuracy, while delivering relatively fast results.

The second approach builds on complexity estimates. Müller-Glaser et al. [34] integrated an area and power estimator into a design planning and management tool. Its input is the expected number of gate equivalences, which is empirically calibrated with respect to design styles, tool flow, and technology to produce area and power estimates. This approach is also used in the spreadsheets presented earlier in this chapter. If the required estimate of the number of gates needed is not available because, for instance, no experience exists for a new application, information theoretic approaches step in.

Their input is the functional I/O behavior of a module. A key indicator of the computational complexity and thus of the energy required, is the entropy. The entropy is a measure of uncertainty. The larger the uncertainty of the function's result, the larger is the effort to compute this value. The entropy of a module output can thus be used as an indicator of its computing power consumption [35–37].

18.4.2.5 Controller Power Estimation

Scheduling, resource allocation, and binding have defined the requirements for the controller. Yet, its structure and implementation are still to be determined. The power consumption of the controller (Figure 18.7) depends on its implementation (i.e., the number of registers and their activity, the implementation of the state-transition and output functions, and their signal probabilities). As with the behavioral-level power estimation, a full controller synthesis will deliver the most accurate controller model, which can be used for power analysis.

To reduce the power estimation time, empirical power models [27] have been proposed, which use regression techniques to generate power models for controllers.

The input parameters for the regression include: the number of states, inputs, outputs, and the state coding, as well as the input signal probabilities, which can be extracted from the schedule, the status, and control signals.

18.4.2.6 Interconnect Power Estimation

So far, we have discussed the algorithmic-level power estimation of software and the various hardware components of an embedded system. These components can be separately analyzed or estimated once they have been allocated and their input activity was captured. The power consumption of the communication between these components and their synchronization by the clock, however, requires physical information of the placement of these components and their interconnect as well as their clock tree. As we will see, it is important to consider the effect of the interconnect on the total power consumption during the different steps of the architecture definition. A power aware interconnect design can significantly reduce the total power consumption of the system. For example, Zhong and Jha [38] and Stammermann [39] report power reductions of more than 20% by an interconnect aware high-level synthesis for low power.

This interconnect aware power optimization requires an estimation technique for the interconnect and its power consumption. The interconnect power models applied so far are based on the switched capacitance of the wires, as formulated in Equation (18.1). For a global power estimate, it is sufficient to estimate the total switched capacitance. Empirical wire models like Rent's Rule [40] can be applied to predict the number and average length of wires; however, because a power estimate of an optimized floorplan is needed, this average figure is too pessimistic. Such a power optimal floorplan will locate components, which are communicating at a high data rate as close together as possible and thus save power. Thus, to steer interconnect power optimization, the capacitance and switching activity of individual wires must be known.

The problem of interconnect power estimation is to estimate the capacitance and activity of each wire of an RT-level architecture. Because the activity can be derived from the activity data of the modules, which have been discussed previously, the remaining problem is to estimate the wire capacitance, which is primarily determined by the wire's length, the physical layers used to implement the wire, and the number of vias. The wire length depends on the location of the modules of the design on the floorplan and the routing structure between the modules. The capacitance of a wire, including the effects of vias and multiple connection layers, typically correlates with the wire's length in a nonlinear way [41]. Thus, the main problem remaining is to calculate the expected length of each wire in a power-optimized floorplan. This requires including floorplanning and routing into the estimation. Because of the large impact of the floorplan on the total power consumption, a separated floorplanning phase, once the architecture is fixed, will create suboptimal solutions. Existing approaches, which consider interconnect power during power analysis and optimization at the system-level and algorithmic-level, attack the problem by integrating floorplanning and routing into the architecture optimization discussed previously.

Traditionally, high-level synthesis consists of the phases: allocation, scheduling, and binding, which are typically performed in a sequential manner. High-level synthesis for low power adds a further step: interconnect optimization. Each of these steps is a NP-complete problem, thus the entire problem is NP -complete (i.e., a guaranteed optimal solution cannot be found in reasonable computation time). An optimal design can further not be achieved by applying these steps sequentially because the optimizations are not independent. Consequently, because power analysis at this level requires a detailed understanding of the target architecture, heuristics are needed to synthesize such architecture in a power-optimized way under simultaneous consideration of allocation, scheduling, binding, and floorplanning.

First approaches to combine several of these tasks of high-level synthesis into one optimization loop have been proposed [38,41,42]. The common feature of these optimization flows is to apply a set of moves on a preliminary design, to evaluate the impact of these moves, and to follow an optimizing heuristic such as simulated annealing, thus applying further moves until a stopping criterion is fulfilled.

Prabhakaran et al. [42] apply moves changing the schedule and the binding. Before evaluating the cost function, they perform a floorplanning step during each iteration. Zhong and Jha [38] use allocation and binding moves followed by a floorplanning step for cost estimation. Stammermann et al. [41] include allocation, binding, and floorplanning moves into their optimization heuristics (see Figure 18.9). The upper part of the figure presents the outer loop of the optimization, during which binding and allocation moves are performed. If, based on a preliminary power estimate, a binding/allocation move is promising, then the floorplan is updated and optimized by several floorplan moves in an inner loop, as presented in the lower part of Figure 18.9.

The floorplan and the allocated registers are also the basis for the generation of a clock tree model, which can be used for clock power prediction.

The result is a power-optimized architecture automatically generated from an algorithmic-level description. The expected power consumption of this architecture is analyzed during the optimization loops. This power figure will have a high relative accuracy and can serve as an estimate of the power consumption of the input algorithm. It can be taken as a guide for optimizing the input algorithm for low power.

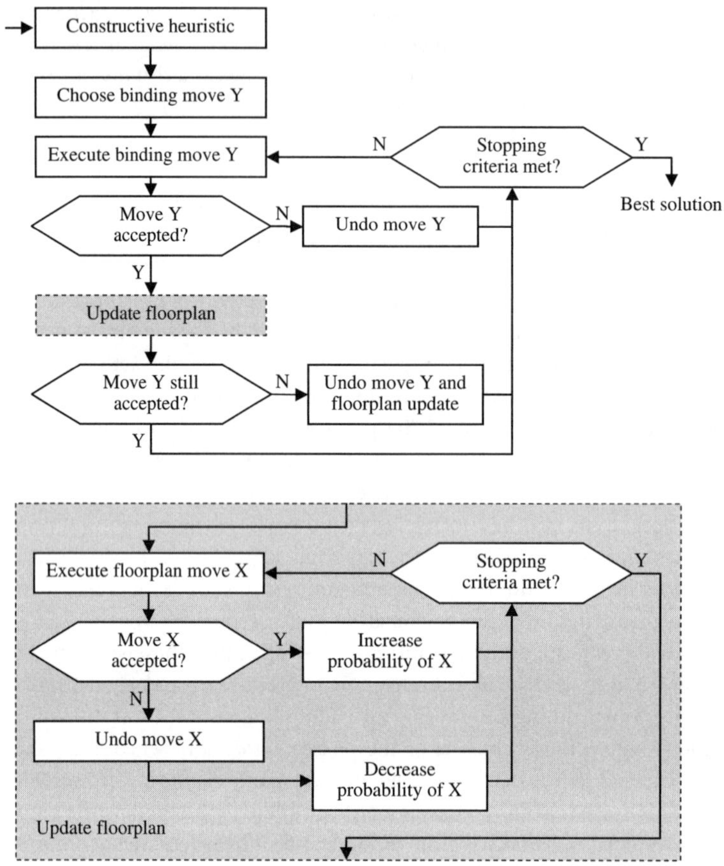

FIGURE 18.9 Simultaneous allocation, binding, and floorplanning for low power.

If, however, the implementation style assumed during the optimization fits the implementation style of the final SoC design, and the architectural parameters (e.g., allocation, scheduling, binding, and floorplanning) of the generated architecture are applied to the final implementation, then the power estimate is a good prediction of the absolute power consumption to be expected for the design.

18.5 ORINOCO: A Tool for Algorithmic-Level Power Estimation

The algorithmic-level power estimation approaches presented so far are results of academic research. Besides these, the ORINOCO tool [3] is commercially available. It is partly based on the research results presented here [15,18,28,29,33,39,41].

Figure 18.10 illustrates the ORINOCO workflow. ORINOCO accepts algorithmic design specifications in the C or SystemC languages. The input description is analyzed and automatically instrumented. The analysis generates the CDFG of the algorithm, which is needed for optimization. The code instrumentation inserts protocol statements, which capture the activity of the algorithm during execution.

In the analysis and optimization phase, presented on the right-hand side of Figure 18.10, the optimization steps described in the previous subsections are applied to generate a power optimized architecture, which is the base of power calculation. The power models are automatically generated by characterization tools, which are part of the ORINOCO tool suite.

Figure 18.11 presents a window of the ORINOCO design browser after power estimation of the vocoder design. The left column lists the various processes of the vocoder; the other columns present parameters of the design and the results of the power analysis. The graphical power reports enable the designer to

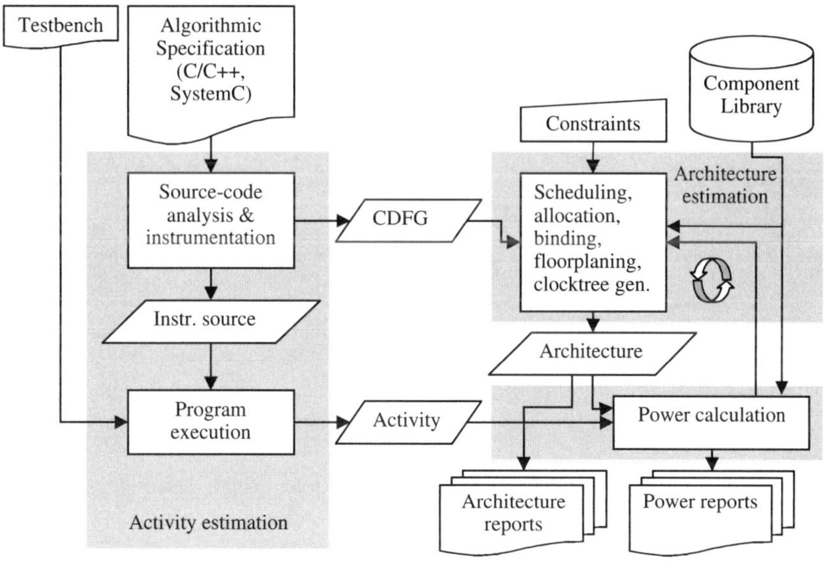

FIGURE 18.10 Workflow ORINOCO Algorithmic-level power estimation.

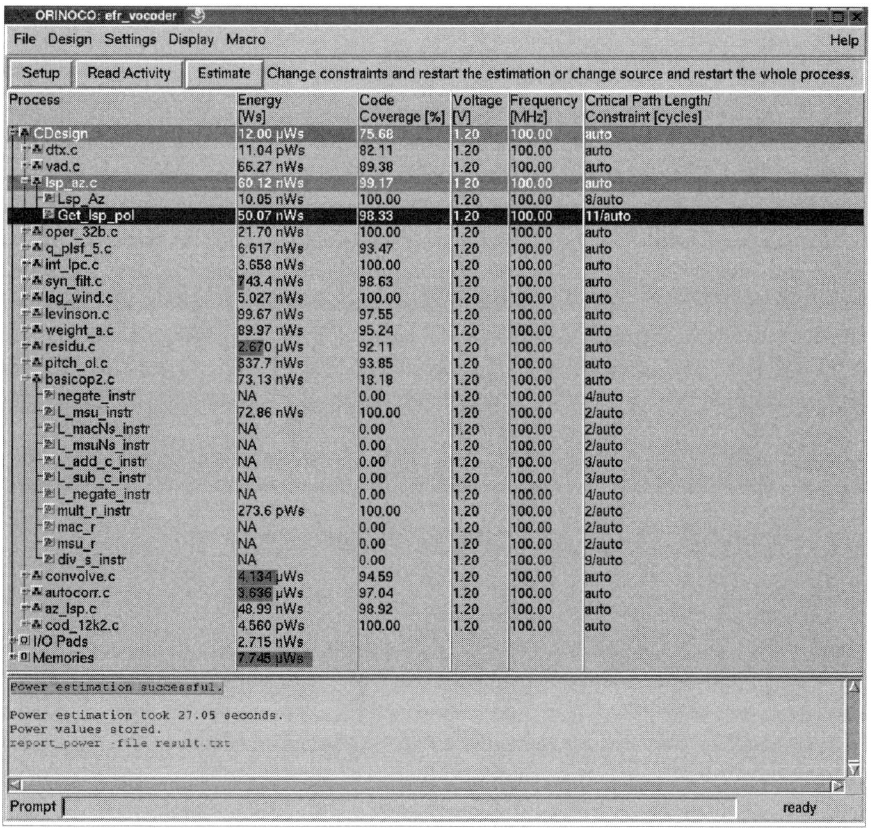

FIGURE 18.11 ORINOCO browser.

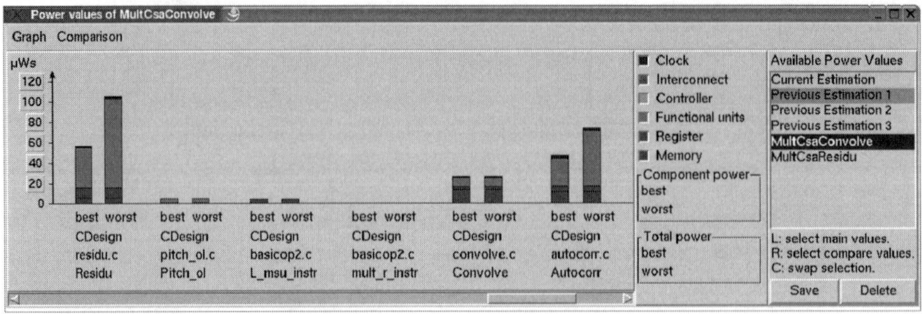

FIGURE 18.12 ORINOCO process power report.

efficiently detect hot spots, the highest potential for power reduction of the design, and the sources of the power consumption. Figure 18.12 is a typical output: the power breakdown of some processes of the vocoder example.

In addition to the power reports, ORINOCO generates outputs, which describe the parameters of the architecture and the floorplan. These outputs can guide the designer to develop a low-power implementation of the current algorithm.

18.6 Conclusion

We described methodologies and techniques used to analyze the power consumption of systems on chip (SoCs) at the earliest possible phase of the design. It is this phase that offers the largest impact on the design, and thus the best chances to optimize the design for performance, power, and cost. Second, performing an early analysis of the expected design properties will help to avoid design iterations from later design phases.

Estimating power at the system-level or the algorithmic-level requires anticipating all the design steps the design will have to go through during the later stages of the design process. It has to predict the impact of the process technology as well as of the design style. It has to consider the application-specific properties of the design, and the performance and cost constraints imposed.

A complete high-level, low-power design flow has been explained with the necessary tools and data to perform power analysis and estimation at the system-level and at the algorithmic-level. This flow is based on the collection, and, if needed, generation as well as the combination of the necessary amount of design data to predict the power consumption of a later implementation in a reliable and sufficiently accurate way.

Nowadays, such a prediction is possible by combining the available techniques into a design and tool flow. The first commercial tools supporting this flow are on the market.

References

[1] 2002 Update of International Technology Roadmap for Semiconductors. Available at http://public.itrs.net/Files/2002Update/Home.pdf.

[2] Hedenstierna, N. and Jeppson, K., CMOS circuit speed and buffer optimization, *IEEE Transactions on CAD*, Vol. 6, No. 3, 1987.

[3] http://www.chipvision.com/dokumente/ORINOCO_WhitePaper.Ext. ChipVision Design Systems AG, *White Paper ORINOCO.*

[4] http://www.bulldast.com/Pee.pdf BullDAST s.r.l., *Power Checker.* Sequence Design, *Power Theater.*

[5] http://www.sequencedesign.com/2_solutions/2b_power_theater.html.

[6] http://www.synopsys.com/products/solutions/galaxy/power/power.html Synopsys, *Galaxy Power Management.*

[7] Rammamoorthy, C.V. and Gonzalez, M.J., Recognition and representation of parallel processable streams in computer programs—II, *Proc. ACM/CSR-ER, Proc. 1969 24th National Conf.*, 1969.
[8] Henkel, J., A low-power hardware/software partitioning approach for core-based embedded systems, *Proc. Design Automation Conf.*, New Orleans, LA, June 1999.
[9] Benini, L. and de Micheli, G., System-level power optimization: techniques and tools, *ACM Trans. on Design Automation of Electronic Syst.*, Vol. 5, No. 2, 115–192, 2000.
[10] Lidsky, D. and Rabaey, J.M., Early power exploration — a World Wide Web application, *Proc. Design Automation Conf.*, Las Vegas, NV, June 1996.
[11] Benini, L., Hodgson, R., and Siegel, P., System-level power estimation and optimization, *Proc. Int. Symp. on Low-Power Electron- and Design,* Monterey, CA, August 1998.
[12] Brooks, D., Tiwar, V., and Martonosi, M., Wattch: a framework for architectural-level power analysis, *Proc. Int. Symp. on Computer Architecture,* Vancouver, Canada, 2000.
[13] Liu, X. and Papefthymiou, M.C., HyPE: hybrid power estimation for IP-based programmable systems, *Proc. Design Automation Conf.*, Anaheim, CA, June 2003.
[14] Catthoor, F., *Custom Memory Management Methodology: Exploration of Memory Organisation for Embedded Multimedia System Design,* Boston, MA: Kluwer Academic Publishers, 1998.
[15] Schmidt, E., von Cölln, G., Kruse, L., Theeuwen, F., and Nebel, W., Memory power models for multilevel power estimation and optimization, *IEEE Transactions on Very Large-Scale Integration Syst.,* Vol. 10, No. 2, 2002.
[16] Fornaciari, W., Sciuto, D., and Silvano, C., Power estimation for architectural exploration of HW/SW communication on system-level buses, *Proc. 7th Int. Workshop on Hardware/Software Codesign (CODES),* Rome, Italy, 1999.
[17] Sarker, B., Nebel, W., and Schulte, M., Low-power optimization techniques in overlap add algorithmus, *Proc. Int. Conf. on Computer, Commn. and Control Technologies: CCCT '03,* Orlando, FL, July/August, 2003.
[18] Stammermann, A., Kruse, L., Nebel, W., Pratsch, A., Schmidt, E., Schulte, M., and Schulz, A., System-level optimization and design space exploration for low power, *Proc. Int. Symp. on System Synthesis,* Montreal, Canada, September, 2001.
[19] Henkel, J. and Li, Y., Energy-conscious HW/SW-partitioning of embedded systems: a case study on an MPEG-2 encoder, *Proc. 6th Int. Workshop on Hardware/Software Codesign (CODES),* Seattle, WA, 1998.
[20] European Telecommunications Standards Institute. Available at http://www.etsi.org/.
[21] Girczyc, E.F. and Knight, J.P., An ADA to standard cell hardware compiler based on graph grammers and scheduling, *Proc. IEEE Int. Conf. on Computer Design,* October, 1984.
[22] Raghunathan, A. and Jha, N.K., Behavioral synthesis for low power, *Proc. IEEE Int. Conf. on Computer Design,* October, 1994.
[23] Brandolese, C., Fornaciari, W., Pomante, L., Salice, F., and Sciuto, D., A multi-level strategy for software power estimation, *Proc. Int. Symp. on System Synthesis,* Madrid, Spain, 2000.
[24] Tiwari, V., Malik, S., and Wolfe, A., Power analysis of embedded software: a first step towards software power minimization, *Proc. Int. Conf. on Computer-Aided Design,* San Jose, CA, November 1994.
[25] Simunic, T., Benini, L., and de Micheli, G., Cycle-accurate simulation of energy consumption in embedded systems, *Proc. Design Automation Conf.,* New Orleans, LA, June 1999.
[26] Hsieh, C.-T., Pedram, M., Mehta, G., and Rastgar, F., Profile-driven program synthesis for evaluation of system power dissipation, *Proc. Design Automation Conf.,* Anaheim, CA, June, 1997.
[27] Mehra, R. and Rabaey, J., Behavioral-level power estimation and exploration, *Proc. 1st Int. Workshop on Low-Power Design,* Napa Valley, CA, April, 1994.
[28] Von Cölln, G., Kruse, L., Schmidt, E., Stammermann, A., and Nebel, W., Power macro-modelling for firm-macros, *Proc. PATMOS,* Göttingen, Germany, September, 2000.

[29] Schmidt, E., von Cölln, G., Kruse, L., Theeuwen, F., and Nebel, W., Automatic nonlinear memory power modelling, *Proc. Design, Automation, and Test in Europe (DATE),* Munich, Germany, March, 2001.

[30] Martin, R.S. and Knight, J.P., Power-profiler: optimizing ASICs power consumption at the behavioral level, *Proc. Design Automation Conf.,* San Francisco, CA, June, 1995.

[31] Musoll, E. and Cortadella, J., Scheduling and resource binding for low power, *Proc. Int. Symp. on System Synthesis,* Cannes, France, September, 1995.

[32] Khouri, K.S., Lakshminarayana, G., and Jha, N.K., Fast high-level power estimation for control-flow intensive designs, *Proc. Int. Symp. on Low-Power Electron. and Design,* Monterey, CA, August, 1998.

[33] Kruse, L., Schmidt, E., Jochens, G., Stammermann, A., Schulz, A., Macii, E., and Nebel, W., Estimation of lower and upper bounds on the power consumption from scheduled data flow graphs, *IEEE Trans. on Very Large-Scale Integration (VLSI) Syst.,* Vol. 9, No. 1, February, 2001.

[34] Müller-Glaser, K.D., Kirsch, K., and Neusinger, K., Estimating essential design characteristics to support project planning for ASIC design management, *Proc. Int. Conf. on Computer-Aided Design,* San Jose, CA, November 1991.

[35] Marculescu, D., Marculescu, R., and Pedram, M., Information theoretic measures of energy consumption at register transfer level, *Proc. Int. Symp. on Low-Power Electron. and Design,* Dana Point, CA, April, 1995.

[36] Nemani, M. and Najm, F.N., High-level area and power estimation for VLSI circuits, *Proc. Int. Conf. on Computer-Aided Design,* San Jose, CA, November, 1997.

[37] Ferrandi, F., Fummi, F., Macii, E., Poncino, M., and Sciuto, D., Power estimation of behavioral descriptions, *Proc. Design, Automation, and Test in Europe (DATE),* Paris, France, March, 1998.

[38] Zhong, L. and Jha, N.K., Interconnect-aware high-level synthesis for low power, *Proc. Conf. on Computer-Aided Design,* San Jose, CA, November, 2002.

[39] Stammermann, A., Helms, D., Schulte, M., and Nebel, W., Interconnect-driven low-power high-level synthesis, *Proc. PATMOS,* Torino, Italy, September, 2003.

[40] Christie, P. and Stroobandt, D., The interpretation and application of Rent's Rule, *IEEE Trans. on VLSI Syst.,* Vol. 8, No. 6, December, 2000.

[41] Stammermann, A., Helms, D., Schulte, M., Schulz, A., and Nebel, W., Binding, allocation and floorplanning in low-power high-level synthesis, *Proc. Int. Conf. on Computer-Aided Design,* San Jose, CA, November 2003.

[42] Prabhakaran, P, Banerjee, P., Crenshaw, J., and Sarrafzadeh, M., Simultaneous scheduling, binding, and floorplanning for interconnect power optimization, *Proc. VLSI Design,* Goa, India, January, 1999.

19

Power Macro-Models for High-Level Power Estimation

Enrico Macii
Politecnico di Torino

Massimo Poncino
Università di Verona

19.1 Introduction

As clearly stated at the beginning of Chapter 38, the addition of the power dimension to the already large area/speed design space dramatically expands the number of available design alternatives. Therefore, in a power-conscious design flow, the ability of estimating the impact of the various choices made by the designers or the effects of automatic optimizations on the final power budget is of utmost importance.

Most of the research on power estimation has initially focused on gate and transistor levels; where, due to the available information on the structure and the macroscopic parameters of the devices, accurate power estimates are expected and satisfactory methods are available.

More recently, techniques for high-level (i.e, register transfer level (RTL) and algorithmic-level) power estimation have been proposed to enable designers to cope with increased design complexity and time-to-market requirements. These approaches are usually based on the construction and the evaluation of abstract power models. This information is supposed to guide the designer in exploring the impact of his or her choices on the quality of the final design.

RTL power estimation is at the transition point between research and industrial applications. Power estimators at the RTL are available as commercial tools [1–3], but they have not yet gained widespread acceptance in the design practice, due to two key technical reasons for this. First, the accuracy gap between gate-level and RTL power estimation has not been fully quantified in an industrial setting. Second, RTL estimators are based on macro-modeling, which requires a preliminary characterization step, where a power macro-model is created for the basic functional components in RTL libraries (e.g., adders and multipliers).

Automated macro-model characterization is a fundamental requirement for the acceptance of RTL power estimation flows in the industrial practice, and this step has been initially overlooked by EDA developers. Fortunately, effective automatic macro-model characterization approaches do exist and are now implemented in the newest commercial tools [3].

As already discussed in Chapter 18, the availability of accurate RTL power models goes beyond the usage in RTL estimation tools; such models are, in fact, also at the basis of the success of power estimators operating at higher levels of abstraction (i.e., algorithmic). This chapter digs deep inside RTL power modeling, and provides a detailed insight to the different facets of the problem of building accurate, yet efficient and easy to characterize RTL power models. Section 19.2 begins by investigating the modeling problem from the theoretical stand-point and by analyzing what are the main dimensions that need to be explored during model definition. In particular, the issues related to the choice of the physical quantities (i.e., the parameters) upon which the models depend are discussed. The semantics associated with the models are also discussed. We conclude by looking at the problem of model representation and storage.

Section 19.3 continues with a discussion of the most relevant steps of a typical macro-modeling flow. Advantages and drawbacks of the macro-modeling technology are outlined, and we demonstrate how such a technology can be exploited for the definition of an RTL power estimation methodology, which is at the foundation of modern EDA tools for low-power design.

Section 19.4 addresses the issues related to the integration of the macro-modeling based approach to RTL power estimation into industry-strength design flows. We discuss how we can deal with RTL descriptions specified through hardware description language (HDLs) and the implications that HDLs may have on the required power modeling capabilities. In addition, we illustrate how models can be enhanced to account for the effects that synthesis and technology may have on the power consumed by the final implementation of the design. Finally, Section 19.5 closes the chapter with some concluding remarks.

19.2 RTL Power Modeling

The problem of power estimation at the RTL amounts to building a power model that relates the power consumption of the target design to suitable quantities. In formula, $P = P(X_1,\ldots,X_n)$, where $X_i, i = 1,\ldots,n$ are the n model parameters.

The construction of a model P implies addressing the following issues:

- The granularity of the model (i.e., to what types of components the power model is referred); this issue implies the definition of the reference architectural model for RTL power estimation.
- The choice of the model parameters (i.e., what and how many parameters upon which the model should depend).
- The semantics of the model (i.e., what is the interpretation of the values returned by the model).
- The way the model is built (i.e., how parameters are put in relation with power); this issue also involves how the model is represented and stored.

The rest of this section discusses the preceding list of four modeling space dimensions, which result in various modeling alternatives.

19.2.1 Model Granularity

In principle, building an RTL power model for an RTL design could be done by considering the design as a monolithic entity. In this case, the model should relate power consumption to properties of the description that can be observed from its RTL I/O behavior.

The choice of a single, monolithic model has several drawbacks:

- Its construction will be extremely time-consuming for RTL designs of realistic size.

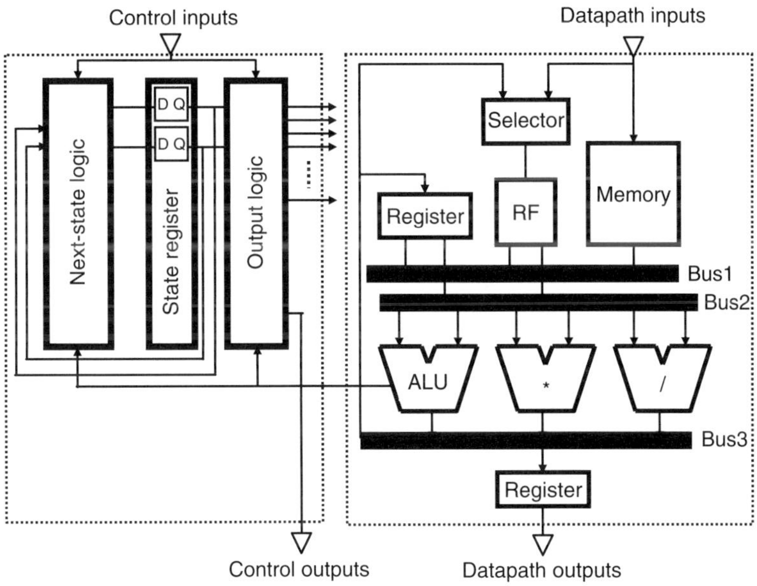

FIGURE 19.1 Architectural template for RTL power estimation.

- Its accuracy will be rather low because we would be trying to express a complex behavior with a single model.
- Its reusability will be quite limited, given its highly customized nature.

These shortcomings prompt for a smaller granularity of the power models. This can be achieved by defining an architectural template and by assuming that all RTL descriptions will map onto that template. The advantage of such a template is that of defining some finer grain objects, for which specific power models can be built. The previous limitations are thus removed by construction: target blocks are smaller, so model construction is simpler and more accurate. Furthermore, because all designs will map onto the same template, reuse will be maximum.

Most of the scientific contributions on RTL power modeling assume (sometimes with minor variants) an architectural template that views an RTL design as the interaction of a datapath and a controller, which fits well the so-called finite state machine with datapath (FSMD) model [4]. Variants to the base FSMD model concern the structure of the controller (e.g., sparse logic implementation vs. wired logic), the structure of the interconnect (e.g., number, type, and size of the buses), or the supported arithmetic operations (e.g., number and type of available functional units). Figure 19.1 depicts an example of a conventional FSMD organization, in which the basic building blocks are explicitly exposed: the controller (shown on the left) and the datapath (shown on the right), with the latter consisting of a register file (and possibly some sparse registers), a memory, various interconnection buses, and some functional units (integer or floating-point).

Fitting an RTL description to this template allows us to restrict the scope and the granularity of the power models to those of the following five main building components: the controller, the registers, the memory, the buses, and the functional units. This approach is followed by most RTL power estimation approaches proposed in the literature [5–12].

19.2.2 Model Parameters

The choice of what parameters should be included in the model is constrained by the fact that they must be quantities that are observable at the RTL. Given the architectural assumptions described in the previous subsection, the traditional model for switching power translates into the following high-level expression:

$$P_{total} = k \cdot \sum_{\forall \; module \; i} A_i C_i \tag{19.1}$$

where A_i and C_i denote the switching activity and the physical capacitance of the generic component i, respectively. k represents the $f \cdot V_{dd}^2$ term, which can be considered as a scaling factor at the RTL because V_{dd} is a technological variable, and the clock frequency is specified up front in the RTL description. Notice that the subscript *total* refers to the fact that power is computed over all components; P_{total} is actually the value of the average power.

Equation (19.1) decouples the problem of building a power model into that of building a model for activity and a model for capacitance, for each type of component. Generally, activity and capacitance models will depend on different parameters because they are affected by different physical quantities.

19.2.2.1 Activity Parameters

Activity is generally easier to model because, even at the RTL, it is a well-defined quantity; however, because RTL simulation is able to monitor activity only at the granularity of the clock cycle, RTL activity is an approximation of the actual one. Therefore, intra-cycle activity, such as that caused by glitches, cannot be extracted from RTL simulation.

Activity models rely on activity parameters, including:

- Bit-wise static and transition probabilities (i.e., quantities referred to specific input or output signals of a component)
- Word-wise static and transition probabilities (i.e., quantities referred to input or output values of a component)

Most activity models proposed in the literature use these probability measures as parameters; the choice between bit- and word-wise quantities allows us to trade off model accuracy for model complexity (i.e., the number of parameters). A n-input, m-output component will require $n + m$ bit-wise parameters, but only two word-wise parameters.

These basic models may be enhanced by using additional activity parameters such as:

- Transition density [13], defined as an average (over time) switching rate. At the RTL, where time is intrinsically discretized to clock edges, density can be regarded as a quantity that incorporates switching and clock frequency information. Density is usually more useful at the gate level, because it allows to capture activity also for nonperiodic (i.e., nonclocked) signals.
- Correlation measures, which take into account spatial correlation between individual inputs or outputs of a component. Spatial correlation, defined as the joint probability of two signals being one, is roughly equivalent to computing the correlation coefficient between these signals. As for probabilities, spatial correlation can be computed bit-wise (one value for each signal pair), or word-wise, by averaging all bit-wise values over the number of signals, to get a single quantity [14]. Notice that transition probabilities already account for temporal correlation (between signals or words).
- Entropy [15,16], which can be used in place of transition probability. In fact, it can be demonstrated that the entropy of a digital signal vs. its static probability p is given by the formula: $p\log_2(1/p)+(1-p)\log_2(1/1-p)$, which closely mimics the behavior of transition probability, with an expression of $2p(1 - p)$. In this sense, entropy does not provide particular advantages over transition probability as a parameter, although it can be used also as a measure of complexity [17].

As a general comment, all activity parameters can be used to represent the switching activity of any of the components of the FSMD model discussed in the previous subsection (i.e., registers, datapath components, memories, buses, and a controller). We can say then that activity parameters are independent of the architecture of the components, and a generic activity model template could be used for all types of components.

19.2.2.2 Complexity Parameters

Modeling physical capacitance is more difficult than modeling switching activity. As a matter of fact, the term "physical" suggests that it is unlikely that we are able to link capacitance to quantities observable at the RTL. For example, the relation between a generic datapath component (e.g., an adder) and its physical capacitance may not be so intuitive. We thus expect capacitance models to be generally less accurate than activity models.

Despite this, RTL capacitance models can be derived with a reasonable degree of accuracy. They all rely on the intuitive observation that capacitance will be roughly related to the number of "objects" (i.e., gates, transistors, or similar lower-level primitives) of the target component. In other words, physical capacitance at the RTL is approximated by complexity, and we thus speak of complexity models, based on complexity parameters.

At the RTL, only a few complexity parameters are available. Those that map onto the basic building blocks of our architectural template are:

- The width of a component, meant as its number of inputs and outputs. This parameter applies to any type of component.
- The number of states. This parameter applies only to the controller, for which the notion of state is explicit.

As mentioned in the previous section, some works have also used entropy as an approximation of complexity [15,16,18], although the relation between the two quantities is quite weak and mostly valid for obsolete circuit technologies.

Any complexity parameter different from the two listed above would require some additional information derived from back-annotation of physical information of previous implementations.

Given the limited choice of parameters, it is quite common that capacitance models exploit information about the "architecture" or the implementation style of a given component to customize the form of the model, and impose some predefined mathematical dependencies. For instance, the model for a N-bit ripple-carry adder might be something like $C_{rpc\_adder} = k_1 \cdot N$, recognizing its "linear" complexity in the number of input bits. Using the same line of reasoning, a $N \times N$ array multiplier might have a model of the type: $C_{arr\_mult} = k_2 \cdot N^2$ [19,20].

This implies that capacitance models for the various components will not fit to a single generic template, but they will have different shapes (e.g., equations) for each type of component.

19.2.3 Model Semantics

So far, the discussion has focused on models for average power, which is normally used as a metric to track battery lifetime or average heat dissipation. In this case, the semantics of the model is that of having a single figure to represent the consumption of the target description. Average power models are called cumulative power models [21].

However, the notion of cycle intrinsic of RTL simulation allows us to obtain a power model with a richer semantics by simply changing the way we collect statistics. The first step in this direction consists of modifying the model of Equation (19.1) as follows:

$$P_{total} = k \sum_{\forall\, cycle\ j} P_j = k \sum_{\forall\, cycle\ j} \sum_{\forall\, module\ i} A_{ij} C_{ij} \tag{19.2}$$

where P_j denotes the power consumption at cycle j, which can be obtained by summing the power consumption for each component (as in Equation (19.1)), this time using activities and capacitances of component i at each cycle j.

The semantics of the model of Equation (19.2)is cycle-accurate, because it allows to track cycle-by-cycle (total) power. Equation (19.2) can be thought of as a particular case of a more general model in which power is computed over a sliding window of size W; in this case, $W = 1$ corresponds to the cycle-accurate model, whereas when W equals the length of the simulation stream we fall into the cumulative model of Equation (19.1) [22,23].

The use of a cycle-accurate model clearly affects the choice of the model parameters. For example, transition or static probabilities are not suitable quantities anymore because, as statistical measures, they are intrinsically "average." Conversely, cycle-accurate models should use cycle-based activity measures, such as the number of bit toggles between consecutive patterns (i.e., the hamming distance) [23–25], or the values of consecutive input patterns [21,22]. In the former case, one single parameter is able to capture the information, while in the latter case, the parameter space consists of all possible word pairs, and may thus become quite large.

A cycle-accurate model provides several advantages over a cumulative one. First, it goes beyond the bare evaluation of average power and can be used to perform sophisticated analysis of power consumption over time, which may be required in some application, such as reliability, noise, or IR drop analysis [23]. In addition, a cycle-accurate model is more accurate than a cumulative one, not just because it provides a series of power values as opposed to a single one. In fact, the relation between input statistics and power is nonlinear: average consumption is usually different from the consumption associated to average input statistics, especially when power consumption varies significantly over time. Therefore, even when average power is the objective, averaging the series of cycle-by-cycle values will yield a more accurate estimate than a model of average power. On the negative side, cycle-accurate models require significant larger storage space than cumulative ones.

19.2.4 Model Construction and Storage

The last modeling issue to be considered concerns the way models are built and stored. In our context, the two dimensions are relatively independent, so that we can analyze them separately.

19.2.4.1 Model Construction

Two options are used to build an RTL power model: a top-down (or analytical) approach, and a bottom-up (or empirical) one [5]. All power models proposed in the literature fall into one of these two categories.

Top-down approaches relate the power consumption (but also the activity and the physical capacitance) of an RTL component to the model parameters through a closed formula. The term "top-down" refers to the fact that the model is derived directly from the RTL description, and it is not based on lower-level information. For this reason, such a formula normally has a physical interpretation. Analytical models are particularly useful in two cases:

1. When dealing with a newly designed circuit, for which no information of previous implementations is available
2. When the implementation of the circuit, even if not available, follows some predictable template, which can be exploited to force some specific relation between the model parameters

Memories are good examples of blocks for which analytical models are particularly suitable. Their internal organization is well-known and relatively fixed (e.g., cell array, bit-lines, word-lines, decoders, MUXs, and sense amps), thus allowing accurate modeling based on various "internal" parameters [26,27].

If we exclude these special cases, however, top-down models are not very accurate because their link to the implementation (e.g., technology, or synthesis constraints) are quite weak. For instance, the analytical models based on entropy [15,16] are totally insensitive to technology and timing information, and are mostly useful for architectural exploration rather than actual estimation.

Bottom-up approaches, conversely, are based on estimating the power consumption of existing implementations, from which the actual power model is derived. Typically, the template of the power model (i.e., the parameters and a set of coefficient used to weigh the parameters) is defined up front; statistical techniques are then used to fit the model template to the measure of power values. This approach is known as macro-modeling, and has proved to be a very accurate and robust methodology for RTL power

Table 19.1 Model Construction Space

	Equation	Lookup Table
Top-Down	√	Not Used
Bottom-Up	√	√

estimation, and can be considered the state-of-the-art solution. Section 19.3 is devoted entirely to the detailed description of the macro-modeling flow and of its role in the definition of an RTL power estimation methodology.

19.2.4.2 Model Storage

The issue of model storage is concerned with the form of the model. Because models express a mathematical relation between power and a set of parameters, the problem amounts to that of representing such a relation. The two options are:

1. Equation-based models
2. Table-based models

The classification is self-explanatory, and it corresponds to the choice of representing a relation as a continuous function (equation-based models), or a discrete-function approximated by points (table-based models).

These two types of models differ in their storage requirements and robustness; the latter is a measure of model sensitivity to the conditions (i.e., the experiments) used for the construction. In that sense, robustness is an issue only for empirical models. Section 19.3 provides further insight on model robustness.

Concerning storage requirements, equation-based models are clearly much more compact than table-based ones. Generally, an equation will only require the storage of the coefficients of the model, as opposed to a full table. In addition, the accuracy of a table-based model is directly proportional to its size (the denser the table, the higher the accuracy), whereas the accuracy of an equation-based model is independent of the model size.

To summarize, the model construction approach and the model shape are substantially independent characteristics, and, in principle, all combinations of options are feasible. As presented in Table 19.1, however, top-down approaches do not resort to table-based models because they naturally try to construct a formula. In this case, approximating the function by points would not bring any particular advantage.

19.2.5 Accuracy Issues

One important point to be addressed concerns the estimation accuracy that RTL power models can guarantee. In the literature, a 20% estimation error with respect to gate- or transistor-level estimates appears to be accepted as the mark that defines "accurate" models; however, power estimation accuracy is affected by so many factors that expressing it as a single figure can have a very poor meaning. Moreover, sometimes even the definition of accuracy itself is not well understood. This section analyzes what affects the evaluation of the estimation accuracy, and provides some hints on how to critically interpret the results available in the literature.

Having accuracy as a target, we refer to empirical models, which are intrinsically more accurate than analytical ones.

19.2.5.1 Accuracy Metrics

In principle, accuracy can be defined as the (absolute or relative) error of the estimate with respect to the "measured" quantity, that is,

$$E = \frac{|P_e - P|}{\max(P_e, P)},$$

where P_e is the power obtained from model evaluation and P the power obtained from gate- or transistor-level simulation. P_e and P refer to a single evaluation of the model (i.e., for one assignment of the parameters) and to a single low-level simulation run, respectively. Therefore, E is a good indicator for a specific execution of the estimation flow.

However, a correct assessment of the accuracy of the model requires that the dependence of power consumption on the input statistics is taken into account. We call robustness the capability of a power model to provide accurate power estimates over a wide range of parameter values (i.e., statistics).

Assuming that S estimation runs have been performed (using different input values), robustness can be computed by averaging the error E over S experiments, that is:

$$E_{avg} = \frac{\sum_{i=1,\ldots,S} E_i}{S}$$

where subscript i denotes the generic experiment.

Sensitivity to input conditions can be assessed by way of the standard deviation of the relative error, that is:

$$SD = \sqrt{\frac{1}{S} \sum_{i=1,\ldots,S} (E_i - E_{avg})}$$

In the case of cycle-accurate models, the root-mean-square (RMS) of the error can be used instead of the average error to track the robustness.

19.2.5.2 Choice of Experiments

Another important element to be considered when evaluating the accuracy of a model is the choice of the set of experiments used to construct the model. This issue is critical in the case of empirical models, where the model is built based on a set of measured points. A common mistake in this case is to evaluate model accuracy on the same set of experiments (usually, the input/output (I/O) statistics) used to build the model. Such error represents the intrinsic error of the model (typically, very low), but it is not a significant measure of the quality of the model under generic conditions, which can be very different from those used for model training.

This is again a matter of model robustness; in fact, we should distinguish between the in-sample accuracy (the intrinsic error of the model) and the out-of-sample accuracy (the error under all other conditions) [28].

The problem of ensuring robustness has been partially solved by resorting to table-based models, where the parameter space is discretized into equivalence classes, and the dependence of the model on the specific values used for model training is weaker. However, a thorough evaluation of even the most robust and (intrinsically) accurate models has demonstrated that estimation errors higher than 100% can be obtained for specific corner cases (namely, input conditions where the activity parameters have very low values [23]).

A realistic evaluation of model accuracy should be carried out by measuring the suitable metrics (e.g., average error or RMS error, and the relative standard deviation) over a very large set of parameters configuration, including pathological cases. Under these assumptions, an average accuracy of 20% is quite hard to achieve, and figures around 30 to 40% are by far more realistic.

19.3 RTL Power Macro-Modeling and Estimation

In statistical analysis, the term macro-model defines models with a “coarse” level of details, which are thus used for overview purposes (as opposed to micro-models, which incorporate finer levels of details).

The definition of macro-model fits well to RTL power models because the latter are employed to relate quantities pertaining to different abstraction levels, such as RTL parameters to the actual power.

In RTL power estimation, the term macro-model has a more restricted meaning because it is used to identify empirical models, without further distinction between other characteristics.

A number of articles dealing with power macro-models have appeared in the literature. Besides those referenced in Section 19.2, other approaches have been proposed recently [29–36], thus leading to a large variety of macro-models spanning many points of the power modeling space. In fact, cycle accurate and cumulative, activity-based and complexity-based, and equation-based and table-based models have been successfully demonstrated.

Despite the differences that do exist between the various macro-models, the design of a macro-model goes through a well-defined sequence of steps; the following describes the macro-modeling flow in detail.

19.3.1 Macro-Modeling Flow

The construction of an accurate macro-model consists of the following four major steps:

1. Choice of model parameters. Although this step applies also to nonempirical models, it is particularly important for macro-models because it defines the parameter space and it affects the complexity of the next macro-modeling steps. Generally, the goal of this phase consists of choosing what and how many parameters X_i will be part of the model.
2. Design of the training set. The training set is a representative subset of the set of all possible pairs of input vectors that will be used to construct the model. The decisions to be made during this phase concern the size of the training set (i.e., the total number of pairs of input vectors) and the statistical distribution of the pattern pairs in the training set. Although the former issue has to do with the total simulation time, the second is more critical; a bad statistical distribution of the training set may easily offset the advantage of a large number of vector pairs.

   What defines a "good" distribution depends on what are the parameters chosen for the model. A general requirement for the training set is that it should span the domain of all the model parameters as much as possible. When one or more domains are not sufficiently covered by the training set, we say that the model is insufficiently trained.

   For instance, if the parameter of the model is the switching activity, the choice of random patterns as a training set would not be a good one because only a very small portion of the activity domain (i.e., the one around a switching activity of 0.5) would be exercised.

   Although the statistical distribution of the training set is important, it is not the only criterion to be used for choosing the training set. In fact, it is also important to consider how the training set is representative of the actual conditions under which the target component for which power is being modeled will be used. For instance, if we take switching activity as a parameter, we should consider that input vectors with low switching activity will be more frequent than those with high switching activity, in normal operating conditions of the component. In this case, a larger number of low-to-medium activity vectors should be included in the training set.

   In the case of table-based models, the generation of the training set is subject to additional constraints. In fact, models stored as lookup tables are defined only for a set of discrete points, corresponding to specific values of the parameters. This implies that the training set is constrained to vector pairs with values of the parameters corresponding to such discrete points only.
3. Characterization. This step entails the usage of the training set to generate a set of points in the power-parameter space. For each element in the training set (i.e., a vector pair), a corresponding value of power is obtained by means of an accurate, low-level power simulator (i.e., a gate-level or a circuit-level simulator).

   More sophisticated schemes introduce an additional averaging of the samples, by grouping a set of vector pairs and associating a single power value to the set (instead of to each vector pair). This solution is often preferred because simulating a set of patterns instead of a single one increases the confidence of the resulting power value.

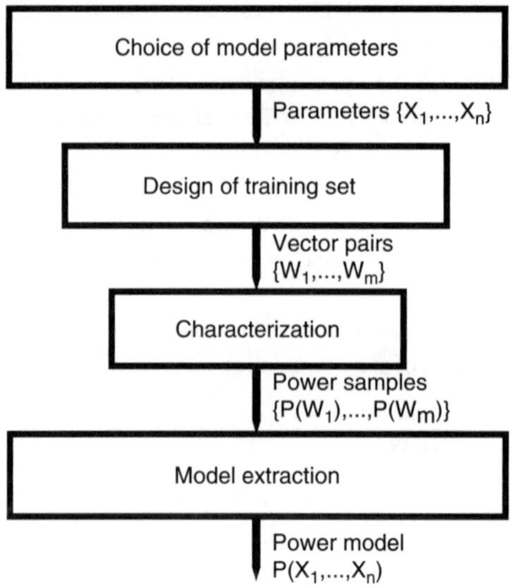

FIGURE 19.2 Summary of macro-modeling flow.

4. Model extraction. This phase consists of deriving the model from the set of power data obtained in the previous step. The actual calculation depends on how the model is stored. For equation-based models, a least-mean-square (LMS) regression engine is applied to the sample generated during characterization. Depending on the options offered by the regression engine, different equation families can be built (e.g., linear, polynomial, logarithmic/exponential).

   For table-based models, the extraction of the model consists of collecting the power values for each of the discrete points of the parameter space. After characterization, each point (i.e., table entry) may contain many power values; the decision on whether to store a single value (e.g., the average of the values) or the complete list of values in each table entry depends on how much room is available for model storage and on how the model will be used for power estimation.

Figure 19.2 summarizes the four steps of the power macro-modeling flow, emphasizing the inputs and the outputs of each phase.

19.3.2 Macro-Modeling Example

Let us apply the macro-modeling flow to the case of a 16-bit ripple-carry adder. To emphasize the difference between equation-based and table-based models, we deal with the two cases separately.

1. Choice of model parameters. For the sake of illustration, we consider a power macro-model containing only one parameter, namely the average switching activity of the inputs S_{in} to the adder. This is a real number between 0 and 1, and it is computed as the number of input transitions between input pattern pairs, divided by the total number of inputs (here, 32). This step is common to both equation-based and table-based models.
2. Design of the training set.
   - Equation-based model. We choose as training set a set $(w_1,\ldots,w_m)$ of $m = 3000$ pattern pairs, with a uniform distribution of S_{in} between 0 and 1.
   - Table-based model. Because the lookup table will have a finite number of entries, we have to choose a proper discretization of the parameter. We use intervals of S_{in} of 0.1; the (one-dimensional) table representing the model will thus have 10 entries. Then, we select 300 pattern pairs for each discrete value of *Sin* (0.1,0.2,…,1.0), for a total of 3000 vector pairs. Notice that, for table-based models, the goodness of the statistical distribution of the training set is

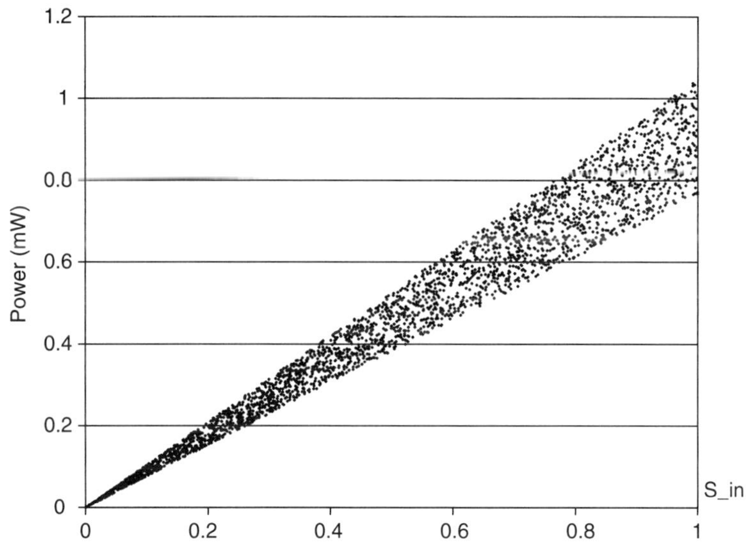

FIGURE 19.3 Scatter plot for the equation-based model.

not an issue. The discretization forces the training set to be designed based on specific values of the parameters.

3. Characterization.
   - Equation-based model. For each vector pair *wi* the corresponding power *P*(*wi*) is determined by means of a low-level simulator. Each simulated value is plotted in the (*P*,*Sin*) space, as depicted in Figure 19.3. Note how the cloud of points illustrates the intuitive trend of an increased power consumption for higher values of the input switching activity.
   - Table-based model. As for the equation-based model, for each vector pair *wi* the corresponding power *P*(*wi*) is computed. Figure 19.4 is a pictorial representation of the set of power data points, which emphasizes the discrete nature of the model.
4. Model extraction.
   - Equation-based model. Given the scatter plot of Figure 19.3, we can derive an equation from it by simply running LMS regression on the set of raw data. Assuming that linear regression is used, and that some corrections are applied to force the intercept with the *y*-axis to be 0, we obtain the following model: $P(S_{in}) = 0.9079 \cdot S_{in} [mW]$.
   - Table-based model. One possible option to store the model is to build a table with one row for each value of S^i_{in}, and a single entry, calculated as the average of the corresponding values $P(S^i_{in})$. With this choice, our table-based model *P* will be as follows:

0.0	0.000
0.1	0.091
0.2	0.181
0.3	0.271
0.4	0.361
0.5	0.451
0.6	0.541
0.7	0.631
0.8	0.721
0.9	0.821
1.0	0.911

   - Using the average as a single representative is based on the assumption of a uniform distribution of the values. More sophisticated solutions analyze the actual distribution of the values and

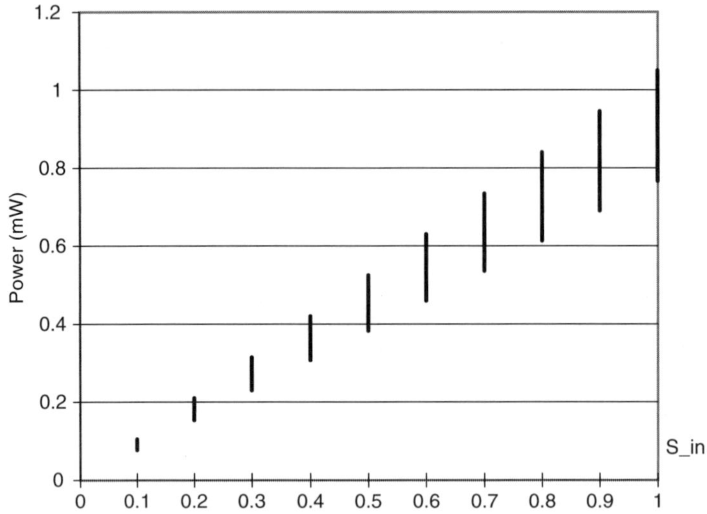

FIGURE 19.4 Plot of power data points for the table-based model.

determine an average value. Some approaches even apply linear regression locally, thus storing an equation instead of a single value [37].

19.3.3 RTL Power Estimation Based on Macro-Modeling

The macro-modeling technology discussed in the previous subsections can be successfully used to enhance state-of-the-art RTL-to-physical design flows with power estimation capabilities.

Assuming that the design to be estimated is described by an FSMD, the estimation procedure consists of the following basic steps [10]:

1. Identification of the individual components in the FSMD. This implies identifying and separating the datapath components from each other and from the finite state machine that represents the control. This is needed to enable the power estimator to generate the power macro-models for each component in the FSMD.
2. Simulation of the FSMD. An RTL simulator is used to trace all the internal signals that define the boundaries between the various components in the FSMD. This is of fundamental importance for the evaluation of the macro-models, which is necessary to complete the estimation procedure.
3. Power estimation. The design hierarchy is traversed at the top level, from the inputs toward the outputs. For each component, the following operations are carried out:
   - Model construction. For each component, the proper model is built using the macro-modeling flow illustrated earlier in this section. A caching strategy can be used to limit the number of times that macro-models are built. More specifically, after the macro-model for a given component is built, it is stored into a cache, so that it can be reused in subsequent runs of the power estimator, should the technology and design constraints not change.
   - Model evaluation. The proper parameter values obtained during the RTL simulation phase are plugged into each model to get the actual values of power consumption.

The total power information for the design is then obtained by summing up the contribution of the model of each component, so that both a total power budget and a power breakdown can be reported to the user. Estimation of the individual components present in the hierarchical description can also be provided. This is possible because components represent the finest level of granularity in the RTL description.

19.4 RTL Power Estimation in Real-Life Settings

Section 19.3 has defined an effective and robust methodology for RTL power estimation based on the macro-modeling paradigm. This section revisits the problem of RTL power estimation from a designer's perspective; and in particular, we try to answer the question of how the macro-modeling approach can be integrated into a standard design flow.

As mentioned in Section 19.2, RTL power macro-modeling relies on a view of an RTL description that fits into the FSMD model. We call this abstraction level structural RTL to distinguish it from the typical designer's view of RTL, usually that of an HDL description containing clocked processes communicating through signals. We call the latter abstraction level cycle-accurate RTL. The gap between these two views of RTL amounts to the presence or absence of structure in the specification, respectively.

In a traditional RTL synthesis flow, the cycle-accurate description undergoes a compilation process that transforms it into an internal format in which structure is made explicit. In VHDL terminology, the compilation process includes the steps of analysis and elaboration; the latter takes care of the delicate task of inferring RTL components from HDL operators.

Therefore, the HDL compilation appears able to fill the gap between cycle-accurate and structural RTL, and, therefore, consistency exists between the two styles. Unfortunately, this consistency is only apparent: HDL compilation builds a structure based on automatic inferencing rules that have limited semantic power. Therefore, the view of structural RTL as an FSMD is ideal. In a realistic design flow, HDL compilation results into the instantiation of four basic types of primitives: RTL components, logic gates, multiplexers, and flip-flops.

At first glance, the difference with the FSMD model is limited to how the controller is specified. Because RTL components are the equivalent of RTL operators, gates, selectors, and memory elements should represent the building blocks that constitute the controller. In practice, however, the degree of inferencing is quite limited, and only the basic HDL operators are inferred ("+", "−", "*",...). For example, registers of the datapath are not instantiated as a single component, and are translated to a set of memory elements. Figure 19.5 depicts how a fragment of an RTL VHDL description is transformed by analysis and elaboration.

The RTL code on the left of Figure 19.5 describes the computation of $z = |x - y|$, which, after HDL compilation, is transformed into the internal database presented on the right of Figure 19.5 (the notation is not bound to any specific synthesis tool). We notice that the "$\geq$" and the "−" operations are translated into the corresponding synthetic operators: One `GTE (component001)` and two `SUB (component002 and component003)`. There is no clear notion of a controller, however, and the rest of the database consists of six gates, one multiplexer (MUX), and four individual flip-flops. Even in this simple example, where the control is limited, a nonnegligible amount of sparse logic exists.

This limited degree of inferencing is due to the fact that the compiled HDL database is meant for being a starting point for synthesis, which will transform primitives into actual library cells. Thus, the instantiation of larger blocks is not so useful to the synthesis tool.

A more detailed analysis on a set of industrial designs confirms the results of the previous example [39]. Although the distribution of the various block types varies significantly across different benchmarks, the relative importance of other primitives with respect to RTL components remains high.

All this discussion demonstrates that, in a realistic design flow, the granularity of a structural RTL design is usually very small, and only a limited number of RTL components is exposed in the RTL internal database. As a consequence, accurate RTL power models for the non-RTL primitives are as important as the macro-models for the RTL components to achieve satisfactory power estimates.

Solutions such as those offered by Synopsys DesignWare [41] represent a step toward a higher degree of inferencing: a library of RTL components is linked to the design, and each RTL component (e.g., a floating-point divider) comes with a VHDL/verilog API that allows designers to replace standard HDL operators (e.g., "/") with a function call (e.g., `DWF_DIVF()`). Designers tend to be uncomfortable with this paradigm, however, because it complicates the reuse of existing designs.

```
ENTITY example IS
GENERIC (W : integer := 4);
PORT(
  clk : IN bit;   -- Global clock
  xin : IN std_logic_vector(W-1 downto 0);
  yin : IN std_logic_vector(W-1 downto 0);
  oup : OUT std_logic_vector(W-1 downto 0));

END example;

ARCHITECTURE rtl OF example IS
BEGIN
  main: PROCESS
  VARIABLE   x   : integer;
  VARIABLE   y   : integer;
  VARIABLE   z   : integer;
  BEGIN
    WAIT UNTIL clk = '1';
    x <= conv_std_logic_vector(xin,W);
    y <= conv_std_logic_vector(yin,W);

    IF (x >= y) THEN
       z := x - y;
    ELSE
       z := y - x;
    END IF;
    oup <= conv_std_logic_vector(z,W);
  END PROCESS main;
END rtl;
```

```
BUF     gate001;
ANDNOT  gate002;
ONE     gate003;
BUF     gate004;
NOT     gate005;
AND2    gate006;
--------------------
MUX     mux001;
--------------------
DFF     ff001_0;
DFF     ff001_1;
DFF     ff001_2;
DFF     ff001_3;
--------------------
GTE     component001;
SUB     component002;
SUB     component003;
```

FIGURE 19.5 Initial RTL VHDL code and internal database after compilation.

The preceding discussion highlights the two main requirements of RTL power estimation in a realistic, HDL-based design flow:

- Power macro-models for the basic RTL components (e.g., "+", "−", "*", "/", "=", "≠", "≥", "≤").
- Power models for the other types of primitives, namely gates, multiplexers, and flip-flops at the RTL. Hereafter, we group these primitives under the name of non-synthetic operators.

Macro-models for RTL components have been extensively discussed in Section 19.2 and Section 19.3. In the sequel, we address the problem of constructing RTL power models for non-synthetic operators.

19.4.1 Power Models of Non-Synthetic Operators

Power estimation of non-synthetic operators at the RTL is different from the case of RTL components. In principle, their power models are quite well understood (e.g., the power model of a NAND gate), and the difficulty arises from two main facts:

- The RTL netlist is different from the actual netlist that will be produced by synthesis, which will optimize the design, possibly under some design constraints. Optimization will reduce the total number of primitives as well as their distribution.
- The RTL netlist is expressed in terms of technology-independent primitives; synthesis, on the contrary, will map primitives onto instances of library cells.

These two facts complicate the problem because our underlying assumption is that RTL power estimation must not rely on RTL synthesis. We usually talk of RTL power estimation as pre-synthesis power estimation.

This leads us to the main issue behind power estimation at the RTL for these types of primitives: estimating the impact of RTL synthesis on power consumption.

Solutions to this problem are not addressed in the literature because the topic is deemed as a practical issue, thus of limited interest from the research point of view.

One exception is the approach followed in [38], which in fact targets an industrial design flow. In that work, the RTL description is decomposed into a set of fine-grain power primitives by a process that combines elaboration and low-effort synthesis. Power primitives are non-synthetic operators for which straightforward power models can be used. The technology independence issue is solved by resorting to fast synthesis.

The rest of this section describes one possible approach that solves both the synthesis estimation and the technology independence issues in an integrated way [39]. This solution, which has been implemented into BullDAST PowerChecker, has provided an estimation accuracy of about 20% with respect to post-synthesis estimates.

One important point to understand is that to achieve acceptable estimation accuracy, we cannot be completely independent of the synthesis tool used in the actual design flow. In the following, without loss of generality, we refer to the Synopsys DesignCompiler [40] flow; similar considerations may apply to any other synthesis flow.

19.4.1.1 Estimating the Effects of Synthesis

Estimating the effects of synthesis consists of relating the pre-synthesis netlist to the postsynthesis one by means of an empirical macro-model, parameterized with respect to complexity and activity parameters.

In this context, complexity parameters are N_R, the number of flip-flops (FF), N_M, the number of MUXs, and N_G, the number of gates in the structural RTL description. The activity parameters extracted from RTL simulation are A_R, the average toggle rate of the FFs, A_M, the average toggle rate of the MUXs, and A_G, the average toggle rate of the gates.

The characterization phase is based on the application of the synthesis flow to a set of RTL benchmarks used as a sample. Regression analysis is applied to the points corresponding to the gate-level power estimation of the synthesized benchmarks, to generate an equation $P = P(N_R, N_M, N_G, A_R, A_M, A_G)$.

The exploration described in Bruno et al. [39] yields a second-order model that is able to provide an average accuracy of about 10%.

19.4.1.2 Achieving Technology Independence

The model described in the previous subsection is technology-dependent because the characterization process refers to a given technology library. The synthesis onto a different technology library (e.g., with different cell types and complexities, or a different feature size) would result in different synthesized implementations, and thus different models.

One way of achieving a model independent of the technology library is that of defining a model scaling mechanism [42]. Starting from a reference technology library L_{ref} (i.e., the one used to build the original power model), a technology scaling factor K is determined for a new technology, such that if P_{ref} is the reference power model, the model for the new library L_{new} will be obtained as $P_{new} = K \cdot P_{ref}$.

The determination of K can be done by defining a sort of "golden" RTL design G to be used for the calibration of the model. G is first synthesized onto L_{ref} and its power consumption P^G_{ref} is evaluated; then, G is synthesized onto L_{new} to determine its power consumption P^G_{new}. K is thus simply obtained as $K = P^G_{new} / P^G_{ref}$.

Concerning the choice of G, a relatively simple description is preferred because it maps onto a small number of non-synthetic operators. For instance, an AND gate is used in Bruno et al. [39], achieving an estimation accuracy of about 20%.

19.5 Conclusions

The ability of accurately characterizing power consumption of complex digital components is at the basis of the setup of power estimation capabilities usable at high levels of abstraction.

This chapter has addressed the problem of building power models for RTL components, and we have discussed how such models can be used for RTL power estimation, a key feature for the enhancement of state-of-the-art RTL-to-physical design flows.

We have illustrated in detail the basic principles of RTL power macro-modeling, which represents today's most advanced technology for RTL power estimation. Both theoretical and practical issues have been considered, thus providing a comprehensive overview of the problem and a review of the solutions that are currently available.

19.6 Acknowledgments

The authors thank BullDAST s.r.l. for the provision of the valuable experimental data included in Section 19.3 and Section 19.4. In particular, the help and support offered by Fabrizio Pro and Maurizio Bruno (BullDAST s.r.l. R&D Division) is acknowledged.

References

[1] Synopsys PowerCompiler, available at http://www.synopsys.com, April 2004.

[2] Sequence PowerTheater, available at http://www.sequencedesign.com, April 2004.

[3] BullDAST PowerChecker, available at http://www.bulldast.com, April 2004.

[4] D.D. Gajski, N.D. Dutt, A.C.-H. Wu, and S.Y.-L. Lin, *High-Level Synthesis: Introduction to Chip and System Design,* Kluwer Academic Publishers, Boston, 1992.

[5] P. Landman, High-level power estimation, *ISLPED-96: ACM/IEEE Int. Symp. on Low-Power Electron. and Design,* pp. 29–35, Monterey, CA, August 1996.

[6] P. Landman and J. Rabaey, Activity-sensitive architectural power analysis, *IEEE Trans. Computer-Aided Design,* Vol. 15, No. 6, pp. 571–587, June 1996.

[7] P. Landman, R. Mehra, and J. Rabaey, An integrated CAD environment for low-power design, *IEEE Design Test of Computers,* Vol. 13, No. 2, pp. 72–82, Summer 1996.

[8] A. Raghunathan, S. Dey, and N. Jha, Register-transfer level estimation techniques for switching activity and power consumption, *ICCAD-96: IEEE/ACM Int. Conf. in Computer-Aided Design,* pp. 158–165, San Jose, CA, November 1996.

[9] S. Katkoori and R. Vemuri, Architectural power estimation based on behavioral profiling, *J. VLSI Design,* Vol. 7, No. 3, pp. 255–270, 1998.

[10] A. Bogliolo, I. Colonescu, R. Corgnati, E. Macii, and M. Poncino, An RTL power estimation tool with on-line model building capabilities, *PATMOS-01: Int. Workshop on Power and Timing Modeling, Optimization and Simulation,* pp. 2.3.1–2.3.10, Yverdon-les-Bains, Switzerland, September 2001.

[11] S. Ravi, A. Raghunathan, and S. Chakradhar, Efficient RTL power estimation for large designs, *IEEE Int. Conf. on VLSI Design,* pp. 431–439, New Delhi, India, January 2003.

[12] D. Helms, E. Schmidt, A. Schulz, A. Stammermann, and W. Nebel, An improved power macro-model for arithmetic datapath components, *PATMOS-02: Int. Workshop on Power and Timing Modeling, Optimization, and Simulation,* pp. 16–24, Sevilla, Spain, September 2002.

[13] F. Najm, Transition density: a new measure of activity in digital circuits, *IEEE Trans. on Computer-Aided Design,* Vol. 12, No. 4, pp. 310–323, April 1993.

[14] S. Gupta and F. Najm, Power macromodeling for high-level power estimation, *DAC-34: ACM/IEEE Design Automation Conf.,* pp. 365–370, Anaheim, CA, June 1997.

[15] M. Nemani and F. Najm, Towards a high-level power estimation capability, *IEEE Trans. on Computer-Aided Design,* Vol. 15, No. 6, pp. 588–598, June 1996.

[16] D. Marculescu, R. Marculescu, M. Pedram, Information theoretic measures for power analysis, *IEEE Trans. on Computer-Aided Design,* Vol. 15, No. 6, pp. 599–609, June 1996.

[17] K.-T. Cheng and V.D. Agrawal, An entropy measure for the complexity of multi-output boolean functions, *DAC-27: ACM/IEEE Design Automation Conf.,* pp. 302–305, Orlando, FL, June 1990.

[18] M. Nemani and F. Najm, High-level area and power estimation for VLSI circuits, *IEEE Trans. on Computer-Aided Design,* Vol. 18, No. 6, pp. 697–713, June 1999.
[19] P. Landman and J. Rabaey, Black-box capacitance models for architectural power analysis, *IWLPD-94: ACM/IEEE Int. Workshop on Low-Power Design,* pp. 165–170, Napa Valley, CA, April 1994.
[20] P. Landman and J. Rabaey, Architectural power analysis: the dual-bit-type model, *IEEE Trans. on VLSI Syst.,* Vol. 3, No. 1, pp. 173–187, March 1995.
[21] Q. Qiu, Q. Wu, C.-S. Ding, and M. Pedram, Cycle-accurate macro-models for RT-level power analysis, *IEEE Trans. on VLSI Syst.,* Vol. 6, No. 4, pp. 520–528, December 1998.
[22] L. Benini, A. Bogliolo, M. Favalli, and G. De Micheli, Regression models for behavioral power estimation, *PATMOS-96: Int. Workshop on Power and Timing Modeling, Optimization, and Simulation,* pp. 125–130, Bologna, Italy, October 1996.
[23] C. Anton, A. Bogliolo, P. Civera, I. Colonescu, E. Macii, and M. Poncino, RTL macromodels for non-stationary workloads, *PATMOS-99: Int. Workshop on Power and Timing Modeling, Optimization, and Simulation,* pp. 313–322, Kos, Greece, October 1999.
[24] H. Mehta, R.M. Owens, and M.J. Irwin, Energy characterization based on clustering, *DAC-33: ACM/IEEE Design Automation Conf.,* pp. 702–707, Las Vegas, NV, June 1996.
[25] S. Gupta and F. Najm, Energy-per-cycle estimation at RTL, *ISLPED-99: ACM/IEEE Int. Symp. on Low-Power Electron. and Design,* pp. 16–17, Monterey, CA, August 1999.
[26] D. Liu and C. Svensson, Power consumption estimation in CMOS VLSI chips, *IEEE J. Solid-State Circuits,* Vol. 29, No. 6, pp. 663–671, June 1994.
[27] E. Schmidt, G. Jochens, L. Kruse, F. Theeuwen, and W. Nebel, Memory power models for multilevel power estimation and optimization, *IEEE Trans. on VLSI Syst.,* Vol. 10, No. 2, pp. 106–109, April 2002.
[28] A. Bogliolo and L. Benini, Robust RTL power macromodels, *IEEE Trans. on VLSI Syst.,* Vol. 6, No. 4, pp. 578–581, December 1998.
[29] A. Bogliolo, L. Benini, and G. De Micheli, Adaptive least mean square behavioral power modeling, *EDTC-97: IEEE European Design and Test Conf.,* pp. 404–410, Paris, France, March 1997.
[30] Z. Chen and K. Roy, Estimation of power dissipation using a novel power macromodeling technique, *IEEE Trans. on Computer-Aided Design,* Vol. 19, No. 11, pp. 1363–1369, November 2000.
[31] M. Barocci, L. Benini, A. Bogliolo, B. Riccò, and G. De Micheli, Lookup table power macro-models for behavioral library components, *IEEE Alessandro Volta Memorial Workshop on Low-Power Design,* pp. 173–181, Como, Italy, March 1999.
[32] A. Bogliolo, E. Macii, V. Mihailovici, M. Poncino, Combinational characterization-based power macro-models for sequential macros, *PATMOS-99: Int. Workshop on Power and Timing Modeling, Optimization, and Simulation,* pp. 293–302, Kos, Greece, October 1999.
[33] G. Jochens, L. Kruse, E. Schmidt, and W. Nebel, A new parameterizable power macro-model for datapath components, *DATE-99: IEEE Design Automation and Test in Europe,* pp. 29–36, Munich, Germany, March 1999.
[34] G. Bernacchia and M.C. Papaefthymiou, Analytical macromodeling for high-level power estimation, *ICCAD-99: IEEE/ACM Int. Conf. on Computer-Aided Design,* pp. 280–283, San Jose, CA, November 1999
[35] A. Bogliolo, L. Benini, and G. De Micheli, Regression-based RTL power modeling, *ACM Trans. on Design Automation of Electronic Syst.,* Vol. 5, No. 3, pp. 337–372, July 2000.
[36] L. Benini, A. Bogliolo, E. Macii, M. Poncino, and M. Surmei, Regression-based RTL power models for controllers, *GLSVLSI-00: ACM/IEEE Great Lakes Symp. on VLSI,* pp. 147–152, Evanston, IL, March 2000.
[37] M. Anton, I. Colonescu, E. Macii, and M. Poncino, Fast characterization of RTL power macro-models, *ICECS-01: IEEE Int. Conf. on Electron., Circuits and Syst.,* pp. 1591–1594, La Valletta, Malta, September 2001.
[38] R. Peset Llopis and K. Goossens, The Petrol approach to high-level power estimation, *ISLPED-98: ACM/IEEE Int. Symp. on Low-Power Electron. and Design,* pp. 130–132, Monterey, CA, August 1998.

[39] M. Bruno, A. Macii, and M. Poncino, A statistical power model for non-synthetic RTL operators, *PATMOS-03: Int. Workshop on Power and Timing Modeling, Optimization, and Simulation*, pp. 208–218, Torino, Italy, September 2003.

[40] Synopsys DesignCompiler, available at http://www.synopsys.com, April 2004.

[41] Synopsys DesignWare Library, available at http://www.synopsys.com, April 2004.

[42] A. Bogliolo, R. Corgnati, E. Macii, and M. Poncino, Parameterized RTL power models for combinational soft macros, *IEEE Trans. on VLSI Syst.*, pp. 880–887, Vol. 9, No. 6, December 2001.

20

Synopsys Low-Power Design Flow

Renu Mehra
Barry Pangrle
Synopsys, Inc.

20.1 Introduction

Design automation tools for analysis and optimization are key enablers for low-power design. With million-gate designs becoming commonplace and design sizes reaching the tens of millions mark, it is impossible to get a power-efficient implementation without appropriate automation. Automated optimization techniques are the fastest way to low-power design and, in fact, sometimes the only way for the highly complex chips of today. For optimization, a comprehensive set of register transfer level (RTL) and gate-level techniques are needed. These include clock gating, operand isolation, and many logic optimizations. In addition, several capabilities that allow both multi-threshold design and multi-voltage design are becoming increasingly important. A good analysis capability provides a basis for understanding the power needs of the design, identifying bottlenecks, and aiding in making correct decisions to reduce power. The power analysis tools need to provide a detailed, time-based power analysis capability for full chips.

Figure 20.1 presents some sources of power dissipation using an example inverter. Dynamic power is consumed by internal or short-circuit switching current that occurs when both transistors are on as when the input is in transition. Also contributing to dynamic power is the switching of the output that causes

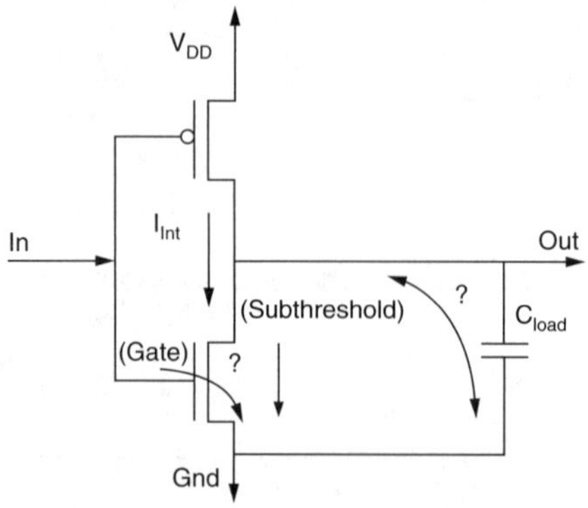

FIGURE 20.1 Example sources of power dissipation.

the charging and discharging of the output capacitive load. Static or leakage power is consumed largely by subthreshold leakage that is becoming more prevalent as threshold voltages are dropping and by gate leakage currents that are on the rise with thinner gate oxides being used.

The rest of this chapter gives an overview of the capabilities needed in today's power optimization and analysis tools. Various automated optimization techniques are discussed in Section 20.2 through Section 20.7. Section 20.8 and Section 20.9 cover the basics of power analysis in computer-aided design (CAD) tools in terms of modeling of the various components and provide an idea of the analysis flow from the designer's point of view.

20.2 Clock Gating

Clock gating involves dynamically shutting off the clock to portions of a design that are idle or are not performing useful computation. This technique is one of the most successful and widely used techniques for power reduction [1–4]. Figure 20.2(b) depicts the concept of clock gating using an AND gate. The basic idea is to AND the clock with an enable signal, so that the register receives a clock signal only when the enable is high.

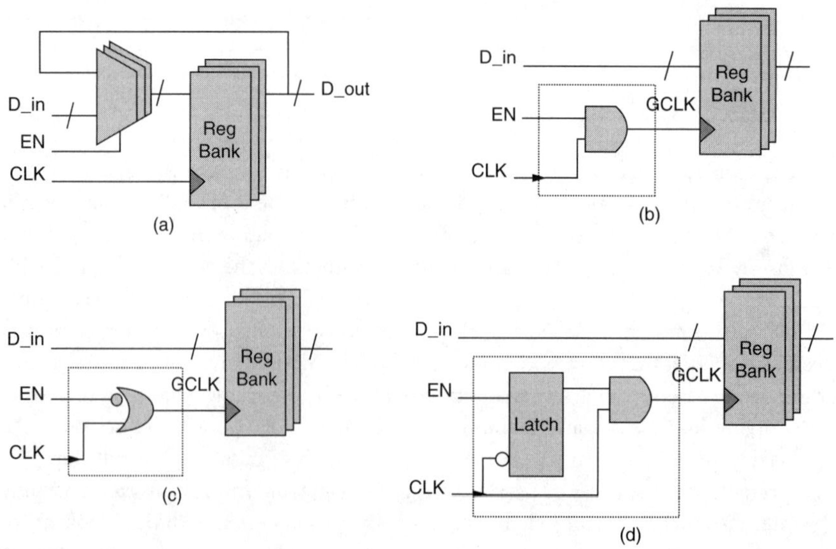

FIGURE 20.2 Clock gating: (a) traditional load-enabled register bank implementation, (b) clock-gated implementation using AND gate only, (c) clock gating using an OR gate, (d) clock gating using a latch and AND gate.

The granularity of the circuit block at which clock gating is applied greatly affects the power savings that can be achieved because gating larger blocks results in higher power savings in the "off" clock cycle, but allows fewer number of "off" clock cycles. Three levels of granularity can be distinguished and are discussed next.

20.2.1 Module-Level Clock Gating

This involves shutting off an entire block or module in the design. Usually, this decision is taken by the system or RTL designer. This method saves large amounts of power when the entire block is not functioning. Perfect cases for this are when a block is used only for a specific mode of operation (e.g., the receiver and transmitter parts in a transceiver may not be active at the same time), and the receiver can be shut off during transmit stages or vice versa. The opportunities for this kind of clock gating are limited and must be identified by the designer and incorporated into the RTL code.

20.2.2 Register-Level Clock Gating

In this method of clock gating, the clock to a single register or set of registers is gated. Synchronous load-enabled registers are usually implemented using a clocked D flip-flop and a recirculating multiplexer as depicted in Figure 20.2(a), with the D flip-flop being clocked every cycle. Clock-gated versions of the same register are depicted in Figure 20.2(b), Figure 20.2(c), and Figure 20.2(d). In the clock-gated versions, the register does not get the clock signal in the cycles when no new data is loaded, thereby saving power. Eliminating the multiplexer also saves power. Gating a single bit register, however, has the associated penalty of power consumption in the clock-gating logic. The key, therefore, is to amortize this penalty over a large number of registers, saving the flip-flop clocking power and the multiplexer power of all of them using a single clock-gating circuit.

Although power saving per clock-gate is much less with register-level clock gating than that obtained with module-level clock gating, this method detects many more opportunities to shut off clocks than would be possible with module-level clock gating. In addition, it lends itself well to automated insertion and can result in very large number of clock-gating cells in the design or massively clock-gated designs. Massively clock-gated designs cause several issues with automated flows most of which are discussed in Section 20.3.

20.2.3 Cell-Level Clock Gating

The cell designer usually introduces cell-level clock gating. For example, a register-bank can be designed such that the registers in the bank receive the clock only when the register is loading new data. Similarly, a memory block may be clocked only during active access cycles. Although this is an easier method of implementing clock gating with no flow issues, it may not be the most efficient from the area and power point of view. It has an area overhead and limits the amount of power savings because all the registers in the design would need to be predesigned with clock gating. In addition, it does not allow the sharing of clock-gating logic across many registers. The previous two methods have additional power savings due to reduced switching in the capacitance of the clock lines from the clock gate to the register, which are not realized with this implementation.

20.3 Automated Clock Gating at the Register Level

In a traditional automated ASIC design flow, all registers are assumed to synchronously read data. This paradigm is violated with register-level clock gating that introduces several challenges in an ASIC design flow. This section presents the challenges of using massively gated clocks in a practical design flow and discuss some of the solutions. For ease of explanation, we assume that the registers being clock gated are positive-edged; however, all concepts will apply, with inverted clock sense, for negative-edged registers.

20.3.1 Practical Gating Circuits

In the simple clock-gating configuration depicted in Figure 20.2(b), glitches on the enable signal that occur when the clock is high are propagated to the clock pin of the register. Although most glitches can be prevented by applying appropriate setup and hold constraints on the enable signal of the AND gate, any spurious changes during runtime (due to coupling with other signals, etc.) can cause wrong values to be latched into the gated register. A slightly safer method is to use an OR gate as depicted in Figure 20.2(c), which holds the output values at logic "1" when the enable signal, *en*, is high. Spurious runtime glitches in this configuration can cause wrong values to be loaded into the flip-flop but the final value loaded at the "valid" rising edge of the clock will be correct.

A better way to avoid this potential problem is to add a level-sensitive, active-low latch on the enable path as depicted in Figure 20.2(d). This freezes the latch output at the rising edge of the clock, and ensures that the new enable signal, *en1*, at the AND gate is stable when the clock is high. In addition, the enable signal can time-borrow from the latch, so that it essentially has the entire clock period to propagate to the latch.

Alternatively, a falling-edge flip-flop can be used instead of the latch to ensure a clean signal to the AND gate. This requires that the enable signal be stable before the falling edge of the clock, however, resulting in a stricter timing constraint on the enable path.

20.3.2 Clock Latency

To maximize the power savings, a single clock gate may be used to gate several flip-flops, if the enable signal is common to all of them; but the clock gate may not have the drive strength required to drive all these registers, requiring a clock tree at its output. If a clock tree is introduced between the clock gate and the registers it gates, the clock signal at the gating logic arrives much before the clock signal at the registers and the enable signal must be ready before the clock arrives at the gating logic. This applies strict timing requirements on the enable signal, which must be addressed during synthesis. Otherwise, this can result in large timing violations after clock-tree synthesis.

One way to address this issue is to specify the clock latency at the clock gate to be smaller than that at the registers themselves during synthesis. The difference in the latencies is the delay of the clock-gating circuit and the clock tree between the clock gate and the registers. This forces the synthesis tool to ensure that the enable signal arrives on time. The difficulty with this approach is that the designer should be able to estimate the latency difference at the two points far in advance of the actual clock-tree synthesis step. The designer can either use a conservative (worst-case) estimate of the clock-tree synthesis delay, or "force" a specific delay by limiting the fanout of each clock-gating cell.

20.3.3 Effect of Clock Skew

Another problem in the latch-based architecture comes from the fact that clock skew between the latch and the AND gate can result in glitches at the gated-clock output. This is explained in Figure 20.3. Figure 20.3(a) illustrates the case when the clock arrives much earlier at the AND gate than at the latch. Here, the clock-skew between the latch and the AND gate should be less than the clock-to-output delay of the latch for the circuit to function properly. Figure 20.3(b) illustrates the case when the clock arrives earlier at the latch. Here, the clock-skew between the AND gate and latch should be less than the sum of the setup time of the latch and the input-to-output delay of the latch to function properly. Therefore, the clock-skew between the latch and AND gate, C_s, should be carefully controlled according to the following equation:

$$-(s + d_{in}) < C_s < d_{clk}$$

where s is the setup time of the latch, d_{in} is the input-to-output delay of the latch, C_s is the difference in clock arrival time between the latch and the AND gate (the clock arrival time at the AND gate minus the clock arrival time at the latch), and d_{clk} is the clock-to-output delay of the latch.

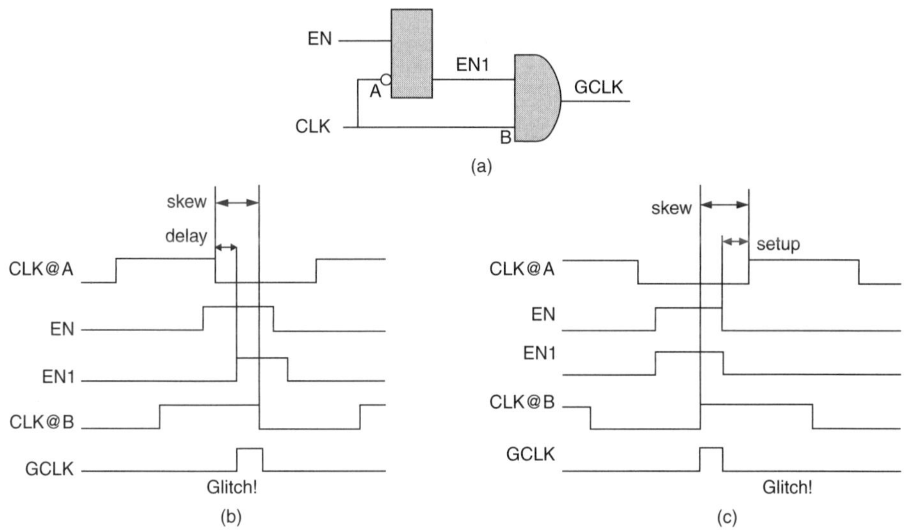

FIGURE 20.3 Clock skew within clock-gating logic: (a) clock-gating logic, (b) glitches due to positive clock skew between the latch and AND gate, (c) glitches due to negative clock skew between the latch and AND gate.

Depending on the relative placement of the latch and the AND gate, these requirements may pose very stringent constraints on the clock-tree synthesis tool.

The best way to control the relative timing of the two clock signals is to keep the entire structure in a single cell, called the integrated clock gating (ICG) cell. The cell should be designed specifically for clock gating, with the explicit requirements discussed previously. Because this cell cannot be modeled either as a combinational or sequential cell, the new "state-table" model in Liberty library format [24] is used for this.

Another way to address this issue is to ensure that the latch and AND gate are close to each other during the placement phase of the design, by placing hard constraints on the distance between them. This makes it simpler for clock-tree synthesis tools to reduce the clock skew between them during clock-routing phase.

20.3.4 Clock-Tree Synthesis

In an automated ASIC design environment, the clock signal typically remains untouched during synthesis, and clock-tree synthesis is done as one of the last steps in the design flow after placement and routing. Because manual (module-level) clock gating introduces only few clock gates in the design, clock-tree synthesis tools can work with these with some manual intervention. In the presence of massively gated clocks, however, clock-tree synthesis tools must automatically address the presence of clock gates on the clock line to be a viable solution. The requirements from the clock tree synthesis are:

- Optimization of the clock-tree in the presence of logic.
- Support for the integrated clock-gating cell on the clock tree. This is a sequential cell, but is not an end point on the clock-tree and, therefore, must be handled in a special way.
- Support for different relative latency requirements at different points in the clock tree.
- Stringent control of clock skew between the latch and the AND gate if the integrated clock-gating cell is not used.

20.3.5 Physical Clock Gating

Physical clock gating simultaneously takes into account the factors mentioned in the three previous subsections, namely latency and clock-tree synthesis issues.

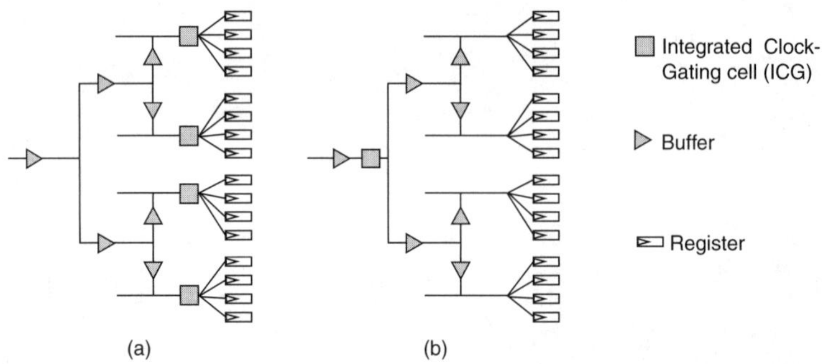

FIGURE 20.4 Clock-gating cell placement: (a) clock gating close to registers, (b) clock gating with post-gate buffering.

There exists a spectrum of clock-gating approaches with regard to the placement of clock-gating cell in the clock tree. Designers often opt to place the clock-gating cells as close as possible to the final placement of their corresponding registers as depicted in Figure 20.4(a). This placement can be enforced during physical synthesis by specifying a bound for the proximity of the clock-gating cells to the registers.

Some advantages of this approach are that it makes it easier to estimate the latency from the clock-gating cell, and it increases the amount of available slack for the arrival of the enable signal. The impact on the clock-tree is minimal because the clock-gating cells are placed close to the registers and can eliminate the need for post clock-gating cell buffer insertion.

A disadvantage to this approach is that it leaves the majority of the clock tree switching even when branches are leading to registers that will have the clock blocked by a clock gate. To save as much power as possible, it is desirable to gate as many buffers on the clock tree as possible. This is difficult for the designer to do without the knowledge of the actual physically induced timing constraints.

In a physically aware clock-gating system, clock-tree synthesis works in conjunction with placement and clock gating to determine an optimized placement and insertion of the clock-gating cells into the clock-tree. This information is used to balance the delay on the enable signal with the amount of potential power saved by placing the clock-gating cell closer to the root of the clock tree as depicted in 20.4(b).

By creating a system that has access to the physical timing information while inserting clock gating and synthesizing the clock-tree, selective clock skewing can also be used to improve timing and peak power characteristics of the design. In general, a tighter clock skew constraint forces more activity into a narrower window of time and that forces peak power higher (see Figure 20.5(a)). If these events are dispersed over a longer period, the peak can be flattened out (see Figure 20.5(b)), thus lowering the impact of peak power and IR drop.

20.3.6 Testability Concerns

Clock gating reduces test-coverage of the circuit because clock-gated registers are not clocked unless the enable signal is high. During test or scan modes, test-vectors need to be loaded into the registers, thus they must be clocked irrespective of the value of the enable signal. One way to address this is to include a control point or control-gate at the enable signal, as illustrated in Figure 20.6(a). This allows the clock-gating signal, *en*, to be overridden during the scanning in or out of vectors by the test mode signal. In this way, during the test clock cycles, the clock signal is not gated by the enable signal, *en*, and the register can be tested to see if it holds the correct state.

Further, the test mode signal is held at logic “1” during test-mode, making any stuck-at faults on the enable signal unobservable. If full observability is required, this signal must be explicitly made observable by tapping it into an observability XOR tree, as illustrated in Figure 20.6(b).

A growing concern around scan-based testing is the power consumed during the scanning in and out of test vectors. [5,6] The changing register values during scanning can create activity levels that are

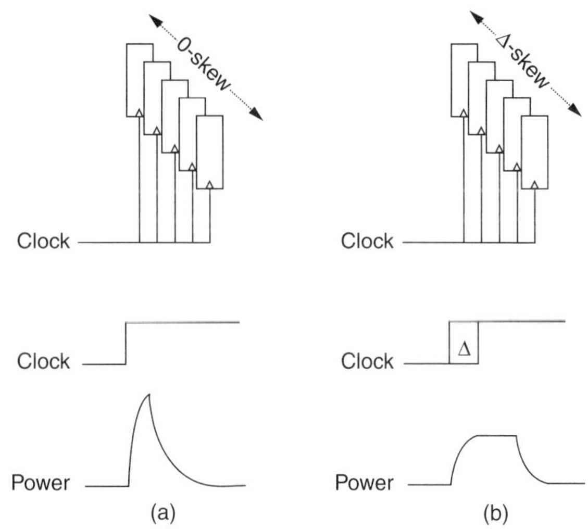

FIGURE 20.5 Selective clock skew to reduce peak power: (a) 0-skew, (b) Δ-skew.

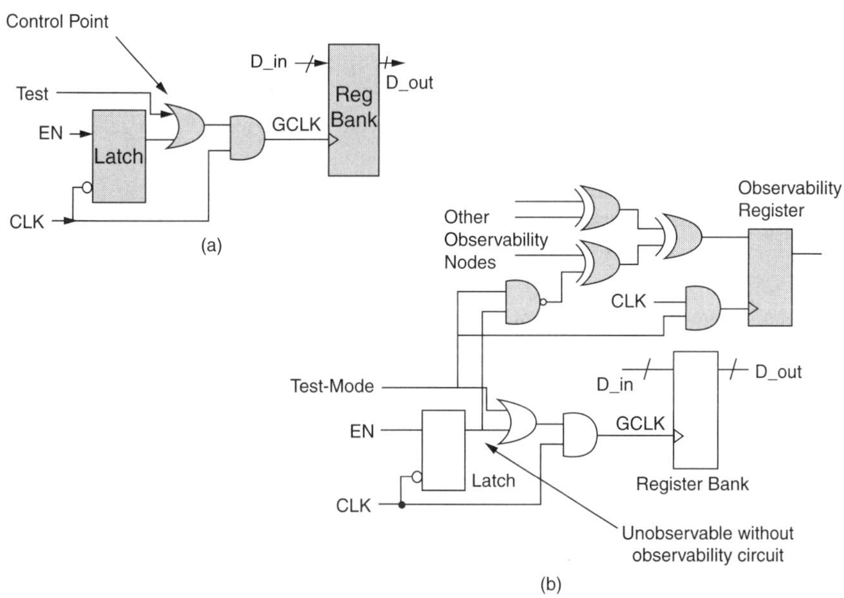

FIGURE 20.6 Testability issues with clock gating: (a) adding controllability, (b) improving observability.

much higher than those experienced during "normal" operation, and can lead to "good" chips failing during testing.

20.4 Operand Isolation

Although clock gating saves power dissipation in the clocked or sequential parts of the design, power savings in the combinational portions are untapped. At the RT level, the most popular technique for power reduction in the combinational parts is operand isolation [7]. Similar to clock gating, the basic concept here is to "shut-off" logic blocks in clock cycles when they do not perform any useful computation.

"Shutting-off" a combinational block involves preventing the activity in the block by not allowing the inputs to toggle in clock cycles in which the block output is not used. The basic concept is depicted in

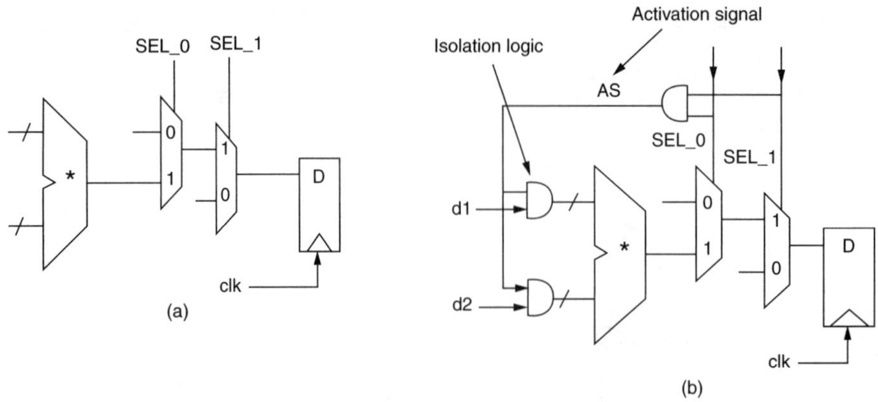

FIGURE 20.7 Operand isolation: (a) original circuit, (b) after operand isolation.

In Figure 20.7(a), we notice that the output of the multiplier is only used when the control signals to the multiplexers, *SEL_0* and *SEL_1*, are both high. In cycles when either of the control signals is low, if the multiplier inputs change, the multiplier performs computation, but its result is not used. The wasted power may be substantial if these idle cycles occur for long periods.

Figure 20.7(b) illustrates the operand isolation applied to this multiplier circuit. First, the activation signal, *AS*, is created to detect the idle cycles of the multiplier. The activation signal is high in the "active" clock cycles when the multiplier output is being used, and low otherwise. This signal is used to isolate the multiplier by freezing its inputs during idle cycles using a set of gates, called isolation logic. In Figure 20.7(b), AND gates are used as isolation logic but OR gates or latches may also be used. Using AND/OR gates avoids the introduction of new sequential elements and reduces the impact on the rest of the flow. In addition, in our experiments, we found that AND/OR gates are cheaper and give better power savings overall.

Operand isolation saves power by reducing switching in the operator being isolated, but it also introduces timing, area, and power overhead from the additional circuitry for the activation signal and the isolation logic. This overhead must be carefully evaluated against the power savings obtained to ensure a net power saving without too much delay or area penalty.

20.5 Logic Optimization

Logic-level optimizations reduce the power of a given circuit by transforming them into a different but functionally equivalent implementation. These transformations include RTL and gate-level techniques. Due to the wide acceptance of logic-synthesis tools in the design market today, these techniques are good candidates for automatic optimizations.

These techniques usually aim at reducing either the dynamic or short-circuit power. Because both of these components occur during switching, the techniques rely heavily on the availability of activity statistics at the input pins of all the cells of the circuit. To provide this information to the optimization tool, the user can either simulate the circuit to obtain the switching statistics at input pins of all cells or simply provide activity statistics at the primary inputs. In the latter case, the optimization tool must propagate the activity from the primary inputs to the internal cell inputs using either binary decision diagram (BDD)-based probabilistic propagation techniques [8] or some internal simulation.

These techniques also require estimates of capacitance values of the nets and input/output pins of cells in the circuit — values that are provided through library models of cells, wires, etc.

The next few subsections discuss some of the most popularly used logic transformations for power reduction.

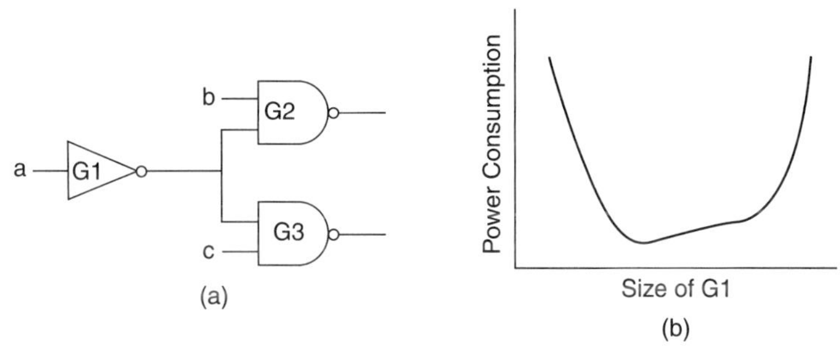

FIGURE 20.8 Gate sizing and power: (a) original circuit, (b) impact of sizing on power dissipation.

20.5.1 Sizing and Buffering

Gate sizes affect the power consumption of the design in the following ways:

1. A larger cell contributes higher capacitance that increases the overall dynamic power dissipation.
2. For a given input transition time, a larger gate has higher short-circuit currents.
3. The output of a gate with high drive has a sharper slope compared to one with lower drive strength, directly decreasing the short-circuit dynamic currents in the fanout gates.

Buffering has a similar effect as sizing because the addition of a buffer contributes both extra capacitance and short-circuit currents but improves the slope of the output signals.

These concepts can be more clearly illustrated using an example. Although we demonstrate the ideas using a sizing example, the points made here also apply to buffering. Consider Figure 20.8 where gate G_1 that fans out to two other gates G_2 and G_3. Let us assume that the input transition times for G_1, G_2, and G_3 are t_i, t_o, and t_o; their gate sizes are W_1, W_2, and W_3; and their output switching frequencies are f_1, f_2, and f_3, respectively. In addition, G_1 has an input switching frequency of f_i and input and output capacitances of C_i and C_o, respectively. Using the equations for the switching and short-circuit power of the gates, the power consumption of the circuit is given by:

$$P = C_o V_{dd}^2 f_1 + C_i V_{dd}^2 f_i + k(W_1 t_i f_1 + W_2 t_o f_2 + W_3 t_o f_3)$$

Let us consider each of the terms. The first term represents the power consumed in switching the capacitance C_o. If we assume that the source/drain capacitance of the gate is a small component of C_o, this term is almost constant and increases slightly with the sizing up of the gate, G_1. The second term represents the switching power at the input of the gate G_1 and increases linearly with the size of G_1. The last three terms represent the short-circuit power of each of the three gates. As the size of G_1 is increased, this short-circuit current of G_1 increases due to an increase in W_1, while the short-circuit current of G_2 and G_3 decrease due to a reduction in the transition time of their input signals. Therefore, as the size of the gate G_1 is increased, the power decreases first and then starts rising, showing a clear minima and a potential for optimization [9].

Sizing and buffering are also used for "path balancing." Different arrival times of the different inputs of a gate cause spurious switching or glitches at the output causing excess power dissipation. These techniques can be used to equalize the arrival times at the inputs of a given gate, thereby reducing glitches.

20.5.2 Technology Mapping

Technology mapping involves optimal mapping of a block of logic using cells from a given library. Technology mapping for low power is driven by three observations:

1. Internal nodes of library cells usually have lower capacitance than external nodes.
2. Gate size has a large impact on power consumption, as discussed in the previous subsection.

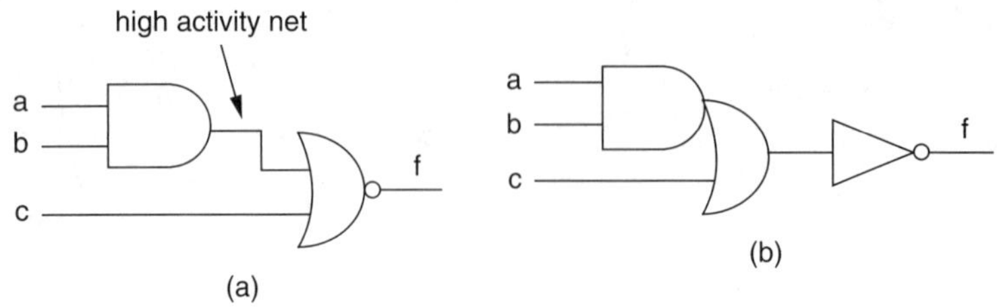

FIGURE 20.9 Technology mapping for low power: (a) original circuit, (b) remapped circuit.

3. The lowest-power mapping does not coincide with either the minimum delay or area mapping [10,11].

The first observation indicates that it is better to map nodes with high switching activity in a design to internal nodes of a cell. This reduces power by reducing the capacitance on the highly active nodes. The second observation indicates that the cell-size selection process must explore the trade-off between the driving capacity and the power consumption of the cells. The third observation indicates that power must explicitly be part of the cost function to achieve a power-sensitive technology mapping solution.

Two implementations of a simple and-or-invert (AOI) are shown in Figure 20.9. We are given that node $x = a \cdot b$ is a high activity node. The circuit in Figure 20.9(b) implements the high activity node, x, as an internal node and therefore dissipates less power.

20.5.3 Phase Assignment

Phase assignment inverts the inputs to an operation and, at the same time, also inverting the output. This transformation reduces power in the following ways. First, because this transformation adds inverters on nets that previously did not have inverters, it creates opportunity for several other transformations: Two inverters next to each other can be merged and removed, and an inverter at the output of a gate may be absorbed into the gate using a composite gate from the library. Second, it can be used to remove inverters from high-activity nets and move them to lower-activity nets.

20.5.4 Algebraic Transformations

Algebraic transformations use algebraic properties to derive equivalent implementations of a given circuit to reduce power. The most popularly used properties for power reduction include commutativity, associativity, and distributivity.

Commutativity is used in a transformation called pin swapping. At the gate level, many Boolean operations, such as AND, OR, NAND, NOR, and XOR, are commutative (i.e., their inputs can be interchanged without affecting the functionality). Based on the capacitance and toggle-rates on pins, the input pins of a gate can be swapped, connecting the lower input-capacitance pin to the net with the higher toggle rate, thus reducing power consumption. The same technique can be used at the RT level with larger commutative blocks such as adders and multipliers. Besides directly reducing power consumption, this technique generates more opportunities for the other techniques and helps to pull the overall algorithm out of local minima.

The associative and distributive properties of gate-level operations, such as AND, OR, and XOR, as well as RTL operators, such as adders and multipliers, are used in a transformation for low power called factoring. Factoring is based on the idea that the power dissipated by an operation and the activity at its outputs depends on the activity at its inputs. Therefore, the power dissipation caused by a certain net depends on the number of operations to which its activity is propagated. The goal of factoring is to reduce the logic depth connected to high-activity nets. This is illustrated in the Figure 20.10, which presents two implementations of the function $ab + bc + cd$. Of the inputs, input b has the highest activity. The thick lines in the figure identify high-activity nets. In this case, the implementation in Figure 20.10(b) dissipates less power because b is propagated only through one gate.

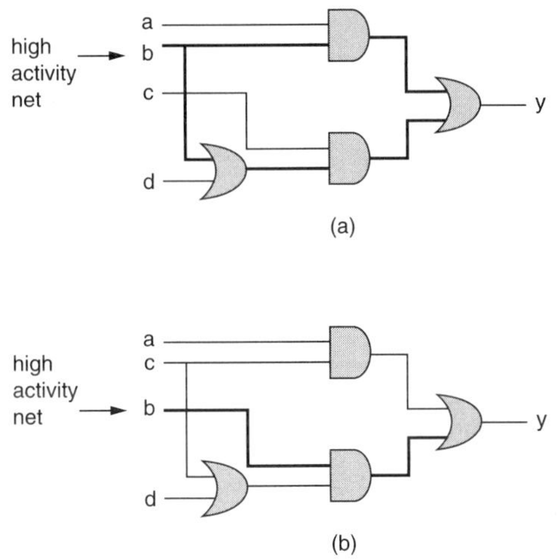

FIGURE 20.10 Factoring for low power: (b) is the lower power implementation.

20.6 Leakage Control — Managing Thresholds

Since the .5-μm technology node, core voltage levels have been scaling at approximately 1 V/0.1 μm. As technologies scale to 100 nm and below, the reduced operating voltages are forcing threshold voltages down to .25 V and below. This has had a major impact on the leakage current of the transistors built in these technologies. For approximately every 65 mV to 85 mV decrease in threshold voltage (depending on temperature), there is an order of magnitude increase in subthreshold leakage current.

As the next equation demonstrates, the subthreshold leakage current grows exponentially as the threshold voltage decreases.

$$I_{sub} = I_0(e^{[-Vth/S]}\ [1 - e^{-qVds/kT}])\ (\text{at } V_{gs} = 0)$$

20.6.1 Multi-Threshold Design

Silicon foundries have started to offer multiple threshold devices at the same process node to address the need to control leakage current and enabling designers to trade off leakage and performance [12]. From the standard V_{th}, a low- and high-V_{th} transistor may be offered. It is not uncommon for the low-V_{th} device to have an order of magnitude higher leakage than the standard V_{th} device and the high-V_{th} device to have leakage characteristics an order of magnitude below the standard. For special applications, a special low-leakage device may exist that will reduce the leakage further by another order of magnitude. This reduction in leakage is not free, however, and it comes at the expense of the speed of the device. There could be a 20% to more than 2× delay penalty between the standard and the high-V_{th} devices and an increase in the cost of the fabrication process.

The challenge for EDA tools is to use the available characteristics of the cells in the design library to create an implementation that will meet the timing constraints, while reducing the leakage current as much as possible.

A simplistic approach is to use the percentage of high- and low-threshold cells in the design as a quality measure. It is important to keep in mind that the goal is to reduce the total leakage. A design with a higher percentage of high-V_{th} cells may appear to be better at first glance, but it could also be inferior to a design with a lower percentage of high-V_{th} cells that uses fewer cells overall. It is for this reason that these trade-offs should be considered early in the synthesis process.

A better approach for synthesis using multi V_{th} cells is to create a post-logic-synthesis routine that replaces low-V_{th} cells with high-V_{th} cells as long as the timing constraints and other design rules and constraints are not violated. This can be a useful "clean-up" step for an existing gate-level netlist design.

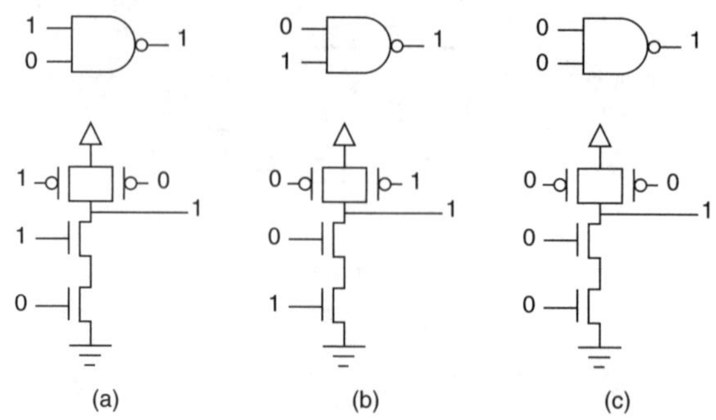

FIGURE 20.11 Leakage varies depending on input values.

The cells may be created to have the same "footprint" to facilitate this type of optimization in a post-placement phase

A more sophisticated approach used during synthesis incorporates the leakage power as a component of the optimization's objective function. Here, it is important to have libraries that have been properly characterized to perform this type of optimization. In particular, state-dependent leakage information can be used to reduce the leakage in the design. It is understood that the amount of current that a cell will leak is dependent on the state of its inputs. In particular, transistors in series will have varying amounts of leakage current depending on the values placed on their gates. This is depicted in Figure 20.11, where the leakage current decreases from (a) to (c) even though no change occurs in the output value of the gate. In this case, as in dynamic power reduction, pin-swapping techniques may also be used to reduce the average leakage component or the leakage for a given state. To perform this optimization, it is necessary for the libraries to include state-dependent leakage information for the cells. Another necessary component is the state probabilities on the inputs. The switching activity interchange format (SAIF) [13] can be used to annotate this information onto a design, which can then be internally propagated and updated for use in optimizing pin selection for leakage reduction. Some details on the SAIF format are presented in Section 20.8.

Another relevant area for multiple V_{th} designs is the impact on noise sensitivity. Higher-V_{th} devices, by their nature, will have a lower sensitivity to noise. Optimization tools that simultaneously consider the signal integrity impact can make use of these cells to improve signal integrity properties as well as leakage.

20.6.2 Variable Threshold Biasing

The threshold voltage and therefore the leakage current in a CMOS transistor are controllable by varying the back biasing. The change in the threshold voltage is roughly proportional to the square root of the back bias voltage. As threshold voltages drop below .25 V, variable back biasing may gain more appeal.

A distinct advantage of this approach is that during periods when heavy processing is needed, the threshold voltage can be reduced, thus speeding up the cells. When the cells are in a slower drowsy or idle mode, the threshold voltage can be raised, thus lowering the leakage.

One significant impact of using variable back biasing is that two new terminals for each cell need to be routed. A common ASIC design practice is to create cells that tie the N-well regions to V_{dd} and the P-well regions to ground. In the physical implementation, these are simply predefined contacts designed into the cell, which are connected as part of the power and ground routes. To enable back-biasing, new voltage lines are routed to control the bias. These can be to individual cells or, more likely, to regions that contain multiple cells sharing the same WELL and a common tie-cell to control the WELL bias [14,15]. Figure 20.12 illustrates the concept of variable body biasing. Figure 20.12(a) is a traditional implementation with wells tied to V_{dd} and V_{ss}, and Figure 20.12(b) presents wells biased for leakage reduction.

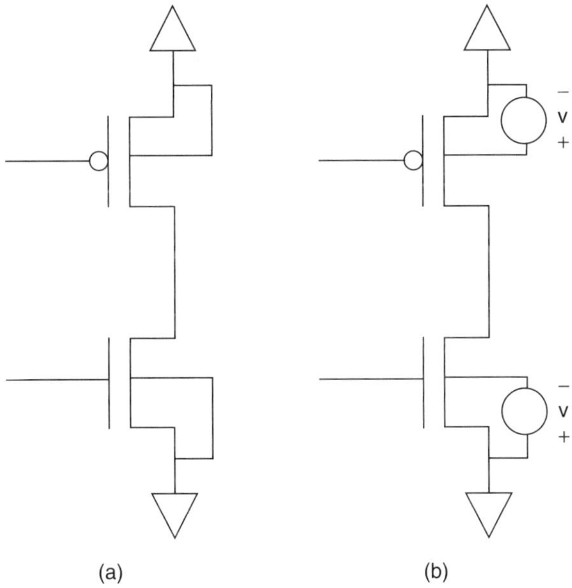

FIGURE 20.12 Using bias to control threshold voltages.

20.7 Voltage Scaling

The switching power consumed in a design is proportional to the square of supply voltage. This relationship of power to voltage makes voltage scaling a prime candidate for reducing the power in CMOS designs.

One approach is to create separate voltage islands that operate at levels that best suit the power and performance levels for each block of logic [16–20]. This can range from having each island run at one voltage and selectively turning blocks of logic completely off, to dynamically varying the voltages supplied to those blocks. These design techniques create interesting dynamics on the chip. Turning the voltage ON or OFF to a block can cause large transients on the power grid, affecting many other blocks on the chip. Further, it is necessary to provide isolation on the outputs of the block that is shut down. It is also possible to use register structures that have a second voltage rail that provides power to retention logic that can be used to save and then restore the state to a block that has had its power shutdown [21,22]. An example of this type of structure is given in Figure 20.13.

A major impact on the design flow when multiple voltages are used is the need to treat the supply line as another variable. For most previous mainstream designs, logical netlists only specified the input and output connections between gates. V_{dd} and V_{ss} were constants, and the V_{dd} pins for all the cells (as well as the V_{ss} pins) were attached to the same net. Figure 20.14 is a simple diagram indicating the need for cells to be able to handle new voltage terminals. A level shifter, for instance, needs to handle two V_{dd} levels as well as separate potential well biases.

The tools have to manage cells that have more than one supply rail and circuitry that can vary or completely shut down the supply voltage to a block. Communication between blocks operating at different voltages requires the insertion of voltage level shifters to transform signals to the appropriate levels. Clock-tree generators need to account for buffers that operate at different voltages to provide clock signals to each block and the router needs to account for buffer placement in the context of different voltage regions on the chip. Routing a feed-through signal via a region may now require the insertion of level shifters to adequately drive the signal. Analysis tools need to understand these different situations — tracking all these new voltage based modes — and provide useful feedback to the designer.

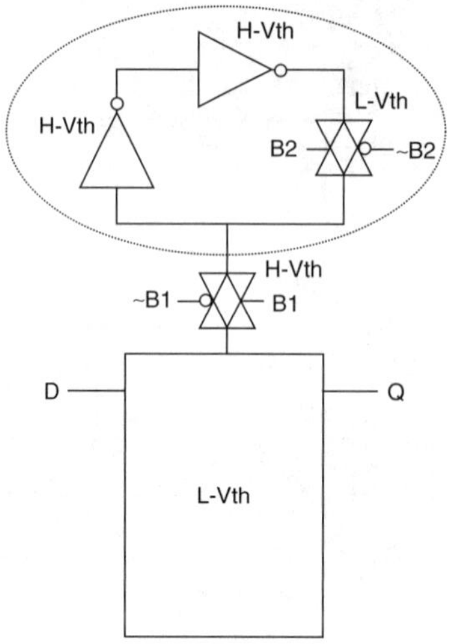

FIGURE 20.13 Low-power state-saving register.

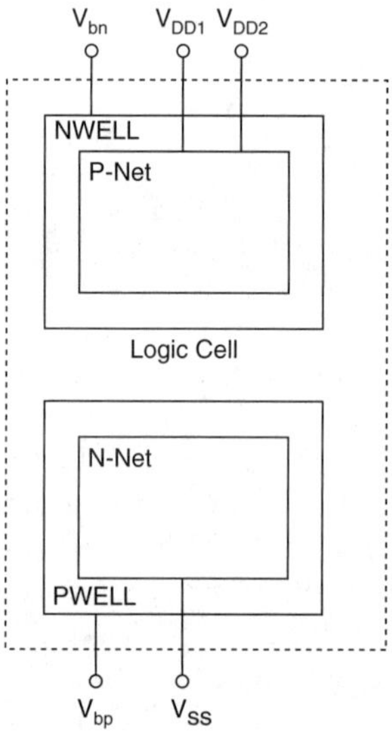

FIGURE 20.14 Handling multiple supply voltages.

Another design implication that optimization and analysis tools must account for is the impact of driving some lines at higher voltages than others. The higher-voltage lines can cause larger spikes in neighboring low-voltage lines than other lower-voltage aggressors, which impacts timing analysis, power, and the routing of lines on the chip. This is presented in Figure 20.15.

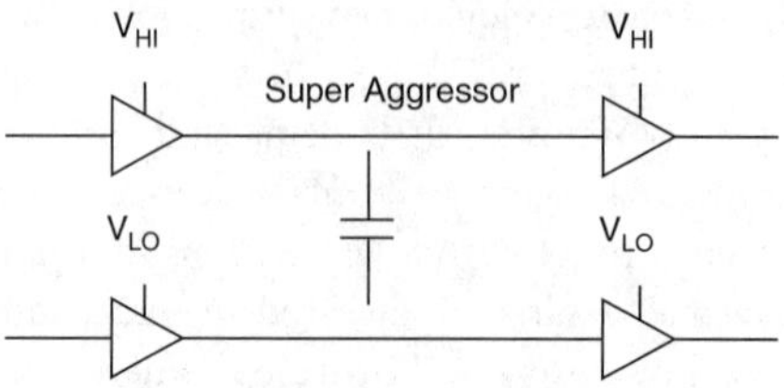

FIGURE 20.15 Signal integrity impact with multiple voltages.

Implementation tools need to account for the effects of the different operating conditions. Instead of designing for the typical *Min* and *Max* conditions, the circuitry now has to deal with a much broader range of conditions. Complicating matters, as the voltages have been decreasing, the current on chip has also been increasing, thus raising the sensitivity to IR-drop and *L di/dt* effects.

20.8 Modeling Basics

The previous sections presented a set of automated power-optimization solutions. To be effective and powerful, these optimization solutions should be supported by robust and accurate power modeling and analysis techniques. Current EDA tools provide comprehensive support for modeling power consumption on logic blocks and for analyzing the power consumption of designs. This section looks at some of the basics of modeling power for gates [24]. Section 20.9 presents typical power analysis flows.

20.8.1 Switching Power

Switching power consumption is computed using the famous CV^2f formula. To compute switching power accurately, the library must provide voltage and capacitance data. Capacitance is computed from the pin capacitances, which are specified on the library cells, and wire capacitances, which may either be back annotated from physical tools or calculated from wire-load models.

20.8.2 Internal Power

The internal power of a cell includes the power consumed due to the switching activity on the internal nodes of the cell and the short-circuit power consumed by the cell, as depicted in Figure 20.1. Internal power of a cell depends on the input rise or fall times and the output capacitances. It is modeled in the library as a lookup table, also called the nonlinear power model (NLPM), which is indexed by two variables: the input rise/fall time and the output capacitance.

A lookup table is specified by a table template that specifies the values of the indices to the table for the different entries. For specifying power values, a specific table template is used, and only the power values at the different points in the table are specified. An example of a one-dimensional table template, *Power_1D*, and a two-dimensional one, *Power_2D*, is presented next. These are used in power models presented later in this section.

```
power_lut_template(power_1D) {
variable_1 : input_transition_time;
index_1 ("1000, 1004, 1005, 1006");
}
power_lut_template(power_2D) {
variable_1 : input_transition_time;
variable_2 : total_output_net_capacitance;
index_1 ("1000, 1001, 1002");
index_2 ("1000, 1001, 1002, 1003, 1004, 1005, 1006");
}
```

A more advanced modeling method that allows multiple voltages and more variables to be modeled is called the scalable polynomial power model (SPPM). This format is explained later in this section.

Internal power can be state-dependent or path-dependent, or both. State dependency captures the fact that the internal power dissipation of a cell can be different based on the states of the inputs or outputs of the cell that are not switching. For example, the power dissipated when the clock pin of a RAM memory block switches depends on whether the RAM is reading, writing, or idle. This kind of effect can be captured by the state-dependency construct “when.” The state-dependent power model for a RAM cell given next demonstrates that the power is separately specified for the three states: read, write, and idle.

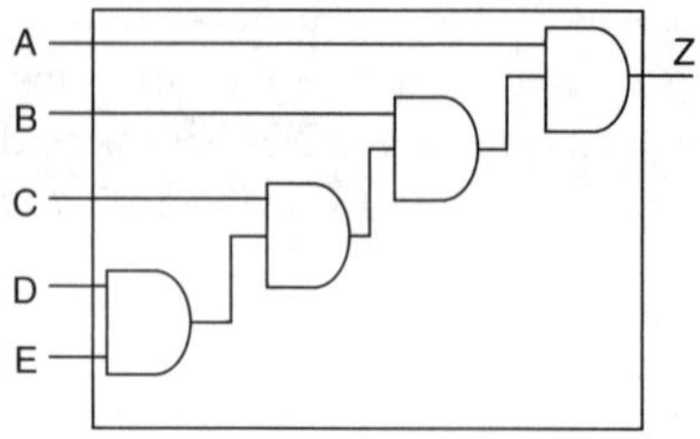

FIGURE 20.16 Path dependency: changes in output caused by a toggle at input E dissipate more power than those caused by changes in input A.

```
cell (RAM)
.........
pin (clk)
internal power {
when (!RD & WR & CS) {/* write state */
rise_power(power_1D)
      values ("1.5 2.5 5.6 7.7") }
fall_power (power_1D)
      values ("1.7 2.4 5.3 7.2") }
when (RD & !WR & CS) {/* read state */
rise_power(power_1D)
      values ("2.5 3.5 6.6 8.7") }
fall_power (power_1D)
      values ("2.7 3.4 6.3 8.2") }
when (!CS) {          /* idle state */
power (power_1D)
values ("0.6 2.0 1.6 4.7") }
}
```

Path dependency captures the fact that the internal power dissipation of a cell is different depending on which input change caused the switching behavior. In the example depicted in Figure 20.16, the power consumption is more if the output change is caused by a change in input D, instead of a change in input A. This effect can be captured by the path-dependency construct "related pin" for internal power presented next.

```
cell (AND5)
pin (Z)
internal power {
related_pin A {
power (power_2D)
values ("0.5 1.5 3.6," "0.6 1.7 4.0,"Ö.)}
ÖÖÖÖ
related_pin D {
power (power_2D)
values ("1.5 2.5 5.6," "1.6 2.7. 5.7,"Ö.) }
Ö......Ö
}
```

20.8.3 Leakage or Static Power Modeling

As discussed in previous sections, leakage may happen in a circuit from several different sources: subthreshold leakage, drain induced barrier lowering, gate induced drain leakage, etc. [23]. Regardless of

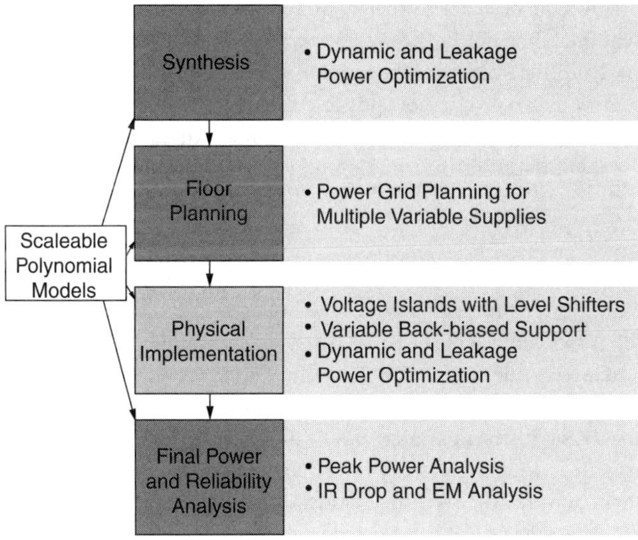

FIGURE 20.17 A flow for energy-efficient design.

the physical reasons for leakage power, library developers can annotate a cell with the total leakage power dissipated by it. The leakage model can be state dependent and can specify different leakage values for different input states of the cell. Here is an example of the leakage model on a cell:

```
cell my_cell() {
leakage_power () {
when: "A & B"
value: 5.5
}
cell_leakage_power: 4.0
}
```

This specifies that the cell `my_cell` consumes 5.5 units of leakage when both A and B inputs are high and consumes 4.0 units otherwise.

20.8.4 Scalable Polynomial Power Models (SPPMs)

It was once possible to create tables based on a few characterization points and use scaling factors, commonly called k-factors, to extrapolate to uncharacterized points in between. With multiple voltages, multiple thresholds, and back biasing, however, that methodology is now failing to provide the accuracy required. More sophisticated techniques are now needed to enable energy-efficient design. In addition, as chip sizes become larger and transistor sizes shrink, intra-chip temperature variations are becoming significant and must be appropriately modeled.

Scalable polynomial models provide an efficient and faster alternative to nonlinear lookup tables. These models capture library characterization information in an accurate equation-based format. Each of the voltages used for supply and back biasing become a variable in an equation, which is stored with the cell's information in the library. This equation-based format allows the tools to obtain precise data on the timing and power characteristics of a cell across a broad operating range of conditions. The cells are characterized for timing using scalable polynomial delay models (SPDMs), for power using SPPMs, and for leakage using scalable polynomial leakage models (SPLMs). Due to their compact representation, they can be used to model much higher degrees of freedom. For example, they allow us to also model the impact of supply voltage and temperature variations without blowing up in size as would happen with lookup table models.

The SPPM syntax allows the designer to specify up to seven variables. For very large data with abrupt changes, a single polynomial may not fit the entire operating range of interest. In this case, the piecewise or adaptive domain polynomial syntax can be used.

Equipped with the new libraries, it is now possible to ensure that the chip functions correctly across the expected operating ranges of process, voltage, and temperature. For this, design optimization and analysis tools must be able to use this information on a per instance basis.

20.8.5 Modeling Activity

The previous subsections discussed the models for the physical components that are required for power consumption, switching power, internal power, and leakage. The other component that needs to be appropriately modeled is switching activity. Based on the analysis flow that is used, switching activity can be modeled in different levels of detail. If a complete time-based power profile view of the chip is desired, the value change dump (VCD) or VCD+ formats can be used to capture detailed switching activity.

If, however, one is only interested in average power dissipation, a much more compact representation for switching activity called switching activity interchange format (SAIF) [13] can be used. SAIF is an open ASCII format and captures the switching statistics for each node in the design in terms of static and dynamic attributes that can be state and path dependent. The attributes captured are listed next:

20.8.5.1 Static Attributes

- T0: time spent in 0 state
- T1: time spent in 1 state
- TX: time spent in unknown X state
- TZ: time spent in floating Z state
- TB: time spent in bus-contention state (two or more drivers simultaneously driving same object)

20.8.5.2 Dynamic Attributes

- TC: number of 0→1 or 1→0 transitions. This can be split into rise and fall transitions.
- TG: number of transport glitches. These are glitches on the output where the output pulse width is more than gate delay. These consume the same power as a full transition.
- IG: number of inertial glitches. These are glitches where the output pulse width is less than gate delay. These do not consume same power as full transition, and a derating factor is used for power dissipation calculation for these.

Besides these basic attributes, the SAIF language provides state and path dependency constructs to capture more specific information about the switching statistics on a particular node or cell. Both the static and dynamic attributes can be state-dependent, thus capturing the switching statistics separately for the different states of a cell. State dependent static attributes are useful for computing state dependent leakage power and for computing dynamic power.

The dynamic attributes can also be path dependent, capturing separate switching statistics based on which input path caused the transition on the pin.

The modeling techniques for the various power components discussed in this section provide the underlying fabric for analysis and optimization tools discussed in this chapter.

20.9 Analysis Flows

Power optimization within power compiler [25] provides average power and uses the SAIF compact representation for switching activity. You can use SAIF from RTL simulation or gate-level simulation for your power analysis. Although not as accurate as gate-level simulation based analysis, RTL-simulation based analysis has the huge advantage in that it avoids the long time-consuming gate-level simulations.

The RTL-simulation based power analysis flow within power compiler is depicted in Figure 20.18. Switching activity is captured via RTL simulation at the synthesis invariant points in the design. These

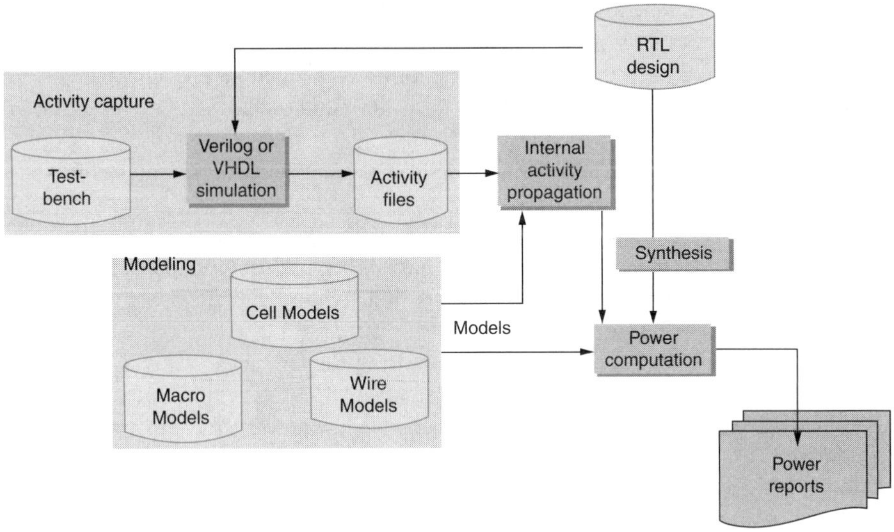

FIGURE 20.18 RTL simulation-based power analysis flow.

include the hierarchy boundaries and sequential elements. Capacitance and power models for wires and gates are taken from the library. The RTL design is synthesized to gate level along with all the constraints. Activity information that is captured at the synthesis invariant points is propagated to all the input pins of all cells in the gate-level design. This information is passed to the power computation engine, which reports the power for the entire design.

The gate-level simulation based flow is similar, except that no internal activity propagation is required because activity is captured at the input pins of all the cells in the gate-level netlist via gate-level simulation. Because this activity is captured in full detail, it is possible to use the state and path dependent information in the library models and in SAIF to perform a more accurate power analysis. In a post-placement or a post-routing netlist, the wire capacitances can be back annotated is for more accuracy.

PrimePower [26] provides a detailed analysis of the power dissipation in a design and relies on the more complete VCD switching activity format. It works on a gate-level netlist with gate-level simulation data and is targeted for full-chip capacity. Along with the average power numbers, it also gives the time-based waveforms of power consumption in different parts of the design. This allows the designer to do more detailed debugging of hot spots in the design.

The analysis tools and flows presented in this section enable the designer to understand and optimize the power of his or her design, and provide the basis for the optimization capabilities discussed earlier in this chapter.

20.10 Conclusion

This chapter described some of the power optimization technology currently available or in development in commercial CAD solutions. Both synthesis and analysis solutions were discussed. Power models that form the basis of such a solution were also presented. The authors thank the Power Compiler and PrimePower teams for their support.

References

[1] M. Gowan, L.L. Biro, and D.B. Jackson, Power considerations in the design of the Alpha 21264 microprocessor, *Proc. Design Automation Conf.*, 1998, pp. 726–731.

[2] A. Correale, Overview of the power minimization techniques employed in IBM PowerPC 4xx embedded controllers, *Int. Symp. on Low-power Design,* 1995, pp. 75–80.

[3] V. Tiwari, D. Singh, S. Rajgopal, G. Mehta, R. Patel, and F. Baez, Reducing power in high-performance microprocessors, *Proc. Design Automation Conf.*, 1998, pp. 726–731.
[4] Z. Khan and G. Mehta, Automatic clock gating for power reduction, *SNUG '99*.
[5] B. Pouya and A. Crouch, Optimization for vector volume and test power, *Proc. Int. Test Conf.*, 2000, pp. 873–881.
[6] J. Saxena, K. Butler, and L. Whetsel, An analysis of power reduction techniques in scan testing, *Proc. Int. Test Conf.*, 2001, pp. 670–677.
[7] M. Muench, B. Wurth, R. Mehra, and J. Sproch, Automating RT-level operand isolation to minimize power consumption in datapaths, *Proc. Design Automation and Test in Europe*, 2000, pp. 624–631.
[8] F. Najm, Transition density, a stochastic measure of activity in digital circuits, *Proc. Design Automation Conf.*, 1991, pp. 644–649.
[9] M. Borah, R. Owens, and M. Irwin, Transistor sizing for low-power CMOS circuits, *Trans. on Computer-Aided Design*, June 1996, pp. 665–671.
[10] O. Coudert and R. Haddad, Integrated resynthesis for low power, *Proc. Int. Symp. on Low-Power Electron. and Design*, 1996, pp. 169–174.
[11] V. Tiwari, P. Ashar, and S. Malik, Technology mapping for low power, *Proc. Design Automation Conf.*, 1993, pp. 74–79.
[12] S. Svilan, J.B. Burr, and G.L. Tyler, Effects of elevated temperature on tunable near-zero threshold CMOS, *Proc. Int. Symp. on Low-Power Electron. and Design*, 2001, pp. 255–258.
[13] Switching Activity Interchange Format (SAIF), http://www.synopsys.com/partners/tapin/saif.html.
[14] K. Flautner, D. Flynn, and M. Rives, A combined hardware-software approach for low-power SoCs: applying adaptive voltage scaling and the vertigo performance-setting algorithms, *Proc. Design Conf.*, 2003.
[15] S. Martin, K. Flautner, T. Mudge, and D. Blaauw, Combined dynamic voltage scaling and adaptive body biasing for lower power microprocessors under dynamic workloads, *Proc. Int. Conf. on Computer-Aided Design*, 2002, pp. 721–725.
[16] D. Tamura, B. Pangrle, and R. Maheshwary, Techniques for energy-efficient SoC design, http://www.eedesign.com/features/exclusive/OEG20030724S0044.
[17] D.E. Lackey, S. Gould, T.R. Bednar, J. Cohn, and P.S. Zuchowski, Managing power and performance for system-on-chip designs using voltage islands, *Proc. Int. Conf. on Computer-Aided Design*, 2002, pp. 195–202.
[18] K. Usami, M. Igarashi, F. Minami, T. Ishikawa, M. Kawakawa, M. Ichida, and K. Nogami, Automated low-power technique exploiting multiple supply voltages applied to media processor, *IEEE J. Solid-State Circuits*, Vol. 33, No. 3, 1998, pp. 463–472.
[19] L. Wei, K. Roy, and V. De, Low-power, low-voltage CMOS design techniques for deep submicron ICs, *Proc. Int. Conf. on VLSI Design*, 2000, pp. 24–29.
[20] F. Ishihara, F. Sheikh, and B. Nikolic, Level conversion for dual supply systems, *Proc. Int. Symp. on Low-Power Electron. and Design*, 2003, pp. 164–167.
[21] S. Shigematsu, S. Mutoh, Y. Matsuya, Y. Tanabe, and J. Yamada, A 1-V high-speed MTCMOS circuit scheme for power-down application circuits, *IEEE J. Solid-State Circuits*, Vol. 32, June 1997, pp. 861–869.
[22] V. Zyuban and S. Kosonocky, Low-power integrated scan-retention mechanism, *Proc. Int. Symp. on Low-power Electron. and Design*, 2002, pp. 98–102.
[23] J. Rabaey, A. Chandrakasan, and B. Nikolic, *Digital Integrated Circuits, 2nd ed.*, 2003, Prentice Hall/Pearson.
[24] *Library Compiler User Guide: Modeling Timing and Power Technology Libraries*, Synopsys.
[25] *Power Compiler Reference Manual*, Synopsys.
[26] *PrimePower Reference Manual*, Synopsys.

21

Magma Low-Power Flow

Ed Huijbregts
Lars Kruse
Eric Seelen
Magma Design Automation

21.1 Introduction

In the case of today's increasingly large and complex digital integrated circuit (IC) and system on chip (SoC) designs, design power closure and circuit power integrity are becoming one of the main drains on engineering resources, thereby impacting the device's total time-to-market.

The shear amount of power consumed by some devices can cause significant design problems. For example, a recently announced CPU consumes 100 A at 1.3 V, which equates to 130 W. This class of device requires expensive packaging and heat sinks, the heat gradient across the chip can cause mechanical stress leading to early breakdown, and the act of physically delivering all this power into the chip is nontrivial. Thus, even in the case of devices intended for use in nonportable equipment where ample power is readily available, power-aware designs can offer competitive advantages with respect to such considerations as the size and cost of the power supply and cooling systems.

The majority of power considerations are exacerbated in the case of low-power designs. The increasing use of battery-powered portable (often wireless) electronic systems is driving the demand for IC and SoC devices that consume the smallest possible amounts of power.

Whenever the industry moves from one technology (i.e., feature size) to another, existing power constraints are tightened and new constraints emerge. Power-related constraints are now being imposed throughout the entire design flow to maximize the performance and reliability of devices. In the case of today's extremely large and complex designs, implementing a reliable power network and minimizing power dissipation have become major challenges for design teams.

Creating optimal low-power designs involves making trade-offs such as timing vs. power and area vs. power at different stages of the design flow. To enable designers to perform these trade-offs accurately and efficiently, it is necessary for low-power optimization techniques to be integrated with — and applied throughout — the entire RTL-to-GDSII flow.

21.1.1 Integrated Tool Suite

A number of very sophisticated power analysis tools are available to designers; however, these tools are typically provided as third-party point-solutions that are not tightly integrated into the main design environment. Either these tools require the use of multiple databases or they combine disparate data models into one database. This means that design environments based on these tools have to perform internal or external data translations and file transfers, making data management cumbersome, time-consuming, and error-prone.

Correlating results from different point-tools can be difficult, which means that problems may be discovered late in the design cycle or may never be detected at all. Perhaps the most significant problem with existing design environments, however, is that power, timing, and signal integrity effects are strongly interrelated in the nanometer domain, but conventional point-solution design tools do not have the capability to consider all of these effects and their interrelationships concurrently.

The lack of integration between power analysis tools and the rest of the environment can result in a tremendous amount of false errors, such as minor voltage drops in portions of the design that will not affect the performance or functionality of the device. Engineers often overcompensate for these false errors and modify the power grid unnecessarily. In turn, this can cause these portions of the design to fail to meet their area constraints and to become congested, and compensating for this can cause ripple effects throughout the rest of the design.

Even worse, the lack of integration between power analysis tools and the rest of the environment — coupled with extremely limited (if any) repair capabilities — means that, when the results from the power analysis are used to locate and isolate timing or signal integrity problems, the act of fixing these problems may introduce new problems into the power network. This can result in numerous, time-consuming design iterations.

Ultimately, using point-solution power analysis tools can result in nonconvergent solutions that prevent designs from achieving their time-to-market windows (or from being realized at all). Thus, a true low-power design environment should have all of the power analysis tools operating concurrently with the implementation tools, including synthesis, place-and-route, clock-tree, extraction, timing, and signal integrity analysis. Furthermore, all the tools in the environment should employ a single, unified database.

The rest of this chapter describes capabilities of Magma's Blast Fusion and Blast Rail* tool. Built on top of Magma's unique data model, they offer an integrated analysis and optimization engine, combining synthesis, place and route engines together with extraction, timing, power, and rail analysis capabilities.

21.2 Power Dissipation

This section discusses only complementary metal oxide semiconductor (CMOS) devices only because this is currently the most prevalent digital IC implementation technology.

21.2.1 Dynamic Power

Dynamic power dissipation occurs in logic gates that are in the process of switching from one state to another. During the act of switching, any internal capacitance associated with the gate's transistors has to be charged, thereby consuming power. Of more significance, the gate also has to charge any external (load) capacitances, which are comprised of parasitic wire capacitances and the input capacitances associated with any downstream logic gates.

*Blast Fusion, Blast Rail, GlassBox are (registered) trademarks of Magma Design Automation, Incorporated.

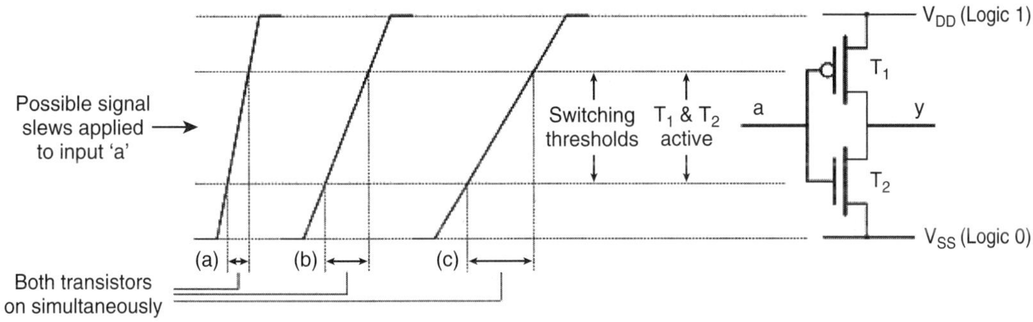

FIGURE 21.1 While the gate is switching, both transistors may be active simultaneously.

Refer to Figure 21.1. Consider a simple inverter gate, in which only one of transistors T_1 and T_2 is usually on at any particular time. When the gate is in the process of switching from one state to another, however, both T_1 and T_2 will actually be on simultaneously for a fraction of a second. This causes a momentary short circuit between the V_{DD} (logic 1, power) and V_{SS} (logic 0, ground) rails, and the ensuing crowbar current results in a transitory power surge.

The amount of time the two transistors are simultaneously active is a function of their input switching thresholds and the slew (slope) of the input signal driving the gate.

For the purposes of this chapter, the amount of dynamic power dissipation may be represented as:

$$\textit{Dynamic Power} \sim af \times C \times V^2 \tag{21.1}$$

where af is the amount of activity as a function of the clock frequency f, C is the amount of capacitance being driven/switched, and V is the supply voltage.

This equation demonstrates that minimizing the circuit activity, reducing the capacitance being driven, or reducing the supply voltage may reduce the dynamic power dissipation.

21.2.2 Static Power

Static power dissipation is associated with logic gates when they are inactive (static); that is, not currently switching from one state to another. In this case, these gates should theoretically not be consuming any power at all. In reality, however, there is always some amount of leakage current passing through the transistors, which means they do consume a certain amount of power.

Even though the static power consumption associated with an individual logic gate is extremely small, the total effect becomes significant when we come to consider today's ICs, which can contain tens of millions of gates. Furthermore, as transistors shrink in size when the industry moves from one technology to another, the level of doping has to be increased, thereby causing leakage currents to become relatively larger. The result is that, even if a large portion of the device is completely inactive, it may still be consuming a significant amount of power. In fact, static power dissipation is expected to exceed dynamic power dissipation for many devices in the near future.

Two key equations need to be considered when it comes to addressing static power dissipation. The first describes the leakage associated with the transistors as:

$$\textit{Leakage} \sim \exp\,(-qV_t/kT) \tag{21.2}$$

where q is the elementary charge, V_t is the transistor's threshold voltage, k is Boltzmann's constant, and T is the temperature.

One important point about this equation is that it shows that static power dissipation has an exponential dependence on temperature. This means that as the chip heats up, its static power dissipation increases exponentially. Furthermore, we see that static power dissipation has an inverse exponential dependence on the switching threshold of the transistors.

The second equation describes the delay (switching time) associated with a transistor affected by its threshold voltage and the supply voltage V_{DD} as:

$$Delay \sim V_{DD} \times (V_{DD} - V_t)^{-\alpha}, \text{ with } 1 < \alpha < 2 \tag{21.3}$$

From this, we see that delay goes up if the threshold increases.

21.3 Power Analysis

Power analysis consists of calculating leakage power, internal power, and switched capacitance power (which accounts for wire and input capacitance) for each cell in the design. Reporting is then done in textual reports of various types and drawing colored power maps in the layout graphical user interface (GUI).

For accurate power analysis, a number of inputs are required:

- Accurate switching activity information for all signals in the design
- Correct capacitance values for interconnect and cell inputs
- Library information for internal and leakage power at the desired operating conditions

21.3.1 Activity

The activity in the circuit has direct impact on the power dissipated by the circuit, as is shown by Equation (21.1). Therefore, it is very important to use correct activity numbers when analyzing and optimizing for power.

The activity of a signal is defined by a pair (*pr*, *tr*), where *pr* denotes the probability that a signal is a logic one and *tr* denotes the toggle rate. Activity can be specified in a number of ways:

- User annotation, where the user specifies the pair for a signal net, a signal pin, or an entire model
- From timing constraints, which specify clock frequencies, and user-specified activity ratios, which relate clock frequencies to data activities for each specific clock domain
- Reading activity obtained via simulation using formats such as value change dump (VCD), global activity format (GAF), and switching activity interchange format (SAIF)
- Automatic activity propagation from primary inputs and flip-flop outputs

21.3.2 Interconnect Modeling

For modern technologies, interconnect capacitance dominates the input capacitance of standard cells and therefore directly affects accuracy of power analysis.

In a synthesis, place and route can flow, and information that is more detailed is added along the way. Mapping, sizing, cell placement, global routing, detailed routing, and metal fill all impact the type of cells and their locations and gradually refine the interconnect estimates up until the point where exact wires and aggressor wires are known. It is only then that detailed extraction can compute wire capacitances with accuracy of within a few percent. At any stage before this point, interconnect capacitance estimation has to be used, exploiting the available information. We therefore distinguish six different interconnect models: constant, wire load, Manhattan, global routing, track routing (i.e., refined global routing), and detailed routing model. Model selection is typically done automatically and different blocks in a design can use different interconnect models depending on where they are in the flow.

21.3.3 Multiple Corner Analysis

A library describes the leakage power and the internal power for all event arcs that are possibly interesting: combinations of any switching input, any switching output, and possibly a set of values for the remaining pins. Depending on the arc type, the internal power is typically described as function of input slew and output load.

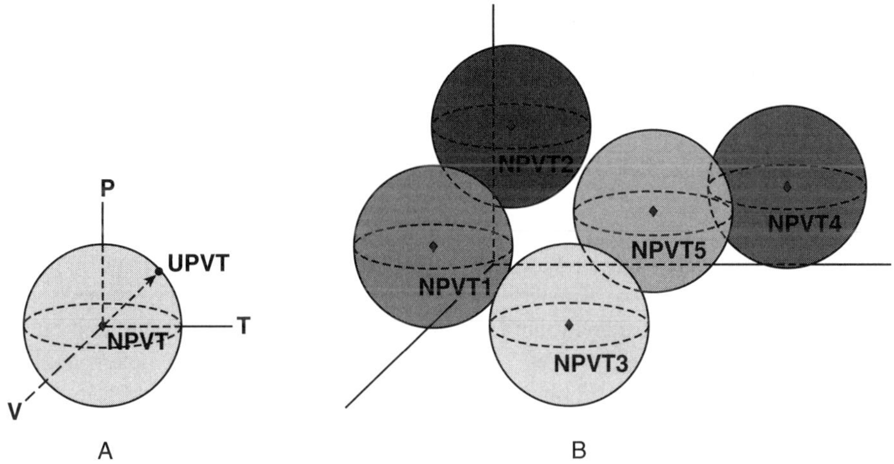

FIGURE 21.2 (a) Region around nominal PVT (NPVT) for which derating to the user defined PVT (UPVT) can be done accurately; (b) increased accuracy by using multiple characterization points.

A typical problem during analysis and optimization for different operating conditions (and, in general, for any scenario where process, voltage, and temperature can change) is the correct selection and derating of the available cell characterization information.

For instance, by changing the voltage of a supply net, the timing and power behavior of all cells supplied by that net changes. If the voltage differs only slightly from the characterization point, then derating might suffice (refer to Figure 21.2(a)). Typically, k-factor derating is then used, although obtaining correct k-factors turns out to be a problem in practice. Better derating models, such as the scalable polynomial delay model (SPDM) and similar models for power (SPPM) and leakage (SPLM), might perform better here.

To improve accuracy or to avoid derating altogether, an approach that characterizes the cell library at many different operating conditions is frequently used (refer to Figure 21.2(b)). All this data is read into the tool. The user or the tool then has to select that characterization data that best fits the desired operating condition.

Therefore, either the user selects the operating condition for which he or she wants to analyze/optimize, or automatic characterization selection is available. The latter takes full advantage of the amount of libraries because it automatically selects the right characterization point (or nearest set of surrounding points) to determine delay and energy values for the individual cells as soon as any process, voltage, or temperature changes. Derating can still be used, but now from the nearest operating condition available.

21.4 Power Optimization

The majority of today's design environments concentrate on analyzing and addressing power considerations toward the back end of the physical portion of the design process. This makes it almost impossible to fix any problems caused by poor decisions made during the early stages of the design.

A key requirement for a true low-power design environment is to provide an early analysis of the effects, such as voltage drop, using whatever data is available at the time, and to successively refine the analysis as more accurate data becomes available. This allows potential problems to be identified and resolved as soon as possible.

Creating optimal low-power designs involves making trade-offs such as timing vs. power and area vs. power at different stages of the design flow. To enable designers to perform these trade-offs accurately and efficiently, it is necessary for low-power optimization techniques to be integrated with, and applied throughout, the entire RTL-to-GDSII flow.

A wide variety of power-aware design optimization techniques can be brought into play. During the early (presynthesis) stages of the design, the RTL can be modified to employ architectural optimizations, such as replacing a single instantiation of a high-powered logic function with multiple instantiations of low-powered equivalents. The design may also be partitioned for implementation in multiple voltage domains (aka voltage islands), and power-aware clock gating techniques can be automatically applied.

In the following paragraphs, all low-power techniques minimize one or more of the factors in Equation (21.1).

During synthesis, power-aware mapping techniques may be used to optimize the netlist. These techniques include mapping highly active nodes into specific cells and mapping highly active input signals onto low-capacitance input pins.

Lowering the supply voltage dramatically reduces a logic gate's power consumption, but this also significantly reduces the switching speed of the gate. One solution is to use multiple voltage domains, allowing different areas of the chip running at different voltages (aka voltage islands). In this case, any performance-critical functions would be located in a higher voltage domain, while noncritical functions would be allocated to a lower voltage domain.

Advanced techniques also enable optimization for power during floorplanning and placement. To correctly implement multiple voltage domains, it is necessary to separate the different power meshes for each domain. The results from early voltage drop analysis can be used to determine better locations for any buffers that are to be inserted. Advanced clustering techniques can also be applied to clock-trees to reduce power consumption.

One way to reduce the amount of switching activity is to reduce the frequency of the system clock. Obviously, this will have a corresponding impact on the performance of the device. Another technique is to employ clock gating, which restricts the distribution of the clock to only those portions of the device that are actually performing useful tasks at that time. It is also possible to minimize local data activity (glitches and hazards) by applying appropriate delay balancing.

The amount of capacitance may be reduced in a number of ways. One approach is to downsize the gates driving overdriven wires, thereby lowering the capacitances associated with these gates. Another technique is to use power-aware cell placement, based on weighing nets according to their activity. The idea is to minimize the total weighed net length to minimize the switched capacitance thereby minimizing dynamic power consumption. Yet another alternative is to exploit technology options such as using low-k dielectric (insulating) materials and low-resistance/capacitance copper (Cu) tracks.

Interesting trade-offs can also be made between functional parallelism and frequency or voltage during the algorithmic and architectural stages of the design flow. For example, replacing one block of logic running at frequency f and voltage V with two copies of that block, each of which performs half of the task, and each of which is running at a lower frequency or a lower voltage. In this case, the total power consumption of this function may be reduced while maintaining performance at the expense of using more silicon real estate.

The following subsections highlight the three techniques that were mentioned previously.

21.4.1 Power Management

One of the frequently used techniques in power management solutions is partitioning the design into different blocks, each operating on block-specific voltages.

Having block specific power supply allows for switching off the supply entirely, to minimize leakage power when the block does not have to perform any logic function. Second, it allows for selecting the block supply voltage(s) that gives just the desired performance, thus minimizing dynamic power. This selection can de done once, at design time, or can be done over time using dynamic voltage scaling (DVS) schemes. Obviously, supply switching techniques and voltage-scaling techniques are orthogonal, and can be combined as such.

This section does not discuss the typical system and architectural issues that need to be addressed, such as the data retention problem for switched blocks and creating the partition of the design into

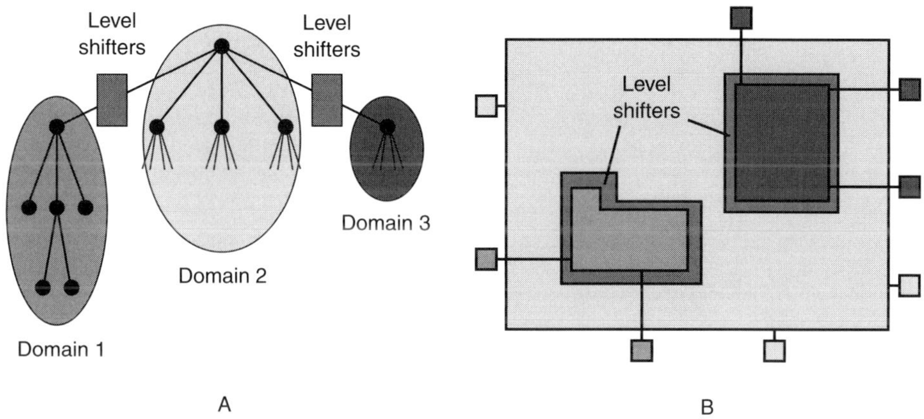

FIGURE 21.3 (a) Design partitioned into electrical domains; (b) design partitioned geometrically into physical floorplans.

separately supplied blocks. Addressed instead are the needs for the synthesis and place-and-route (P&R) flow to correctly handle such a design, such that the electrical behavior of the design is guaranteed.

We describe three types of design partitioning: functional partitioning using logical hierarchy, electrical partitioning in so-called domains, and physical partitioning in so-called (sub)floorplans (refer to Figure 21.3). Domains and floorplans can be introduced early in the flow, at or before RTL input.

A logical hierarchy starts with the top model. Each model contains model pins, cells, and nets. Nets connect to model pins and cell pins. A cell is an instantiation of a model, thus allowing for hierarchical design descriptions.

Clearly, a model represents a group of cells. The same holds for domains and floorplans.

A domain defines the operating conditions for a group of cells and the supply nets together with the recipe how to hook up each library cell to these supply nets. Domain membership is enough to fully define a cell's operating conditions and pin voltages. This information will be used during any analysis (e.g., delay, timing, power, and rail analysis) and optimization (e.g., mapping, sizing, cloning, and buffering). For instance, at the beginning of the design cycle, many cells that will be in the final design still need to be inserted. Knowing to what domain they belong is enough to do analysis correctly.

A floorplan describes a rectilinear area of cell rows in which all cells associated with the floorplan need to be placed. This information is used to guide P&R. Furthermore, each floorplan is required to be associated with one domain, from which power routing deduces what supply nets need to be routed as the rails in the cell rows, and what supply connections needs to be made by point-to-point routing.

Domains can specify more than one power and one ground net. This is required to connect complex cells, such as level shifters, which oftentimes have two power and one ground connection, cells with supply as well as bias lines, and macros. Per net, a voltage level is specified (actually, per analysis case as well) as well as the supply type. The supply type describes the behavior of the net's voltage level over time. The four supply types are:

- Constant. The voltage level is constant over time.
- Switched constant. Constant, but it may be become floating (undefined).
- Variable. The voltage level varies over time.
- Switched variable. Variable, and it may become floating (undefined).

Apart from voltage levels, a domain can also describe the process and temperature for its cells.

Assuming a cell is supplied by a specific power and ground net, its output will carry the voltage of the power (ground) net when driven to a logical one (zero). Likewise, each of its inputs will assume the power (ground) supply voltage to drive it to a logical one (zero). We extend this to multi-supplied cells, where we associate a supply line with each input and output.

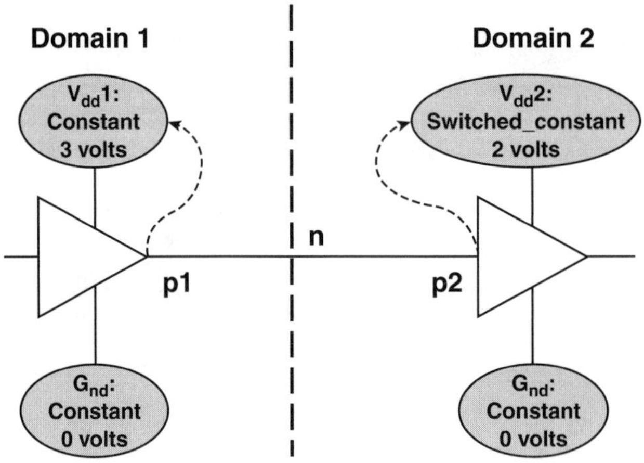

FIGURE 21.4 Net n needs level shifter with isolation capabilities.

Now, suppose we have a net that is connected to a driver cell (source) and a number of cell input pins (sinks). With each of the connected cell pins, we can associate a supply net. These supply nets have their voltage and supply-type defined on the domains. Using this information we can determine what nets need level-shifters and or isolation cells. Level–shifters are used to up-shift (and sometimes explicitly down-shift) a signal's voltage level to overcome the difference in source and sink swing. Isolation cells are cells that keep their output at a predefined logical level when the input becomes floating, and are typically required on signals that travel from switched to unswitched blocks and vice versa. Refer to Figure 21.4 for an example.

Any net crossing domain boundaries needs to be inspected for special situations described previously. In addition, however, cells within one domain can be connected to different supply nets and, therefore, should be checked as well. This, however, depends on the setup of your domains.

21.4.2 Gate Sizing

Refer to Figure 21.1. One of the factors controlling the slew of the signal being presented to the inverter's input is the size of the transistors forming the logic gate driving this signal. These need to be sufficiently large such that the signal transitions fast enough to keep the amount of time the inverter's transistors are both active to a reasonable level (Figure 21.1(b)).

Now consider what happens if the driving gate's transistors are too large and the driving gate is overpowered. In this case, the power savings achieved by minimizing the time where the inverter's transistors are both on (Figure 21.1(a)) will be negated by the driving gate having to charge the increased capacitance associated with its oversized transistors, thereby consuming excessive amounts of power. Furthermore, the high speed of the signal's transitions will also cause signal integrity problems in the form of noise, overshoot, undershoot, and cross talk.

By comparison, if the driving gate's transistors are too small and the driving gate is underpowered, the inverter's transistors will both be on for a significant amount of time (Figure 21.1(c)), thereby causing the inverter to consume unwarranted amounts of power (the under-driven input signal will also be susceptible to noise and cross talk coupling effects from other signals).

The gain-based scenario that we use selects minimum gate sizes, subject to maximum slew limits and maximum load limits. Minimizing gate size is good for cell area and cell power dissipation, since it implies minimization of cell parasitic capacitances. In doing so, all paths are automatically made as slow as possible, thereby balancing the delay of the paths as much as possible. This is good for glitch and hazard power as well.

21.4.3 Multiple Thresholds

To address low-power designs, IC foundries offer multiple V_t libraries, in which each type of logic gate is available in two (or more) flavors:

1. With low-threshold transistors that switch quickly but have higher leakage and consume more power
2. With high-threshold transistors that have lower leakage and consume less power but switch more slowly

One option is to use low-V_t transistors only on timing-critical paths and to use high V_t transistors on noncritical paths, the so-called dual V_t approach. Another solution is to use multiple voltage domains to implement MTCMOS (i.e., to selectively power-down leaking blocks using nonleaking transistors whenever those portions of the device are not required), for example, when those portions are placed in a standby mode. These two solutions may of course be used in conjunction.

Given the possibility to use multiple thresholds and (dynamic) voltage selection for blocks makes that the engineers have to perform a complicated balancing act. For instance, lowering the supply voltage reduces the amount of heat being generated, which, in turn, lowers the static power dissipation; however, lowering the supply voltage also increases gate delays. By comparison, lowering the transistors' switching thresholds speeds them up, but this exponentially increases their leakage and therefore their static power dissipation. In addition, switching entire blocks on and off can cause dramatic current surges and thus generate *Ldi/dt* voltage drops, which may require the use of additional circuitry to provide a soft (staged) power on/off for these blocks.

21.5 Rail Analysis

Deep submicron (DSM) and ultra-deep submicron (UDSM) devices are prone to voltage drop effects, which are caused by the resistance associated with the network of wires used to distribute power and ground from the external pins to the internal circuitry.

Purely for the purposes of providing a simple example, consider a chain of inverter gates connected to the same power and ground tracks. Refer to Figure 21.5. Every power and ground track segment has a small amount of resistance associated with it. This means that the logic gate closest to the IC's primary power or ground pins (gate G1 in this example) is presented with the optimal supply. The next gate in the chain (G2 in this example) will be presented with a slightly degraded supply, and so on down the chain.

The problem is exacerbated in the case of transient or alternating current (AC) voltage drop effects. These occur when gates are switching from one value to another or, even worse, when entire blocks are switched on and off. This causes transitory power surges, which momentarily reduce the voltage supply to gates farther down the power supply chain. The simple example circuit shown in Figure 21.5 consists

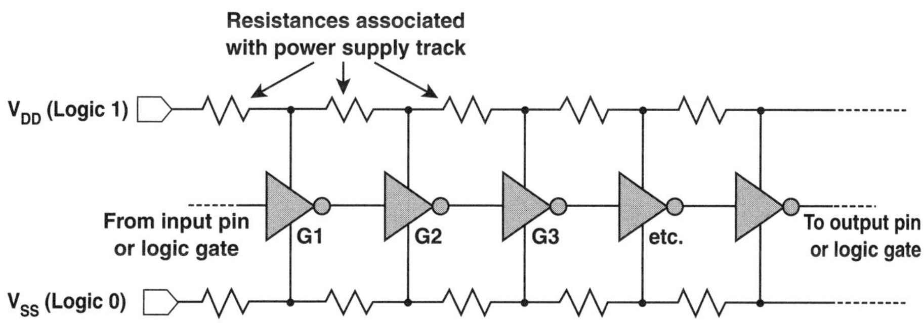

FIGURE 21.5 A chain of invertors connected to the same power and ground tracks.

only of inverter gates, but a real design typically contains tens of thousands of register (storage) elements triggered by a clock signal. The clock can cause large numbers of register elements to switch simultaneously, resulting in significant glitches in the power supply. To analyze and address these effects, it is necessary to consider resistive, inductive, and capacitive effects.

Voltage drop caused by resistive effects is often referred to as IR drop and, erroneously, associated with DC currents only. Compare this with the inductive voltage drop effects caused by transient currents through inductors. Of course, IR drop also occurs in this last case.

The reason voltage drop effects are so important is that the input-to-output delays across a logic gate increase as the voltage supplied to that gate is reduced, which can cause the gate to miss its timing specifications. An increase in the interconnect delays associated with wires driven by underpowered gates also occurs. Furthermore, a gate's input switching thresholds are modified when its supply is reduced, which causes that gate to become more susceptible to noise.

21.5.1 Analysis Flow

Blast Rail is a full-chip voltage drop analysis and repair tool completely integrated into the RTL-to-GDSII design environment. The tight integration makes error-prone data transfer between tools superfluous. A voltage drop analysis based on power estimation (refer to Figure 21.6) consists of six steps:

1. Layout extraction of power and ground nets
2. Determination of current sources from power analysis
3. Creation of an electrical network suitable for voltage drop computation
4. Matrix solving
5. Reporting
6. Update cell timing information using computed cell voltages

For static DC voltage drop analysis, the electrical network model of the power and ground nets consist of electrical nodes, resistances that connect nodes, and current as well as voltage sources. Resistance values are derived from the geometries of the extracted segments and the layer specific sheet resistance, which might be width dependent. Voltage sources are due to connected supply pads or bump cells in flip chip designs. All other cells, such as standard cells, macros, or I/O pads, result in current sources in the electrical network. Their values are derived from a power analysis. The position of the current sources within the network depends on when the analysis is performed within the design flow. After detail placement, the position is easily derived from the placement location of the corresponding cells and macros. A preplaced design is analyzed by assuming a uniform distribution of the current sources while placement blockages are honored.

A transient analysis additionally requires the extraction or specification of inductances and capacitances. On-chip wire inductances are negligible for current technologies; however, bonding wires have inductances that must be taken into account. Capacitive effects are mainly due to decoupling capacitors. Decoupling capacitors are placed close to high switching cells to act as a charge supply. The result is a low-pass filtering of current spikes.

21.5.2 Voltage-Drop-Induced Analysis

The flow described previously assumes that the current sources have known values. These values are derived from a power consumption analysis of the entire design; however, the initial power and timing analysis is based on a fixed voltage supply for all cells and macros (i.e., voltage drop is not considered at this stage). This results in optimistic slews and delays but a pessimistic power consumption behavior of the cells and macros. Blast Rail offers the feature to feed back the voltage drop values for timing and power analysis. The user can then iterate the process of timing, power, and voltage drop analysis. It turns out that the analysis results converge very quickly after only a few iterations (two to three iterations). Since Blast Rail is integrated into Magma's unified data model, the updated timing information can also be used for timing driven optimizations such as buffer sizing and placement.

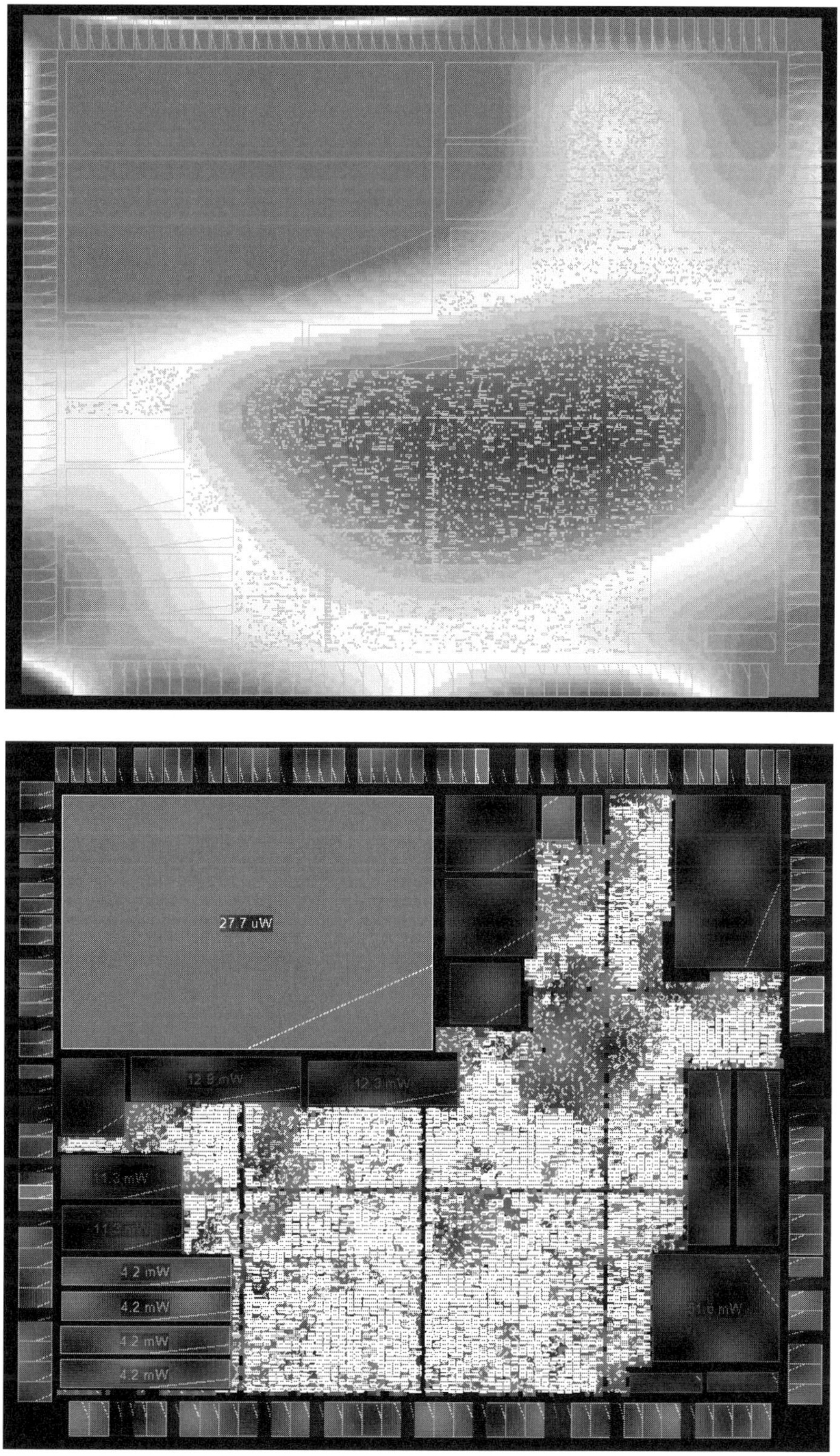

FIGURE 21.6 (Top) Power map showing the power dissipation for each cell of a design after power analysis and (Bottom) the corresponding voltage-drop oil map after rail analysis.

21.5.3 Abstraction

To handle large SoCs (such as 10 million gates or beyond), we offer our GlassBox abstraction technique for hierarchical chip design. The basic idea is to abstract away from a block as much data as possible. Only boundary information is kept such that for instance accurate top-level timing analysis and optimization is still possible. Blast Rail also supports the GlassBox abstraction technique. The power and ground layout information is deleted, while a reduced electrical network is kept. The reduced network mimics the same electrical behavior of the block at the top-level as the original block. Typical reduction of up to 95% is obtained with accuracy loss of less than 1%.

After reduction, the voltage drop inside the block cannot be observed while doing a top-level analysis. An excessive drop inside the block due to a poor supply network might remain undetected. Three different solutions can be applied to circumvent this drawback:

1. Assume pessimistic voltages at the boundaries while doing block level analysis.
2. The user can specify regions (aka view ports) inside the block, which are not reduced and thus stay observable. This combines the best of both, namely observability in combination with high-reduction factors.
3. The electrical network of the GlassBox is not reduced at all (but all other abstractions are applied).

These hierarchical analysis techniques offer the designer to control the trade-off between memory reduction and observability of voltages and currents within a block.

21.5.4 What-If Analysis

Designing the power and ground distribution networks amounts to trading off the chip area used by these networks and the area used by signal and clock-net routing. A robust supply network ideally has a dense mesh with as many via connections to rails and rings as possible. It should consist of wide wires and a lot of connected power pads or bumps; however, such a network results in highly congested routing layers, making it harder or impossible to route signal- and clock-nets.

What-if analyses support the designer to deal with this trade-off. It allows for analyzing different power network configurations without generating the corresponding layout. The designer can add or delete resistances, capacitances, inductances, current sources, and voltage sources directly in the electrical network. An analysis of the modified network can then be done without performing an extraction. Adding a resistance means that a connection is made in the geometric domain, such as dropping a via between the mesh and a rail. Replacing an extracted resistance by a smaller resistance corresponds to widening a wire or creating a larger via array. Introducing or removing capacitances and inductances are used to model different decoupling alternatives or chip package models respectively. Adding current sources allows for testing the robustness of the network. For example, the designer can distribute a certain amount of current over a region of the chip where a macro will be placed later on in the design flow. Adding or removing voltage sources allows for instance to play with a different number of supply pads or different pad locations.

21.5.5 Partial Grids

Manually adding resistances to an extracted electrical network becomes a tedious and error prone task if entire power routing steps are to be simulated. Blast Rail offers the feature of partial grid analysis, which automatically creates resistances to mimic the routing steps of via dropping and pin tapping. Pin tapping is the task to create the proper connections of macro pins to rings and meshes during power routing.

The automatic resistance creation reduces design as well as extraction time. It is typically used when designing the power rings and meshes, when the designer is focusing on the global aspects of the power grid and does not want to be disturbed by detailed connectivity issues such as pin tapping and via dropping.

In this phase of the flow, positioning and connectivity of current sources also play a role. During preplacement phases, such as floorplanning and power grid planning, unplaced cells are modeled as current sources that are uniformly distributed over the unused standard cell area as defined by the floorplan.

21.5.6 Electromigration

Electromigration occurs when the current density (current per cross-sectional area) in tracks is too high. In the case of power and ground tracks, electromigration effects are DC-based. The so-called electron wind induced by the current flowing through a track causes metal ions in the track to migrate. This migration creates voids in the upwind direction, while metal ions can accumulate downwind to form features called hillocks and whiskers. The increased track resistance associated with a void can result in a corresponding voltage drop and thus might cause timing problems or even functional errors due to undersupplied logic. Major functional errors can also occur when the voids eventually lead to open circuits or when the hillocks and whiskers may cause short circuits to neighboring wires.

Electromigration rules are defined as a maximum current per width of a wire (and not per area) because the height of each wire is constant and a parameter of the applied process technology (neglecting intra-die variation effects on layer thickness). Modern process technologies require the definition of width dependent electromigration rules for each layer. These rules are usually staircase functions where wider wires have a higher current density limit. Vias are treated differently compared with routing wires. Via electromigration rules define a maximum current per via cut.

21.6 Power Grid Synthesis

To accommodate variations in operating temperature and supply voltage, designers have traditionally been obliged to pad device characteristics and design margins; however, creating a device's power network using excessively conservative design practices consumes valuable silicon real estate and results in performance that is significantly below the silicon's full potential. This is simply not an option in today's highly competitive marketplace.

Voltage drop effects are becoming increasingly significant, because the resistance of the power and ground tracks rises as a function of decreasing feature sizes (e.g., track widths). Increasing the width of power and ground tracks can minimize these effects, but can cause routing congestion problems. To solve these problems, the logic functions have to be spaced farther apart, which increases delays (and power consumption) due to longer signal tracks. Thus, implementing an optimal power network requires the balancing of many diverse factors.

21.6.1 Grid Synthesis

The process of designing the power distribution network should be based on the results of early rail analysis performed when the power grids are still incomplete (refer to Section 21.5.5, "Partial Grids"). Correct distribution of dissipating elements across the chip can avoid hot spots and local voltage drop problems, and special wire-widening algorithms can be used to address voltage drop and electromigration issues.

Voltage drop problems can be fixed by wire widening, via insertion, and by adapting mesh frequencies. Support for automatic determination of the mesh parameters, namely mesh frequency and wire width, is available as part of automatic power grid synthesis. Mesh optimization is done for all meshes simultaneously and can be guided by the user.

Appropriate on-chip decoupling capacitors should be added to minimize the inductive voltage drop effects caused by off-chip current variations over time. The transient effects should be kept to a minimum (i.e., the charge gets to the cells in time). Thus, the voltage-drop reduction problem is made as *DC* as possible. To lower the current-per-pad and bond-wire inductance, many pads are allocated for power and ground, thereby making the analysis of pad placement a nontrivial task. Flip-chip packaging technologies can be used to increase the number of pads connected to the power and ground supplies, thereby lowering the current-per-pad and lowering the inductance.

Electromigration violations in power and ground tracks are fixed by widening the wires, which cause the violations. Electromigration rules for vias are defined as current per via cut. This means that fixing an electromigration problem in a via translates to increasing the number of via cuts. If the via already consists of the maximum number of via cuts for the overlapping top and bottom routing wires, then the wires have to be widened as well. The electromigration report generated by Blast Rail prints for each violated wire the required width and the number of required via cuts per violated via, and may apply these suggestions.

21.6.2 Packaging Considerations

Power consumption — both static and dynamic — increases a device's operating temperature. In turn, this may require engineers to employ expensive device packaging and external cooling technology.

Yet, another consideration is that the on-chip temperature gradient (i.e., the difference in temperatures at different portions of the device caused by unbalanced power consumption) can produce mechanical stress, which may degrade the device's reliability.

When it comes to power distribution, the first problem is to get the power from the outside world, through the device's package, to the silicon chip itself. Typically, the amount of current per pin is limited requiring many pins for supply. The wires used to distribute power throughout the chip have resistances associated with them — the longer the wires, the larger the resistance, and the larger the resistance, the greater the associated voltage drops. This means that traditional packaging technologies based on peripheral power pads are no longer an acceptable option in the case of today's extremely large and complex designs.

One solution is to use a flip-chip packaging technology, in which pads located across the face of the die are used to deliver power from the external power supply directly to the internal areas of the chip. In addition to being able to support many more power and ground pads, this minimizes the distance the power has to travel to reach the internal logic. Furthermore, the inductance of the solder bumps used in flip-chip packages is significantly lower than that of the bonding wires used with traditional packaging techniques.

21.7 Conclusion

Addressing the problems associated with DSM and UDSM devices requires power design and analysis tools that work throughout the entire RTL-to-GDSII design flow. Identifying and resolving power problems late in the flow may result in expensive, time-consuming iteration cycles. What is required is to identify and resolve these problems throughout the flow and to be able to "forget" issues once they have been addressed and made "safe."

To handle complex interrelationships between diverse effects, it is necessary for all of the power tools to be fully integrated with each other, and also with other analysis engines in the flow, including synthesis, place-and-route, timing, and signal integrity analysis. This requires that all design and analysis tools have concurrent access to a single, unified design database, and that any changes made by one tool are immediately tested and validated by the others. This results in a convergent algorithm that quickly determines optimal solutions with a minimum of time-consuming iterations.

22

Sequence Design Flow for Power-Sensitive Design

Jerry Frenkil
Sequence Design

22.1 Introduction

Integrated circuit (IC) power consumption has become a significant issue for most applications. For wireless and battery-powered applications, the key issue is that of maximizing battery life. This issue is exacerbated by the continuously increasing amounts of computing required for advanced functionality, such as color displays and full-motion video. For tethered applications, where batteries do not limit the power budgets, power is also an issue because it directly affects critical manufacturing parameters, such as die size, packaging choices, and unit cost.

These issues and concerns are not new. Various forms of low-power design have been practiced for decades, but what is new is the criticality of effectively addressing the various facets of power consumption. Simply put, power threatens to derail the constant progress of Moore's law [1]. In many ways, the advances delivered by process engineers in the form of smaller line widths and thinner oxides are now creating as many problems as they solve. Moreover, some of the problems, such as transistor leakage, appear unlikely to be solved in the near future by the processing advances, so it is the design community that will have to step up with solutions to the power problem.

The earliest forms of low-power design involved the basic practice of reducing the power supply voltage, either in the entire design or in certain parts. The attractiveness of this approach was, and still is, the quadratic relationship between the supply voltage and the resulting power consumption. An additional method involved the use of more advanced semiconductor processes with narrower transistors and shorter wires; all else being equal, the reduction in parasitic capacitances due to the smaller geometries resulted in less dynamic power.

Today, these approaches are no longer enough and, in some cases, such as subthreshold leakage, they are counterproductive. Thus having employed the most basic approaches, designers have turned to more design-oriented techniques in the architecture, logic, and physical design spaces. In doing so, designers have found it necessary to employ various forms of design automation tools. In some cases, the design tools enable the designer to intelligently choose between various design alternatives based on power characteristics. In other cases, the tools evaluate the choices and make the decisions automatically.

22.2 Design Flow Overview

The vast majority of digital ICs designed today are built in complementary metal-oxide semiconductor (CMOS) technology. Once viewed as a low-power technology, CMOS chips can consume as little as a few microwatts or as much as 100 W. Where a chip's power consumption characteristics fall on this continuum depend on a large number of variables, not the least of which is the amount of attention paid to power consumption during the design process.

22.2.1 CMOS Power Consumption

Generally, power-efficient CMOS design involves the minimization of one or more of the terms in the basic power consumption equation

$$P = C_L V_{dd}^2 f + V_{dd} I_{dd} \tag{22.1}$$

or, in its more detailed version

$$P = C_L V_{dd} V_{swing} f + V_{dd} Q_{sc} f + V_{dd} I_{lkg} + V_{dd} I_{through} \tag{22.2}$$

where P represents the total power consumed, V_{dd} represents the supply voltage, I_{dd} represents the static current drawn from the supply, C_L represents the load capacitance, and f represents the switching frequency. The VI term represents the static, or DC power consumption, while the CV^2f term represents the dynamic power consumption. In the more detailed version, V_{swing} represents the signal voltage swing (which for CMOS is usually equal to V_{dd}), Q_{sc} represents the charge consumed due to the short-circuit momentary current (also known as crowbar current) drawn from the supply during switching events, I_{lkg} represents the parasitic leakage current, and $I_{through}$ represents the (by design) quiescent static current. The first two terms of this equation represent the dynamic power consumption, while the latter two represent the static power consumption.

Until relatively recently, the design and design automation communities viewed low-power design as being primarily focused on the CV^2f component; however, with chips inexorably becoming bigger, faster, and more power-hungry, it has become clear that the problems, as well as their solutions, are much more complicated.

22.2.2 Power-Sensitive Design Challenges

Today, what is commonly known as low-power design actually comprises two different but related design activities. The first is power minimization — the reduction of the power consumption characteristics of the design. The second is power integrity management — managing the delivery of power to the various portions of the design as well as the effects of a nonideal power source on the design's timing and functionality. Together, these two design activities are sometimes referred to as power-aware design or power-sensitive design.

Power minimization seeks to reduce power consumption, be it average power or instantaneous power or both. It may be directed at all modes of operation, or only a particular power mode, such as standby or sleep mode. It may focus on only dynamic power, only on leakage power, or on the total. By comparison, power integrity management seeks to illuminate and minimize the effects of power on the design. These effects include timing, noise, reliability, and cost.

The determination and optimization of these various effects becomes more complex as designers go to greater lengths to control the effects. For example, the reduction of power supply voltages over the last several years, from the long standing standard of 5 V to around 1 V, exacerbates already challenging issues such as large on-chip supply currents and miniscule noise margins. The use of multiple voltages to obtain higher performance or to interface to devices running at higher voltages creates additional issues in physical design.

The implications of these challenges are that a variety of design tools are needed to address the various power issues — the power problem is sufficiently critical and complex that a "one size fits all" approach is inadequate. Fortunately, a number of design automation solutions are available now. For example, power can be calculated early in the design process by utilizing tools that estimate power from a high-level register-transfer level (RTL) description, while later in the design process, detailed analyses of dynamic supply currents flowing through on-chip power distribution networks can be obtained via the use of power rail analyzers. In between, power can be calculated and optimized at the gate level, after synthesis and before physical design.

22.2.3 Feed Forward Design Flow

Given an appropriate variety of tools, effective use is often dependent upon a well-structured design flow. For example, for power optimization as for other parameters, such as performance and cost, it is critical to architect the system properly at the beginning and successively refine it as the project proceeds. Such a multilevel approach increases the likelihood of meeting design goals by providing both early visibilities into critical issues as well as multiple opportunities for mitigation.

Much of digital design is performed today utilizing a top-down or modified top-down design flow. Here top refers to the higher levels of design abstraction, such as the system, behavior, and register-transfer (RT) levels, and time flows downward toward the lower levels of design abstraction, such as the gate and transistor levels. In this case, flow refers to the sequence of tasks; however, the flow of detailed design information is somewhat less clear.

In conventional practice, detailed design information tends to follow a feedback design flow, wherein information about particular power characteristics does not become available until the design has progressed to the lower abstraction levels. A feedback design flow features a relatively lengthy feedback loop from the analysis results obtained at the gate or transistor level back up to the design tasks at the RT-level and above. Thus, information about the design's power characteristics is not obtained until quite late in the design process. Once this information is available, it is fed back to the higher abstraction levels to be used in determining how to deal with the power issues of concern. The farther the lower-level power analysis results exceed the target specification, the higher the abstraction level in which the design must be changed.

By comparison, a feed forward approach, illustrated in Figure 22.1, replaces these lengthy, cross-abstraction feedback loops with more efficient abstraction specific loops. Thus, the design that is fed forward to the lower abstraction levels is much less likely to be fed back for reworking, and the analysis performed at the lower levels becomes essentially a verification task. The key concept is to identify, as early as possible, the design parameters and trade-offs that are required to meet the project's power specs. This helps to ensure that the design being fed forward is fundamentally capable of achieving the power targets. Later in the design flow, optimizations at the lower levels can be used to further minimize the power as desired.

The feed forward flow is enabled by a high-level analysis tool, such as PowerTheater [2], which can accurately predict power characteristics. These early, high-level analysis capabilities are employed to make informed trade-offs, such as which algorithms and architectures to employ, without having to resort to detailed design efforts or low-level implementations to assess performance against the target power

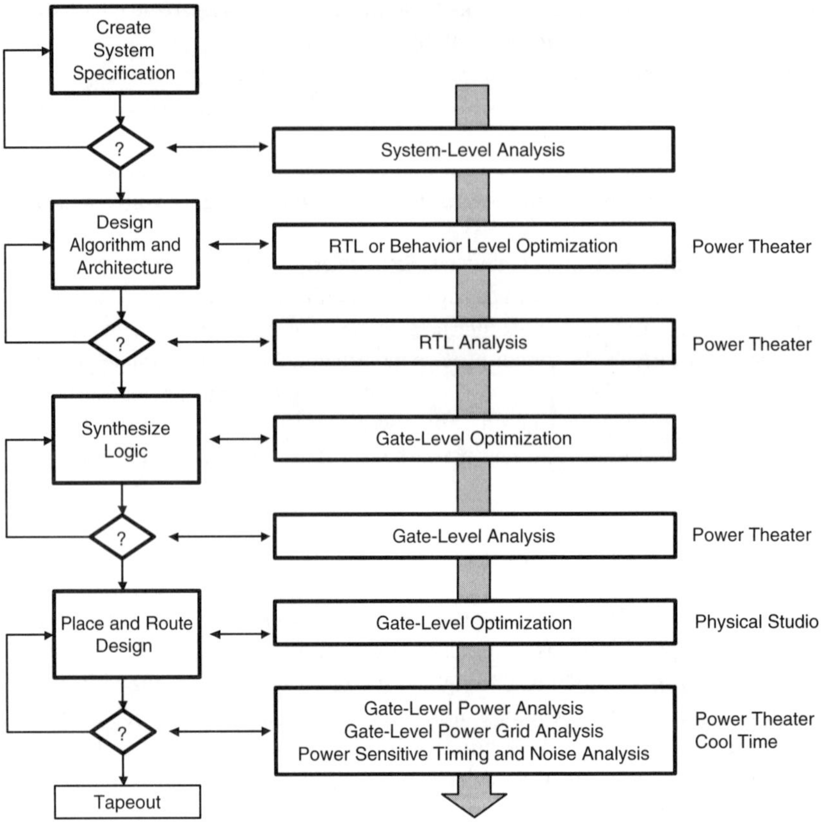

FIGURE 22.1 Feed forward design flow.

specification. Compared with the traditional top-down methods, the key difference and advantage is added by the early prediction technology.

Proceeding in parallel with, or sometimes ahead of, the architecture development is the design of the library macro functions and custom elements, such as datapath cells. These are used in the subsequent implementation phase in which the RTL design is converted into a gate-level netlist. At this point, appropriate optimizations are performed again, and power is reestimated with more detailed information, such as floorplanned wiring capacitances. The power grid is planned and laid out using this power data.

Once the design has been synthesized into a technology mapped gate-level netlist, lower level power optimizations can be employed, using a tool such as PhysicalStudio [3], to further reduce dynamic or leakage power consumption. Specific goals or issues, such as battery life or noise margin repair, will determine the particular optimizations employed.

These optimizations can be performed either before (using estimated wiring parasitics) or after routing (using extracted wiring parasitics). In either case, after the design has been routed and optimized, a final tape-out verification and electrical verification check is performed with an electrical sign-off tool such as CoolTime [4]. In this step, power is calculated and used to compute and validate key design parameters, such as total power consumption in active and standby modes, junction temperatures, power supply droop, noise margins, and signal delays.

Thus, power is analyzed and optimized multiple times, at each abstraction layer following the feed forward approach. Each analysis is successively refined from the previous analysis by using information fed forward from prior design decisions along with new details produced by the most recent design activities. Each optimization, at the various abstraction layers, results in more efficient logic structures to feed forward to the downstream design tasks, thereby successively squeezing out the wasted power.

This approach encourages design efforts to be spent up front, at the higher abstraction levels, where design efforts are most effective in terms of minimizing and controlling power [5]. In addition, because power-sensitive issues are tracked from the beginning to the end, the likelihood of a late surprise issue is minimized.

22.3 Sequence Tools for Power-Sensitive Design

Sequence design provides several different tools for use in a comprehensive power-sensitive design flow. These tools are described next, in the order that they would be utilized in a feed forward methodology.

22.3.1 PowerTheater

PowerTheater is a mixed-level power analysis tool that analyzes RTL designs, gate-level designs, and mixed- — RTL and gate — level designs. It reads in the design description, technology libraries, and environmental data, such as power supply values and external loadings, and activity information. It produces a detailed report listing the amount of power consumed by the various portions of the design along with additional power debug information.

PowerTheater's architecture is illustrated in Figure 22.2. Its front end includes a language parser and inference engine that reads designs described in Verilog, VHDL, or mixed Verilog/VHDL, translates the design into an internal representation, and loads the internal database. The power calculation engine reads the internal database, and, using the specified simulation activities, environmental data, and technology libraries, calculates the power for each portion of the design [2].

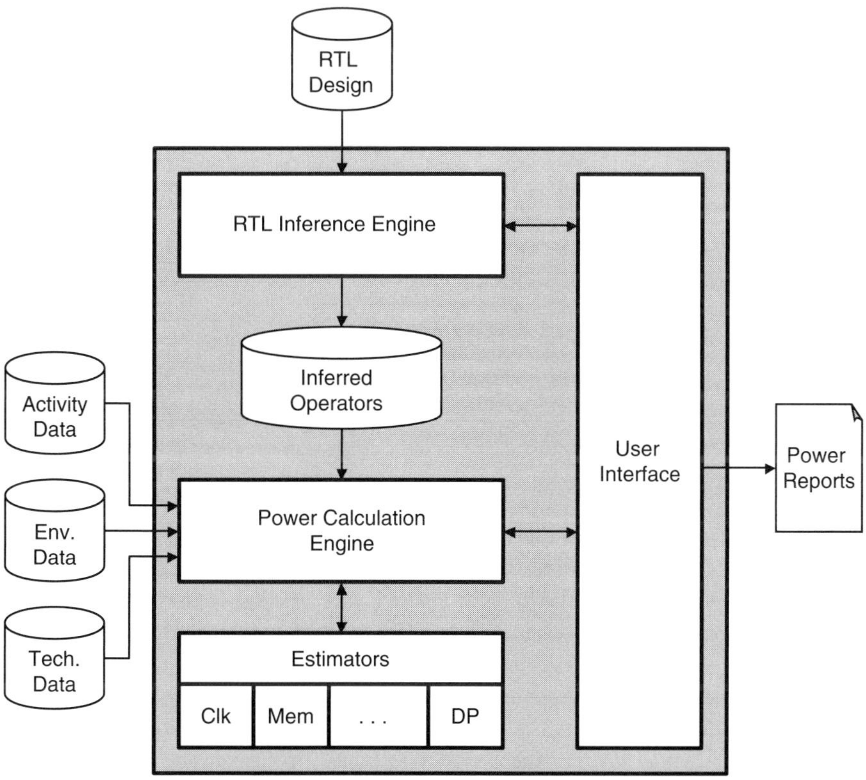

FIGURE 22.2 PowerTheater architecture.

The technology library describes the electrical characteristics of the circuit primitives to be employed in the design. Data, such as input capacitances, state dependent static and dynamic power consumption, logical function, and physical size, are all used during various computations leading up the final power calculation. The data can be read in any of several industry standard formats, such as Liberty, advanced library format (ALF), or open library architecture (OLA) [6–8].

Upon encountering RTL code, PowerTheater infers the hardware that would be produced by a synthesizer for that code; however, unlike a full logic synthesizer, PowerTheater does not produce a gate-level netlist. Instead, it produces a micro-architectural netlist that contains technology unmapped, inferred operators, such as multi-bit registers, arithmetic units, multiplexers, and decoders. Although similar to the first steps executed by a conventional gate-level synthesizer, this approach results in faster execution time, a smaller internal database, and an ability to easily cross-reference the original RTL code with the inferred operators — the latter, especially, being a key advantage to those writing the RTL code. By contrast, when encountering instantiated objects, such as compiled memories or gate-level primitives, PowerTheater passes these objects directly to the internal database.

Once the inferencing step is completed and the internal database is loaded, the design is ready for analysis. The calculation engine loads the database, along with activity data (which can come from a simulation trace, in the form of a value change dump (VCD) file, or from a vectorless activity specification), technology data, and environmental data. For gate-level power analyses, the calculations are relatively straightforward: for each instance, PowerTheater determines the particular stimulus from the activity data, looks up the power characteristics for that stimulus in the technology library, and computes the instance's power using the specified environmental data.

For RTL analyses, the operation is similar but with a key difference: instead of processing gate-level instances, the engine calculates power for each inferred instance. This is accomplished by elaborating a parameterized model for each inferred instance. The elaboration process involves evaluating a built-in parameterized power equation for each instance utilizing power information from the technology library along with the activity and environmental data. Each inferred operator has its own unique power equation, thus enabling PowerTheater to separately calculate power for the different design structures and objects that are found in the RTL. For example, datapath operators are inferred and evaluated separately from control and clock operators. This divide and conquer approach enables faster and more accurate calculations (more accurate than the "one size fits all" algorithm common to gate-level tools) along with a reporting format that is closely linked to the RTL source code. For example, clock power, input/output buffer (I/O) power, register power, and random logic power are all reported separately. Power consumed in driving wiring capacitances, and cell internal power can be reported separately.

The various reporting styles and mechanisms are key features enabling PowerTheater to be used as a design tool. Given the objective of writing power-efficient RTL code, effective analysis, and debugging tools, capable of pinpointing power problem areas, are critical for identifying power minimization opportunities.

22.3.2 Using PowerTheater

PowerTheater is primarily used for two different purposes: power minimization and power verification. In the former case, the objective is to minimize power consumption by writing power-efficient RTL code and by providing early visibility into the design's power characteristics. In the latter case, the objective is to check or verify that the design, whether at the RT or gate level, is within an acceptable power consumption limit.

Producing power-efficient RTL code requires the ability to identify the amount and source of power consumption along with its underlying causes. For example, considering Equation (22.2) above, isolating the largest contributing factors enables the designer to focus on the largest opportunities for power reduction. Although many of these factors, such as V_{dd} or C_L, are fixed by either technology or environmental dictates, others, such as nodal switching frequency, can be affected by coding styles.

Conventional Code:
64x32 Register File using Flip-Flops

```
input web, oe,clk;
input [31:0] di;
input [5:0] aadr, badr;
output [31:0] do;

// define storage array
reg[31:0] array [63:0];
reg[31:0] do;

// Write Cycle: edge trigger
//              implies flops
always @ (posedge clk) begin
    if (web == 0) array[aadr] = di;
end

// Read Cycle - a sync read
always @ (badr or oe) begin
    if (oe == 1) do = array[badr];
end
```

Low Power Code:
64x32 Register File using Latches

```
input web, oe,clk;  // clk not needed
input [31:0] di;
input [5:0] aadr, badr;
output [31:0] do;

// define storage array
reg[31:0] array [63:0];
reg[31:0] do;

// Write Cycle: level trigger
//              implies latches
always @ (aadr or web or di) begin
    if (web == 0) array[aadr] = di;
end

// Read Cycle - a sync read
always @ (badr or oe) begin
    if (oe == 1) do = array[badr];
end
```

FIGURE 22.3 Register file RTL code.

In a power minimization methodology, PowerTheater is used to estimate power as the RTL code is written, thus enabling the designer to quickly understand the impact of design decisions and to optimize the code during the creation process, which is the essence of low-power design. If a simulation test bench is available, then the design is simulated to collect activities for the subsequent power calculation. If a simulation test bench is not available, then PowerTheater's vectorless activity function is used to generate activity and state information for use in the power calculations. Although the accuracy of the resulting calculations is much better with simulated data, vectorless activity specification can be used to determine which of several coding alternatives will consume the least power when simulation data is not yet available.

A comparative example of different RTL codes for a given function is listed in Figure 22.3. Consider a register array organized as 64 words by 32 bits. Such a storage function can be coded in several different ways, two of which are presented here. Although the amount of code is the same, the power difference is substantial: the latch-based array consumes only about half as much power as the conventionally coded register array.

In this case, the power reduction principle at work is the use of latches instead of flip-flops — latches being more power-efficient than flip-flops. This targets the $V_{dd}Q_{sc}f$ term in Equation (22.2) — latches, implemented with fewer transistors than flip-flops, consume less internal current; however, a more common method of power reduction is the minimization of effective switching frequencies, targeting the two frequency dependent terms in Equation (22.2), $C_L V_{dd} V_{swing} f$ and $V_{dd}Q_{sc}f$. Here, the concept is to reduce unwanted or unnecessary toggles, which in turn reduces the dynamic power consumption.

The classic example of reducing effective switching frequency is the use of gated clocks to inhibit the clocking of storage elements when it is known that stale data would be latched. Thus, inhibiting the clock edge saves power without changing overall functionality. A subtler example involves operator isolation. Consider the case of a multiplexer that must select between an operand and a multiplied version of that operand as presented in the code in Figure 22.4. In the original version, the multiplier multiplies all the time, whether the multiplexer is set to select the multiplier's result or not — the multiplier multiplies any time a change occurs on either of its inputs. In the modified code, however, the multiplier only multiplies when its output results will be used, thus the effective switching output frequency is reduced because data is not allowed to flow into the multiplier unless its output will be selected.

PowerTheater provides two different methods of finding such opportunities, among others, for power reduction. The first method involves the use of the graphical user interface (GUI) for interactive power

Conventional Code: Selecting Operand or Multiplied Operand	Low Power Code: Selecting Operand or Multiplied Operand

`assign muxout = sel ? A : A*B;`	`assign isoA = sel ? 0 : A;`
	`assign isoB = sel ? 0 : B;`
	`assign muxout = sel ? A : isoA*isoB;`

FIGURE 22.4 Operator isolation RTL code.

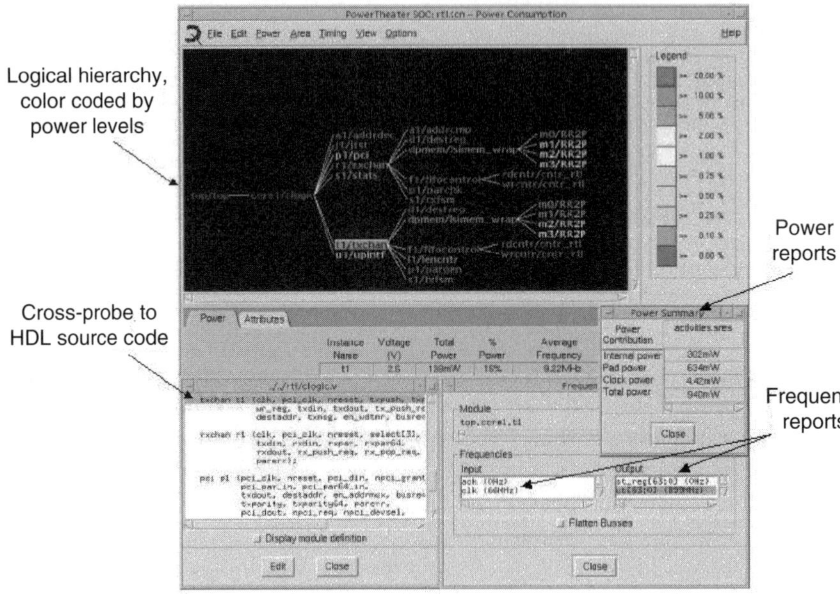

FIGURE 22.5 PowerTheater GUI.

debugging. As presented in Figure 22.5, the PowerTheater GUI displays a hierarchical representation of the design's inferred operators. The display is colorized to facilitate discovery of the design's power-hungry portions via a quick visual inspection. Once a particular "hot spot" is identified, debugging proceeds by clicking on the inferred operator of interest, which will bring up a display of the particular lines of RTL code along with the power consumed by those lines of code. The root cause of that consumption can be investigated by displaying the nodal frequencies of the operator under consideration.

PowerTheater's second method for finding power waste in a design is fully automatic. This method utilizes design search modules known as "WattBots" that walk the inferred netlist looking for various types of power inefficient structures. In addition to finding several different types of clock gating opportunities, the WattBots will also identify operator isolation opportunities, inefficient memory structures, bus conflicts and floating busses, and glitchy nets and control signals. The WattBots will also estimate the amount of power wasted in each instance so that the designer can make informed decisions as to which changes will produce the largest power savings. This is especially important in the case of clock gating because gating too many clocks presents significant problems later during physical design with respect to clock skew management. Thus, it becomes critical to identify which clocks are worth gating (the ones that result in the largest power savings) as well as the ones that should not be gated (the ones that result in the least power savings). The WattBot reports support this type of decision making by presenting the potential power savings for each identified opportunity. A sample WattBot report is pictured in Figure 22.6.

PowerTheater is also used at the RT level for power verification. For many designs, gate-level simulation is impractical due the number of gates and the slowness of simulation speed relative to the operational

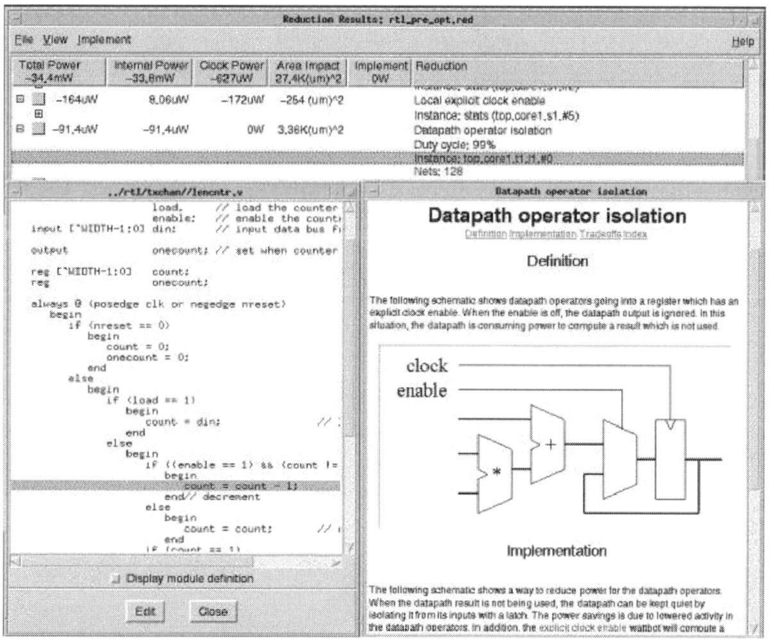

FIGURE 22.6 WattBot report.

speed — significant numbers of operational cycles cannot be run in a reasonable period of time. For these designs, power analysis of the RTL code is the only practical method of obtaining accurate full-chip power estimates because roughly an order of magnitude more cycles can be simulated in a given period using an RTL representation as opposed to a gate-level representation. Another motivation for RTL power verification is the desire to check a design's conformance to its power specification as early as possible. This, in particular, is a key methodology step in a feed forward design flow.

Nevertheless, some designs are verified at the gate level, and PowerTheater is used for this task as well. The use model for gate-level verification is identical to that for RTL verification: the (gate-level) design is simulated, and the resulting activities and the design itself are loaded into PowerTheater for analysis. The analysis can be performed using either estimated or extracted wiring capacitances. If an RT level analysis was previously run, then the same setup and command files can be used to run the gate-level analysis.

To aid in choosing the most appropriate simulation activities for power analysis, PowerTheater provides a simulation activity viewer, which enables the user to view aggregate activity over time. This viewer is used to find, for example, the point in the simulation in which the activities reach a sustained maximum — the most appropriate period for calculating worst-case average power. The viewer can also be used for power debugging, such as searching for those modules that should be inactive but instead are toggling unexpectedly.

22.3.4 PhysicalStudio

PhysicalStudio is a cell-level physical design closure tool for analyzing and optimizing power, timing, and signal integrity issues concurrently in both pre- and post-route designs. It includes a static timing analysis (STA) engine, a delay calculator, and a signal integrity (SI) analyzer for both coupling-delay effects and glitching, along with placement aware optimizations. The combination of the various analysis capabilities, operating off a single database, enables concurrent optimizations in which individual power optimizations are implemented only if they do not break either timing or noise margin limits [3].

PhysicalStudio produces as output an optimized physical design in LEF/DEF format [9]. The optimizations may be performed on a preroute database, in which case the output is placement optimized DEF and corresponding Verilog netlist, or the optimizations may be performed post-route, in which case the output is placement optimized DEF and corresponding Verilog netlist along with a routing ECO file.

PhysicalStudio loads design information and technology libraries to build an internal database from which analyses and optimizations are launched. The required design information includes the design netlist, the placement description, and timing constraints; for routed designs, the required design information includes extracted wiring parasitics. The technology information includes cell libraries for logical, timing, power, signal integrity, and layout views. PhysicalStudio reads the same libraries as PowerTheater, in either Liberty or ALF formats, but it also requires layout definitions in the LEF format.

Central to PhysicalStudio's optimizations are the delay calculator and STA engine. Delays are computed using full three-dimensional coupling capacitances, including aggressor and victim analyses. Similarly, glitch injection and propagation are modeled to evaluate effective noise margins for all instances. Subsequently, during the optimization phase, any potential transform is evaluated against how it affects timing and noise margins. The transform is committed to the database only if it does not violate any of the other constraints; otherwise, the transform is discarded and subsequent optimizations are considered.

PhysicalStudio employs two different types of power optimizations, one aimed primarily at reducing dynamic power consumption and the other aimed at leakage power reduction. Both of the optimizations employ the same fundamental trade-off: speed is traded for power on those paths that have positive slack timings.

This trade-off is implemented as follows. All timing paths are analyzed and slack timings (timing margins) are computed. For each path with positive slack timing, PhysicalStudio replaces individual cells along the path with lower power equivalents as long as the slack timing remains positive. If a potential cell replacement causes the slack timing to become negative, the replacement is not implemented. The overall post-optimization result is that fewer paths exhibit substantial positive slack and a commensurately larger number exhibit less slack, thus indicating that those paths have become slower. Nevertheless, no path is allowed to exceed the clock period timing constraint so that the circuit will still function as desired, but with reduced power consumption.

The specific power reductions that PhysicalStudio employs to trade-off slack timing for power are cell resizing and dual-Vt cell swapping. In the former case, the $C_L V_{dd} V_{swing} f + V_{dd} Q_{sc} f$ terms in Equation (22.2) are targeted, and cell drive strengths (or sizes) are reduced as far as possible without introducing timing or noise problems. The reduction in size reduces several parameters: occupied area, cell crowbar current, and most importantly capacitive loading for the fan-in logic. The dual-Vt optimization targets the $V_{dd} I_{lkg}$ term in Equation (22.2) and cells employing a high-Vt threshold implant, resulting in reduced leakage albeit with lengthier delays, are substituted for low-Vt cells wherever timing slack permits [10].

22.3.4 Using PhysicalStudio

PhysicalStudio is used in two different optimization modes: one before routing and one after routing. The preroute mode is used to prepare the design to avoid or prevent timing, noise, or power problems from arising after routing. The post-route mode is used to fix any problems that persist after the physical design and routing have been completed.

In preroute mode, PhysicalStudio reads the placement and estimates route lengths from the placement. Wire parasitics are estimated from these route lengths and the parasitics are, in turn, used to calculate delays. PhysicalStudio then performs a static timing analysis and a noise analysis to determine available timing slack and noise margins after glitching effects have been considered. Timing constraints are used to guide the static timing analysis. If either timing violations or noise violations are detected, PhysicalStudio will repair the violations using timing and noise optimizations, respectively. Once the design is timing and noise violation free, physical power optimizations are applied.

Dynamic power is optimized by utilizing the resizing command in the optimization script, along with the target minimum amount of timing slack. PhysicalStudio will then examine the sizings of all instances, and decrease the instances' sizes wherever possible, so long as the minimum timing slack parameter is not violated.

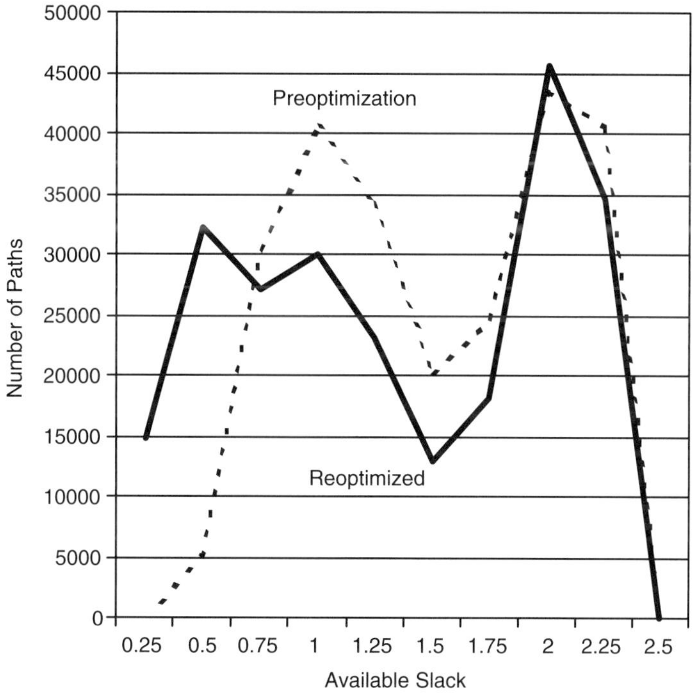

FIGURE 22.7 Power optimization and delay redistribution.

PhysicalStudio supports two different leakage power reduction flows. The first flow works similarly to dynamic power optimization with one significant difference — a second library is required. This second library is functionally and physically equivalent to the first, but each cell has different timing and leakage characteristics due to the use of a different transistor threshold than that in the first library. PhysicalStudio loads both libraries — the primary low-Vt as well as the second high-Vt version to be used for replacement — and automatically establishes functional equivalencies between them. After this point, the flow and use model are identical to that of resizing. PhysicalStudio replaces a low-Vt cell in the original netlist with a high-Vt cell wherever possible as long as the replacement does not violate the minimum timing slack parameters. Leakage power is reduced because the high-Vt cells exhibit much less leakage than the low-Vt cells. An example of this type of optimization is plotted in Figure 22.7. In this particular case, the design consisted of 255,000 cells, and the dual-Vt optimization resulted in 84% of the low-Vt cells being swapped for high-Vt cells. Leakage power was reduced by 43% without changing any of the critical timing.

The second leakage power reduction flow can be thought of as a timing closure flow that starts with a design utilizing all high-Vt cells, but later selectively substitutes low-Vt cells to repair paths with negative timing slack [11]. In this flow, the design is initially synthesized and placed using the slower, high-Vt library as the primary target, and the timing is optimized as much as possible with this single library. For those paths that cannot meet timing using high-Vt cells alone, PhysicalStudio uses low-Vt cells to replace as many of the high-Vt cells as necessary to meet the specified timing constraints.

The second flow generally results in less leakage power consumption, but may require more area as the synthesizer will attempt to close timing with the single, slower library by adding additional logic or buffering for heavily loaded nets.

PhysicalStudio supports both of the leakage reduction flows as well as the dynamic power reduction capability in the post-route mode in addition to the preroute mode described previously. The only difference between the two modes is that in post-route mode the wiring parasitics are known exactly instead of being estimated. This knowledge enables PhysicalStudio to reduce both dynamic and leakage power even further because no uncertainty exists regarding the timing slack (with the exception of on-chip process variation). Both dynamic power and leakage power can be optimized together, although priority must be given to one over the other.

22.3.5 CoolTime

CoolTime is a cell-based electrical integrity analysis tool for analyzing the effects of power on timing, noise, and reliability. Similar to PhysicalStudio, CoolTime includes a STA engine, a delay calculator, and a signal integrity (SI) analyzer for both coupling-delay effects and glitching. It additionally includes a power rail parasitic extractor, a power calculator for both average and instantaneous power, a power rail voltage solver, and several power rail display capabilities [4]. The CoolTime architecture is shown in Figure 22.8.

CoolTime also employs a cell electrical modeler, ElMo, to characterize each cell for its timing and glitching characteristics under various voltage conditions. ElMo reads Liberty and SPICE models, and then runs numerous SPICE simulations to create a set of glitch models and voltage derating factors for each characterized cell. These models and derating factors are subsequently used by CoolTime to compute the response of each individual instance to the calculated voltage variations.

CoolTime produces reports, both textural and graphical, of the power rail voltage variation across the design. This variation data is used to calculate the effects of the power rail voltages and currents upon timing and noise.

CoolTime loads the same design and technology information as PhysicalStudio along with additional information, such as extraction rule definitions, decoupling capacitor definitions, and package parasitics. Once the power rails are extracted, CoolTime uses a combination of internal estimation algorithms, along

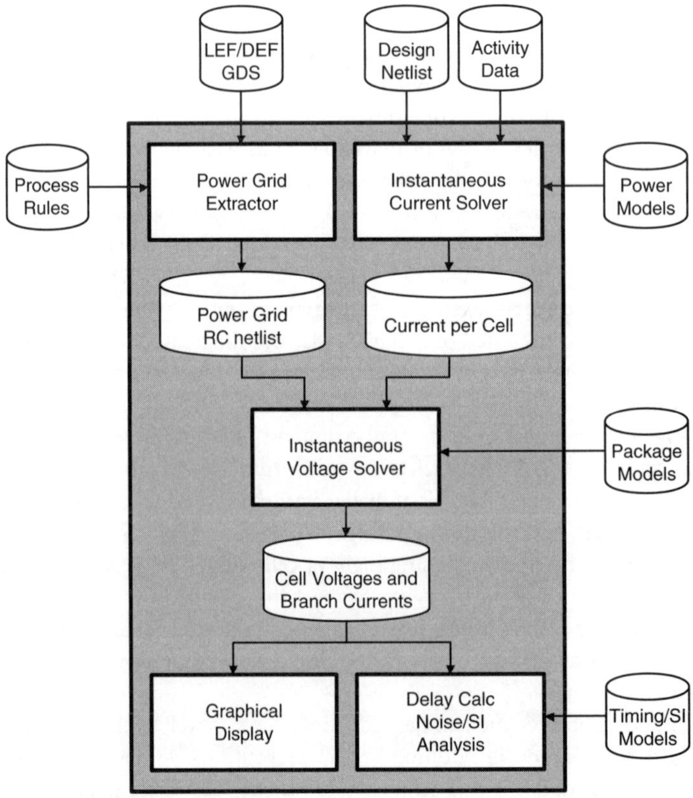

FIGURE 22.8 CoolTime architecture.

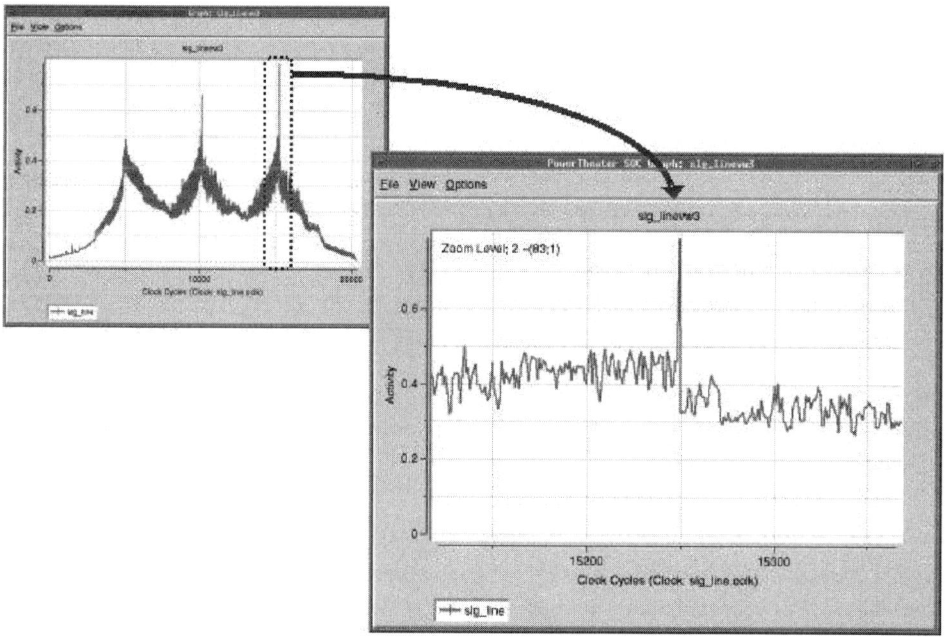

FIGURE 22.9 Finding a simulation cycle to use in CoolTime.

with timing constraints to compute current waveforms for each node in the power rail network. These current waveforms are then fed, along with the extracted power rail network and package parasitics, to a matrix solver to solve for nodal voltages. The resulting output is a set of time varying voltage waveforms for each node in the power rail network. These voltages are subsequently used to recalculate delays and noise margins for each instance.

CoolTime utilizes two different approaches for computing power. The first approach is a vectorless approach in which no external stimulus is needed. The second is a simulation-based approach that relies upon a PowerTheater analysis of the simulation's activities to find the simulation cycle with the most activity as pictured in Figure 22.9. Once found, the state points in that selected cycle are fed to CoolTime as a seed vector from which CoolTime determines which nodes switch and when. By contrast, the vectorless approach requires no simulation data at all but instead makes nodal switching activity determinations based upon the input design constraints and a topological analysis of the netlist. The advantages of the vectorless approach are that no logic simulations are required and that the resulting current and power estimates will be conservative. The advantage of the simulation-based approach is that the results will represent the actual conditions for a particular cycle of interest, although the simulation time required to reach that cycle may be excessive, and as with all simulation results, there can be no assurance that the results represent a worst-case condition.

Either of these two power calculation approaches can be used to compute both average and instantaneous power. For average power, CoolTime computes an average current for one cycle, I_{avg}, which is then used to compute a time-averaged voltage drop:

$$V_{avg} = I_{avg} R \tag{22.3}$$

where I_{avg} is the sum of all currents consumed during the period of interest divided by the length of time of that period.

CoolTime also computes the time varying, instantaneous voltages, $V(t)$, according to the following equation:

$$V(t) = (I(t) + Cdv/dt)R + Ldi/dt \tag{22.4}$$

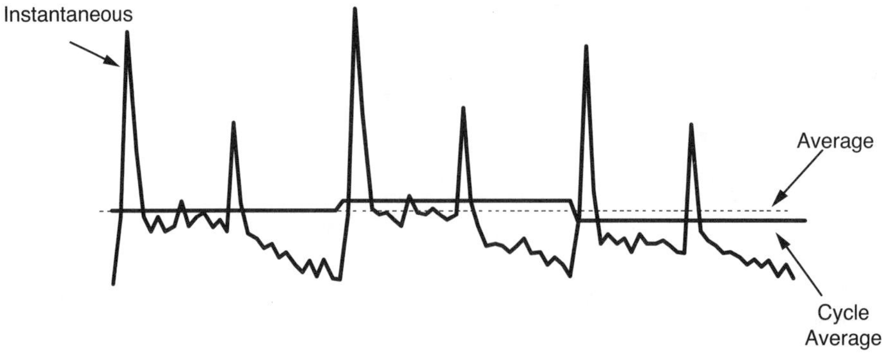

FIGURE 22.10 Current waveform comparison.

where C represents the sum of various parasitic capacitances such as power rail capacitances, well capacitances, and explicitly inserted decoupling capacitances, and L represents the sum of various parasitic inductances such as power rail inductance, bond-wire inductance, and package pin inductance.

Figure 22.10 illustrates the significance of using time-averaged or instantaneous currents for the voltage drop calculations. Not only are the magnitudes materially different, but also the rapid magnitude changes due to spiking can lead to large inductive voltage fluctuations. In effect, the use of time-averaged currents negates the $(Cdv/dt)R$ and Ldi/dt terms in Equation (22.4), and is thus inappropriate for detailed analyses of the power rail network.

The time varying currents and voltages can be viewed using CoolTime's voltage and current recorder (VCR) as depicted in Figure 22.11. This visualization capability enables the user to single step through time, either forward or backward, to see where the "hot spots" occur in the layout along with when they occur. The CoolTime VCR contains four panes, one each for V_{dd} voltage and current as well as V_{ss} voltage and current, providing the ability to view animations of the dynamic current and voltage variations on both V_{dd} and V_{ss}.

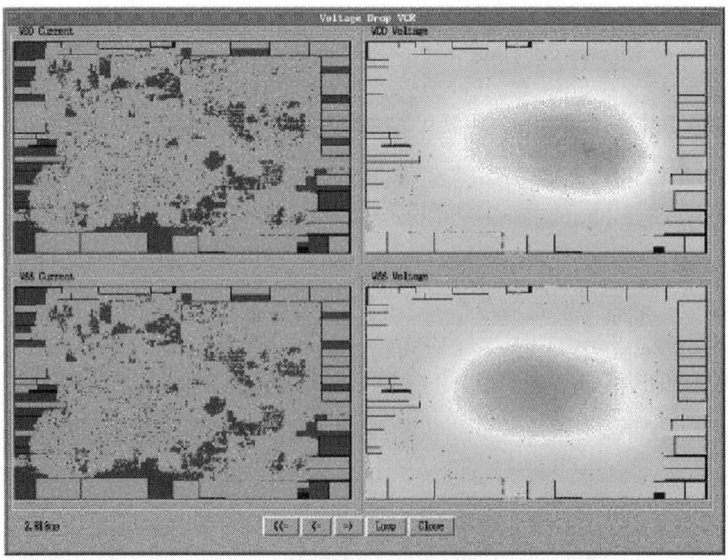

FIGURE 22.11 CoolTime VCR.

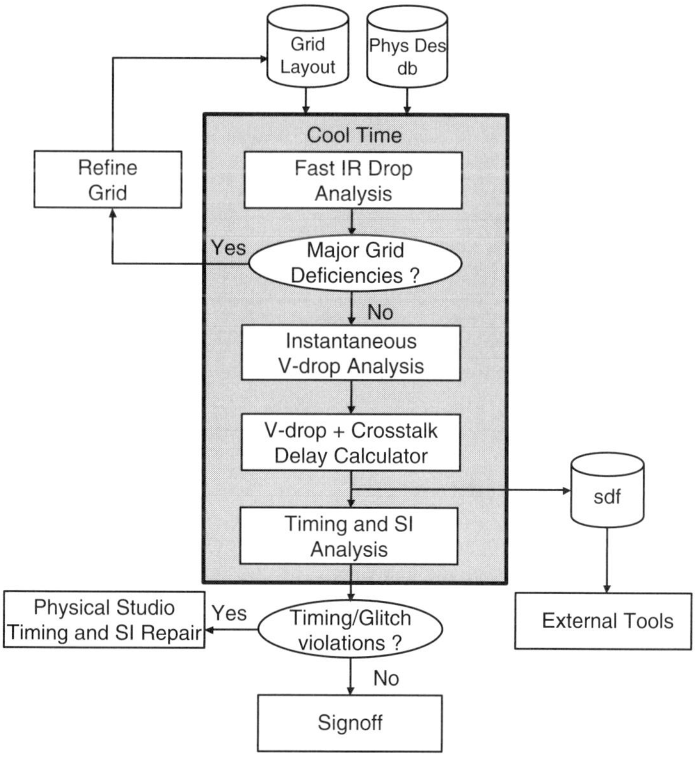

FIGURE 22.12 CoolTime methodology for power grid integrity.

22.3.6 Using CoolTime

CoolTime is primarily intended to be used as an electrical verification and sign-off tool, however it also serves a role in the design and implementation process. A guide to CoolTime's usage in the overall IC development flow is depicted in Figure 22.12.

During initial power rail design and layout, CoolTime is used to check basic power rail sizes, using its fast average IR drop calculator. The intent at this point is to identify any basic power rail deficiencies before proceeding to more detailed physical design. To perform this analysis, the power rails are extracted and the power is calculated, either from a simulation-based analysis or from a vectorless analysis. In either case, the average power computation method is selected because the intent is to isolate any resistive issues in the rails' design. If excessive IR drop is found at any point in the power rail network, the rail sizes or topologies are adjusted to reduce the IR drop to acceptable levels.

Once the physical design has been refined and nears completion, CoolTime is used to analyze instantaneous current and voltage drop effects to verify timing and reliability for sign-off. The power grid and signal parasitics are extracted. Pad voltages and package parasitic models are specified along with decoupling capacitor definitions. As before, power and current consumption are calculated; however, in this case, instantaneous calculations are employed to incorporate the effects of the various capacitances and inductances, which are needed to verify timing and reliability with the highest confidence. The results are a set of instantaneous voltage waveforms for all nodes in the power rail network. These instance-specific, time-varying voltages are fed to the delay calculator, along with fully coupled signal parasitics, to compute voltage derated delays, where the voltage derating incorporates the effects of driver and receiver voltages being different than the library characterization conditions represented in the cell library. These voltages are also used by the signal integrity analyzer to adjust the noise margins according to the nonideal V_{dd} and V_{ss} voltages that power each cell.

CoolTime also evaluates electromigration sensitivity at this point. Using the extracted power rail network, along with the computed branch currents, CoolTime will calculate the current density in the various power rail network branches. The results can be displayed by the GUI for a visual inspection; violations will be highlighted as well as written to a violation report file.

The conclusion of all these analyses is a set of timing, power, and reliability reports that include the effects of power on each parameter. If no violations are found, then the design is ready for electrical sign-off.

22.4 A Design Example

The usage of the sequence tools in a feed forward power-sensitive design flow is perhaps best illustrated by example. Consider the design of an application-specific digital signal processor slated for use in a wireless device.

22.4.1 High-Level Design

Following a feed forward design flow, the first steps involve the system specification and design which includes setting the power supply voltage(s) as low as possible within the constraints of performance, battery availability, and the voltage requirements of other chips in the system. A power budget is set for the entire design and apportioned for the various modules or design components. Once the system specification solidifies, RTL design begins.

As the RTL coding proceeds, PowerTheater is used to evaluate the RTL code's power characteristics. Whenever available, activities derived from RTL simulations, such as those of the inner loops of the most frequently used DSP algorithms, are used as the input stimulus. Power debugging proceeds by using PowerTheater to identify the design constructs that consume the largest portions of the power budget, as well as those that are unexpectedly high. Debugging continues by using PowerTheater's reporting mechanisms, especially the frequency reports, to isolate the causes behind any excessive power consumption. Once the root causes have been identified and rectified by code modifications, the code-simulate-analyze local loop is repeated until the power consumption target is met. PowerTheater's automatic power linting utilities are also employed at this point to highlight any overlooked opportunities for code optimization.

During this process, particular attention is paid to clocking, datapaths, and memories. Clock power can be reduced by incorporating clock gating, although large numbers of gated clocks often pose problems in clock skew management by downstream physical tools. PowerTheater's WattBots are used to identify the most effective clock gating opportunities so that the total number of gated clocks can be minimized, thus avoiding later issues in clock tree synthesis and layout.

Inspecting the switching frequencies of intermediate nodes identifies power waste in datapaths. Code with quiescent outputs and active inputs represent opportunities for improvement by moving gating logic as far upstream as possible Conversely, code with glitchy control inputs is rearranged so that the glitches are prevented or blocked. If this is not possible, then the function should be coded such that the glitchy inputs are fed as deeply into the datapath logic as possible to limit the amount of logic through which the glitches could propagate.

Memories and data storage structures warrant particular attention as they often consume the largest portion of the total power. Because PowerTheater computes memories' read-and-write access rates, inspection of PowerTheater's frequency reports can uncover inadvertent memory accesses, which waste power. More aggressive power consumption goals may dictate revising the entire data storage architecture to minimize the total number of accesses; in this case, PowerTheater provides the mechanism by which to evaluate which architecture will be the most power efficient for that particular application. One example of this situation is the consideration of asynchronous vs. synchronous memories — which type of memory will be more power efficient depends not only on its internal structure, but also on the details of how it is employed within the particular target system.

22.4.2 Physical Design

Once the RTL has been demonstrated to meet the target power specification, the design is synthesized or otherwise converted into a cell level netlist suitable for placement. After placement and timing closure, PhysicalStudio is used to minimize the dynamic power consumption through resizing and, if desired, minimize leakage power through dual-Vt cell swapping. The output of these operations is a timing-closed, power-minimized physical design ready for initial power rail sizing. CoolTime can now be used to evaluate the basic integrity of the power rail sizing by analyzing the design's IR drop. If this analysis indicates a sufficiently small amount of IR drop, then the design is fed to the router. On the other hand, if the IR drop is judged as excessive then mitigating steps are undertaken to rectify the issues. The rails may be resized or even redesigned altogether, after which CoolTime is used again to verify the IR drop. Once the IR drop is within the target spec the design is fully routed.

After routing PhysicalStudio is employed again to further squeeze both dynamic and leakage power using similar transformations to those employed preroute. Additional reductions are usually possible at this point because extracted parasitics are used in lieu of estimated wire lengths and the timing slacks are known with much greater certainty. Any changes other than simple cell swaps will necessitate another route, but once the design has been routed after the last optimization, the design is ready for electrical sign-off verification using CoolTime.

22.4.3 Electrical Sign-Off

At this point, the completed physical design is reextracted (if the layout changed at all, otherwise the original extracted data is used) to produce both signal and power rail parasitics, which are loaded into CoolTime along with the package parasitics definition. CoolTime then computes the amount of supply droop and bounce along with their effects on timing and noise margin degradation. This data is, in turn, used by the static timing analysis and noise analysis engines to verify electrical performance and noise immunity in the presence of power rail voltage variations. Electromigration limits are also calculated in all power rail branches. If none of the limits for timing, noise, or reliability are violated, then the design is ready for tape-out.

22.5 Conclusion

Power-sensitive design has become an essential focus in this age of wireless and multimedia computing, but it is no longer directed at simply reducing the amount of power consumption. Power-related issues now directly affect many facets of design and these issues are sufficiently complex as to require significant amounts of design automation.

Sequence design develops advanced tools to address these critical issues and views power-sensitive design as a multilevel endeavor. To that end, PowerTheater supports RTL design and analysis early in the design process, PhysicalStudio optimizes power, timing, and noise characteristics at the physical level, and CoolTime verifies cell-based electrical characteristics for verification and sign-off. As presented in this chapter, these tools are used in a comprehensive methodology for developing robust, power-efficient designs.

References

[1] Wilson, R. and Lammers, D., Grove calls leakage chip designers' top problem, *EE Times,* December 13, 2002.

[2] Sequence Design, *PowerTheater Reference Manual,* Sequence Design Inc., Santa Clara, CA, 2003.

[3] Sequence Design, *PhysicalStudio Reference Manual,* Sequence Design Inc., Santa Clara, CA, 2003.

[4] Sequence Design, *CoolTime User Manual,* Sequence Design Inc., Santa Clara, CA, 2003.

[5] Landman, P. et al., An integrated CAD environment for low-power design, *IEEE Design and Test of Computers,* vol. 13, Summer 1996, pp. 72–82.

[6] Synopsys, *Liberty User Guide, Version 2001.08,* Synopsys, Inc., Mountain View, CA, 2001.
[7] IEEE, *P1603/D9, A Draft Standard for Advanced Library Format (ALF),* IEEE, New York, 2003.
[8] IEEE, *1481, Standard for Delay & Power Calculation Language Reference Manual,* IEEE, New York, 1999.
[9] Cadence Design Systems, *LEF/DEF Language Reference,* Cadence Design Systems, Inc., San Jose, CA, January 2003.
[10] Lee, W. et al., A 1V DSP for Wireless Communications, *Proc. ISSCC,* February 1997, pp. 92–93.
[11] Wang, Q. and Vrudhula, S.B.K., Algorithms for minimizing standby power in deep submicrometer, dual-Vt CMOS circuits, *IEEE Trans. on Computer-Aided Design of Integrated Circuits and Syst.,* vol. 21, no. 3, March 2002, pp. 306–318.

Index

A

D

E

F

G

M

Q

R

S

T

U

V

W

X

Y

Z